The Geometry of Heisenberg Groups

With Applications in Signal Theory, Optics, Quantization, and Field Quantization

Mathematical
Surveys
and
Monographs

Volume 151

The Geometry of Heisenberg Groups

With Applications in Signal Theory, Optics, Quantization, and Field Quantization

Ernst Binz
Sonja Pods

with an Appendix by
Serge Preston

American Mathematical Society
Providence, Rhode Island

2000 *Mathematics Subject Classification.* Primary 22B05, 22E70, 43A40, 43A65, 46L05, 46L65, 53D55, 57R25, 78A05, 80M99.

For additional information and updates on this book, visit
www.ams.org/bookpages/surv-151

Library of Congress Cataloging-in-Publication Data

Binz, Ernst, 1939–
The geometry of Heisenberg groups : with applications in signal theory, optics, quantization, and field quantization / Ernst Binz, Sonja Pods ; with an appendix by Serge Preston.
p. cm. – (Mathematical surveys and monographs ; v. 151)
Includes bibliographical references and index.
ISBN 978-0-8218-4495-3 (alk. paper)
1. Heisenberg uncertainty principle–Mathematics. 2. Quantum theory–Mathematics. I. Pods, Sonja, 1974– II. Title.

QC174.17.H4B53 2008
512′.482–dc22   2008030289

---

10 9 8 7 6 5 4 3 2 1   13 12 11 10 09 08

Contents

Introduction

The notion of a three-dimensional Heisenberg group is an abstract algebraic formulation of a geometric phenomenon in everyday life. It occurs if we select a plane in our three-dimensional space. For example, the page you are now reading is such a plane and produces a Heisenberg group. If you take a photograph you are in the middle of Heisenberg group theory; you have transmitted information along a line and encoded it in a plane and hence you have established a Heisenberg group. These few remarks indicate already some aspects of our program encoded in the title. We will study Heisenberg groups and Heisenberg algebras as mathematical objects in detail, identify them in several physical and mathematical areas and thereby exhibit close relationships among them even though they look quite different at the beginning of our discourse. This is to say we will build bridges between fields by means of Heisenberg groups and Heisenberg algebras.

This short introductory look at our program hints that we like to invite both graduate students of mathematics and mathematicians with some interest in physics as well as graduate students of physics and physicists with some interest in mathematics on our journey through the very mathematics and physics of Heisenberg groups.

Now we go into more detail. Looking at a Euclidean oriented three-dimensional space E as a Heisenberg group or at a Heisenberg algebra amounts to a splitting of E into an oriented plane F and an oriented real $\mathbb{R}\cdot a$ line, say, orthogonal to F. Here $a \in S^2$. The orientation on E shall be made up by the orientations of the plane and the real line, respectively. A constant symplectic structure on the plane determines its orientation while the real line is oriented by a vector in it yielding a direction of the plane's oriented rotations. These geometric ingredients can be encoded in detail by a specific non-commutative group operation on the Euclidean space, yielding a Heisenberg group, a Lie group. Its center is the one-dimensional subspace. Its Lie algebra is called a Heisenberg algebra. In fact any $(2n+1)$-dimensional Euclidean space admits a Heisenberg group structure for any integer n bigger than zero. However, with the applications we have in mind we mainly concentrate on three-dimensional Heisenberg groups and Heisenberg algebras. However, in field quantization we have to pass over to infinite dimensional ones.

Heisenberg groups have a very remarkable property: By the Stone–von Neumann theorem any Heisenberg group up to equivalence admits only one irreducible unitary representation on an infinite dimensional Hilbert space, if its action is specified on the center, i.e. on the real line introduced above. The rather simple looking Schrödinger representation of a Heisenberg group on the L_2-space of the real line (the Hilbert space) is unitary and irreducible. Therefore, up to equivalence, the Schrödinger representation is uniquely determined by its action on the center. To define the Schrödinger representation, a coordinate system is needed turning the plane F into a phase space.

Together with the notion of time, requiring a fourth dimension, the oriented Euclidean space immediately determines the skew field of quaternions $\mathbb{H}$. This skew field provides us with a convenient mathematical structure to treat three-dimensional Heisenberg groups in a larger context. In fact, any Heisenberg algebra structure on E emanates from the multiplication of the quaternions. Moreover, the various ways of turning the oriented Euclidean space into a Heisenberg group determine the skew field $\mathbb{H}$, its three-sphere, i.e. the spin group $SU(2)$, and the Hopf bundles on the two-sphere. These Hopf bundles fibring S^3 over S are uniquely described by their respective Chern numbers. In addition, the skew field structure yields a natural Minkowski metric (which can be rescaled to meet the needs of special relativity) on $\mathbb{H}$ which intertwines Minkowski geometry with Euclidean geometry. This is nicely seen by computing the rotation angles of the inner automorphisms of $\mathbb{H}$ since any inner automorphism amounts to an oriented rotation on the Euclidean space. In fact, any Minkowski metric on an oriented four-dimensional linear space emanates from the natural Minkowski metric on $\mathbb{H}$.

Now it is conceivable that the modeling based on Heisenberg groups naturally involves the mathematical structures just described above. Let us shortly review three applications of the Heisenberg group, each of which reveals clearly this group as underlying mathematical background.

We begin these presentations by signal theory. As formulated in the seminal book of Groechenig [**47**], for example, the general framework of signal theory consists of three main steps, namely analysis, processing and synthesis of a signal. Since already the first step, namely signal analysis, is a rather huge field, for our purpose we need to restrict our scope. To show where Heisenberg groups and Heisenberg algebras appear in signal analysis, we focus on one of its branches, namely on time-frequency analysis, which from a mathematical point of view is a branch of harmonic analysis. Later, we describe in a rather rough fashion what time-frequency analysis focuses on by means of an analogy (cf. [**47**]) and consider a musical score. Time behavior is encoded horizontally whereas the frequency information is expressed vertically. The score represents an analysis of the signal in terms of time-frequency information. Playing the music is the synthesis or reconstruction of the signal. Truncation is a form of signal processing, for example. Expressed in a more abstract fashion, the time axis and the frequency axis generate the time frequency plane and a signal is given by a quadratically integrable, complex-valued function on the real line, the time axis, say. Its values give time information. Frequency information in the signal is visible in the Fourier transform of the function.

Methods of the analysis of a signal involve various technical tools, in particular various sorts of transformations of signals which exhibit specific properties. One of them is the ambiguity function (defined on the time-frequency plane), a fundamental tool, as for instance in radar engineering or in geometric optics in terms of its Fourier transform, the Wigner function. The ambiguity function compares two signals with each other, for example an outgoing with an incoming one. This situation is typical for radar. In the case of a plane the outgoing signal hitting the object is known in detail. The incoming signal contains information on position and velocity. Formulated in a simplified fashion one can extract this information from the ambiguity function built up by the outgoing and the reflected signal.

As a first highlight in this discourse let us demonstrate the appearance of the Heisenberg group in time frequency analysis by means of the ambiguity function. From a technical point of view it is a simple but beautiful insight based on the Schrödinger representation first mentioned by W. Schempp in [**71**]: The first Fourier coefficient of the Schrödinger representation of the three-dimensional Heisenberg group is nothing more than the ambiguity function. Similar results on other tools can be found in [**47**]. The relation of the Heisenberg group with such a fundamental tool in time-frequency plane shows that this group is omnipotent in time-frequency analysis, as is also expressed in [**47**].

The time frequency plane is the Cartesian product of two axes, the time axis and the frequency axis. One goal of time-frequency analysis seems to be to resolve a point in this plane arbitrarily well by means of signals treated by various tools. Surprisingly, such a resolution process is not possible since time and frequency information of a signal are encoded in the signal itself, respectively in its Fourier transform. Uncertainty relations are obstructing it. These relations can be derived from a Heisenberg algebra and its infinitesimal Schrödinger representation, too.

In geometric optics in 3-space, the effect of an optical system placed in between two parallel planes can nicely be described by the Wigner function exhibiting a symplectic transformation A acting on the first plane, i.e. by a linear map of this plane which preserves a symplectic structure, as shown for example in [**24**]. The same linear map can be found by arguing in terms of wave optics as done in [**41**].

In our presentation of the appearance of the Heisenberg group in geometric optics we partly follow [**50**] since this approach naturally carries on to the quantization of homogeneous quadratic polynomials. To describe geometric optics a little more precisely, let the plane F be mapped by light rays to another plane F' parallel to F. In between these two planes an optical system is placed. A symplectic structure on F is caused by the (2+1)-splitting of the Euclidean 3-space E initiated by the choice of F and its orthogonal complement in E. The choice of a coordinate system turns the symplectic plane into a phase space. As mentioned above, the image in F' caused by the light rays passing the optical system is described by a symplectic transformation A of F. Vice versa, any map in the group $\mathrm{Sp}(F)$, consisting of all symplectic transformations of F, corresponds to an optical system. What happens with a light distribution on F? Associated with a light distribution on F is a phase distribution. The image of this phase distribution on F' caused by the optical system is computed by Fresnel integrals, a tool in wave optics (cf. [**41**]). These Fresnel integrals are very closely related to metaplectic representations, i.e. representations of the metaplectic group $\mathrm{Mp}(F)$ of F with the L_2-space of the real line as representation space (cf. [**50**]). The metaplectic group $\mathrm{Mp}(F)$ is a twofold covering of the symplectic group $\mathrm{Sp}(F)$ of the plane. Back to the Wigner function mentioned above, it detects A also and is the Fourier transform of the ambiguity function. Therefore, geometric optics are based on a Heisenberg group as well. It is determined by the (2+1)-splitting of E produced by the symplectic plane F and its orthogonal complement in E. This is to say the array naturally yields a Heisenberg group $\mathcal{H}$. The symplectic map A on F (characterizing the optical system) is extended to all of $\mathcal{H}$ by the identity on the center of $\mathcal{H}$ (still called A). It is a Heisenberg group isomorphism and hence determines a new Heisenberg group, namely $A(\mathcal{H})$. Due to the famous Stone–von Neumann theorem, its Schrödinger representation $\rho \circ A$ is equivalent to the Schrödinger representation ρ of $\mathcal{H}$ and causes the Wigner function on F describing the optical system. This is the content of chapter 8, so you can see we are already right in the middle of our manuscript.

Now let us pass on to ordinary quantum mechanics. There are several different looking approaches to it. For example, one way of formulating quantum mechanics is by concentrating more on analytic aspects such as Wigner functions and operator theory (cf. [**4**], [**37**], [**47**] and [**32**]). Another one is based directly on symplectic geometry (cf. [**50**] and [**75**]); operator theory enters here via representation theory. In the very beautiful book [**32**] it becomes clear how these approaches are intertwined. Since we concentrate on the appearance of the Heisenberg group and other geometric structures based on them, it is natural to focus on a geometric basis of Quantum mechanics. In doing so, in our investigations we adopt the view point taken in [**50**].

Keeping the role of time and frequency in time-frequency analysis in mind, in classical mechanics the analogous object of the time-frequency plane is the phase space of a line, which is a plane F, say. F is equipped with a coordinate system in which one coordinate axis is identified with the line, on which a point is thought to move. Let us call it the q-axis. At any instant this point has a position q and a momentum p, say, visualized on the second coordinate axis, the p-axis. Hence the pair (q, p) of coordinates characterizes a point in phase space F. As in time-frequency analysis, F is equipped with a symplectic structure, hence determines a Heisenberg group structure on E.

Quantization of position and momentum in classical mechanics is achieved by means of the infinitesimal Schrödinger representation $d\rho$ multiplied by $-i$ where i is the imaginary unit of the complex plane. Hence $-i \cdot d\rho$ converts each element of the Heisenberg algebra into a self-adjoint operator acting on the L_2-space of the real line, a Hilbert space. It consists of all quadratically integrable, complex-valued functions of the real line. The quantization of q and p yields two non-commuting operators, obeying Heisenberg's uncertainty relations.

Kinetic energy (a classical observable) of the moving point is a homogeneous quadratic polynomial. The quantization of this type of polynomials (called classical observables) defined on the phase space F is in a sense an infinitesimal version of geometric optics. Here is why: The infinitesimal metaplectic representation multiplied by $-i$ represents the Lie-algebra of the metaplectic group $\mathrm{Mp}(F)$ in the space of the self-adjoint operators of the L_2-space of the real line. This Lie algebra is identical with the Lie algebra $\mathrm{sp}(F)$ of $\mathrm{Sp}(F)$ and the Lie algebra (F) made up of all trace-free linear maps of F. Now the Lie algebra $sl(F)$ is naturally isomorphic to the Poisson algebra of homogeneous quadratic polynomials on the plane F, an algebra of classical observables of the moving point. Thus any homogeneous quadratic polynomial on F is converted to a self-adjoint operator acting on the L_2-space of the real line. Of course the definition of a homogeneous quadratic polynomial requires the coordinate system on F. This resembles the situation of the Schrödinger representation, and in fact, the metaplectic representation can be constructed out of the Schrödinger representation. Representing all quadratic polynomials of the plane requires a representation of the semidirect product of the Heisenberg group with the metaplectic group and the Lie algebra of this product. At this stage we point out that the collection of quantized homogeneous quadratic polynomials together with the identity allows the reconstruction of the field of quaternions and hence of a Minkowski space. This is found in chapter 9.

Of a quite different nature is the quantization associated with a vector field in 3-space, elaborated in chapter 10. The goal here is to specify a collection of classical observables of the vector field and to associate field operators (on some infinite dimensional Hilbert space) to them. But first let us analyze the vector field in order to single out a collection of classical observables.

Given a vector field on a possibly bounded three-dimensional submanifold in the oriented Euclidean space E we may cut out all its singularities and obtain a singularity free vector field X on a smaller topological space of which we assume that it is a manifold M in E with or without boundary, say. A complex line bundle $\mathbb{F}$ on M is obtained by taking the orthogonal complement in E of each field vector as fibres. The points in $\mathbb{F}$ are called internal variables of the vector field. Each fibre of $\mathbb{F}$ admits a constant symplectic form determined by inserting the respective field vector into the volume form of E. Thus at each point in M, the fibre together with the line containing the field vector yields a three-dimensional Heisenberg group and hence in total a Heisenberg group bundle $\mathbf{H}^{\mathbf{a}}$ as well as a Heisenberg algebra bundle on M. The bundle $\mathbf{H}^{\mathbf{a}} \subset M \times E$ determines the vector field and vice versa. Passing on to the collection $\Gamma\mathbf{H}^{\mathbf{a}}$ of all Schwartz sections and integrating up the fibrewisely given symplectic forms yields an infinite dimensional commutative Weyl algebra, a C^*-algebra whose involution sends a section into its negative and which in addition contains a natural Poisson algebra. The elements of this Poisson algebra $P^{\mathbf{a}}$ are called the classical observables of the vector field. This natural Poisson algebra determines the vector field and vice versa, as shown in chapter 6.

Now we begin to describe the quantization procedure of the vector field X as done in chapter 10. Here this procedure is split up into two steps, namely into a prequantization and into the specification of the physical observables reached by representations. The Poisson algebra $P^{\mathbf{a}}$ is the domain of the quantization map $Q_\hbar$, called here the prequantization. The quantization map $Q_\hbar$ represents the $*$-algebra $P^{\mathbf{a}}$ on the C^*-algebra $\mathcal{W}^\hbar\Gamma\mathbf{H}^{\mathbf{a}}$, involving a parameter $\hbar$ varying on the real line. It may in particular assume the value of Planck's constant. If this parameter differs from zero, the multiplication (an $\hbar$ dependent deformed convolution) of the Weyl algebra $W^\hbar\Gamma\mathbf{H}^{\mathbf{a}}$ is non-commutative. This construction yields a real parameterized family of Weyl algebras $W^\hbar\Gamma\mathbf{H}^{\mathbf{a}}$ with parameter $\hbar$ and the Poisson algebra $P^{\mathbf{a}}$ for vanishing $\hbar$. In fact, $Q_\hbar$ is a strict and continuous deformation quantization in the sense of Rieffel (cf. [**55**]). The range of the map $Q_\hbar$ can be reproduced from a C^*-group algebra $C^*\mathcal{H}^\infty$ of the infinite dimensional Heisenberg group $\mathcal{H}^\infty := \Gamma\mathbf{H}^{\mathbf{a}} + \mathbb{R}\cdot e$. This group is characteristic for the vector field, too. On $\mathbb{R}\cdot e$ in $\mathcal{H}^\infty$ varies the deformation parameter $\hbar$ mentioned above. If this parameter approaches 0 the Heisenberg group deforms to $W^0\Gamma\mathbf{H}^{\mathbf{a}}$ containing the Poisson algebra $P^{\mathbf{a}}$. This family of Weyl algebras mentioned determines a C^*-algebra of so-called Weyl fields, $*$-isomorphic to $C^\infty\mathcal{H}^\infty$.

A representation of $W^\hbar\Gamma\mathbf{H}^{\mathbf{a}}$ or in some cases also of $C^*\mathcal{H}^\infty$ represents these C^*-algebras on the C^*-algebra $\mathbf{B}\mathcal{H}$, the C^*-algebra of all bounded operators of $\mathcal{H}$. The respective images are called the collection of physical quantum observables.

From here we construct the field operators and derive the canonical commutation relations (CCR). This construction is called the Weyl quantization of the vector field.

Thus the infinite dimensional Heisenberg group $\mathcal{H}^\infty$ governs the Weyl quantization and allows a classical limit as $\hbar$ tends to 0 in a continuous rigorous fashion.

We close chapter 10 by studying the influence of the topology of the three-manifold M to the quantization of the vector field X defined on M.

Now we have already alluded to the content of the later chapters of these notes. But let us start from the beginning. The first two chapters collect and prepare the mathematical material for the later ones. We intend to show that the quaternions $\mathbb{H}$ are a convenient tool to describe the geometry in three- and four-space naturally hidden in $SU(2)$. In particular, we investigate the automorphisms of $\mathbb{H}$; these automorphisms provide a link between the natural Minkowski geometry on $\mathbb{H}$ and the Euclidean one on E. These studies open the doors to the Hopf bundles on S^2. We hence pay a little more attention to this skew field than the mere application of Heisenberg groups and Heisenberg algebras would require.

The C^*-quantization associated with a singularity free vector field in 3-space requires the notion of a Heisenberg algebra bundle associated with it. These bundles naturally contain the complex line bundles of such vector fields. The geometry of these line bundles is treated in chapter 3. A classification of them, in terms of homotopy theory and Chern classes, is the goal of chapter 4. It prepares the effect of the topology to the field quantization done in chapter 10.

In chapter 5 Heisenberg groups and their Lie algebras are introduced. We need them in the quantization of homogeneous polynomials in two variables. We observe that the skew field of quaternions is determined by only one Heisenberg group or one Heisenberg algebra inside of $\mathbb{H}$. These groups and algebras link Euclidean and Minkowski geometry. The close ties of $\mathbb{H}$ and Heisenberg algebras with Minkowski geometry are exhibited and group theoretically formulated. Here the symplectic group and $\mathrm{SL}(2,\mathbb{C})$ reproduce isometry groups of three- and four-dimensional Minkowski spaces.

The main tools of the quantization of vector fields in 3-space are their Heisenberg group bundles and the C^*-algebras of sections of them. The infinite dimensional C^*-Heisenberg group as well as a natural C^*-Weyl algebra emanate from the vector fields. Both are characteristic for the field. To show this is the topic of chapter 6.

The Schrödinger representation of Heisenberg groups and the metaplectic representation are the basic topics of chapter 7. These representations are essential for the quantization of inhomogeneous quadratic polynomials. Both representations influence signal analysis and geometric optics fundamentally.

The notes end with a remarkable appendix by Serge Preston. The deep relations between information theory and thermodynamics are well recognized and utilized as documented in the references of the appendix. Therefore, in the spirit of the approach of this monograph, one might expect the Heisenberg group to play some prominent role in geometrical structures of thermodynamics. This is beautifully presented in this appendix.

Hence after the applications in Chapters 8 to 10 described above, the appendix "Thermodynamics, Geometry and Heisenberg group" provides an answer to the following question: The energy-phase space (P, θ, G) of a homogeneous thermodynamical system, together with its contact structure θ and natural indefinite metric G introduced by R. Mrugala, is isomorphic to the Heisenberg group $\mathcal{H}_n$ endowed with a right-invariant contact structure and the right-invariant indefinite metric G. Different properties of these structures are studied in terms of curvature and isometries of the metric G. Geodesics of the metric G are closely related with the three-dimensional Heisenberg subgroups $\mathcal{H}_1$ of the group $\mathcal{H}_n$. A natural compactification $(\hat{P}, \hat{\theta}, \hat{G})$ of the triple (P, θ, G) with its stratification by the subgroups of the type $R_k \times \mathcal{H}_{n-k}, k = 0, \ldots, n$ is investigated.

The above outlines of the chapters shall be complemented by a short description of the main interdependencies which are graphically visualized in the following diagram:

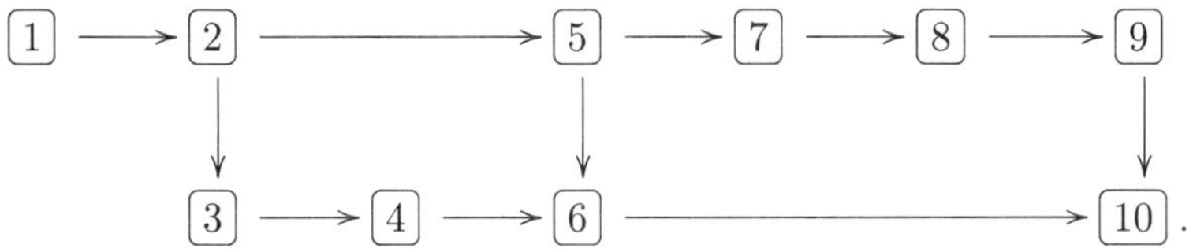

The first chapter provides the basis for the second one and presents techniques used throughout the book. Results of both of them are applied to smooth vector fields in 3-space by means of complex line bundles in the third chapter. The classification of these vector fields in terms of complex line bundles in chapter 4 is based on the earlier chapters, however, new technical means are introduced and interrelated with earlier ones in order to understand and formulate the classifications mentioned. The techniques provided by the first two chapters are used in chapter 5 to interrelate the concept of a Heisenberg group and a Heisenberg algebra with Euclidean and Minkowski as well as with symplectic geometry. The Schrödinger and the metaplectic representations introduced in chapter 7 form the basis for all the later ones. Chapter 6 lays the foundation for chapter 10 and uses among newly introduced techniques the ones provided in the first three chapters. The Chapters 8 and 9 use the representation theory presented in chapter 7 and require material from chapter 5. The last chapter on field quantization is based on chapter 6 and part of the study of the topological influence to this quantization relies on chapter 4.

Finally, a word on the prerequisites: The first three chapters as well as chapter 5 can easily be read with a background in linear algebra and elementary differential geometry as provided by senior undergraduate or low level graduate courses. Technically more involved are chapters 4, 6 and 7. With the prerequisites mentioned for chapter 4, the reader has to invest some time reading the topological part involving the classification of complex line bundles and the section concerning the mapping degree. In chapter 6 different concepts are introduced and studied; it is partly self-contained. However, it is technically more advanced than earlier ones. In chapter 7 some knowledge in representation theory would be helpful. The literature referenced contains all the technicalities which are used. Technically less involved are the self-contained parts of chapters 8 and 9. Other, more advanced parts are complemented by references to standard literature. Some basic knowledge in functional analysis would make the reading easier. Both chapters, however, require knowledge from earlier ones, in particular from chapter 7 on representation theory. More knowledge from the earlier chapters and from the literature referenced is needed to follow the last chapter. Together with the appendix on thermodynamics it is the most complex one.

Many valuable discussions with mathematicians and physicists have been the basis of these notes. We are particularly indebted to Karl-Heinz Brenner, Maurice de Gosson, Basil Hiley, Reinhard Honegger, Andrei Khrennikov, Alfred Rieckers and Hartmann Römer (in alphabetical order). Sincere thanks go to all of them. Very special thanks go to Serge Preston who read carefully through the manuscript, made many very helpful remarks and suggestions, and wrote the appendix.

Our colleague Serge Preston would like to thank Professor Michel Goze from the Université de Haute Alsace for the attention to his work and useful information and references.

Sincere thanks go to Mrs. S. Braak and Mrs. G. Nusser for typing many parts of this manuscript in LaTeX.

CHAPTER 1

The Skew Field of Quaternions

For most of us, the first time we learned about the quaternions may have been in our introductory master class at university serving as an example for a skew field. However, once you start working with the quaternions, you find that they are a very powerful tool. This chapter is about that tool.

Looking from later chapters one might say that the first chapter provides us with the fundamental tools. But this is not all. In taking a little time we will notice that the structure of the quaternions is extremely rich and hence is much more than a starting point and a basis for our investigations. Furthermore, the quaternions allow us to combine many fields that do not seem to be connected at first sight. Hence in the following sections we will occasionally invite the reader to take a little tour into some more involved features of the quaternions.

But now let us start right at the beginning, i.e. with the definition of the quaternions.

1.1. Definition of the Field of Quaternions $\mathbb{H}$ and Elementary Formulae

1.1.1. The Definition. To treat the geometry of a *three-dimensional, oriented Euclidean space* E in a more algebraic manner, it is convenient to enlarge E by one dimension since the enlarged $\mathbb{R}$-linear space naturally admits the structure of a *skew field.* This field will be denoted by $\mathbb{H}$ and will be called the *skew field of quaternions.* We will come to this picture of $\mathbb{H}$ a little later. First we will present a specific $\mathbb{R}$-linear subspace of the $\mathbb{C}$-linear space $\mathrm{End}_{\mathbb{C}}\mathbb{C}^2$ which under addition and composition forms a skew field naturally isomorphic to $\mathbb{H}$. Here $\mathrm{End}_{\mathbb{C}}\mathbb{C}^2$ denotes the $\mathbb{C}$-algebra of all $\mathbb{C}$-linear endomorphisms of $\mathbb{C}^2$ under the natural linear operations and composition.

The geometries of the four- and the three-dimensional spaces will thus be treated from one common point of view. This is especially important since $\mathbb{H}$ naturally encompasses a whole variety of fundamental notions such as $\mathrm{SU}(2)$, Heisenberg groups, Heisenberg algebras, Jordan algebras and Minkowski spaces, etc., all of which will be of use in this work.

As mentioned above, we use a realization of $\mathbb{H}$ in $\mathrm{End}_{\mathbb{C}}\mathbb{C}^2$. The complex four-dimensional space $\mathrm{End}_{\mathbb{C}}\mathbb{C}^2$ carries a natural *Hermitian structure* (cf. [**43**]) given by

$$(1) \qquad (h|k) := \frac{1}{2} \cdot \mathrm{tr}\ h \cdot \tilde{k} \qquad \forall\, h, k \in \mathrm{End}_{\mathbb{C}}\mathbb{C}^2.$$

Now, the skew field $\mathbb{H}$ is described as follows (cf. [**43**]):

DEFINITION 1. *Given $\mathbb{C}^2$ together with the standard Hermitian inner product,*

$$\mathbb{H} := \{h \in \text{End}_{\mathbb{C}}\mathbb{C}^2 \mid h + \tilde{h} = (\text{tr } h) \cdot id\}$$

is a skew field under the addition and composition in $\text{End}_{\mathbb{C}}\mathbb{C}^2$. The map $\tilde{h}$ is the adjoint of h and $\cdot$ denotes the composition and the multiplication by scalars. Let us write e instead of id.

1.1.2. Scalar Product and Norm. Any $h \in \mathbb{H}$ has a *real trace*, and $\mathbb{H}$ is an $\mathbb{R}$-algebra. Hence

$$< h, k > := \frac{1}{2} \cdot \text{tr } h \cdot \tilde{k} \qquad \forall\, h, k \in \mathbb{H}$$

is a ($\mathbb{R}$-valued) *scalar product* on $\mathbb{H}$.

For any $h \in \mathbb{H}$ the *norm* $|h|$ is derived from the scalar product as usual by

$$|h| := < h, h >^{\frac{1}{2}} .$$

1.1.3. Matrix Notation. The matrix $M(h)$ of any $h \in \mathbb{H}$ formed with respect to the complex basis $(1,0), (0,1)$ is given by

$$M(h) = \begin{pmatrix} \alpha & \beta \\ -\bar{\beta} & \bar{\alpha} \end{pmatrix}$$

with $\alpha, \beta \in \mathbb{C}$. From the structure of $M(h)$ we immediately see that

$$\dim_{\mathbb{R}} \mathbb{H} = 4.$$

If no confusion arises, we will identify h with its matrix. Its determinant is

$$\det h = \alpha \cdot \bar{\alpha} + \beta \cdot \bar{\beta},$$

a real number. The norm $|h|$ obviously satisfies

$$|h|^2 = \det\ h \tag{2}$$

for any $h \in \mathbb{H}$ and hence

$$|h \cdot k| = |h| \cdot |k| \qquad \forall\, h, k \in \mathbb{H}.$$

If $M_{\mathbb{C}}(2)$ denotes the collection of all two by two matrices with complex entries, then

$$M : \mathbb{H} \longrightarrow M_{\mathbb{C}}(2), \qquad h \longmapsto M(h)$$

is an isomorphism onto the skew-field $M(\mathbb{H})$ in $M_{\mathbb{C}}(2)$.

1.1.4. $\mathbb{C}^2$ and $\mathbb{H}$. The linear space $\mathbb{C}^2$ itself can be equipped with a multiplication turning it into a skew field isomorphic to $\mathbb{H}$. This multiplication can be derived from the matrix multiplication above and is given by

$$(\alpha_1, \beta_1) \cdot (\alpha_2, \beta_2) := \big(\alpha_1 \cdot \alpha_2 - \beta_1 \cdot \bar{\beta}_2\ ,\ \alpha_1 \cdot \beta_2 + \beta_1 \cdot \bar{\alpha}_2\big)$$

for any $(\alpha_s, \beta_s) \in \mathbb{C}^2$ with $s = 1, 2$. Hence the assignment

$$\text{Mat} : \mathbb{C}^2 \longrightarrow M(\mathbb{H}) \subset M_{\mathbb{C}}(2), \qquad \text{Mat}(\alpha, \beta) := \begin{pmatrix} \alpha & \beta \\ -\bar{\beta} & \bar{\alpha} \end{pmatrix}$$

is an isomorphism between skew fields.

The *skew field of quaternions* has so far the three realizations as $\mathbb{H}$, $M(\mathbb{H})$ and $\mathbb{C}^2$. We will make use of all these realizations throughout these notes.

1.1.5. E and $\mathbb{H}$. So far we neglected the fact that an oriented Euclidean space causes a skew field if enlarged by one dimension, as mentioned at the beginning of this section. This extra dimension of $\mathbb{R} \cdot e$ provides the unity element and hence the center of the skew field. This aspect shall now be prepared.

By construction the quaternions naturally split into

$$\mathbb{H} = \mathbb{R} \cdot e \oplus E$$

where $E := e^{\perp}$. Since $\operatorname{tr} h = 2 \cdot < h, e >$ for any $h \in \mathbb{H}$, we have $h \in E$ iff it is traceless. For any $u \in E$,

$$M(u) = \begin{pmatrix} \gamma & \beta \\ -\bar{\beta} & -\gamma \end{pmatrix}$$

with $\gamma \in \mathbb{R}$ and $\beta \in \mathbb{C}$, saying that $M(u)$ is *skew Hermitian.* Hence E can be identified with the $\mathbb{R}$-linear space su(2) of all skew Hermitian elements in $\operatorname{End}_{\mathbb{C}} \mathbb{C}^2$. Therefore,

$$u^2 = -|u|^2 \cdot e$$

for any $u \in E$ and, in particular,

$$(3) \qquad u^2 = -e \quad \text{for} \quad |u| = 1.$$

The linear space E is a real vector space of dimension three and carries the scalar product on $\mathbb{H}$ restricted to E. As we will see below, E is oriented and hence an *oriented Euclidean space.*

1.1.6. Inverse Elements. To compute h^{-1} for any $h = \lambda \cdot e + u \in \mathbb{H}$ we use the splitting $\mathbb{H} = \mathbb{R} \cdot e \oplus E$ and introduce the conjugate $\bar{h}$ given by

$$\bar{h} = \overline{\lambda \cdot e + u} = \lambda \cdot e - u.$$

Hence the element $u \in E$ is also called the *imaginary part* of h. Then

$$(4) \qquad h^{-1} = \frac{\bar{h}}{|h|^2} \quad \forall\, h \in \mathbb{H} \qquad \text{and} \qquad u^{-1} = -\frac{u}{|u|^2} \quad \forall\, u \in E.$$

Let us point out here that

$$(5) \qquad (\lambda \cdot e + u) \cdot \overline{(\lambda \cdot e + u)} = \lambda^2 + |u|^2 \qquad \forall\, \lambda \in \mathbb{R} \text{ and } \forall\, u \in E.$$

Moreover, any $h \in \mathbb{H}$ satisfies

$$h + \bar{h} = 2 \cdot \lambda \cdot e,$$

implying that

$$(6) \qquad \bar{\bar{h}} = \bar{h} \qquad \forall\, h \in \mathbb{H}.$$

Since $\overline{u_1 \cdot u_2} = \bar{u}_2 \cdot \bar{u}_1$ for all $u_1, u_2 \in E$, the equation

$$\overline{h_1 \cdot h_2} = \bar{h}_2 \cdot \bar{h}_1 \qquad \forall\, h_1, h_2 \in \mathbb{H}$$

holds true.

If $h \neq 0$ and $k \cdot h^{-1} = h^{-1} \cdot k$, then we will write it as $\frac{k}{h}$ in the following.

1.1.7. Commutator. Given any two $h, k \in \mathbb{H}$, its commutator $[h, k]$ is defined by

$$[h, k] := h \cdot k - k \cdot h.$$

We have $< [h, k], e >= 0$ and thus $[h, k] \in E$.

Together with the commutator we consider $h \cdot \tilde{k} + k \cdot \tilde{h}$ and obtain

$$h \cdot \tilde{k} + k \cdot \tilde{h} = 2 \cdot < h, k > \cdot e,$$

as for example can be shown by performing the matrix calculations. This implies

$$h \cdot k = \frac{1}{2} \cdot [h, k] - < h, k > \cdot e \qquad \forall\, h, k \in E.$$

1.1.8. The Special Unitary Group in $\mathbb{H}$. It is a remarkable fact that the unit sphere $S^3 \subset \mathbb{H}$ admits a group structure. This is due to (2) since

$$|h \cdot k| = |h| \cdot |k| = \det h \cdot \det k \qquad \forall\, h, k \in \mathbb{H}$$

and thus the collection of all $h \in \operatorname{End} \mathbb{C}^2$ with $\det h = 1$ forms a group called $\mathrm{SU}(2)$. Obviously,

$$S^3 = \mathrm{SU}(2).$$

This group $\mathrm{SU}(2) \subset \mathbb{H}$ of all *unitary* mappings in $\operatorname{End}_{\mathbb{C}} \mathbb{C}^2$ with determinant one is called the *special unitary group*. Its *Lie algebra* $\mathrm{su}(2)$ is a complement to $\mathbb{R} \cdot e$, i.e.

$$\mathbb{H} = \mathbb{R} \cdot e \oplus \mathrm{su}(2),$$

with half of the commutator as Lie product. This explains the earlier notation of $\mathrm{su}(2) = E$.

1.1.9. Orientation on E and on $\mathbb{H}$. The commutator $[\ ,\ ]$ in $\mathbb{H}$ yields an orientation

$$\mu : \overset{3}{\times} E \longrightarrow \mathbb{R}$$

on E given by

$$\mu(u_1, u_2, u_3) :=< \frac{1}{2} \cdot [u_1, u_2], u_3 > \qquad \forall\, u_1, u_2, u_3 \in E$$

on the threefold Cartesian product $\overset{3}{\times} E$ of E. Hence

$$u_1 \times u_2 := \frac{1}{2} \cdot [u_1, u_2] \qquad \forall\, u_1, u_2 \in E \tag{7}$$

is the *cross product* associated with μ and $<\ ,\ >$. Therefore, the product of any two $u_1, u_2 \in E$ in $\mathbb{H}$ can be rephrased as

$$u_1 \cdot u_2 = u_1 \times u_2 - < u_1, u_2 > \cdot e \tag{8}$$

and the cross product is the *Lie product* in $\mathrm{su}(2)$. Thus

$$\mu(u_1, u_2, u_3) =< u_1 \cdot u_2, u_3 > \qquad \forall\, u_1, u_2, u_3 \in E.$$

This orientation is extended to all of $\mathbb{H}$ by choosing a normed volume form $\mu_{\mathbb{H}}$ such that

$$\mu_{\mathbb{H}}(e, u_1, u_2, u_3) :=< \frac{1}{2} \cdot [u_1, u_2], u_3 > \qquad \forall\, u_1, u_2, u_3 \in E.$$

Since both orientations can be constructed from each other, we will denote both the orientation on $\mathbb{H}$ as well as the one on E by μ. Together with the scalar product $<\ ,\ >$, the space E will be our standard Euclidean space; its orientation will occasionally also be called μ_E.

Of course we may now start with an $\mathbb{R}$-linear three-dimensional, oriented Euclidean space E, choose the unit element e and define the product $u_1 \cdot u_2$ in $\mathbb{R} \cdot e + E$ of any two elements $u_1, u_2 \in E$ by (8). The product

$$(\lambda_1 \cdot e + u_1) \cdot (\lambda_2 \cdot e + u_2) \\ := (\lambda_1 \cdot \lambda_2 - < u_1, u_2 >) \cdot e + \lambda_1 u_2 + \lambda_2 u_1 + u_1 \times u_2 \tag{9}$$

for all $\lambda_1, \lambda_2 \in \mathbb{R}$ and any $u_1, u_2 \in E$ yields a skew field called $\mathbb{H}$ again. Hence

$$\mathbb{H} = \mathbb{R} \cdot e + E.$$

1.1.10. Minkowski Metric. On $\mathbb{H}$ there is a natural *Minkowski metric*. To see this let $h_s \in \mathbb{H}$ for $s = 1, 2$ be split into $h_s = \lambda_s \cdot e + u_s$. Setting

$$g_M(h_1, h_1) := \frac{1}{2} \cdot \operatorname{tr}(h_1 \cdot h_2) = \lambda_1 \cdot \lambda_2 - < u_1, u_2 > \tag{10}$$

yields a *Minkowski metric* g_M of signature $+---$. Hence if $h = \lambda \cdot e + u \in \mathbb{H}$,

$$g_M(h, h) = \lambda^2 - |u|^2 \tag{11}$$

(cf. (3) and (9)). If h is in the fore-cone or in the past-cone, i.e. if $g_M(h, h) < 0$, we set

$$|h|_M^2 := -g_M(h, h).$$

1.1.11. Minkowski Space in Physics. The Minkowski space often used in physics is of the following type: The four-dimensional $\mathbb{R}$-linear space $\mathbf{M}$ of all *self-adjoint* mappings in $\operatorname{End}_{\mathbb{C}}\mathbb{C}^2$ is of the form

$$\mathbf{M} = \mathbb{R} \cdot e \oplus i \bullet E.$$

Here $\bullet$ denotes the pointwise multiplication of the endomorphism u_s by the scalar $i \in \mathbb{C}$. A natural Minkowski metric on $\mathbf{M}$ is given by

$$g_{\mathbf{M}}(V_1, V_2) = \frac{1}{2} \cdot (\operatorname{tr}\, (V_1 \cdot \tilde{V}_2) - \operatorname{tr}\, V_1 \cdot \operatorname{tr}\, \tilde{V}_2) \qquad \forall\, V_1, V_2 \in \mathbf{M}.$$

Hence

$$g_{\mathbf{M}}(V, V) = \det V \qquad \forall\, V \in \mathbf{M}$$

due to the Cayley-Hamilton theorem (cf. [**43**]). It coincides with the scalar product on E introduced above and is a Minkowski metric on $\mathbf{M}$ with signature $+\ -\ -\ -$.

1.1.12. Minkowski Space and $\mathbb{H}$. Looking at the definition of the Minkowski space $\mathbf{M}$, it is natural to look for a map linking $\mathbb{H}$ and $\mathbf{M}$. In fact, $\mathbb{H}$ and $\mathbf{M}$ are related by the $\mathbb{R}$-linear map

$$\operatorname{quat} : \mathbb{H} \longrightarrow \mathbf{M}, \qquad \operatorname{quat} h := \frac{1-i}{2} \bullet e \cdot \operatorname{tr} h + i \bullet h, \tag{12}$$

an isomorphism. Its inverse quat^{-1} is

$$\operatorname{quat}^{-1}(V) = \frac{1+i}{2} \bullet e \cdot \operatorname{tr} \operatorname{quat} V - i \bullet \operatorname{quat} V \qquad \forall\, V \in \mathbf{M}.$$

As shown in [**43**],

$$< \operatorname{quat} h, \operatorname{quat} k >= -\frac{1}{2} \operatorname{tr} h \cdot k \qquad \forall\, h, k \in \mathbb{H},$$

showing that quat is an isometry of $\mathbb{H}$ (equipped with g_M) to $\mathbf{M}$. For any $h \in E$,

$$\operatorname{quat} h = i \bullet h$$

holds true.

1.1.13. Lorentz group. Having introduced the Minkowski space $\mathbf{M}$, let us determine the *Lorentz group* $O_0(1,3)$ of $\mathbf{M}$, i.e. the group of all *proper* (i.e. having determinant one, *orthochronous*) $\mathbb{R}$-linear maps of $\mathbf{M}$ leaving $g_{\mathbf{M}}$ invariant. To this end we consider $\mathrm{SL}(2,\mathbb{C}) \subset \mathrm{End}_{\mathbb{C}}\, \mathbb{C}^2$ consisting of all $\mathbb{C}$-linear transformations of $\mathbb{C}^2$ with determinant one.

Given $S \in \mathrm{SL}(2,\mathbb{C})$ and $V \in \mathbf{M}$, let

$$\Phi_L(S)(V) := S \circ V \circ \tilde{S} \qquad \forall\, V \in \mathbf{M},$$

with $\tilde{S}$ being the adjoint of S. We see that $\Phi_L(S)$ is a Lorentz transformation:

$$\begin{aligned} g_{\mathbf{M}}(\Phi_L(S)(V), \Phi_L(S)(V)) &= \det \Phi_L(S \circ V \circ \Phi_L(\tilde{S})) \\ &= \det S^2 \det V \\ &= -\det S^2 \cdot g_{\mathbf{M}}(V,V). \end{aligned}$$

Thus we have a surjective map

$$\Phi_L : \mathrm{SL}(2,\mathbb{C}) \longrightarrow O_0(1,3)$$

with e and $-e$ as kernel (cf. [**43**], [**78**]). In fact, $O_0(1,3)$ is the connected component of id in $O(1,3)$. Hence $\mathrm{SL}(2,\mathbb{C})$ is a *twofold covering* of $O_0(1,3)$.

1.1.14. Jordan Product and Lie Product. The multiplication in $\mathbb{H}$ can be split in a quite natural way introducing a Jordan product. We write the product $k_1 \cdot k_2$ as

$$k_1 \cdot k_2 = \frac{1}{2}(k_1 \cdot k_2 + k_2 \cdot k_1) + \frac{1}{2}(k_1 \cdot k_2 - k_1 \cdot k_2). \tag{13}$$

The first term on the right-hand side is the *Jordan product* denoted by $k_1 \odot k_2$, the second one is half of the commutator (cf. (7)). Thus

$$k_1 \cdot k_2 = k_1 \odot k_2 + \frac{1}{2}[k_1, k_2] = k_1 \odot k_2 + u_1 \times u_2 \tag{14}$$

for $k_s = \lambda_s \cdot e + u_s$ with $\lambda_s \in \mathbb{R}$ and $u_s \in E$ where $s = 1, 2$. Hence the product $k_1 \cdot k_2$ is the amalgam of the Jordan product and the Lie product on E. Clearly, the Jordan product is the polarization of the *squaring operator* on $\mathbb{H}$ which sends each $k \in \mathbb{H}$ into k^2.

The natural splitting of k_s yields

$$k_1 \cdot k_2 = (\lambda_1 \cdot \lambda_2 - < u_1, u_2 >) \cdot e + \lambda_1 \cdot u_2 + \lambda_2 \cdot u_1 + u_1 \times u_2$$

or

$$k_1 \cdot k_2 = g_M(k_1, k_2) + \lambda_1 \cdot u_2 + \lambda_2 \cdot u_1 + u_1 \times u_2$$

and hence

$$k_1 \odot k_2 = (\lambda_1 \cdot \lambda_2 - < u_1, u_2 >) \cdot e + \lambda_1 \cdot u_2 + \lambda_2 \cdot u_1. \tag{15}$$

In this decomposition of $k_1 \odot k_2$ one can easily see that the *Minkowski metric* g_M on $\mathbb{H}$ enters naturally, and (15) turns into

$$k_1 \odot k_2 = g_M(k_1, k_2) \cdot e + \lambda_1 \cdot u_2 + \lambda_2 \cdot u_1. \tag{16}$$

Clearly, the Jordan algebra structure can be transferred to the Minkowski space $\mathbf{M}$ by the map quat in section 1.1.12. This algebraic structure reproduces the Minkowski metric again.

1.2. Embeddings of $\mathbb{C}$ into the Quaternions and Natural Unitary Groups

Some properties of $\mathbb{H}$ such as $u^2 = -|u|^2 \cdot e$ for all $u \in E$ or the existence of the conjugation, mapping any h in $\mathbb{H}$ into $\bar{h}$ suggest that we may embed the field of complex numbers $\mathbb{C}$ into the quaternions. However, there is no canonical embedding of $\mathbb{C}$ into $\mathbb{H}$ but rather one embedding for each a in the unit sphere S^2 of E. In this section we will study these embeddings of $\mathbb{C}$ into $\mathbb{H}$ and their parametrization by S^2 as well as some implications in more detail.

1.2.1. Complex Numbers in $\mathbb{H}$. Any $a \in S^2$ satisfies $a^2 = -e$ (cf. (3) in section 1.1.5) and thus may be interpreted as an *imaginary unit.* In fact, the real two-dimensional subspace

$$\mathbb{C}^a := \{\lambda \cdot e + \xi \cdot a \,|\, \lambda, \xi \in \mathbb{R}\}$$

is a *commutative* subfield of $\mathbb{H}$ called a *realization of* $\mathbb{C}$ *in* $\mathbb{H}$ and

$$i^a : \mathbb{C} \longrightarrow \mathbb{C}^a \subset \mathbb{H}, \qquad \lambda + \xi \cdot i \longmapsto \lambda \cdot e + \xi \cdot a \tag{17}$$

is an isomorphism of fields. There are as many embeddings i^a as there are elements $a \in S^2$.

Since any $k \in \mathbb{H}$ splits into

$$k = \lambda \cdot e + u = \lambda \cdot e + |u| \cdot \frac{u}{|u|},$$

we observe that $k \in \mathbb{C}^{\frac{u}{|u|}}$, saying that k can be interpreted as a complex number in $\mathbb{C}^{\frac{u}{|u|}}$ by using the isomorphism $i^{\frac{u}{|u|}}$ in (17).

Fixing some $a \in S^2$, the isomorphism i^a provides $\mathbb{H}$ with a $\mathbb{C}$-algebra structure; in fact, $\mathbb{H}$ is even a C^* algebra (cf.[**3**]) with the conjugation as involution. The splitting (14) of the product into a symmetric and skew-symmetric part is a typical feature of a C^*-algebra (cf. [**3**]).

1.2.2. Unitary Group and Exponential Map. The field $\mathbb{C}^a$ for $a \in S^2 \subset E$ gives rise to the introduction of the unitary group $U^a(1)$ as the unit circle in $\mathbb{C}^a$. We set

$$U^a(1) := i^a(U(1)).$$

Every element $z^a \in U^a(1)$ is of the form

$$z^a = e \cdot \cos t + a \cdot \sin t,$$

which we write as $z^a = \cos t + a \cdot \sin t$ wherever the meaning is clear. Hence

$$z^a = i^a(\cos t + i \cdot \sin t).$$

Moreover, the series

$$\exp^a(t \cdot a) := \text{id} + \sum_{n=1}^{\infty} \frac{(t \cdot a)^n}{n!}$$

converges uniformly in $\mathbb{H}$ for any $t \in \mathbb{R}$; its limit is also denoted by $e^{t \cdot a}$. As one easily verifies,

$$e^{t \cdot a} = \cos t + a \cdot \sin t = i^a \left(e^{t \cdot i}\right). \tag{18}$$

The conjugate of $e^{t \cdot a} = \cos t + a \cdot \sin t$ is given by $e^{-t \cdot a} = \cos t - a \cdot \sin t$.

1.2.3. The 2-by-2-Splitting of $\mathbb{H}$. Let $a \in S^2$. If we denote the (two-dimensional) orthogonal complement of $\mathbb{C}^a$ in $\mathbb{H}$ by F^a, the skew field $\mathbb{H}$ can be split into

$$\mathbb{H} = \mathbb{C}^a \oplus F^a, \tag{19}$$

called the 2-*by*-2-*splitting* of $\mathbb{H}$, referred to often as the $2+2$-splitting. In terms of this splitting, any $k \in S^3 \equiv \mathrm{SU}(2)$ is uniquely written as $k = z + h$ with $z \in \mathbb{C}^a$ and $h \in F^a$ satisfying

$$|z|^2 + |h|^2 = 1.$$

If $z \in U^a(1)$ and if h is an element of the unit circle $K_1 \subset F^a$, then $k = r \cdot z + r' \cdot h$ for some $r, r' \in \mathbb{R}$ for which $r^2 + r'^2 = 1$. Hence $\mathrm{SU}(2)$ is fibred into tori

$$\mathrm{SU}(2) = \bigcup_{r^2+r'^2=1} r \cdot U^a(1) \oplus r' \cdot K_1, \tag{20}$$

some of which are degenerate to circles, of course (cf. [**68**]).

1.2.4. Operation of $\mathbb{C}^a$ on F^a. By the multiplication in $\mathbb{H}$, we observe:

LEMMA 2. *Let $a \in S^2$ and $z^a = \lambda \cdot e + \xi \cdot a \in \mathbb{C}^a$. Then with $\bar{z}^a = \lambda \cdot e - \xi \cdot a$*

$$z^a \cdot v = v \cdot \bar{z}^a \qquad \forall v \in F^a. \tag{21}$$

PROOF. To verify this, let $z^a = |z^a| \cdot e^{t \cdot a} = |z^a| \cdot (e \cdot \cos t + a \cdot \sin t)$ for $t \in \mathbb{R}$. For any $v \in F^a$,

$$\begin{aligned} z^a \cdot v &= |z^a| \cdot (e \cdot \cos t + a \cdot \sin t) \cdot v \\ &= |z^a| \cdot (v \cdot \cos t + a \cdot v \cdot \sin t) \\ &= |z^a| \cdot (v \cdot \cos t - v \cdot a \cdot \sin t) \\ &= v \cdot \bar{z}^a, \end{aligned}$$

as proposed. □

Using the structure of the Euclidean space

$$E = \mathbb{R} \cdot a \oplus F^a \tag{22}$$

we deduce from (9) and by the properties of the cross product $a \cdot v \in F^a$ that

$$z^a \cdot v \in F^a \qquad \forall\, z^a \in \mathbb{C}^a \text{ and } \forall v \in F^a.$$

Moreover,

$$v \times w = (|v| \cdot |w| \cdot \sin \beta) \cdot a \qquad \forall v, w \in F^a,$$

where β is the oriented angle between v and w. Thus

$$v \cdot w = v \times w - < v, w > \cdot e \in \mathbb{C}^a.$$

This allows us to verify that the product in $\mathbb{H}$ induces the following left and right actions:

$$\begin{aligned} \mathbb{C}^a \cdot \mathbb{C}^a &= \mathbb{C}^a, \\ \mathbb{C}^a \cdot F^a &= F^a, \qquad (23) \\ F^a \cdot \mathbb{C}^a &= F^a, \qquad (24) \\ F^a \cdot F^a &= \mathbb{C}^a. \end{aligned}$$

One easily verifies that F^a is a left and right $\mathbb{C}^a$-linear space.

From these remarks it follows that the choice of a also turns $\mathbb{H}$ into the *Clifford algebra* of $(F^a, -<\ ,\ >)$ (cf. [**44**]).

1.2.5. Rotations in $\mathbb{C}^a$ and $\mathbb{F}^a$. To interpret the exponential map as a rotation in $\mathbb{H}$ for each $t \in \mathbb{R}$ and $a \in S^2$, we consider the linear map

$$\ell_{\exp^a(t)} : \mathbb{H} \longrightarrow \mathbb{H}, \qquad \ell_{\exp^a(t)}(k) := e^{t \cdot a} \cdot k$$

denoting the left multiplication by $e^{t \cdot a}$. More generally, we may replace a by $|a| \cdot \frac{a}{|a|}$ to define $\exp_a t$ for all $a \in \dot{E} := E \backslash \{0\}$.

Due to (18) we observe that

$$|\ell_{\exp^a}(k)| = |k| \qquad \forall\, k \in \mathbb{H}$$

exhibiting $\ell_{\exp^a}$ as a rotation on $\mathbb{H}$. Furthermore, we observe

$$\ell_{\exp^a t}(z) \in \mathbb{C}^a \qquad \forall\, z \in \mathbb{C}^a$$

and

$$\ell_{\exp^a t}(h) \in F^a \qquad \forall\, h \in F^a.$$

Thus

$$\ell_{\exp^a t}|_{\mathbb{C}^a} \qquad \text{and} \qquad \ell_{\exp^a t}|_{F^a}$$

are rotations in $\mathbb{C}^a$ and F^a, respectively. None of these rotations in a plane admits an axis of rotation, of course.

Let $k = \lambda \cdot a + u \in E$. Then $\ell_{\exp^a t}(k)$ is not in E, in general. To produce a rotation in E we set for all $k \in E$,

$$r_{\exp^a t}(k) = r_{\exp^a t}(\lambda \cdot a + u) := \lambda a + \ell_{\exp^a t}(u).$$

Obviously, $r_{\exp^a t}$ is a rotation with $\mathbb{R} \cdot a$ as axis of rotation and t as oriented angle of rotation. Since the plane F^a is perpendicular to a, any $h \in F^a$ is rotated by the angle t. Therefore, we call F^a the *plane of rotation*.

1.2.6. Hermitian Product on $\mathbb{H}$. From the above we see that the choice of a turns both F^a and $\mathbb{H}$ into $\mathbb{C}^a$-left and right-linear spaces. Moreover, $\mathbb{H}$ carries a natural *Hermitian product* (or Hermitian structure) as follows: Having fixed $a \in \mathbb{C}^a$ with $|a| = 1$ we define the Hermitian product $(\ |\ )_a$ by

$$(k_1|k_2)_a := z_1^1 \overline{z_2^1} + z_1^2 \cdot \overline{z_2^2} \tag{25}$$

provided $k_s = z_s^1 \cdot e + z_s^2 \cdot a$ for $s = 1, 2$. It is easily seen that e and a are orthonormal. Of course $\mathbb{H}$ is complete, i.e. a complex Hilbert space.

1.3. ℂ-linear Structures, Symplectic Structures and Orientation, Pauli Elements

1.3.1. ℂ-Linear Structures on F^a. Let $a \in \dot{E}$ where $\dot{E} := E \backslash \{0\}$. The orthogonal complement of a formed in E is the $\mathbb{R}$-linear space F^a. Instead of $\mathbb{C}^{\frac{a}{|a|}}$ we just write $\mathbb{C}^a$ in the sequel. As shown by the equations (23) and (24), F^a carries both a ℂ-left and a ℂ-right-linear structure formulated pointwisely by

$$v \cdot z := v \cdot i^a(z) \qquad \text{and} \qquad z \cdot v := i^a(z) \cdot v$$

for all $z \in \mathbb{C}$ and all $v \in F^a$. By equation (21)

$$\bar{z} \cdot v = v \cdot z \qquad \forall z \in \mathbb{C} \text{ and } \forall v \in F^a.$$

Remark. *If we refer to F^a as a $\mathbb{C}$-linear space or equivalently to a* complex line, *we always mean the* $\mathbb{C}$-right-*linear structure on* F^a.

1.3.2. F^a as a Complex Field. There is no canonical complex field structure on F^a. It depends on the choice of a *unit vector* in F^a. Given such a unit vector $v \in F^a$, the $\mathbb{R}$-linear space F^a inherits the left and right $\mathbb{C}$-linear structures from

$$F^a = v \cdot \mathbb{C}^a = \mathbb{C}^a \cdot v.$$

The right (left) $\mathbb{C}$-linear structures turn F^a into a field isometrically isomorphic to $\mathbb{C}^a$.

Alternatively formulated the *special rotation group* $\mathrm{SO}(F^a)$ parameterizes the totality of all complex field structures of F^a, isometrically isomorphic to $\mathbb{C}$.

1.3.3. Orientation on F^a. The two different $\mathbb{C}^a$-linear structures on F^a are in direct correspondence to the two orientations $\mu\left(\frac{a}{|a|}, \dots, \dots\right)$ and $\mu\left(-\frac{a}{|a|}, \dots, \dots\right)$ where μ is the fixed orientation (cf. section 1.1.9) on E. This is due to

$$v \cdot i = v \times \frac{a}{|a|} \qquad \text{and} \qquad i \cdot v = \frac{a}{|a|} \times v,$$

both holding for all $v \in F^a$. These equations yield for any two $v, w \in F^a$,

$$< v \cdot i, w >= \mu\left(v, \frac{a}{|a|}, w\right) = \mu\left(-\frac{a}{|a|}, v, w\right)$$

and

$$< i \cdot v, w >= \mu\left(\frac{a}{|a|}, v, w\right).$$

Thus the conjugation in $\mathbb{C}$ switches from one orientation on F^a to the other.

Since we will use the right action of $\mathbb{C}^a$ on F^a, we will orient F^a by μ_{F^a} as we set

$$\mu_{F^a}(h_1, h_2) := \mu\left(h_1, \frac{a}{|a|}, h_2\right).$$

1.3.4. F^a as a Phase Space. The choice of some $v \in F^a$ with $v \neq 0$ allows another interpretation of F^a which we will use in chapter 8: The splitting

$$F^a = \mathbb{R} \cdot \frac{v}{|v|} \oplus \mathbb{R} \cdot \frac{v}{|v|} \times a$$

with $a \in S^2$ turns F^a into a phase space. $\mathbb{R} \cdot \frac{v}{|v|} \times \; a$ is regarded as the tangent space at each point $\lambda \cdot \frac{v}{|v|}$ in the configuration space $\mathbb{R} \cdot \frac{v}{|v|}$. Thus $\mathbb{R} \cdot \frac{v}{|v|} \times \mathbb{R} \cdot a$ is the space of all velocities at each point in $\mathbb{R} \cdot \frac{v}{|v|}$. Hence $F^a = T(\mathbb{R} \cdot \frac{v}{|v|})$, the (trivial) tangent bundle of $\mathbb{R} \cdot \frac{v}{|v|}$.

By using the scalar product on F^a, each velocity at each point can uniquely be converted into a $\mathbb{R}$-valued linear map, called a *momentum* at the same point. The collection of all momenta at each point is denoted by $T^*(\mathbb{R} \cdot \frac{v}{|v|})$ and is called the cotangent bundle of $\mathbb{R} \cdot \frac{v}{|v|}$. In mechanics it is referred to as the *phase space*. We can elementwisely identify $T(\mathbb{R} \cdot \frac{v}{|v|})$ with $T^*(\mathbb{R} \cdot \frac{v}{|v|})$ and in turn with F^a. Thus F^a can be viewed as a phase space.

Accordingly, if $\mathbb{R} \cdot \frac{v}{|v|}$ is the configuration space of a periodic motion on $\mathbb{R}$ with period T, and length 2π, the space of all velocities is $\mathbb{R} \cdot \frac{v}{|v|} \times a$. A position x in $\mathbb{R}$ at time t is $t \cdot w$ where w is the velocity at the position x. The frequency ν is by definition $\frac{2\pi}{T}$. Hence the pair (x, w) is uniquely determined by the pair (t, ν). The plane F^a can, therefore, be identified as the *time-frequency plane*, playing the role of a phase space in signal analysis as used in chapter 8.

1.3.5. Symplectic Structures on F^a. The non-vanishing vector $a \in E$ determines two *symplectic structures* ω^a and $\bar{\omega}^a$, say, on F^a defined for all $h_1, h_2 \in F^a$ by

$$\omega^a(h_1, h_2) := < h_1 \times a, h_2 > \qquad \text{and} \qquad \bar{\omega}^a(h_1, h_2) := < a \times h_1, h_2 >,$$

respectively. Obviously,

$$|a| \cdot \mu_{F^a} = \omega^a \qquad \text{and} \qquad -|a| \cdot \mu_{F^a} = \bar{\omega}^a.$$

In this way, $\dot{\mathbb{R}}$ parameterizes the collection of all symplectic structures on F^a.

1.3.6. Pauli Matrices. The goal of this section is to relate the $2+2$-splitting of $\mathbb{H}$ with the Minkowski space $\mathbf{M}$ in section 1.1.11. The Pauli matrices in $\mathbf{M}$ will be identified with particular elements in $\mathbb{H}$, the Pauli elements. In section 1.3.2 we showed that we can represent $\mathbb{H}$ as

$$\mathbb{H} = \mathbb{C}^a \oplus \mathbb{C}^a \cdot q_0$$

for some unit vector $q_0 \in F^a$. For simplicity let $a \in S^2$ for the rest of this section.

For any two $h = \alpha \cdot e + \beta \cdot q_0$ and $k = \gamma \cdot e + \delta \cdot q_0$ with $\alpha, \beta, \gamma, \delta \in \mathbb{C}^a$ the equation

$$h \cdot k = (\alpha \cdot \gamma - \bar{\beta} \cdot \delta) \cdot e + (\alpha \cdot \delta + \bar{\beta} \cdot \gamma) \cdot q_0$$

holds true. Therefore, the map

$$j^l_{q_0} : \mathbb{C}^a \oplus \mathbb{C}^a \cdot q_0 \longrightarrow M_{\mathbb{C}}(2), \qquad j^l_{q_0}(\alpha \cdot e + \beta \cdot q_0) := \begin{pmatrix} \alpha & \beta \\ -\bar{\beta} & \bar{\alpha} \end{pmatrix}$$

is an isomorphism of $\mathbb{H}$ onto the skew field $M(\mathbb{H}) \subset M_{\mathbb{C}}(2)$. It is important to note that $j^l_{q_0}$ is *not* a $\mathbb{C}$-linear map with respect to the pointwisely defined $\mathbb{C}$-linear structure on $M_{\mathbb{C}}(2)$.

$\mathbb{H} = \mathbb{C}^a \oplus q_0 \cdot \mathbb{C}^a$ is a $\mathbb{C}$-linear space, as well. The multiplication on this $\mathbb{R}$-linear space is represented as

$$(\alpha \cdot e + q_0 \cdot \beta) \cdot (\gamma \cdot e + q_0 \cdot \delta) = (\alpha \cdot \gamma - \bar{\beta} \cdot \delta) \cdot e + q_0 \cdot (\overline{\alpha} \cdot \delta + \beta \cdot \gamma)$$

for all $\alpha, \beta, \gamma, \delta \in \mathbb{C}^a$. Moreover, $\mathbb{H}$ together with this multiplication can be converted into a skew subfield of matrices $M_{\mathbb{C}^a}(2)$. The injective map is given by

$$\begin{aligned} j^r_{q_0} : \mathbb{C}^a \oplus q_0 \cdot \mathbb{C}^a &\longrightarrow M_{\mathbb{C}^a}(2), \\ j^r_{q_0}(\alpha \cdot e + q_0 \cdot \beta) &:= \begin{pmatrix} \alpha & -\beta \\ \overline{\beta} & \overline{\alpha} \end{pmatrix}. \end{aligned}$$

$j^r_{q_0}$ is an anti-isomorphism onto the skew field $j^r_{q_0}(\mathbb{H})$ (cf. section 1.1.4). This is to say that

$$j^r_{q_0}((\alpha \cdot e + q_0 \cdot \beta)(\gamma \cdot e + q_o \cdot \delta)) = j^r_{q_0}(\gamma \cdot e + q_0 \cdot \delta) \cdot j^r_{q_0}(\alpha \cdot e + q_0 \cdot \beta).$$

In terms of $\mathbf{M} \subset M_{\mathbb{C}^a}(2)$ the Pauli matrices have the form

$$e = \begin{pmatrix} 1 & 0 \\ 0 & 1 \end{pmatrix}, \qquad \sigma_1 := \begin{pmatrix} 0 & 1 \\ 1 & 0 \end{pmatrix},$$
$$\sigma_2 := \begin{pmatrix} 0 & -i \\ i & 0 \end{pmatrix}, \qquad \sigma_3 := \begin{pmatrix} 1 & 0 \\ 0 & -1 \end{pmatrix}$$

(cf. [**43**]). Hence

$$\text{quat}^{-1}\begin{pmatrix} 1 & 0 \\ 0 & 1 \end{pmatrix}, \qquad \text{quat}^{-1}\begin{pmatrix} 0 & 1 \\ 1 & 0 \end{pmatrix},$$
$$\text{quat}^{-1}\begin{pmatrix} 0 & -i \\ i & 0 \end{pmatrix}, \qquad \text{quat}^{-1}\begin{pmatrix} 1 & 0 \\ 0 & -1 \end{pmatrix}$$

are matrices representing endomorphisms in $\mathbb{H}$. We call these quaternions the *Pauli elements* in $\mathbb{H}$.

In terms of the representations

$$\mathbb{H} = \mathbb{C}^a \oplus \mathbb{C}^a \cdot q_0 \qquad \text{and} \qquad \mathbb{H} = \mathbb{C}^a \oplus q_0 \cdot \mathbb{C}^a$$

of $\mathbb{H}$, the Pauli elements take the form

$$e, a, -q_0 \text{ and } -q_0 \cdot a \qquad \text{and} \qquad e, -a, -q_0, q_0 \cdot a,$$

respectively. The Lie algebra $\text{su}(2) \subset \mathbb{H}$ of $\text{SU}(2)$ (cf. section 1.1.8) is generated by

$$a, q_0 \text{ and } q_0 \cdot a.$$

Hence, the commutation relations are

$$\begin{aligned} [a, q_0] &= a \cdot q_0 - q_0 \cdot a &&= -2 \cdot q_0 \cdot a, \\ [a, q_0 \cdot a] &= a \cdot q_0 \cdot a - q_0 \cdot a \cdot a &&= 2 \cdot q_0, \\ [q_0, q_0 \cdot a] &= q_0 \cdot q_0 \cdot a - q_0 \cdot a \cdot q_0 &&= -2 \cdot a. \end{aligned}$$

In terms of the representations $\mathbb{H} = \mathbb{C}^a + \mathbb{C}^a \cdot q_0$ and $\mathbb{H} = \mathbb{C}^a \oplus q_0 \cdot \mathbb{C}^a$ the analogues of the Pauli matrices are

$$\begin{pmatrix} 1 & 0 \\ 0 & 1 \end{pmatrix}, \quad \sigma_1^a = \begin{pmatrix} 0 & 1 \\ 1 & 0 \end{pmatrix}, \quad \sigma_2^a = \begin{pmatrix} 0 & -a \\ a & 0 \end{pmatrix}, \quad \sigma_3^a = \begin{pmatrix} 1 & 0 \\ 0 & -1 \end{pmatrix}$$

and

$$\begin{pmatrix} 1 & 0 \\ 0 & 1 \end{pmatrix}, \quad {}^a\sigma_1 = \begin{pmatrix} 0 & -1 \\ 1 & 0 \end{pmatrix}, \quad {}^a\sigma_2 = \begin{pmatrix} 0 & -a \\ -a & 0 \end{pmatrix}, \quad {}^a\sigma_3 = \begin{pmatrix} 1 & 0 \\ 0 & -1 \end{pmatrix},$$

respectively. We use the symbols $\sigma_1^a, \sigma_2^a, \sigma_3^a$ and ${}^a\sigma_1, {}^a\sigma_2$ and ${}^a\sigma_3$, respectively, to denote also the respective endomorphisms in $\text{End}_{\mathbb{C}}\, \mathbb{C}^2$.

1.4. Inner Automorphisms of $\mathbb{H}$

1.4.1. Inner Automorphisms. The space $\text{End}_{\mathbb{R}}\mathbb{H}$ can be naturally described by the canonical injective map

$$\begin{aligned} \Phi : \mathbb{H} \otimes \mathbb{H} &\longrightarrow \text{End}_{\mathbb{R}}\mathbb{H} \\ \Phi(h \otimes k)(s) &:= h \circ s \circ \bar{k} \qquad \forall\, h, k \in \mathbb{H} \text{ and } \forall\, s \in \mathbb{H}. \end{aligned}$$

The restriction of Φ to the diagonal $dia = \{h \otimes h \mid h \in \mathbb{H}\}$ of $\mathbb{H} \otimes \mathbb{H}$ embeds $\mathbb{H}$ into $\text{End}\,\mathbb{H}$. Each element $\frac{1}{|k|^2} \cdot \Phi(k \otimes k) \in \text{End}\,\mathbb{H}$ is called an *inner automorphism* determined by $k \in \mathbb{H}$ and is denoted by τ_k.

For any $k \in \dot{\mathbb{H}} := \mathbb{H}\setminus\{0\}$ this inner automorphism τ_k is the conjugation by k given by

$$(26) \qquad \tau_k : \mathbb{H} \longrightarrow \mathbb{H}, \quad \tau_k(h) = k \cdot h \cdot k^{-1} \qquad \forall\, h \in \mathbb{H}.$$

Since any automorphism of the skew field $\mathbb{H}$ is an inner automorphism,

$$\tau : \dot{\mathbb{H}}/\dot{\mathbb{R}} \cdot e \longrightarrow \text{Aut}\ \dot{\mathbb{H}}, \quad k \mapsto \tau_k$$

is a bijection (cf. [**84**]). Thus τ yields a natural action of $\dot{\mathbb{H}}$ on $\mathbb{H}$.

Obviously,

$$\tau_k(h_1 \odot h_2) = \tau_k(h_1) \odot \tau_k(h_1) \qquad \forall\, h_1, h_2 \in \mathbb{H}$$

and

$$\tau_k(u_1 \times u_2) = \tau_k(u_1) \times \tau_k(u_2) \qquad \forall\, u_1, u_2 \in E$$

as is easily seen from (7), (13) and (31). In conjugation with (14) we find for $h_s = \lambda_s e + \nu_s$ with $s = 1, 2$:

$$\tau_k(h_1 \cdot h_2) = \tau_k(h_1 \odot h_2 + u_1 \times u_2) = \tau_k(h_1) \odot \tau_k(h_1) + \tau_k(u_1) \times \tau_h k(u_2).$$

This is a refinement of the preservation of the multiplication in $\mathbb{H}$ by τ.

1.4.2. Properties. Obviously, τ_k is an $\mathbb{R}$-algebra isomorphism of $\mathbb{H}$ satisfying

$$(27) \qquad \tau_k(\mathbb{R} \cdot e) = \mathbb{R} \cdot e \quad \text{and} \quad \tau_k(\text{su}(2)) = \text{su}(2).$$

The latter is due to $< e, h >= \text{tr}\, h = 0$ for all $h \in \text{su}(2)$ (cf. section 1.1.8).

As in every Lie group the inner automorphism τ_k with $k \in \text{SU}(2) = S^3$ determines an $\mathbb{R}$-linear isomorphism on the Lie algebra $\text{su}(2) = E$. It is the tangent map of τ_k at the identity $e \in \text{SU}(2) \subset \mathbb{H}$, i.e.

$$T_e\tau_k(h) = \frac{d}{dt}\left(k \cdot h(t) \cdot k^{-1}\right)\Big|_{t=0} = k \cdot h \cdot k^{-1} \qquad \forall\, h \in E$$

for a smooth family $h(t) \in S^3$ with $h(0) = e$ and $\dot{h}(0) = h$. Thus

$$(28) \qquad T_e\tau_k = \tau_k \qquad \forall\, k \in \text{SU}(2) = S^3$$

due to the linearity of τ_k. Instead of $T_e\tau_k$ the notation Ad_k is common. $\text{Ad}_k \in \text{Aut}\,\text{su}(2)$ is called the *adjoint representation* of k. From (28) we see that

$$\text{Ad} = \tau$$

where $\text{Ad}(k) = \text{Ad}_k$ for all $k \in \text{SU}(2)$.

Another property of τ_k is the fact that $\tau_k \in \text{SO}(E)$ for all $k \in \text{SU}(2) = S^3$, as can be seen from

$$\begin{aligned}
< \tau_k(h), \tau_k(h') > &= \frac{1}{2} \cdot \text{tr}\left(\tau_k(h) \cdot \widetilde{\tau_k(h')}\right)\\
&= \frac{1}{2} \cdot \text{tr}\left(k \cdot h \cdot k^{-1} \cdot \overline{k^{-1}} \cdot \overline{h'} \cdot \overline{k}\right)\\
&= \frac{1}{2} \cdot \text{tr}\left(k \cdot h \cdot k^{-1} \cdot k \cdot \overline{h'} \cdot \overline{k}\right)\\
&= \frac{1}{2} \cdot \text{tr}\left(h \cdot \overline{h'}\right)\\
(29) \qquad &= < h, h' >
\end{aligned}$$

due to (4) and (6).

1.4.3. The Inner Automorphism as a Rotation. In this section we will first show that Ad is a surjection. Together with $\operatorname{Ad}_k = \tau_k$ this allows us to interpret τ_k as a rotation. Let us recall from section 1.1.9 that both E and $\mathbb{H}$ are oriented in a compatible fashion.

In more detail we see that because of (28) and (29) the assignment

$$k \mapsto \operatorname{Ad}_k = \tau_k \qquad \forall\, k \in \mathrm{SU}(2)$$

determines the representation

$$\operatorname{Ad} : \mathrm{SU}(2) \longrightarrow \mathrm{SO}(E).$$

To see that Ad is surjective, we expand $\tau_k(h)$ for $k = \lambda \cdot e + u \in \mathrm{SU}(2)$ and any $h \in E$ into

$$\begin{aligned} \tau_k(h) &= \frac{1}{|k|^2}(k \cdot h \cdot \bar{k}) \\ &= (\lambda e + u) \cdot h \cdot (\lambda e - u) \\ &= (\lambda^2 \cdot h + \lambda \cdot [u, h] - u \cdot h \cdot u). \end{aligned} \tag{30}$$

Hence

$$\tau_k(u) = (\lambda^2 \cdot u - u^2 \cdot u) = u \tag{31}$$

and the *rotation axis* of τ_k is, therefore, $\mathbb{R} \cdot u \subset E$. Thus the rotation τ_k can be described in $u^\perp \subset E$, the orthogonal complement of u in E. The plane $u^\perp$ is called the rotation plane of τ_k. For a unit vector $h \in u^\perp$ we have

$$< \tau_k(h), h >= \frac{1}{|k|^2} \cdot (\lambda^2 - |u|^2).$$

Thus, if ϑ_k denotes an angle of rotation (cf. [**43**]),

$$\cos \vartheta_k =< \tau_k(h), h >= \frac{\lambda^2 - |u|^2}{\lambda^2 + |u|^2}. \tag{32}$$

Of course $\cos \vartheta_k$ is not affected by the orientation of the rotation.

The knowledge of the rotation axis of τ_k allows us to rewrite the action of τ_k on its plane of rotation in terms of the action of $U^a(1)$ on F^a for some $a \in S^2$. Indeed, let $k \in \mathrm{SU}(2)$ be such that the axis of rotation of τ_k is $\mathbb{R} \cdot a$ with $a \in S^2$. Hence

$$k = \lambda \cdot e + \mu \cdot a.$$

Thus k is of the form $k = z^k$ for a complex number $z^k := \lambda e + \mu \cdot a$ in $\mathbb{C}^a$. In fact, $z^k \in U^a(1)$ since $k \in \mathrm{SU}(2)$ and $\tau_k = \ell_{\exp^a t}$ for some $t \in \mathbb{R}$ (cf. section 1.2.5). Vice versa, if $k = \lambda e + \mu \cdot a$, then $\mathbb{R} \cdot a$ is the axis of rotation (cf. (31)). Thus we have:

LEMMA 3. *Given a two-dimensional subspace F^a of the Euclidean space E,*

$$\tau_k \in \mathrm{SO}(F^a) \qquad \Longleftrightarrow \qquad k = \lambda \cdot e + \mu \cdot a$$

for $\lambda, \mu \in \mathbb{R}$ where $a \in S^2$ is perpendicular to F^a.

Now let us determine the *oriented rotation angle* of τ_k. The action of τ_k with $k = \lambda \cdot e + \mu \cdot a$ on F^a is expressed by (30). This is to say that for any $h \in F^a$,

$$\tau_k(h) = \frac{1}{\lambda^2 + \mu^2}(\lambda^2 \cdot h + 2 \cdot \lambda \cdot \mu \cdot a \cdot h - \mu^2 \cdot h)$$

which is rewritten as

$$\tau_k(h) = \left(\frac{\lambda^2 - \mu^2}{\lambda^2 + \mu^2} \cdot e + \frac{2 \cdot \lambda \cdot \mu \cdot a}{\lambda^2 + \mu^2} \right) h.$$

Therefore, we may state:

LEMMA 4. *Given $\lambda, \mu \in \mathbb{R}$ and $a \in S^2$, the action of τ_k on F^a is of the form*

$$\tau_k(h) = z_k \cdot h \qquad \forall\, h \in F^a$$

where $z_k \in U^a(1)$ is given by

$$z_k = \frac{\lambda^2 - \mu^2}{\lambda^2 + \mu^2} \cdot e + \frac{2\lambda \cdot \mu}{\lambda^2 + \mu^2} \cdot a.$$

In terms of the oriented rotation angle ϑ_k (cf. (32)), z_k is expressed as

$$z_k = e^{\vartheta_k \cdot a}. \tag{33}$$

Hence

$$\tau_k(h) = e^{\vartheta_k \cdot a} \cdot h \qquad \forall\, h \in F^a. \tag{34}$$

(Here the orientation has entered via the multiplication in $\mathbb{H}$; the multiplication from the left is directly related to $-\mu_{F^a}$ (cf. section 1.3.3)). Since $k \in U^a(1)$ can be assumed to be of the form $k = e^{\theta_k \cdot a}$ with $a \in S^2$, due to

$$\tau_k(h) = k \cdot h \cdot k^{-1} = e^{2\theta_k \cdot a} \cdot h, \tag{35}$$

we see that

$$\vartheta_k = 2 \cdot \theta_h. \tag{36}$$

By (33), (35) and (36) we finally conclude that Ad is surjective.

Moreover,

$$\tau_{tk} = \tau_k \qquad \forall\, t \in \dot{\mathbb{R}} := \mathbb{R} \setminus \{0\},$$

expressing the projective nature of τ_k, which means that τ_k depends only on the line $\mathbb{R} \cdot k$ and not on k itself. $\tau_k|_{\mathrm{su}(2)}$ is an isometry of su(2) (cf. (28)). Thus we state:

LEMMA 5. *Let $\dot{\mathbb{H}}$ denote the multiplicative group $(\dot{\mathbb{H}}, \cdot)$ of $\mathbb{H}$. Then the surjection*

$$\tau : \dot{\mathbb{H}} \longrightarrow \mathrm{SO}(E)$$

yields the surjection

$$\begin{aligned} \tau : \mathrm{SU}(2) \quad &\longrightarrow \quad \mathrm{SO}(E), \\ k \quad &\longmapsto \quad \tau_k, \end{aligned}$$

a Lie group homomorphism with kernel $\{e, -e\}$, the center of SU(2). Therefore,

$$\mathrm{SO}(E) = \mathbb{R}P\mathbb{H} \tag{37}$$

and SU(2) is the twofold covering of SO(E); here $\mathbb{R}P\mathbb{H}$ denotes the real projective space of $\mathbb{H}$.

In treating the geometry of E by means of the skew field $\mathbb{H}$, it is clear that we have to use all possible embeddings of $\mathbb{C}$ into $\mathbb{H}$. This is obvious from (35).

1.4.4. SU(2) and SL(2, $\mathbb{C}$). Let us point out that a rotation in $F^a = a^{\perp_E}$ about the angle $2t$ can be expressed as

$$\tau_{e^{t\cdot a}}(u) = u \cdot e^{-2\cdot t\cdot a} = u \cdot \exp_{\mathrm{SU}(2)} -2\cdot t\cdot a \tag{38}$$

for all $t \in \mathbb{R}$, for all $a \in S^2$ and all $u \in \mathrm{su}(2)$. This is due to (19). The figure 2 in this formula expresses the fact that SU(2) is the twofold covering of SO(E).

Note that the map τ extends to the conjugation map Φ_L of the Minkowski space on $\mathbf{M}$ by the elements of SL(2, $\mathbb{C}$) (cf. sections 1.1.11 and 1.1.12). Since $\tilde{h} = h^{-1}$ for any $h \subset \mathrm{SU}(2)$ we have the following commutative diagram:

$$\begin{array}{ccc} \mathrm{SL}(2,\mathbb{C}) & \supset & \mathrm{SU}(2) \\ \downarrow{\Phi_L} & & \downarrow{\tau} \\ O_0(1,3) & \supset & \mathrm{SO}(3). \end{array} \tag{39}$$

Here $O_0(1,3)$ denotes the proper (orthochronous) Lorentz group of $\mathbf{M}$. The kernel of τ is in both cases $\{-\mathrm{id}, \mathrm{id}\}$. This diagram shows that SU(2) is naturally related to special relativity. We will come back to it in chapter 5.

The relation between SU(2) and SL(2, $\mathbb{C}$) is partly expressed by the relation between their Lie algebras su(2) and sl(2) namely by

$$\mathrm{sl}(2,\mathbb{C}) = \mathrm{su}(2) \otimes \mathbb{C}. \tag{40}$$

The tensor product on the right-hand side is formed over the reals. The reason this equality holds is that any traceless matrix in SL(2, $\mathbb{C}$) can be written as $A + i\, B$ with A and B real or as $C + i \cdot D$ with C and D in SU(2) (cf. [**78**]).

The brackets are

$$[A_1 + i \cdot B_1, A_2 + i \cdot B_2] = [A_1, A_2] - [B_1, B_2] + i \cdot ([A_1, B_2] + [B_1, A_2]).$$

Therefore, we present a (well-known) construction of a natural basis of $sl(2,\mathbb{C})$ out of the one for su(2) presented above. In matrix notation it is (cf. section 1.3.6)

$$X_1 = \frac{i}{2} \cdot \sigma_1, \quad X_2 = \frac{i}{2} \cdot \sigma_2, \quad \text{and} \quad X_3 = \frac{i}{2} \cdot \sigma_3.$$

In intrinsic terms X_1, X_2 and X_3 read

$$X_1 = \frac{i}{2}, \quad X_2 = \frac{1}{2}, \quad \text{and} \quad X_3 = \frac{i}{2}.$$

Then the basis of $sl(2,\mathbb{C}) \subset M \oplus \mathrm{su}(2)$ reads as follows:

$$\begin{aligned} E_1 : &= -i \cdot X_1 - X_2, \\ E_2 : &= -i \cdot X_1 - X_2, \\ E_3 : &= -i \cdot X_3. \end{aligned}$$

The matrices of E_1, E_2, E_3 are

$$M(E_1) = \begin{pmatrix} 0 & 1 \\ 0 & 0 \end{pmatrix}, \quad M(E_2) = \begin{pmatrix} 0 & 0 \\ 1 & 0 \end{pmatrix}, \quad M(E_3) = \frac{1}{2}\begin{pmatrix} 1 & 0 \\ 0 & -1 \end{pmatrix}.$$

1.4.5. Tangent Maps and Lie Algebra Isomorphisms. Let us turn back to the map τ and Ad (cf. section 1.4.2). To compute the tangent map $T\mathrm{Ad}$, we consider $k(t) \in \mathrm{SU}(2)$ varying smoothly in the real parameter t which runs through a small interval centered about 0. Denoting $k(0)$ by u, the tangent map $T\mathrm{Ad}$ at any $u \in \mathrm{SU}(2)$ is given by

$$\begin{aligned} T_u\mathrm{Ad}\,(\dot{k}(0))(h) &= \frac{d}{dt}\left(k(t)\cdot h\cdot k(t)^{-1}\right)\Big|_{t=0} \\ &= \dot{k}(0)\cdot h\cdot k(0)^{-1} - k(0)\cdot h\cdot k^{-1}(0)\cdot\dot{k}(0)\cdot k^{-1}(0) \\ &= \dot{k}(0)\cdot k^{-1}(0)\cdot k(0)\cdot h\cdot k^{-1}(0) \\ &\quad -k(0)\cdot h\cdot k^{-1}(0)\cdot\dot{k}(0)\cdot k^{-1}(0) \\ &= [\dot{k}(0)\cdot k^{-1}(0), \tau_{k(0)}(h)] \end{aligned}$$

and hence the general formula reads

$$T_u\mathrm{Ad}\,(l)(h) = [l\cdot u^{-1}, \tau_u(h)] \qquad \forall\, h \in E \tag{41}$$

where $k(0) = h \in \mathrm{SU}(2)$ and $\dot{k}(0) = l \in E$; the bracket is formed in $\mathbb{H}$ (cf. section 1.1.7).

In case $u = e$ the value of the tangent map $T_e\mathrm{Ad}\,(l)$ is replaced by ad_l. In terms of this notion formula (41) is

$$\mathrm{ad}_l(h) = [l,h] = 2\cdot l\times h \qquad \forall\, l,h \in E, \tag{42}$$

a commutator in $\mathbb{H}$. This formula allows the following interpretation: Let

$$\mathrm{so}(E) := \{A \in \mathrm{End}_{\mathbb{R}}\, E \mid A \text{ skew-adjoint}\}.$$

Moreover, given $A_1, A_2 \in \mathrm{so}(E)$ then the *commutator* $\mathrm{com}\,(A_1,A_2)$ defined by

$$\mathrm{com}\,(A_1,A_2) := A_1\circ A_2 - A_2\circ A_1$$

obviously is a skew-adjoint linear map on E. Thus $\mathrm{so}(E)$ together with the commutator com is a Lie algebra. The two Lie algebras $\mathrm{su}(2)$ and $\mathrm{so}(E)$ are linked by the tangent map

$$\mathrm{su}(2) \xrightarrow{T_e\tau} \mathrm{so}(E)$$

taken at $e \in \mathrm{SU}(2)$, a Lie algebra isomorphism due to

$$T_e\tau(l)(h) = [l,h] = 2\cdot l\times h \qquad \forall\, l,h\in E.$$

(We will frequently replace $T_e\tau$ by $d\tau$.) This is seen as follows: Obviously,

$$T_e\tau(l) = \mathrm{ad}_l : E \longrightarrow E \qquad \forall\, l \in E = \mathrm{su}(2)$$

is a skew-adjoint map, due to (42). On the other hand, for any $A \in \mathrm{so}(E)$ there is a vector a_A, say, such that

$$A(x) = a_A \times x \qquad \forall\, x \in E.$$

Hence, $\mathrm{so}(E)$ is a three-dimensional $\mathbb{R}$-linear space. Since $a_{A_1\circ A_2} = a_{A_1}\times a_{A_2}$, the commutator $\mathrm{com}\,(A_1,A_2)$ can be rewritten as

$$\begin{aligned} \mathrm{com}\,(A_1,A_2)(x) &= A_1\circ A_2(x) - A_2\circ A_1(x) \\ &= (a_{A_1}\times a_{A_2})\times x - (a_{A_2}\times a_{A_1})\times x \qquad \forall\, x\in E, \end{aligned}$$

showing that

$$\mathrm{com}\,(A_1,A_2)(x) = (2\cdot a_{A_1}\times a_{A_2})\times x = [a_{A_1}, a_{A_2}]\times x \qquad \forall\, x \in E, \tag{43}$$

and hence

$$\mathrm{com}\,(A_1, A_2) = [a_{A_1}, a_{A_2}] = \mathrm{ad}_{\,a_{A_1}} a_{A_2}.$$

Moreover,

$$T_e\tau([l_1, l_2])(x) = ([l_1, l_2]) \times x = \mathrm{com}\,(A_{l_1}, A_{l_2})(x) \qquad \forall\, x \in E,$$

and thus

$$T_e\tau([l_1, l_2]) = \mathrm{com}\,(A_{l_1}, A_{l_2}) \qquad \forall\, l_1, l_2 \in E,$$

saying that

$$T_e\tau : \mathrm{su}(2) \longrightarrow \mathrm{so}(E)$$

is a Lie algebra isomorphism.

Both $\mathrm{so}(E)$ and $\mathrm{su}(2)$ carry two Lie algebra structures, one with the commutator as Lie product and one with the cross product as Lie algebra structure. $T_e\tau$ is a Lie algebra isomorphism with respect to both structures. Here the number 2 in equations (42) and (43) again expresses the fact that $\mathrm{SU}(2)$ is the twofold covering of $\mathrm{SO}(E)$ with τ as the covering map.

1.5. The Oriented Rotation Angle of the Inner Automorphism with Respect to the Natural Minkowski Metric

1.5.1. The Oriented Rotation Angle. In preparing the geometric use of inner automorphisms of $\mathbb{H}$ we investigate $\tau_k \in \mathrm{SO}(E)$ more closely by means of its rotation plane and its oriented rotation angle for any $k \in \mathbb{H}$ (cf. (34)). In doing so we will vary the approach in section 1.4.3 to a more geometric point of view. In particular, we will find the natural Minkowski metric g_M again (cf. (11)), determined by the oriented angle of rotation (cf. (33) and (34)). Moreover, we will relate this angle to the latitude on S^3. In this picture the quaternions appear once more as a unifying entity of Euclidean and Minkowski geometry.

To this end let $k = \lambda \cdot e + u$ with $\lambda \in \mathbb{R}$ and a non-vanishing $u \in E$. Because of (27) we can concentrate on the component in E.

The unit vector $\frac{u}{|u|}$ yields the splitting $E = \mathbb{R}\cdot\frac{u}{|u|} \oplus F^{\frac{u}{|u|}}$ where $F^{\frac{u}{|u|}}$ is the orthogonal complement of $\frac{u}{|u|}$ in E. Hence any $h \in E$ is written as $h = \zeta \cdot \frac{u}{|u|} + h'$ with $h' \in F^{\frac{u}{|u|}}$ and $\zeta =< h, \frac{u}{|u|} >$ yielding

$$\tau_k(h) = \zeta \cdot \frac{u}{|u|} + \tau_k(h')$$

since $\tau_k(u) = u$ according to (30) in section 1.4. The image $\tau_k(h')$ of h' is computed as

$$\tau_k(h') = k \cdot h' \cdot k^{-1} = \frac{k}{|k|^2} \cdot h' \cdot (\lambda \cdot e - u) = \frac{k^2}{|k|^2} \cdot h' \tag{44}$$

since $h' \cdot (\lambda \cdot e - u) = k \cdot h'$. Thus

$$\tau_k(h) = \zeta \cdot \frac{u}{|u|} + \frac{k^2}{|k|^2} \cdot h' \qquad \text{with} \quad h = \zeta \cdot \frac{u}{|u|} + h' \quad \text{and} \quad h' \perp u. \tag{45}$$

The rotation factor $\frac{k^2}{|k|^2}$ can be represented in terms of any $k \in \dot{\mathbb{H}}$ as

$$\frac{k^2}{|k|^2} = \frac{k}{\overline{k}}$$

due to $|k|^2 = k \cdot \bar{k} = \bar{k} \cdot k$ for all $k \in \mathbb{H}$. Therefore, $\tau_k(h)$ can also be written as

$$\tau_k(h) = \zeta \cdot \frac{u}{|u|} + \frac{\lambda \cdot e + u}{\lambda \cdot e - u} \cdot h'$$

for $h = \zeta \cdot \frac{u}{|u|} + h'$.

The cosine of the *oriented* rotation angle ϑ_k (called *rotation angle* in the sequel) of τ_k is determined by the equation

$$< \tau_k(h'), h' > = < \frac{k^2}{|k|^2} \cdot h', h' > = |h'|^2 \cdot \cos\vartheta_k$$

which shall be evaluated next.

1.5.2. The Oriented Rotation Angle in Terms of the Minkowski Metric. We observe that for any $k \in \mathbb{H}$ its square k^2 can be expressed as

(46)
$$k^2 = (\lambda \cdot e + u)^2 = (\lambda^2 - |u|^2) \cdot e + 2 \cdot \lambda \cdot u,$$

yielding

(47)
$$\frac{k^2}{|k|^2} = \frac{\lambda^2 - |u|^2}{\lambda^2 + |u|^2} \cdot e + 2 \cdot \frac{\lambda \cdot |u|}{|k|^2} \cdot \frac{u}{|u|}.$$

Since $\frac{k^2}{|k|^2} \in S^3$, equation (47) can be rewritten as

(48)
$$\frac{k^2}{|k|^2} = e \cdot \cos\vartheta_k + \frac{u}{|u|} \cdot \sin\vartheta_k = e^{\vartheta_k \cdot \frac{u}{|u|}}$$

with $\cos\vartheta_k = \frac{\lambda^2 - |u|^2}{\lambda^2 + |u|^2}$ (according to (32)) and hence

(49)
$$\frac{k}{|k|} = e^{\frac{\vartheta_k}{2} \cdot \frac{u}{|u|}}.$$

In terms of the Minkowski metric g_M introduced in section 1.1, equation (47) reads

(50)
$$\frac{k^2}{|k|^2} = \frac{g_M(k,k)}{< k, k >} \cdot e + 2 \cdot \frac{\lambda \cdot |u|}{|k|^2} \cdot \frac{u}{|u|},$$

yielding

(51)
$$g_M(k,k) = |k|^2 \cdot \cos\vartheta_k,$$

which emphasizes the fact that $\tau : \dot{\mathbb{H}} \to \operatorname{Aut} \mathbb{H}$ determines the natural Minkowski metric (cf. (32)). Hence if k is a unit vector, $k^2 = \cos\vartheta_k \cdot e + 2\lambda \cdot u$. From (32) we conclude

$$\cos\vartheta_k = 0 \quad \Leftrightarrow \quad |\lambda| = |u|.$$

To identify the geometric meaning of the rotation angle ϑ_k in terms of the geometry of S^3 let us consider the plane spanned by $\lambda \cdot e$ and u. This plane is nothing more than $\mathbb{C}^{\frac{u}{|u|}}$. In fact, k, $\lambda \cdot e$ and u form a right triangle. The oriented angle between k and $\lambda \cdot e$ shall be denoted by β_k. Since $\frac{2\lambda \cdot |u|}{|k|^2} = \sin\vartheta_k$ (cf. (47) and (48)),

$$2 \cdot \lambda \cdot |u| = 2 \cdot |k| \cdot \sin\beta_k \cdot |k| \cos\beta_k = |k|^2 \cdot \sin\vartheta_k$$

and we conclude $2 \cdot \vartheta_k = \beta_k$ up to an integer multiple of $2 \cdot \pi$, and hence

(52)
$$\vartheta_k = \frac{\beta_k}{2}.$$

Since $\frac{\pi}{2} - \beta_k$ is the latitude of $\frac{k^2}{|k|^2} \in S^3$, the angle $\frac{\pi}{2} - 2 \cdot \vartheta_k$ is the latitude of $\frac{k^2}{|k|^2}$. In other words:

LEMMA 6. *The oriented rotation angle ϑ_k of τ_k determines the latitude of $\frac{k^2}{|k|^2}$ on S^3 and vice versa.*

If we fix the rotation angle ϑ_k, the unit vector $\frac{k^2}{|k|^2}$ varies on all of the two-dimensional sphere $S^2_{\vartheta_k} \subset S^3$ given by

$$S^2_{\vartheta_k} := \left\{ e \cdot \cos \vartheta_k + \frac{u}{|u|} \cdot \sin \vartheta_k \;\middle|\; \frac{u}{|u|} \in S^2 \right\}.$$

The two-sphere $S^2_{\vartheta_k}$ is the sphere of constant latitude $\frac{\pi}{2} - \vartheta_k$. This is to say that the rotation angle ϑ_k of τ_k fixes the latitude of the unit vector $\frac{k^2}{|k|^2} \in S^3$ within S^3 by (52). Two inner automorphisms with the same rotation angles may have different rotations axes and thus may rotate in different tangent planes of S^2. The geometry of S^3 will be discussed in more detail in chapter 2.

1.6. Link between Space-Time Geometry and Euclidean Geometry on the Quaternions

1.6.1. A Basis of the Two-Dimensional Minkowski Space. We continue the study of the two-dimensional Minkowski geometry on $\mathbb{C}^a \subset \mathbb{H}$, in particular, to understand the interplay between Lorentz transformations and rotations in 3-Space. In doing so we use $\bar{g}_M := -g_M$ (cf. section 1.1.10) in this section. The reason is that $\bar{g}_M$ on E coincides with the scalar product $< \, , \, >$. This simplifies the comparison of the Minkowski geometry with the Euclidean one.

By definition a light vector $k \in \mathbb{C}^a$ satisfies $\bar{g}_M(k,k) = 0$. The collection of all light vectors in $\mathbb{C}^a$, that is the *light cone*, intersects the unit circle in four points. Any light vector thus has the form

$$k = |k| \cdot e^{m \cdot \frac{\pi}{4} \cdot a} \quad \text{for} \quad m = 1, 3, 5, 7$$

(up to an integer multiple of 2π, of course). A convenient basis for the two-dimensional Minkowski space $\mathbb{C}^a$ is formed by

$$k_1 = e^{\frac{\pi}{4} \cdot a} \quad \text{and} \quad k_2 = e^{3 \cdot \frac{\pi}{4} \cdot a}$$

For $\vartheta_{k_1} = \frac{\pi}{2}$ and $\vartheta_{k_2} = \frac{3\pi}{2}$ the two vectors obey

$$\bar{g}_M(k_s, k_s) = \cos \vartheta_{k_s} = 0 \quad \text{for} \quad s = 1, 2$$

and

$$\bar{g}_M(k_1, k_2) = -1.$$

Obviously, any Lorentz transformation ℓ on $\mathbb{C}^a$ with $\det \ell = 1$ acts on k_1 and k_2 by

$$\ell(k_1) = r \cdot k_1 \qquad \text{and} \qquad \ell(k_2) = \frac{1}{r} k_2$$

for some positive real number r (cf. [**43**]) instead of ℓ. The inverse of ℓ_r is $\ell_{\frac{1}{r}}$.

Next, we generalize g_M to g_M^c used in physics: Let the constant c be a positive real (the speed of light in special relativity) and $h_1, h_2 \in \mathbb{C}^a$ be split into

$$h_s = \lambda_s \cdot e + \xi_s \cdot a \quad \text{with} \quad s = 1, 2$$

for $\lambda_s, \xi_s \in \mathbb{R}$ and $a \in S^2$. The Lorentz metric $\bar{g}_M^c$ is then defined by

$$\bar{g}_M^c(h_1, h_2) := -\lambda_1 \cdot \lambda_2 \cdot c^2 + \xi_1 \cdot \xi_2.$$

If ℓ_r is extended to all of $\mathbb{H} = \mathbb{C}^a \oplus F^a$ by letting

$$\ell_r(F^a) = F^a,$$

ℓ is called a *Lorentz boost* or just a *boost.*

For the generalized Minkowski metric $\bar{g}_M^c$ the new light vectors k_1^c and k_2^c are determined by

$$k_1^c := \frac{1}{\sqrt{2}} \cdot \left(\frac{e}{c} + a\right) \quad \text{and} \quad k_2^c := \frac{1}{\sqrt{2}} \cdot \left(\frac{e}{c} - a\right).$$

Thus

$$\frac{e}{c} = \frac{1}{\sqrt{2}} \cdot (k_1^c + k_2^c) \quad \text{and} \quad a = \frac{1}{\sqrt{2}} \cdot (k_1 - k_2).$$

Hence the values of l_r on e and a are

(53) $$l_r\,(a) = \frac{1}{2} \cdot \left(r - \frac{1}{r}\right) \cdot \frac{e}{c} + \frac{1}{2} \cdot \left(r + \frac{1}{r}\right) \cdot a$$

and

$$l_r\left(\frac{e}{c}\right) = \frac{1}{2} \cdot \left(r + \frac{1}{r}\right) \cdot \frac{e}{c} + \frac{1}{2} \cdot \left(r - \frac{1}{r}\right) \cdot a,$$

i.e.

(54) $$l_r(e) = \frac{1}{2} \cdot \left(r + \frac{1}{r}\right) \cdot e + \frac{1}{2} \cdot \left(r - \frac{1}{r}\right) \cdot c \cdot a.$$

Splitting $l_r(k)$ for $k = \lambda \cdot e + \xi \cdot a$ into

$$l_r(k) = \lambda_r \cdot e + \xi_r \cdot a$$

yields the relations

(55) $$\lambda_r = \frac{\lambda}{2}\left(r + \frac{1}{r}\right) + \frac{\xi}{2}\left(r - \frac{1}{r}\right) \quad \text{and} \quad \xi_r = \frac{\xi}{2}\left(r + \frac{1}{r}\right) + \frac{\lambda c}{2}\left(r - \frac{1}{r}\right).$$

Introducing

(56) $$\frac{v}{c} := -\frac{r^2 - 1}{r^2 + 1}$$

turns the equations in (55) into

(57) $$\lambda_r = \frac{\lambda - \xi \cdot \frac{v}{c}}{\sqrt{1 - \frac{v^2}{c^2}}} \quad \text{and} \quad \xi_r = \frac{\xi - \lambda \cdot v}{\sqrt{1 - \frac{v^2}{c^2}}}.$$

(Clearly, (56) exhibits c as an absolute velocity). In terms of v the real r is given by

(58) $$r = \left(\frac{1 - \frac{v}{c}}{1 + \frac{v}{c}}\right)^{\frac{1}{2}}.$$

The Lorentz transformation l_r is thus characterized by v and vice versa.

Independently of the constant c, there is a real number θ_{l_r} for which

$$\cosh\theta_{l_r} = \frac{1}{2}\cdot\left(r+\frac{1}{r}\right) \quad\text{and}\quad \sinh\theta_{l_r} = \frac{1}{2}\cdot\left(r-\frac{1}{r}\right) \tag{59}$$

and hence

$$\tanh\theta_{l_r} = \frac{r^2-1}{r^2+1} = \frac{\bar{g}_M\left(l_r(e),a\right)}{\bar{g}_M\left(l_r(a),e\right)}.$$

Let $k \in \mathbb{C}^a \subset \mathbb{H}$ be split into $k = \lambda\cdot e + a$ with $a \in E$. The effect of l_r onto k is

$$l_r(k) = \cosh\theta_{l_r}\cdot k + \sinh\theta_{l_r}\cdot\overline{k}\cdot a.$$

In analogy to the oriented Euclidean angle of rotation, θ_{l_r} is here called the *hyperbolic (oriented) rotation angle.* It is determined by

$$\bar{g}_M^c(l_r(k),k) = \frac{1}{2}\cdot\left(r+\frac{1}{r}\right)\cdot\bar{g}_M^c(k,k)$$

which yields

$$\tanh\theta_{l_r} = \frac{r^2-1}{r^2+1}. \tag{60}$$

Therefore, by the definition of the hyperbolic functions sinh and cosh the value $\tanh\theta_{l_r}$ is expressed as

$$\frac{e^{2\cdot\theta_{l_r}}-1}{e^{2\cdot\theta_{l_r}}+1} = \frac{r^2-1}{r^2+1},$$

because of which the hyperbolic rotation angle in terms of v is

$$\theta_{l_r} = \log|r| = \frac{1}{2}\cdot\log\frac{1+v}{1-v}. \tag{61}$$

With the notation of (59), equations (53) and (54) read

$$l_r(a) = \frac{1}{c}\cdot\sinh\theta_r\cdot e + \cosh\theta_r\cdot a \quad\text{and}\quad l_r(e) = \cosh\theta_r\cdot e + c\cdot\sinh\theta_r\cdot a.$$

By introducing the imaginary angle $i\cdot\theta_{l_r}$ the following relations hold true:

$$\begin{aligned} \sin i\cdot\theta_{l_r} &= i\cdot\sinh\theta_{l_r}, & \sinh i\cdot\theta_{l_r} &= i\cdot\sin\theta_{l_r},\\ \cos i\cdot\theta_{l_r} &= \cosh\theta_{l_r}, & \cosh i\cdot\theta_{l_r} &= \cos\theta_{l_r}. \end{aligned}$$

These equations imply

$$\tan i\cdot\theta_{l_r} = i\cdot\tanh\theta_{l_r} \quad\text{and}\quad \tan\theta_{l_r} = -i\cdot\tanh i\cdot\theta_{l_r};$$

and hence (60) becomes

$$i\cdot\tan i\cdot\theta_{l_r} = \frac{1-r^2}{1+r^2}.$$

These equations also hold true in $\mathbb{C}^{\frac{u}{|u|}}$ with i replaced by $\frac{u}{|u|}$. In terms of complex rotation angles the Lorentz transformations (53) and (54) hence read

$$l_r(a) = a\left(\cos\frac{u}{|u|}\theta_{l_r} - \sin\frac{u}{|u|}\theta_{l_r}\right) \quad\text{and}\quad l_r(e) = e\left(\cos\frac{u}{|u|}\theta_{l_r} + \sin\frac{u}{|u|}\theta_{l_r}\right).$$

1.6.2. The Exponential Map. In section 1.4.4 we have shown that a rotation in $F^a = a^{\perp_E}$ about the angle $2t$ can be expressed by

$$\tau_{\exp^a(t)}(u) = u \cdot \exp^a(2 \cdot t) \tag{62}$$

(cf. (38)). There is a hyperbolic analogue, i.e. a *boost* in the plane $\mathbb{C}^a$ of the Minkowski space $\mathbb{H}$. To obtain it in a simple fashion let us work with $\mathbf{M}$ as defined in section 1.1.11, i.e. $\mathbf{M} = \mathbb{R} \cdot e \oplus i \bullet E$. We consider

$$\text{quat}(\mathbb{C}^a) = \text{span}\,(e, i \bullet a)$$

where quat is also given as in (12) in section 1.1.11.

The exponential series

$$\sum e^{\theta \cdot i \bullet a}$$

converges uniformly in norm on $\mathbf{M}$ and splits into

$$e^{\theta \cdot i \bullet a} = \sum_{n=0}^{\infty} \frac{(\theta \cdot i \bullet a)^{2n}}{(2n)!} + \sum_{n=-1}^{\infty} \frac{(\theta \cdot i \bullet a)^{2n+1}}{(2n+1)!}$$

Since for all $n \geq 1$,

$$(i \bullet a)^{2n} = i^{2n} \bullet a^{2n} = e$$

and

$$(i \bullet a)^{2n+1} = i^{2n+1} \bullet a^{2n+1} = i \bullet a,$$

we conclude that

$$\sum_{n=1}^{\infty} \frac{(\theta \cdot i \bullet a)^{2n}}{(2n)} = e \cdot \cosh \theta$$

and

$$\sum_{n=1}^{\infty} \frac{(\theta \cdot i \bullet a)^{2n+1}}{2n+1} = i \bullet a \cdot \sinh \theta,$$

therefore,

$$e^{\theta \cdot i \bullet a} = e \cdot \cosh \theta + i \bullet a \sinh \theta \qquad \forall t \in \mathbb{R}.$$

Coming back to a boost we recall the Lorentz transformation on the four-dimensional Minkowski space $\mathbf{M}$ in the form

$$\ell(V) = S \cdot V \cdot \bar{S} \qquad \forall V \in \mathbf{M}$$

for $S \in \text{SL}(2, \mathbb{C})$ (cf. section 1.1.13). Here $\bar{S}$ is the defined conjugate of S. A smoothly parameterized family $S(t)$ for $t \in \mathbb{R}$ yields

$$\dot{\ell}(0)(V) = S(0) \cdot V + V \cdot \bar{S}(0)$$

where $\bar{S}(0)$ is a *trace-free* $\mathbb{C}$-linear map on $\mathbb{C}^2$. If the conjugation by $S(0)$ is a *boost* in quat $\mathbb{C}^a$ with hyperbolic angle θ_r,

$$\ell_r(V) = e^{\frac{\theta_r}{2} i \bullet a} \cdot V \cdot e^{-\frac{\theta_r}{2} i \bullet a}$$

for all $V \in \mathbf{M}$, any $r \in \mathbb{R}$ and any $a \in S^2$. Another way of writing this on $\mathbb{C}^a$ is

$$\ell_r(z) = (\text{quat})^{-1}(e^{\theta_r i \bullet a})(z) \qquad \forall\, z \in \mathbb{C}^a.$$

CHAPTER 2

Elements of the Geometry of S^3, Hopf Bundles and Spin Representations

Chapter 1 provided us with an insight into the structures of the skew field of quaternions. Two rather important tools were the 1-by-3 and the 2-by-2-splittings of $\mathbb{H}$ into $\mathbb{R} \cdot e + E$ and $\mathbb{C}^a \oplus F^a$, respectively, for some $a \in S^2$ as well as the inner automorphism τ_k linking the two different splittings.

These two elements will accompany us throughout this chapter. We will study the geometry of the unit spheres S^3 and S^2 as well as fibrations relating to those spheres.

The first section deals with the one-parameter groups of SU(2) (being identical with $S^3 \subset \mathbb{H}$) and SO(E) allowing us to look at the parallels of latitude and the meridians on S^3 in the next section.

From there on we are equipped with the tools necessary to construct the Hopf bundles, in particular the Hopf fibration. All these bundles link the three-sphere with its equator contained in the quaternions and hence connect the geometry in four-space with the geometry in 3-space. Here the inner automorphism also plays a central role.

Sections 2.4 and 2.5 take a different point of view in focussing on spin representations, in particular the spin $\frac{1}{2}$-representation and the infinitesimal spin $\frac{1}{2}$-representation. These representations offer insights into the two sorts of geometries in the light of representation theory.

The reader will notice that all of the geometry of $\mathbb{H}$ can be recovered by the spin $\frac{1}{2}$-representation.

2.1. One-Parameter Groups of SU(2) and SO(E)

2.1.1. The Exponential Maps of SU(2) and SO(E). Let us consider the plane $\mathbb{C}^{\frac{a}{|a|}} \subset \mathbb{H}$ containing $\mathbb{R} \cdot e$ and a non-vanishing vector, say $a \in E$ (cf. section 1.2). Thus $\mathbb{C}^{\frac{a}{|a|}} = \text{span}\left(e, \frac{a}{|a|}\right)$. In section 1.2 we have defined the exponential map on $\mathbb{C}^{\frac{a}{|a|}}$. Here we will go into more detail to comprise also SU(2) and SO(E). The *unitary group* $U^a(1)$ as defined in section 1.2.2 is represented here as

$$U^a(1) = \mathbb{C}^{\frac{a}{|a|}} \cap S^3$$

and shall be parameterized by (cf. section 1.2.2)

$$\exp^a(t) := \cos|a|t + \frac{a}{|a|}\sin|a|t \qquad \forall\, t \in \mathbb{R}. \tag{63}$$

As we learned in section 1.4.3 any quaternion $k \in \mathrm{SU}(2) = S^3$ is of the form

$$k = \exp^{\frac{u}{|u|}}(\vartheta_k)$$

where $k = \lambda \cdot e + u$ with $u \in E$. Therefore, we define

$$\exp_{\mathrm{SU}(2)} : \mathrm{su}(2) \longrightarrow \mathrm{SU}(2) \tag{64}$$

by

$$\exp_{\mathrm{SU}(2)}(u) = \exp^{\frac{u}{|u|}}(|u|) \qquad \forall\, u \in \mathrm{su}(2) = E.$$

If we want to emphasize dependence on the unit vector $a \in S^2$ we write $\exp^a_{\mathrm{SU}(2)}$ instead of $\exp_{\mathrm{SU}(2)}$. Obviously,

$$\exp^a_{\mathrm{SU}(2)} : \mathbb{R} \longrightarrow \mathbb{C}^{\frac{a}{|a|}} \subset \mathrm{SU}(2)$$

is a group homomorphism and is based on the convergent series in $\mathbb{H}$ we looked at in section 1.2.2. It is called a *one-parameter subgroup* of SU(2) or just a one-parameter group of SU(2). Its image is the unit circle in the plane $\mathbb{C}^{\frac{a}{|a|}}$. Clearly, $\exp_{\mathrm{SU}(2)}$ is surjective.

Let us point out that there is a unique one-parameter group on $U^a(1)$ determined by equation (63).

The image of $\exp^a_{\mathrm{SU}(2)}$ under τ, that is

$$\tau \circ \exp^a_{\mathrm{SU}(2)} : \mathbb{R} \longrightarrow \mathrm{SO}(E),$$

is a one-parameter group, as well. This one-parameter group is easily investigated more closely. At first we observe that

$$\tau_{e^{t\cdot a}}(a) = e^{t\cdot a} \cdot a \cdot e^{-t\cdot a} = a \qquad \forall\, t \in \mathbb{R}$$

since a commutes with $e^{t\cdot a}$. Thus $\mathbb{R} \cdot a$ is an axis of rotation of $\tau_{e^{t\cdot a}}$ and therefore $\tau_{e^{t\cdot a}}|_{F^a} \in \mathrm{SO}(F^a)$ for all $t \in \mathbb{R}$. Hence

$$\tau : U^a(1) \to \mathrm{SO}(F^a)$$

is a *surjective homomorphism.* Since

$$\tau_{e^{t\cdot a}}(h) = e^{2\cdot t\cdot a} \cdot h \qquad \forall\, h \in F^a \text{ and } \forall\, t \in \mathbb{R}$$

as can easily be seen from equation (21), the homomorphism τ is *two-to-one*, saying that $U^a(1)$ is the *twofold covering* of $\mathrm{SO}(F^a)$.

What we have done here can be linked to the geometry of E more closely:

Combining τ and $\exp_{\mathrm{SU}(2)}$ we conclude that

$$\tau \circ \exp_{\mathrm{SU}(2)} : E \longrightarrow \mathrm{SO}(E)$$

is *surjective*, as well. Let us determine the image of $\tau \circ \exp^a_{\mathrm{SU}(2)}$ for a fixed $a \in S^2$.

2.1.2. The Image of $\tau \circ \exp^a_{\mathrm{SU}(2)}$. To do so we consider the eight-dimensional unit sphere $S_{\mathrm{End}_{\mathbb{R}} E} \subset \mathrm{End}_{\mathbb{R}} E$ defined by the Killing form $<\ ,\ >_{\mathrm{End}_{\mathbb{R}} E}$. This *scalar product* is defined by

$$(65) \qquad < A, B >_{\mathrm{End}_{\mathbb{R}} E} := \frac{1}{3} \cdot \mathrm{tr}\, A \cdot \tilde{B} \qquad \forall\, A, B \in \mathrm{End}_{\mathbb{R}} E.$$

Again, $\tilde{B}$ is the adjoint of B. Thus

$$(66) \qquad < \tau_h, \tau_h >_{\mathrm{End}_{\mathbb{R}} E} = 1 \qquad \forall\, h \in \mathbb{H},$$

saying that $\tau(\mathrm{SU}(2)) = \mathrm{SO}(E)$ is a part of $S_{\mathrm{End}_{\mathbb{R}} E}$, the unit sphere in $\mathrm{End}_{\mathbb{R}} E$.

Now for any $h \in E$,

$$\tau_{\exp^a_{\mathrm{SU}(2)}}(t)(h) = \left(h \cdot \cos |a|\ t + \frac{a}{|a|} \cdot h \cdot \sin |a|\ t\right) \cdot \left(\cos |a|\ t \cdot e - \frac{a}{|a|} \cdot \sin |a| t\right)$$

for all $t \in \mathbb{R}$ since

$$\exp^a_{\mathrm{SU}(2)} \cdot \overline{\exp^a_{\mathrm{SU}(2)}} = \mathrm{id} \qquad \forall\, t \in \mathbb{R}.$$

Here the bar denotes the conjugation in $\mathbb{H}$ (which restricts to the conjugation in $\mathbb{C}^{\frac{a}{|a|}}$). Hence

$$\tau_{\exp^a_{\mathrm{SU}(2)}} h = h \cdot \cos 2|a| t + \frac{a}{|a|} \cdot h \cdot \sin 2|a|\ t \qquad \forall\, t \in \mathbb{R} \text{ and } \forall\, h \in E,$$

and therefore,

$$(67) \qquad \tau_{\exp^a_{\mathrm{SU}(2)}}(t)(h) = R_e \cdot \cos 2 \cdot |a| \cdot t + R_{h \frac{a}{|a|}} \cdot \sin 2 \cdot |a| \cdot t$$

where $R_{k'}$ is the right translation in $\mathbb{H}$ which, applied to any $k \in \mathbb{H}$, is $R_{k'}(k) = k \cdot k'$ for all $h \in \mathbb{H}$. Since R_e and $R_{\frac{a}{|a|}}$ are in $\mathrm{End}_{\mathbb{R}}\, \mathbb{H}$,

$$\tau \circ \exp^a_{\mathrm{SU}(2)}(\mathbb{R}) \subset \mathrm{End}_{\mathbb{R}}\, \mathbb{H}$$

is a circle of diameter two in the linear space span $\left(R_e, R_{\frac{a}{|a|}}\right)$ due to (67). It is contained in $\mathrm{SO}(E)$. Thus $\tau \circ \exp^a_{\mathrm{SU}(2)}$ parameterizes the circle $\tau \circ \exp^a_{\mathrm{SU}(2)}(\mathbb{R})$ in $\mathrm{SO}(E)$. Clearly,

$$(68) \qquad \tau : \exp^a_{\mathrm{SU}(2)}(\mathbb{R}) \longrightarrow \mathrm{im}\, \tau_{\exp^a_{\mathrm{SO}(E)}(\mathbb{R})}$$

is a *twofold covering* map due to lemma 5 in section 1.4.3. This is reflected in the following also:

For each $a \subset E$ the map

$$(69) \qquad \begin{array}{rcl} A : E & \longrightarrow & E \\ A^a(x) & := & a \times x \end{array}$$

is skew-adjoint. In fact,

$$(70) \qquad \mathrm{im}\, \tau \circ (\exp^a_{\mathrm{SU}(2)}(\mathbb{R})) = \mathrm{im}\, e^{\mathbb{R} \cdot A^a}$$

where $e^{t \cdot A^a}$ is formed in $\mathrm{End}_{\mathbb{R}} E$ for any $t \in \mathbb{R}$.

2.2. Parallels of Latitude and Meridians on S^3

Closely related to the one-parameter groups studied in the last section, we take a brief look at the parallels of latitude of the three-sphere as well as its meridians.

2.2.1. Parallels of Latitude. Next we study the parallels of latitude of S^3. Since $\mathbb{H} = \mathbb{R} \cdot e \oplus E$ each affine three-dimensional space $\lambda \cdot e \oplus E$ with $|\lambda| \leq 1$ intersects S^3 in a two-sphere $S^2_{r(\lambda)}$ with radius $r(\lambda) = 1 - |\lambda|$, a *parallel of latitude* of S^3. This is easily seen by choosing an orthonormal basis $v_1, v_2, v_3 \in E$ with which $S^2_{r(\lambda)}$ is determined by the equation

$$x_1^2 + x_2^2 + x_3^2 = (1 - |\lambda|)^2.$$

Clearly, the two degenerated spheres $S^2_{r(1)} = e$ and $S^2_{r(-1)} = -e$ are the two poles of S^3. The equator is $S^2_{r(0)}$, the unit sphere in E.

2.2.2. Meridians on S^3. To describe the meridians on S^3 we first make this notion precise in terms of a geometric definition: A *meridian* on S^3 is the intersection of a plane $C \subset \mathbb{H}$ containing the line $\mathbb{R} \cdot e$ with S^3. Thus a meridian is a great circle on S^3 containing the poles e and $-e$. Since C is of the form span (e, a) with $a \in S^2 \subset E$,

$$C = \mathbb{C}^a.$$

Therefore, the sets underlying the groups $U^a(1)$ with $a \in S^2$ provide us with all meridians, and hence:

LEMMA 7. *The oriented meridians are in a one-to-one correspondence with the unitary groups $U^a(1)$ with $a \in S^2$.*

Given any two unitary groups $U^a(1)$ and $U^b(1)$ with $a, b \in S^2$,

$$b = \tau_u(a) \qquad \text{for some} \quad u \in S^3,$$

and thus

$$U^b(1) = u \cdot U^a(1) \cdot u^{-1}.$$

In other words, the meridians are conjugate to each other, which implies

$$\mathbb{C}^b = u \cdot \mathbb{C}^a \cdot u^{-1}.$$

2.3. One-Parameter Subgroups of SU(2) and Hopf Bundles

In the descriptions of the various phenomena throughout these notes the so-called Hopf fibration and more general the Hopf bundles play a central role. They relate the group $S^3 = \mathrm{SU}(2)$ with its equator $S^2 = S^3 \cap E$. These bundles are fibrations of S^3 over S^2. To construct them we first study the image of a circle under τ_k.

2.3.1. Cosets and Circles. The geometry of S^3 allows a partition into circles centered about $0 \in \mathbb{H}$. Using the group structure of S^3 this partition is constructed as follows: Let $a \in S^2$ be fixed. The one-parameter group generated by a circle

$$U^a(1) = \mathbb{C}^a \cap S^3$$

(centered about zero) passes the poles e and $-e$ of S^3. Given any $k \in S^3$ the right coset $U^a(1) \cdot k$ (as well as the left coset) is a circle centered about zero since the multiplication with k from the right is an isometric $\mathbb{R}$-linear map m_k, say, and thus

$$U^a(1) \cdot k = m_k(\mathbb{C}^a) \cap S^3$$

holds true. Obviously, the collection

$$\{U^a(1) \cdot k \mid k \in S^3\} \tag{71}$$

of right cosets partitions S^3. Clearly, not every circle centered about 0 belongs to this partition as the example $U^b(1)$ with $b \in S^2$ different from a manifests.

On the other hand, let $K \subset S^3$ be a circle centered about zero. For $k \in K$ the set $K \cdot k^{-1}$ is a circle centered about zero with

$$e \in K \cdot k^{-1} \quad \text{and} \quad -e \in \quad K \cdot k^{-1}. \tag{72}$$

Thus

$$K \cdot k^{-1} = U^b(1) \quad \text{for some} \quad b \in S^2 \tag{73}$$

yielding

$$K = U^b(1) \cdot k. \tag{74}$$

Therefore,

$$K = U^a(1) \cdot k \quad \text{and} \quad \tau_k(K) = k \cdot U^a(1) \tag{75}$$

for any fixed $k \in K$ which relates left with right cosets (cf. [**68**]). Clearly,

$$\tau_k(K) \subset \mathrm{SO}(E).$$

Equation (75) motivates, therefore, the following construction.

2.3.2. The Image of a Circle under Conjugation. In this section we are given a circle $K \subset S^3$ not necessarily centered about $0 \in \mathbb{H}$. The aim is to identify the image of K under the map

$$\tau|_K : K \longrightarrow \mathrm{SO}(E)$$

and to study the curve $\tau|_K(h)$ for some $h \in E$. As we would expect from the real projective space $\mathbb{RP}\mathbb{H} = \mathrm{SO}(E)$ of $\mathbb{H}$, the image of $\tau|_K$ is a circle. However, let us study this image by algebraic means: The unit quaternion $\frac{k}{|k|}$ obviously satisfies

$$\frac{k^2}{|k|^2} = -e.$$

Therefore, we conclude with (32), (35) and the splitting of k in section 1.4.3 that for $k \in K$,

$$\tau_k(< h, u > \cdot u + h') =< h, u > \cdot u + \tau_k(h')$$

and

$$\tau_k(h') = \frac{k^2}{|k|^2} \cdot h'$$

(cf. (44) in section 1.5.1). Thus

$$\tau_k(h) =< h, u > \cdot u + \frac{k^2}{|k|^2} \cdot h'$$

yielding

$$\tau_k(h) =< h, u > \cdot u - h' = h - 2 \cdot h'$$

with $h' \in u^\perp$ and $h = \zeta \cdot u + h'$. Therefore, τ_k is decomposed into

$$\tau_k = \mathrm{id}_E - 2 \cdot \mathrm{pr}_u \tag{76}$$

where $\mathrm{pr}_u : E \longrightarrow u^{\perp_E}$ is the orthogonal projection and id_E is the identity on E. Since due to (65),

$$|\mathrm{pr}_u|^2_{\mathrm{End}_{\mathbb{R}} E} = \frac{1}{3} \cdot \mathrm{tr}\, \mathrm{pr}_u \circ \widetilde{pr_u} = \frac{2}{3} \qquad \forall\, u \in E,$$

the collection

$$\{\mathrm{pr}_u \mid u \in E\} \text{ for } k \subset S^2$$

(cf. (76)) forms the two-sphere $S^2_{\frac{2}{3}}$ in $\mathrm{End}_{\mathbb{R}} E$ of radius $\frac{2}{3}$. Thus we have

$$\tau|_K(K) \subset \mathrm{id}_E(K) - 2 \cdot S^2_{\frac{2}{3}}.$$

Next, let $v \in S^2$ be orthogonal to all elements in K. Thus

$$\mathrm{tr}(\mathrm{pr}_v \circ \mathrm{pr}_u) = \sum_{i=0}^{3} < \mathrm{pr}_v(\mathrm{pr}_u(e_i)), e_i >$$

where $e_0 := v$, $e_1 := u$, and $e_2 \in S^2$ is orthogonal to both v and u. Hence

$$< \mathrm{pr}_v, \mathrm{pr}_u >_{\mathrm{End}_{\mathbb{R}} E} = \frac{1}{3} \cdot \mathrm{tr}\, \mathrm{pr}_v \circ \mathrm{pr}_u = \frac{1}{3}.$$

$\{\mathrm{pr}_u \mid u \in K\}$ is a circle (on a two-sphere) centered about pr_v in $\mathrm{End}_{\mathbb{R}} E$ and hence pr_u varies on a circle on $S^2_{\frac{2}{3}}$. This shows that the image

$$\tau|_K(K) \text{ of } \tau|_K : K \longrightarrow \mathrm{SO}(E)$$

is a circle in $\mathrm{SO}(E)$.

For any h in a plane in E the vector $\tau_k(h)$ varies in a plane in E, too, since the evaluation map

$$\mathrm{ev}_h : \mathrm{End}_{\mathbb{R}} E \times E \longrightarrow E,$$

defined by $\mathrm{ev}_h(l) := l(h)$ for all $l \in \mathrm{End}_{\mathbb{R}} E$, is linear. Therefore, the image of $\mathrm{ev}_h \circ \tau|_K$ is a circle due to $|\tau_k(h)| = |h|$ for all $k \in K$.

For the general case of a circle $K \subset \mathrm{SU}(2)$ we observe that

$$K = K' \cdot k'$$

for $K' \subset S^2$ and $k' \in \mathrm{SU}(2)$. Hence

$$\tau(K) = \tau(K') \circ \tau_{k'}.$$

By equation (65) we hence conclude that $\tau(K) \subset \mathrm{End}_{\mathbb{R}} E$ is a circle, as well.

Together with the study of T_τ in section 1.4.5 we conclude:

LEMMA 8. *For any circle K on $\mathrm{SU}(2) = S^3$ the image of K under the map*

$$\tau|_K : K \longrightarrow \mathrm{SO}(E)$$

is a circle in $\mathrm{SO}(E)$. In addition, for any fixed $h \in E$,

$$ev_h \circ \tau_k : K \longrightarrow E$$

is a circle, as well. Moreover,

$$T_k\tau(\dot{k})(h) = 2 \cdot \dot{k} \times h$$

is a constant where $\dot{k}$ varies as a tangent vector along the circle parameterized by a multiple of the arc length.

2.3.3. Actions of $U^a(1)$ and SU(2). The aim here is to introduce the geometry of the *Hopf fibration* of SU(2) over S^2 and the principal bundle SO(E) over S^2, interpreted as the bundle of oriented, orthonormal frames of TS^2 (cf. section 2.3.5). These bundles are special cases of Hopf bundles over S^2, as we will see.

To this end we first consider the natural (right) action

$$\begin{array}{rcl} \psi : \mathrm{SU}(2) \times U^a(1) & \longrightarrow & SU(2) \\ \psi(k, e^{t\cdot a}) & := & k \cdot e^{t\cdot a} \end{array} \tag{77}$$

of $U^a(1)$ on $\mathrm{SU}(2) \subset \mathbb{H}$ for some fixed $a \in S^2$. Clearly, the coset $k \cdot U^a(1)$ is a circle for each $k \in \mathrm{SU}(2)$ (cf. section 2.3.1). To view the coset space $\mathrm{SU}(2)/U^a(1)$ as a diffeomorphic image of the sphere S^2 we form $k \cdot U^a(1) \cdot k^{-1} = \tau_k(U^a(1))$, which is a circle, and observe that

$$\tau_k(U^a(1)) = U^{\tau_k(a)}$$

(cf. section 1.4.1). The map

$$\begin{array}{rcl} \mathrm{pr}_{\mathrm{Hopf}\,1} : \mathrm{SU}(2) & \longrightarrow & S^2 \\ \mathrm{pr}_{\mathrm{Hopf}\,1}(k) & := & \tau_k(a) \end{array} \tag{78}$$

is smooth and surjective. This projection fibres SU(2) into circles and is called the *Hopf fibration*. It is now a routine exercise to verify that $\mathrm{pr}_{\mathrm{Hopf}\,1}$ yields the diffeomorphism

$$\begin{array}{rcl} \mathrm{SU}(2)/U^a(1) & \longrightarrow & S^2 \\ U^a(1) \cdot k & \longmapsto & \tau_k(a). \end{array} \tag{79}$$

Hence S^2 is diffeomorphic to the homogeneous space of all cosets $\mathrm{SU}(2)/U^a(1)$ (cf. section 2.3.1). We recall that by section 1.2.1 to any element in $U^a(1)$ there corresponds a unique element in $U(1)$ via $(i^a)^{-1}$. Here $a \in S^2$. Thus the fibre $k \cdot U^a(1)$ may be identified with $k \cdot U(1)$; hence we write $\mathrm{SU}(2)/U(1)$ instead of $\mathrm{SU}(2)/U^a(1)$. Moreover, SU(2) is a $U(1)$-principal bundle over S^2 (as we will see below). In quadruple notation this bundle reads $(\mathrm{SU}(2), S^2, \mathrm{pr}_{\mathrm{Hopf}\,1}, U(1))$. If no confusion arises we will call it $\mathbb{P}_1$. For a different, detailed treatment we refer to [**62**] and [**63**].

The second action we have in mind is

$$\begin{array}{rcl} \underline{\psi} : \mathrm{SU}(2) \times S^2 & \longrightarrow & S^2 \\ \underline{\psi}(k, u) & := & \tau_k(u) \end{array} \tag{80}$$

of SU(2) on the unit sphere $S^2 \subset E$. In terms of $\exp_{\mathrm{SU}(2)}$ this action is rewritten as

$$\underline{\psi}(\exp_{\mathrm{SU}(2)} t \cdot a, u) = e^{t\cdot a} \cdot u \cdot e^{-t\cdot a} = \tau_{e^{t\cdot a}}(u) \qquad \forall\, t \in \mathbb{R}, \forall\, a \in S^2 \text{ and } \forall\, u \in S^2$$

for a suitable $a \in S^2$. Thus SU(2) acts on S^2 by means of the natural action of $\tau(\mathrm{SU}(2)) = \mathrm{SO}(E)$ on S^2 (cf. lemma 5 in section 1.4.3). The stability group of the action (80) is therefore $U^a(1)$.

Next, we compute the *stability group* in SO(E) of $u \in S^2$, a closed subgroup of SO(E) (and hence a Lie group). It is determined by the equation

$$\tau_{e^{t\cdot a}}(u) = u,$$

whose infinitesimal version is

$$a \times u = u \times a.$$

This equation holds iff $a = \pm u$. The stability group in $\mathrm{SO}(E)$ at u is thus the commutative subgroup $\mathrm{SO}(F^u)$ of $\mathrm{SO}(E)$.

Let us fix a for the rest of this paragraph. For each $b \in S^2$ there is some $g \in \mathrm{SO}(E)$ such that

$$b = g(a).$$

Thus

$$\mathrm{SO}(F^b) = g \circ \mathrm{SO}(F^a) \circ g^{-1}$$

for all $b \in S^2$ and $g \in \mathrm{SO}(E)$ with $b = g(a)$. Hence $\mathrm{SO}(F^b) \subset \mathrm{SO}(E)$ is the stability group of $b \in S^2$. Thus we have the projection

$$\begin{array}{rcl} \mathrm{pr}_{\mathrm{Hopf}\,2} : \mathrm{SO}(E) & \longrightarrow & S^2 \\ g & \longmapsto & g(a). \end{array}$$

A reformulation of $\mathrm{pr}_{\mathrm{Hopf}\,2}$ is thus

$$\mathrm{pr}_{\mathrm{Hopf}\,2}(\tau_k) = \tau_k(a) \tag{81}$$

for any $k \in \mathrm{SU}(2)$. Here a acts as a reference point.

The fibre $\mathrm{pr}^{-1}_{\mathrm{Hopf}\,2}(\tau_k(a)) \subset \mathrm{SO}(E)$ is $\mathrm{pr}^{-1}_{\mathrm{Hopf}\,2}(\tau_k(a)) = \tau(k \cdot U^a(1))$, to which the action of $U^a(1)$ on $\mathrm{SU}(2)$ descends. Hence $\mathrm{pr}_{\mathrm{Hopf}\,2}$ fibres $\mathrm{SO}(E)$ into circles on which $U^a(1)$ acts naturally.

In fact, $(\mathrm{SU}(2), S^2, \mathrm{pr}_{\mathrm{Hopf}\,2}, U(1))$ is a principal bundle with structure group $U(1)$ (cf. [**62**] and [**63**]). We call it simply $\mathbb{P}_2$. (Clearly, $\mathrm{SO}(E)$ is a principal bundle over S^2 with structure group $\mathrm{SO}(F^a)$, too).

Here is a proof for this principal bundle structure of $\mathbb{P}_2$: Let us consider the two open sets

$$S^{\pm} := S^2 \backslash \{\pm a\} \tag{82}$$

in S^2 with the respective stereographic projections onto $F^{-a}\backslash(0)$ and $F^a\backslash(0)$. These projections are both bijective. Of course, as linear spaces $F^a = F^{-a}$, however, their orientations ω^a and ω^{-a} are different (cf. section 1.3.5). The preimage of the two sets S^+ and S^- under $\mathrm{pr}_{\mathrm{Hopf}\,2}$ are $\mathrm{SU}(2)^+$ and $\mathrm{SU}(2)^-$, say. For each of S^+ and S^- we choose smooth sections

$$s^{\mp} : S^{\pm} \to \mathrm{SU}(2)^{\pm}$$

of $\mathrm{pr}_{\mathrm{Hopf}\,2}$. For each k in $\mathrm{SU}(2)^{\pm}$ let $\sigma_{\pm}(k)$ be the unique element in $U(1)$ for which

$$k = s^{\pm}(\mathrm{pr}_{\mathrm{Hopf}\,2}(k)) \cdot \sigma_{\pm}(k) \qquad \forall\, k \in \mathrm{SU}(2)|_{S^{\pm}}.$$

Each of the smooth maps

$$\begin{array}{rcl} \Phi^{\pm a} : \mathrm{SU}(2)^{\pm} & \longrightarrow & S^{\pm} \times U(1) \\ k & \longmapsto & (\mathrm{pr}_{\mathrm{Hopf}\,2}(k), \sigma_{\pm}(k)) \end{array}$$

is a bundle isomorphism. The transition map

$$\mathrm{SU}(2)^+ \xrightarrow{\Phi^{-a} \circ (\Phi^{+a})^{-1}} \mathrm{SU}(2)^-$$

is of the form

$$\Phi^{-a} \circ (\Phi^{+a}(k))^{-1} = k \cdot e^{-2\vartheta(\mathrm{pr}_{\mathrm{Hopf}\,2}(k)) \cdot i} \qquad \forall\, k \in \mathrm{SU}(2)^+$$

where $\vartheta(\mathrm{pr}_{\mathrm{Hopf}\,2}(k))$ is the smallest non-negative angle in $[0, 2\pi)$ for which

$$s^-(\mathrm{pr}_{\mathrm{Hopf}\,2}) = s^+(\mathrm{pr}_{\mathrm{Hopf}\,2}) \cdot e^{-2\vartheta(\mathrm{pr}_{\mathrm{Hopf}\,2}(k))\cdot i} \qquad \forall\, k \in \mathrm{SU}(2)^+.$$

We hence have the map

$$g : S^+ \cap S^- \longrightarrow U(1)$$

defined by

$$g(x) := e^{-2\vartheta(x)\cdot i} \qquad \forall\, x \in S^+ \cap S^-.$$

It is called the *cocycle* of SU(2) over S^2.

The principal bundle structure of $(\mathrm{SU}(2), S^2, \mathrm{pr}_{\mathrm{Hopf}\,1}, U(1))$ is verified accordingly.

The bundle $(\mathrm{SU}(2), S^2, \mathrm{pr}_{\mathrm{Hopf}\,2}, U(1))$ and the Hopf fibration are special cases of the so-called *Hopf bundles* which will be introduced next.

2.3.4. Hopf Bundles. After studying the Hopf fibrations $\mathbb{P}_1$ and $\mathbb{P}_2$ we will now extend our investigation on general Hopf bundles (cf. [**62**] and [**63**]). To obtain these principal bundles we fix some $a \in S^2$ and construct them as follows: Given an integer n there is a natural action of $\mathbb{Z}_n := \mathbb{Z}/_{n\cdot\mathbb{Z}}$ on S^3. To present it let us realize $\mathbb{Z}_n$ as a finite subgroup of $U(1)$. More precisely, let

(83) $$C_n := \{e^{\frac{z}{n}\cdot i} | z \in \mathbb{Z}\}.$$

Clearly, $\mathbb{Z}_n$ and C_n are naturally isomorphic. $z_n \in C_n$ corresponds to some $m \in \mathbb{Z}$ with $0 \leq m \leq n-1$. In order to multiply any $k \in \mathrm{SU}(2) \equiv S^3$ with $z_n \in C_n$ we notice that

$$k = e^{\vartheta\cdot b}$$

for some $b \in S^2$ and some $\vartheta \in \mathbb{R}$. Clearly, $C_n^a := \{e^{\frac{z}{n}} \cdot a | z \in \mathbb{Z}\}$ is isomorphic to C_n, as well. Transferring $e^{\frac{z}{n}\cdot i}$ into $e^{\frac{z}{n}\cdot a}$, the multiplication is defined by

$$k \cdot z_n := e^{\vartheta\cdot b} \cdot e^{\frac{z}{n}\cdot a}.$$

This multiplication yields an action

$$\begin{array}{ccc} \mathrm{SU}(2) \times \mathbb{Z}_n & \longrightarrow & \mathrm{SU}(2) \\ (k, z_n) & \longmapsto & k \cdot z_n. \end{array}$$

Here we have used the isomorphism i^a again (cf. (17) in section 1.2.1). The quotient

(84) $$\mathbb{P}_n := \mathrm{SU}(2)/\mathbb{Z}_n$$

of the above action is a smooth manifold. In fact, SU(2) is an n-fold covering of $\mathbb{P}_n$ with the surjective bundle homomorphism

(85) $$\mathrm{pr}_n : \mathrm{SU}(2) \longrightarrow \mathbb{P}_n$$

as the *covering map*. The fibres of pr_n have the form $k \cdot C_n^a$. Obviously, SU(2) is a principal $\mathbb{Z}_n$ bundle over $\mathbb{P}_n$. The compact subgroup $U^a(1) \subset \mathrm{SU}(2)$ is mapped by pr_n onto the compact group $U^a(1)/C_n^a \subset \mathbb{P}_n$. Hence $U^a(1)$ is the n-fold covering of $U^a(1)/C_n^a \subset \mathrm{SU}(2)/C_n^a$.

To verify that $\mathbb{P}_n$ is a principal bundle over S^2 with structure group $U(1)$ we first define the projection

$$\begin{array}{rcll} \mathrm{pr}_{\mathrm{Hopf}\,n} : \mathbb{P}_n & \longrightarrow & S^2 & \\ \mathrm{pr}_{\mathrm{Hopf}\,n}(k \cdot C_n^a) & := & \tau_k(a) & \forall\, k \in \mathrm{SU}(2). \end{array}$$

The fibre over $\tau_k(a)$, i.e. the preimage $(\mathrm{pr}_{\mathrm{Hopf}\,n})^{-1}(\tau_k(a))$ of $\tau_k(a)$ is

$$(\mathrm{pr}_{\mathrm{Hopf}\,n})^{-1}(\tau_k(a)) = (k \cdot U^a(1)) \cdot C_n^a = k \cdot C_n^a \cdot U^a(1).$$

Thus $U^a(1)$ acts transitively on the fibres of $\mathbb{P}_n$.

The charts on which the *Hopf bundle* $\mathbb{P}_n$ trivializes are $S^{\pm}$ (cf. (82)). The transition map of $\mathbb{P}_n$ over S_2 is provided by

$$gs_{\pm}^n(x) := e^{-2n\cdot\vartheta(x)\cdot a} \qquad \forall\, x \in S_{\pm}.$$

As an example let $n = 2$. Then $\mathbb{P}_2 = \mathrm{SO}(E)$. The manifold $\mathbb{P}_2$ is three-dimensional, as $\mathbb{P}_n$ is for all n. If $n = 1$, then $\mathbb{P}_n$ is the Hopf fibration, $\mathbb{P}_1$.

LEMMA 9. *For each $n \in \mathbb{Z}$ the manifold $\mathbb{P}_n$ is a $U(1)/\mathbb{Z}_n$-principal bundle over S^2. The natural map*

$$\mathrm{pr}_{\mathrm{Hopf}\,n} : \mathbb{P}_n \longrightarrow S^2 \tag{86}$$

is surjective and the diagram

$$\begin{array}{ccc} \mathrm{SU}(2) & \xrightarrow{\;pr_n\;} & \mathbb{P}_n \\ & \searrow^{\mathrm{pr}_{\mathrm{Hopf}\,1}} \quad {}^{\mathrm{pr}_{\mathrm{Hopf}\,n}}\swarrow & \\ & S^2 & \end{array} \tag{87}$$

commutes. As a three-sphere SU(2) *is simply connected and hence is the universal covering of $\mathbb{P}_n$.*

2.3.5. Example: The Bundle of Oriented, Orthonormal Frames of the Two-Sphere. As an example of a Hopf bundle we consider the bundle of oriented, orthonormal frames of the oriented sphere S^2 and will identify its total space as $\mathbb{P}_2$. More precisely, in the following we will show that $\mathrm{SO}(E) = \mathbb{P}_2$ is the total space of oriented, orthonormal frame bundles of $S^2 \subset E$, a principal bundle with $SO(2)$ as the structure group.

To relate $\mathbb{P}_2$ with the geometry of the two-sphere more closely we consider the collection

$$\mathbb{S} := \bigcup_{x \in S^2} S_x \subset TS^2$$

of unit circles S_x formed with respect to the Riemannian metric which the sphere S^2 inherits from the inclusion $S^2 \subset E$. Clearly, S_x parameterizes all oriented, orthonormal frames of the tangent space T_xS^2. Hence $\mathbb{S}$ manifests the collection of all oriented, orthonormal frames in TS^2. This collection is in fact a principal bundle as shown next.

Let us fix some v_0 in S_x for an $x \in S^2$. This gives rise to the action

$$\begin{array}{rcl} \Phi_{\mathbb{S}} : \mathrm{SO}(E) & \longrightarrow & \mathbb{S} \\ \Phi_{\mathbb{S}}(g) & := & g(v_0) \end{array}$$

for all $g \in \mathrm{SO}(E)$. Clearly,

$$\begin{aligned} \Phi_{S^2} : \mathrm{SO}(E) &\longrightarrow S^2 \\ g &\longmapsto g(a) \end{aligned}$$

is compatible with $\Phi_{\mathbb{S}}$ in the sense that

$$\mathrm{pr}_{S^2} \circ \Phi_{\mathbb{S}}(g) = \Phi_{S^2}(g) \qquad \forall\, g \in \mathrm{SO}(E)$$

where $\mathrm{pr}_{S^2} : TS^2 \longrightarrow S^2$ is the canonical projection. The map $\Phi_{\mathbb{S}}$ is a bijection and yields a manifold structure on $\mathbb{S}$. Hence $\Phi_{\mathbb{S}}$ is a diffeomorphism. Moreover, $\mathrm{SO}(T_a S^2)$ maps S^1_x into itself. Thus the diagram

$$\begin{array}{ccc} \mathrm{SO}(E) & \xrightarrow{\Phi_{\mathbb{S}}} & \mathbb{S} \\ & {\scriptstyle \mathrm{pr}_{\mathrm{Hopf}\,2}} \searrow \quad \swarrow {\scriptstyle \mathrm{pr}_{S^2}} & \\ & S^2 & \end{array}$$

commutes. Since $\mathrm{SO}(T_a S^2)$ is isomorphic to $\mathrm{SO}(2)$, the above diagram shows that

$$(\mathrm{SO}(E), S^2, \mathrm{pr}_{\mathrm{Hopf}\,2}, \mathrm{SO}(2)) \quad \text{and} \quad (\mathbb{S}, S^2, \mathrm{pr}_{S^2}, \mathrm{SO}(2))$$

are smoothly isomorphic as principal SO(2)-bundles (cf. [**45**]). Hence $\mathbb{P}_2$ is related to $\mathbb{P}_2 = \mathbb{S}$ by the commuting diagram

$$\begin{array}{ccccc} \mathrm{SU}(2) & \xrightarrow{\tau} & \mathrm{SO}(E) = \mathbb{P}_2 & \xrightarrow{\Phi_{\mathbb{S}}} & \mathbb{S} \\ & {\scriptstyle \mathrm{pr}_{\mathrm{Hopf}\,1}} \searrow & \downarrow {\scriptstyle \mathrm{pr}} & \swarrow {\scriptstyle \mathrm{pr}_{S^2}} & \\ & & S^2 & & \end{array} \tag{88}$$

where pr replaces $\mathrm{pr}_{\mathrm{Hopf}\,2}$. This diagram shows that the Hopf fibration is a *spin structure* of S^2.

In closing this section we point out that S_x also parameterizes the collection of all normed complex frames of $T_x S^2$, a complex linear space under the (right) operation of $\mathbb{C}^x$ (cf. section 1.3.1). The totality of all these frames in TS^2 is $\mathbb{P}_2$ with $U(1)$ as the structure group.

2.3.6. Complex Line Bundles on S^2. Each of the Hopf bundles presented in section 2.3.4 naturally yields a complex line bundle on S^2. These bundles are constructed by means of *associated bundles*. In fact, this construction yields all complex line bundles on S^2 (cf. [**45**]). One of them, namely TS^2, will play a fundamental role in several of the forthcoming chapters. At first we sketch the notion of a bundle associated to a principal bundle.

We begin with a simple situation. Given is a Lie subgroup H of a Lie-group G, a finite dimensional vector space V and a representation r of H on V. The principal bundle G over G/H (with its natural manifold structure) with H as the structure group yields the *joint action*

$$\begin{aligned} (G \times V) \times H &\longrightarrow G \times V \\ ((g, v), h) &\longmapsto (g \cdot h, h^{-1} \cdot v) \end{aligned}$$

where $h^{-1} \cdot v := r(h^{-1})(v)$. The quotient of this action is denoted by $G \times_H V$ and is a *homogeneous vector bundle* over G/H. This is to say that each $g \in G$ acts naturally on the vector bundle $G \times_H V$ associated to H (cf. [**45**]) as well as on

G/H (natural action). These actions commute with the projection from $G \times_H V$ to G/H. The vector bundle $G \times_H V$ is associated to the principal bundle G/H.

Now we extend this notion to any principal bundle P with a Lie group H as the structure group, say, over a manifold M. The group H is assumed to operate on V by means of a representation r. As above the joint action is given by

$$\begin{array}{rcl} (P \times V) \times H & \longrightarrow & G \times V \\ ((g, v), h) & \longmapsto & (g \cdot h, h^{-1} \cdot v) \end{array}$$

for any $g \in P$, any $h \in H$ and any $v \in V$. The quotient of this action is a vector bundle denoted by $P \times_H V$. (More general bundles are obtained by replacing the vector space V by a manifold on which the group H operates).

As an example for each n the Hopf bundle $\mathbb{P}_n$ provides us with the complex line bundle

$$\mathbb{F}_n := \mathbb{P}_n \times_{U(1)} F \tag{89}$$

for a complex linear space F. Among these bundles is

$$\mathbb{F}_2 = \mathbb{P}_2 \times_{U(1)} F \tag{90}$$

which is naturally isomorphic to TS^2, as is easily seen by the diagram (88).

To provide a little more insight to the construction of $\mathbb{P}_n \times_{U(1)} F$ let us take a look at some of its details. The goal is to construct an atlas on the quotient $\mathbb{P}_n \times_{U(1)} F$ and to show that it is a vector bundle. The basic ingredients of the construction of $\mathbb{P}_n \times_{U(1)} F$ are the one-dimensional $\mathbb{C}$-linear space F, called *typical fibre* and a *representation*

$$r : U(1) \longrightarrow GL(F).$$

The bundle $\mathbb{P}_n \times_{U(1)} F$ emanates out of the joint action

$$\Phi : (\mathbb{P}_n \times F) \times U(1) \longrightarrow \mathbb{P}_n \times F$$

defined by

$$\Phi(k_x, v, z) := (k_x \times z, r(z^{-1})(v))$$

for any $x \in S^2$, any $k_x \in \mathbb{P}_n$, any $v \in F$ and any $z \in U(1)$. The quotient of this action is $\mathbb{P}_n \times_{U(1)} F$. The manifold structure on $\mathbb{P}_n \times_{U(1)} F$ is obtained by an explicit atlas to be established as follows. The projection

$$\mathrm{pr}_r : \mathbb{P}_n \times F \longrightarrow \mathbb{P}_n \times_{U(1)} F$$

shall be the natural projection onto the quotient. Moreover,

$$\mathrm{pr} : \mathbb{P}_n \times_{U(1)} F \longrightarrow S^2$$

is defined such that the diagram

$$\begin{array}{ccc} \mathbb{P}_n \times F & \xrightarrow{\ \mathrm{pr}_r\ } & \mathbb{P}_n \times_{U(1)} F \\ {\scriptstyle \mathrm{pr}_1}\downarrow & & \downarrow{\scriptstyle \mathrm{pr}} \\ \mathbb{P}_n & \xrightarrow{\ \mathrm{pr}_{S^2}\ } & S^2 \end{array}$$

commutes. Here pr_1 is the projection of the first factor of the Cartesian product. The charts of $\mathbb{P}_n \times_{U(1)} F$ rely on the charts of $\mathbb{P}_n$. To present them let U_α and U_β

be trivializing charts of $\mathbb{P}_n$ together with the two smooth sections s_α and s_β. These sections are related by the structural function:

$$g_{\alpha\beta} : U_\alpha \cap U_\beta \longrightarrow U(1) \subset \mathbb{C}$$

as

$$s_\beta = s_\alpha \cdot g_{\alpha\beta}$$

(cf. [**46**]). The map

$$\varphi_\alpha : U_\alpha \times F \longrightarrow \mathrm{pr}^{-1}(U_\alpha)$$

given by

$$\varphi_\alpha(x, v) := \mathrm{pr}_r(s_\alpha(x), v) \qquad \forall\, x \in S^2 \text{ and } \forall\, v \in F$$

satisfies

$$\mathrm{pr}(\varphi_\alpha(x, v)) = x.$$

Hence φ_α can be restricted to a map

$$\varphi_{\alpha,x} : F \longrightarrow \mathrm{pr}^{-1}(x) \qquad \forall\, x \in S^2.$$

To each orbit $\mathrm{pr}^{-1}(x)$ there is some $w \in r(U(1))$ such that it passes through $(s_\alpha(x), w)$. Hence φ_α, x is bijective. Let φ_β and $\varphi_{\beta,x}$ be constructed accordingly. Due to

$$\mathrm{pr}_r(k \cdot z, v) = \mathrm{pr}_r(k, z \cdot v)$$

we obtain

$$\varphi_\alpha^{-1} \circ \varphi_\beta(x, v) = (x, g_{\alpha\beta}(x) \cdot v) \qquad \forall\, x \in U_\alpha \cap U_\beta \text{ and } \forall\, v \in F.$$

Thus an atlas of $\mathbb{P}_n$ yields an atlas of $\mathbb{P}_n \times_{U(1)} F$, which hence is a vector bundle with typical fibre F.

2.3.7. A Reverse Problem of the Hopf Fibration. In this section we will turn back to the Hopf fibration (cf. section 2.3.3). The purpose here is to determine the fibre in SU(2) over b for any b in S^2. This requires us to study the link between $b = \tau_k(a)$ and k from the point of view of Euclidean geometry. Here b and a are points in S^2 and k is an element of SU(2). To do so we first determine some $k' \in k \cdot U^a(1)$ out of the two data a and b for which the axis of the rotation $\tau_{k'}$ is perpendicular to a. Thereafter we reconstruct the fibre over a given $b \in S^2$.

The *reference point* needed is still $a \in S^2$ (cf. section 2.3.3). To analyse the fibres with respect to our purpose, we split $k \in S^3$ into

$$k = \lambda \cdot e + \mu \cdot u \tag{91}$$

with $\lambda, \mu \in \mathbb{R}$ and $u \in S^2$. Obviously,

$$\lambda^2 + \mu^2 = 1.$$

According to (31) the real line $\mathbb{R} \cdot u$ is the axis of rotation of τ_k. To see how a is rotated by τ_k about $\mathbb{R} \cdot u$, we decompose a into the components

$$a_u := < a, u > \cdot u \qquad \text{and} \qquad a_\perp := a - < a, u > \cdot u.$$

Clearly, $a_\perp \in F^u$ is the component of a rotated by τ_k while a_u is left invariant by τ_k. We observe that

$$\tau_k(a_\perp) = e^{\vartheta \cdot u} \cdot a_\perp$$

since F^u is the plane of rotation of τ_k. The angle ϑ is the (oriented) angle of rotation of τ_k (cf. (49)).

Moreover,

$$< \tau_k(a), u >=< a, \tau_k^{-1}(u) >$$

and therefore

$$< b, u >=< a, u >=: \cos\beta, \tag{92}$$

for some $\beta \in \mathbb{R}$, since u is left invariant by τ_k. Thus

$$b - a \in F^u.$$

Equation (92) expresses the fact that a and b are on the same *cone* with vertex in $0 \in E$, the real number β as *cone angle* and $\mathbb{R} \cdot u$ as axis of rotation of the surface.

Applying τ_k to a yields

$$e^{\vartheta \cdot u} \cdot a_\perp + < a, u > \cdot u = b. \tag{93}$$

Defining b_u and $b_\perp$ in analogy to a_u and $a_\perp$ yields

$$\tau_k(a_u) = b_u$$

and, therefore,

$$\tau_k(a_\perp) = b_\perp.$$

From (49) we obtain for $k^2 = e^{\vartheta \cdot u}$,

$$e^{\vartheta \cdot u} = b_\perp \cdot a_\perp^{-1}$$

or

$$e^{\vartheta u} = -\frac{1}{|a_\perp|^2}(b_\perp \times a_\perp - < b_\perp, a_\perp > \cdot e)$$

due to

$$a_\perp^{-1} = \frac{1}{|a_\perp|^2} \cdot a_\perp.$$

Moreover, since $|a_\perp| = |b_\perp|$, we easily verify

$$e^{\vartheta \cdot u} =< \frac{a_\perp}{|a_\perp|}, \frac{b_\perp}{|a_\perp|} > \cdot e + \frac{a_\perp}{|a_\perp|} \times \frac{b_\perp}{|b_\perp|}.$$

Thus k^2 is entirely determined by the unit vectors $\frac{a_\perp}{|a_\perp|}$ and $\frac{b_\perp}{|b_\perp|}$ in F^u.

Therefore, the components of k^2 formed with respect to the splitting $\mathbb{H} = \mathbb{R} \cdot e + E$ are

$$(\lambda^2 - \mu^2) \cdot e =< \frac{b_\perp}{|b_\perp|}, \frac{a_\perp}{|a_\perp|} > \cdot e \tag{94}$$

and

$$2\lambda \cdot \mu \cdot u = \frac{a_\perp}{|a_\perp|} \times \frac{b_\perp}{|b_\perp|} \tag{95}$$

which are directly related to the rotation angle ϑ of τ_k as seen from lemma 3 in chapter 1.

Clearly, the right-hand side of (95) determines the rotation axis of τ_k by means of a and b. Since $b_\perp$ is given by u, the spatial part of k, we may try to find some k' such that the rotation axis of $\tau_{k'}$ is perpendicular to a. This simplification is our next goal.

As mentioned in section 2.3.1, the coset $K := k \cdot U^a(1)$ is a great circle inclined from $\mathbb{R} \cdot e$ if $k \neq e$. At this point let us remark that there is some $k' \in K$ for which the rotation axis $\mathbb{R} \cdot u'$ is *perpendicular* to a. Here $u' \in S^2$. To see this we represent any element in K with respect to a fixed $k \in K$ as

$$k(t) := k \cdot e^{t \cdot a}$$

and split k according to (91). Thus

$$k(t) = \lambda e \cos t + \lambda a \cdot \sin t + \mu u \cdot \cos t + \mu \cdot u \cdot a \sin t.$$

According to (31) the rotation axis of $\tau_{k(t)}$ is

$$\mathbb{R} \cdot (\lambda a \cdot \sin t + \mu u \cdot \cos t + \mu u \cdot a \cdot \sin t).$$

If this axis is to be perpendicular to a for some $t_0 \in \mathbb{R}$,

$$\lambda \cdot \sin t_0 + \mu \cdot < u, a > \cdot \cos t_0 = 0$$

has to hold. If $\lambda \neq 0$ the solution t_0 is hence determined by

$$\tan t_0 = -\frac{\mu}{\lambda} \cdot < a, u >$$

(Clearly, there is some $k \in K$ for which $\lambda \neq 0$.) Since $k = e^{\frac{t}{2} \cdot u}$, the parameters λ and μ are identified as $\lambda = \cos \frac{t}{2}$ and $\mu = \sin \frac{t}{2}$ and we finally find

$$\tan t_0 = -\tan \frac{t}{2} \cdot < a, u > \qquad \text{or} \qquad \tan t_0 = -\tan \frac{t}{2} \cos \beta.$$

Therefore, we may state:

LEMMA 10. *Given a right coset $k \cdot U^a(1)$ for some $k \in \mathrm{SU}(2)$ for a fixed vector $a \in S^2$ there is some $k' \in k \cdot U^a(1)$ such that the axis of rotation of $\tau_{k'}$ is perpendicular to a. If, moreover,*

$$b := \tau_{k'}(a) = \mathrm{pr}_{\mathrm{Hopf}\,1}(k)$$

the axis of rotation of $\tau_{k'}$ is $\mathbb{R} \cdot \frac{a \times b}{|a \times b|}$.

With this we can come back to the problem posed at the beginning of this section: Given $b \in S^2$ find $k \in \mathrm{SU}(2)$ such that

$$\mathrm{pr}_{\mathrm{Hopf}\,1}(k) = b.$$

Again the one-parameter group in SU(2) to be factored out shall be $U^a(1)$ with $a \in S^2$.

The solution is rather easy now. Without loss of generality, we may assume that the rotation axis of τ_k (where k is still to be constructed) is perpendicular to a due to lemma 3. Thus k has to be of the form

$$k = e^{\frac{\vartheta}{2} \frac{a \times b}{|a \times b|}}. \tag{96}$$

The rotation angle ϑ of τ_k is obtained up to 2π from $|a \times b|$ as

$$\sin \frac{\vartheta}{2} = |a \times b|$$

due to the formula

$$|a \times b| = |a| \times |b| \cdot \sin \vartheta = \sin \vartheta.$$

We therefore have the following theorem:

THEOREM 11. *For each fixed $a \in S^2$ the projection*

$$\begin{aligned} \mathrm{pr}_{\mathrm{Hopf}\,1} : \mathrm{SU}(2) &\longrightarrow S^2 \\ \mathrm{pr}_{\mathrm{Hopf}\,1}(k) &:= \tau_k(a) \end{aligned}$$

is a (smooth) surjection with

$$\mathrm{pr}_{\mathrm{Hopf}\,1}^{-1}(\tau_k(a)) = k \cdot U^a(1).$$

Any coset $k \cdot U^a(1)$ with $k \in \mathrm{SU}(2)$ contains some k' such that the axis of rotation of $\tau_{k'} \in \mathrm{SO}(E)$ is perpendicular to $\mathbb{R} \cdot a$. Hence given $b \in S^2$ the coset $\mathrm{pr}_{\mathrm{Hopf}\,1}^{-1}(b) \subset \mathrm{SU}(2)$ is of the form

$$\mathrm{pr}_{\mathrm{Hopf}\,1}^{-1}(b) = e^{\frac{\vartheta}{2} \cdot \frac{a \times b}{|a \times b|}} \cdot U^a(1)$$

where ϑ satisfies

$$|a \times b| = \sin \vartheta.$$

The element k in (96) does not smoothly depend on b. In fact, a and $-a$ even cause a discontinuity. However, if the set $\{a, -a\} \subset S^2$ is deleted, the map

$$\begin{aligned} s_{\mathrm{Hopf}\,1} : S^2 \backslash \{a, -a\} &\longrightarrow S^3 \\ s_{\mathrm{Hopf}\,1}(b) &:= e^{\frac{\vartheta}{2} \cdot \frac{a \times b}{|a \times b|}} \end{aligned} \tag{97}$$

is smooth. It is certainly a section of $\mathrm{pr}_{\mathrm{Hopf}\,1}$, i.e.

$$\mathrm{pr}_{\mathrm{Hopf}\,1} \circ s_{\mathrm{Hopf}\,1} = \mathrm{id}_{S^2 \backslash \{a,-a\}}. \tag{98}$$

Therefore, the restriction of the principal bundle $(S^3, U^a(1), S^2, \mathrm{pr}_{\mathrm{Hopf}\,1})$ to $S^2 \backslash \{a, -a\}$ is *trivial*.

2.4. Spin Representations

In the next two sections we will leave the context of the Hopf fibration and instead will look at the representation of the special unitary group SU(2) (cf. [**78**]). After looking at the general construction for irreducible unitary representation of SU(2), we will study the spin $\frac{1}{2}$-representation in more detail. Apart from its fundamental role in physics, the spin $\frac{1}{2}$-representation is of interest in the description of the geometry of $S^3 \subset \mathbb{H}$: It describes the action of $S^3 = \mathrm{SU}(2)$ on $\mathbb{H}$ and hence gives a further view on the geometry of S^3. In contrast to the Hopf fibration where we have partitioned S^3 into circles we consider here the totality $\{U^a(1) | a \in S^2\}$ and observe that the action of each of these groups on $\mathbb{H}$ is subsumed under the action of SU(2) on $\mathbb{H}$ by the spin $\frac{1}{2}$-representation.

2.4.1. Spin Representations.

Let us consider an irreducible unitary representation

$$r : \mathrm{SU}(2) \longrightarrow \mathrm{GL}(W)$$

acting on a finite dimensional complex Hermitian vector space W, whose elements are called *spinors*.

To determine the character χ_r of r, i.e. $\operatorname{tr} r(z)$ for $z \in \mathrm{SU}(2)$, let $U^a(1)$ be the one-parameter group passing through z for some $a \in S^2$. Thus

$$r(z) : W \longrightarrow W$$

is a unitary isomorphism. Hence W decomposes into a direct sum of eigenspaces of $r(z)$ as

$$W = \bigoplus_{j=1}^{n} W_{\lambda_j}$$

where the λ_j are the eigenvalues (counted with their multiplicity). Both λ_j and $\bar{\lambda}_j$ occur as eigenvalues. Let $r|_{W_{\lambda_j}}$ be denoted by r_j. Clearly, $r_j(W_{\lambda_j}) = W_{\lambda_j}$ for all j. Therefore, the trace of

$$r_j : U^a(1) \longrightarrow U(W_{\lambda_j})$$

mapping any $z = e^{t \cdot a} \in U^a(1)$ into $r_j(z)$ is

$$\operatorname{tr} r_j(z) = e^{\nu_j t \cdot i} \qquad \forall\, t \in \mathbb{R}.$$

For some integer ν the value of the character χ_r of r at $z = e^{2t \cdot a}$ is thus

$$\chi_r(z^a) = \sum_{\vartheta} \exp \nu \cdot t \cdot i \qquad \forall\, t \in \mathbb{R}.$$

As it is well known (cf. [**78**]), the general form of the character χ_r of the representations r is

$$\chi_r(e^{t \cdot a}) = \exp(-s \cdot ti) + \exp(-s \cdot 1)t \cdot i) + \cdots + \exp(s \cdot ti) \tag{99}$$

for any $t \in \mathbb{R}$ and any $a \in S^2$. Here $s = \frac{n-1}{2}$. For a general s let us write r_s instead of r and χ_{r_s} instead of χ_r. Most of the time we will study the *spin $\frac{1}{2}$-representation*, i.e. the case $s = \frac{1}{2}$ or $n = 2$.

2.4.2. The Spin$\frac{1}{2}$-Representation. To study $r_{\frac{1}{2}}$ adapted to $\mathbb{H}$ as representation space we consider the left action

$$\begin{aligned} \ell_m : \mathbb{H} &\longrightarrow \operatorname{End}_{\mathbb{R}} \mathbb{H} \\ \ell_m(k)(k') &:= k \cdot k' \qquad \forall\, k, k' \in \mathbb{H}. \end{aligned} \tag{100}$$

If $k \in \dot{\mathbb{H}} = \mathbb{H} \backslash \{0\}$, the $\mathbb{R}$-endomorphism $\ell_m(k)$ is invertible. Hence

$$\ell_m : \mathbb{H} \longrightarrow \mathrm{GL}(\mathbb{H})$$

is a representation of the group of invertible elements in $\mathbb{H}$ on the real linear space $\mathbb{H}$. If we restrict ℓ_m further to

$$\ell_m : \mathrm{SU}(2) \to \mathrm{GL}(\mathbb{H}), \tag{101}$$

we find a real representation of $\mathrm{SU}(2)$.

To turn it into a *unitary representation* we need to equip $\mathbb{H}$ with the structure of a *complex linear vector space*. This is done by splitting $\mathbb{H}$ into $\mathbb{H} = \mathbb{C}^a \oplus F^a$ for some $a \in S^2$ and fixing some unit vector $q_0 \in F^a$ yielding

$$F^a = q_0 \cdot \mathbb{C}^a.$$

Hence $\mathbb{H}$ is a $\mathbb{C}^a$-right linear space (cf. section 1.2.3). Moreover, $\mathbb{H}$ is equipped with the Hermitian product $(\ |\ )$, defined by

$$(z_1 + q_1 \cdot z_1' \mid z_2 + q_0 \cdot z_2') = z_1 \cdot \bar{z}_2 + z_1' \cdot \bar{z}_2' \qquad \forall\, z_1, z_2 \in \mathbb{C} \qquad \text{for } s = 1, 2$$

(cf. (25) in section 1.2.6). Since, for any $u \in \mathrm{SU}(2)$,

$$|u \cdot k'| = |u| \cdot |k'| = |k'| \qquad \forall\, k' \in \mathbb{H},$$

the representation (101) can be regarded as a representation

$$\ell_m : \mathrm{SU}(2) \longrightarrow \mathrm{SO}(\mathbb{H}).$$

Obviously, ℓ_m is $\mathbb{C}^a$-linear and

$$\ell_m(u)((z' + h') \cdot z) = (u \cdot (z' + h')) \cdot z$$

for any choice of $u \in \mathrm{SU}(2)$, any $z, z' \in \mathbb{C}^a$ and any $h' \in F^a$. The map $\ell_m(u)$ preserves the norm of the Hermitian product since this norm coincides with the norm given by the scalar product $< \, , \, >$. Thus $\ell_m|_{\mathrm{SU}(2)}$ can be regarded as a *unitary representation*

$$\ell_m : \mathrm{SU}(2) \to U(\mathbb{C}^a \oplus F^a).$$

To compute the character χ_{ℓ_m} of $\ell_m(u)$ we represent u as $u = e^{tb}$ for some $b \in S^2$. The eigenspaces of ℓ_m are $\mathbb{C}^b$ and F^b, as is easily seen. Therefore, the trace in $\mathbb{C}^b$ is

$$\mathrm{tr}\, \ell_m(u) = e^{t \cdot b} + e^{-t \cdot b}$$

(cf. (21)). Hence the trace in $\mathbb{C}$ satisfies

$$\chi_{\ell_m} = \chi_{r_{\frac{1}{2}}} \tag{102}$$

and we can conclude

$$\ell_m = r_{\frac{1}{2}}. \tag{103}$$

The (unitary) spin $\frac{1}{2}$-representation is geometrically easier to understand if we set for any $a \in S^2$

$$r_{\frac{1}{2}}|_{U^a(1)} : U^a(1) \longrightarrow U(\mathbb{C}^a \oplus F^a) = U(\mathbb{H})$$

defined by

$$\begin{aligned} r_{\frac{1}{2}}(z)(z' + h) &:= z' \cdot z + z \cdot h \\ &= z' \cdot z + h \cdot z^{-1} \end{aligned} \tag{104}$$

for all $z \in U^a(1)$, for all $z' \in \mathbb{C}^a$ and for all $h \in F^a$. The unitary operator $r_{\frac{1}{2}}(z)$ for $z \in \mathbb{C}$ is called a *half spin operator* or, more precisely, spin $\frac{1}{2}$*-operator*.

If $z = e^{t \cdot a}$ for $t \in \mathbb{R}$, then

$$r_{\frac{1}{2}}(e^{t \cdot a})(z' + h) = e^{t \cdot a} \cdot z' + h \cdot e^{-t \cdot a}.$$

For some $k \in \mathbb{H}$ the automorphism τ_k maps $\mathbb{C}^a$ and F^a, respectively, into $\mathbb{C}^b$ and F^b. Hence for each $b \in S^2$ and any $z^b \in U^b(1)$ there is a quaternion k depending on b such that

$$r_{\frac{1}{2}}(z^b)\Big(\tau_k(z') + \tau_k(h)\Big) = \tau_k(z^b) \cdot z' \tau_k(h) \cdot (z^b)^{-1}$$

for $z^b = \tau_k(z)$. This is to say that

$$r_{\frac{1}{2}}(e^{t \cdot b}) = \Big(\tau_k^{-1} \circ r_{\frac{1}{2}} \circ \tau_k\Big)(e^{t \cdot a})$$

for any $e^{t \cdot a} \in U^a(1)$. Clearly, k is not unique at all.

Obviously, the spin $\frac{1}{2}$-representation restricted to $U^b(1)$ for any $b \in S^2$ preserves the tori fibration of SU(2) as described in (20).

The character $\chi_{r_{\frac{1}{2}}}$ is given by

$$\chi_{r_{\frac{1}{2}}}(e^{t\cdot a}) = e^{ti} + e^{-ti} \qquad \forall\, t \in \mathbb{R}.$$

If $\mathbb{H}$ is regarded as the representation space of SU(2), we call the elements of $\mathbb{H}$ *spinors*, as well.

2.5. The Infinitesimal Spin $\frac{1}{2}$-Representation

In this final section of this chapter, we will pay attention to the differential of the spin $\frac{1}{2}$-representation, this is to say to the infinitesimal spin $\frac{1}{2}$-representation, and we will take a short look at how it intertwines with the geometry of $\mathbb{H}$.

2.5.1. The Infinitesimal Spin $\frac{1}{2}$-Representation. By definition the *infinitesimal* spin $\frac{1}{2}$-representation is the differential $dr_{\frac{1}{2}}$ at $e \in \mathrm{SU}(2)$. More precisely,

$$dr_{\frac{1}{2}} : \mathrm{su}(2) \longrightarrow \mathrm{su}(\mathbb{C}^a \oplus F^a)$$

since $T_e\,\mathrm{SU}(2) = \mathrm{su}(2)$. By $\mathrm{su}(\mathbb{C}^a \oplus F^a)$ we denote the collection of all skew Hermitian linear maps of the $\mathbb{C}$-linear space $\mathbb{C}^a \oplus F^a$ equipped with its natural Hermitian product $(\;|\;)$ (again cf. (25) in section 1.2.6).

To compute $dr_{\frac{1}{2}}$ at $e \in \mathrm{SU}(2)$ we consider

$$\frac{d}{dt} r_{\frac{1}{2}}(e^{t\cdot a})\Big|_{t=0}(z+h) = \frac{d}{dt}(e^{t\cdot a}\cdot z + h\cdot e^{-t\cdot a})\Big|_{t=0} \tag{105}$$

where $a \in \mathrm{su}(2)$, $z \in \mathbb{C}^a$ and $h \in F^a$. Hence

$$\frac{d}{dt} r_{\frac{1}{2}}(e^{t\cdot a})\Big|_{t=0}(z+h) = a\cdot z - h\cdot a. \tag{106}$$

Since, on the other hand,

$$\frac{d}{dt} r_{\frac{1}{2}}(e^{t\cdot a})\Big|_{t=0}(z+h) = dr_{\frac{1}{2}}(a), \tag{107}$$

(106) turns for each $z \in \mathbb{C}^a$ and any $h \in F^a$ into

$$dr_{\frac{1}{2}}(a)(z+h) = a\cdot z - h\cdot a.$$

Writing $\dot r_{\frac{1}{2}}$ instead of $dr_{\frac{1}{2}}$,

$$\dot r_{\frac{1}{2}}(a)(z+h) = a\cdot z - h\cdot a \qquad \forall\, z \in \mathbb{C}^a \text{ and } \forall\, h \in F^a. \tag{108}$$

In order to compute its matrix we choose $a \in \mathrm{su}(2)$ and fix a unit vector $q_0 \in F^a$. Then the infinitesimal spin $\frac{1}{2}$-representation $\dot r_{\frac{1}{2}}$ acts on the orthonormal basis $(e, a, q_0, q_0\cdot a)$ of the $\mathbb{R}$-linear space $\mathbb{H}$ by

$$\begin{array}{rclrcl} \dot r_{\frac{1}{2}}(a)(e) & = & a, & \dot r_{\frac{1}{2}}(a)(q_0) & = & -q_0\cdot a, \\ \dot r_{\frac{1}{2}}(a)(a) & = & -e, & \dot r_{\frac{1}{2}}(a)(q_0\cdot a) & = & q_0. \end{array} \tag{109}$$

Thus the matrix $M(\dot{r}_{\frac{1}{2}})(a)$ in the above basis is

$$M(\dot{r}_{\frac{1}{2}}(a)) = \begin{pmatrix} 0 & -1 & 0 & 0 \\ 1 & 0 & 0 & 0 \\ 0 & 0 & 0 & 1 \\ 0 & 0 & -1 & 0 \end{pmatrix}. \tag{110}$$

This is directly seen from

$$\frac{d}{dt} r(e^{a\cdot t})(e)|_{t=0} = a, \qquad \frac{d}{dt} r(e^{a\cdot t})(q_0)|_{t=0} = -q_0 \cdot a,$$
$$\frac{d}{dt} r(e^{a\cdot t})(a)|_{t=0} = -e, \qquad \frac{d}{dt} r(e^{a\cdot t})(q_0 \cdot a)|_{t=0} = q_0.$$

2.5.2. Hermitian Operators. Multiplying $\dot{r}_{\frac{1}{2}}(a)(z+h)$ by a from the right ($\mathbb{C}^a$ acts on F^a from the right) yields the Hermitian $\mathbb{C}$-linear map denoted by $a\cdot\dot{r}_{\frac{1}{2}}$ which is given by

$$(a \cdot \dot{r}_{\frac{1}{2}})(a)(z+h) := \dot{r}_{\frac{1}{2}}(a)(z+h) \cdot a, \tag{111}$$

which rewritten is

$$a \cdot \dot{r}_{\frac{1}{2}}(a)(z+h) = -z+h$$

due to (109). Thus $a \cdot \dot{r}_{\frac{1}{2}}$ acts on the above orthonormal basis as

$$\begin{array}{lclclcl} a \cdot \dot{r}_{\frac{1}{2}}(a)(e) & = & -e, & \quad & a \cdot \dot{r}_{\frac{1}{2}}(a)(q_0) & = & q_0, \\ a \cdot \dot{r}_{\frac{1}{2}}(a)(a) & = & -a, & \quad & a \cdot \dot{r}_{\frac{1}{2}}(a)(q_0 \cdot a) & = & q_0 \cdot a, \end{array} \tag{112}$$

and we hence obtain the diagonal matrix

$$M(a \cdot \dot{r}_{\frac{1}{2}}(a)) = \begin{pmatrix} -1 & 0 & 0 & 0 \\ 0 & -1 & 0 & 0 \\ 0 & 0 & 1 & 0 \\ 0 & 0 & 0 & 1 \end{pmatrix}. \tag{113}$$

In physics literature it is common to set

$$\frac{d}{dt}\frac{1}{2} \cdot r_{\frac{1}{2}}(e^{ta})\bigg|_{t=0} (z+h) = \frac{1}{2}\dot{r}_{\frac{1}{2}}(a)(z+h)$$

yielding the matrices

$$M\left(\frac{1}{2} \cdot \dot{r}_{\frac{1}{2}}(a)\right) = \begin{pmatrix} 0 & \frac{1}{2} & 0 & 0 \\ -\frac{1}{2} & 0 & 0 & 0 \\ 0 & 0 & 0 & -\frac{1}{2} \\ 0 & 0 & \frac{1}{2} & 0 \end{pmatrix} \tag{114}$$

and

$$M\left(\frac{a}{2} \cdot \dot{r}_{\frac{1}{2}}(a)\right) = \begin{pmatrix} -\frac{1}{2} & 0 & 0 & 0 \\ 0 & -\frac{1}{2} & 0 & 0 \\ 0 & 0 & \frac{1}{2} & 0 \\ 0 & 0 & 0 & \frac{1}{2} \end{pmatrix}. \tag{115}$$

Taking $\{e, q\}$ as a basis in the $\mathbb{C}$-linear space $\mathbb{H}$,

$$a \cdot \dot{r}_{\frac{1}{2}}(a)(e) = -e \quad \text{and} \quad a \cdot \dot{r}_{\frac{1}{2}}(a)(q) = q \tag{116}$$

is obvious. Thus the matrix with respect to $\{e, q\}$ of the Hermitian operator $a \cdot \dot{r}(a)$ is

$$M(a \cdot \dot{r}(a)) = \begin{pmatrix} -1 & 0 \\ 0 & -1 \end{pmatrix}. \tag{117}$$

The contragradient operator has

$$M((a \cdot \dot{r}(a))^*) = \begin{pmatrix} 1 & 0 \\ 0 & -1 \end{pmatrix} \tag{118}$$

as its matrix. Hence the respective eigenvalues of $a \cdot \dot{r}_{\frac{1}{2}}(a)$ and $(a \cdot \dot{r}(a))^*$ are ∓ 1 and ± 1. The Hermitian operator $\frac{1}{2} \cdot a \cdot \dot{r}_{\frac{1}{2}}(a)$ and its contragradient thus have the eigenvalues $\mp\frac{1}{2}$ and $\pm\frac{1}{2}$, respectively.

To relate $\dot{r}$ with the Lie bracket on SU(2) first we observe that

$$\begin{aligned} [a, q_0] &= 2a \times q_0 = -2 \cdot \omega^a(q_0, q_0 \cdot a) \cdot q_0 \cdot a = -2 \cdot q_0 \cdot a, \\ [a, q_0 \cdot a] &= 2a \times q_0 \cdot a = -2 \cdot \omega^a(q_0 \cdot a, q_0) \cdot q_0 = 2 \cdot q_0, \\ [q_0, q_0 \cdot a] &= 2q_0 \times q_0 \cdot a = -2 \cdot \omega^a(q_0, q_0 \cdot a) \cdot a = -2a. \end{aligned}$$

Since for any two $h_1, h_2 \in F$,

$$[h_1, h_2] = 2 \cdot \omega^a(h_1, h_2) \cdot a \quad \text{and} \quad \frac{1}{2} \cdot \dot{r}_{\frac{1}{2}}([h_1, h_2]) = -\omega^a(h_1, h_2)e,$$

we see the reason for the factor $\frac{1}{2}$. Therefore, on $\mathbb{H}$,

$$\begin{aligned} \frac{1}{2} \cdot \dot{r}([a, q_0]) &= -q_0, \\ \frac{1}{2} \cdot \dot{r}([a, q_0 \cdot a]) &= a, \\ \frac{1}{2} \cdot \dot{r}([q_0, q_0 \cdot a]) &= -e \end{aligned} \tag{119}$$

hold true. The representation $\frac{\dot{r}}{2}$ naturally extends to $\mathrm{SU}(2) \otimes \mathbb{C}^a$ (cf. [**78**]).

Notice that the right-hand side of (119) consists of Pauli elements in $\mathbb{H}$ associated with the splitting $H = \mathbb{C}^a + q_0 \cdot \mathbb{C}^a$ (cf. section 1.3). Thus $\frac{a}{2} \cdot \dot{r}(2a)$ acts on $\mathbb{C}^a$ by conjugation and on $q_0 \cdot \mathbb{C}^a$ by $-\mathrm{id}_{q_0 \cdot \mathbb{C}^a}$.

For any $k \in E$ there is a unique $h \in \mathrm{SU}(2)$ such that

$$\frac{a}{2} \cdot \dot{r}(a)(2h) = k \tag{120}$$

as can immediately be seen by setting

$$h = -\eta \cdot a + h_{F^a},$$

provided $k = \eta \cdot a + k_{F^a}$ with h_{F^a} and k_{F^a} being the F^a-component of h and k, respectively. h is called a *classical spin vector*.

By definition the infinitesimal representation

$$\frac{\hbar \cdot a}{2} \cdot \dot{r} : su(2, \mathbb{C}) \longrightarrow \mathrm{End}_{\mathbb{C}} \mathbb{H}$$

determined by (111) and having Hermitian operators as values provides the totality of all the quantum mechanical spin $\frac{1}{2}$-elements associated with the axis $\mathbb{R} \cdot a$.

Note that in the quantum mechanical literature in the splitting

$$E = \mathbb{R} \cdot e \oplus F^a$$

the line $\mathbb{R} \cdot a$ corresponds to the z-axis while $R \cdot q_0$ and $\mathbb{R} \cdot q_0 \cdot a$ correspond to the x- and y-axis, respectively.

2.5.3. Spin $\frac{1}{2}$-Representation and Symplectic Structure. The spin $\frac{1}{2}$-representation acting on $\mathbb{H} = \mathbb{C}^a \oplus F^a$ for some $a \in S^2$ naturally defines a symplectic structure on F^a: The infinitesimal representation

$$\mathrm{dr}_{\frac{1}{2}}(a) = \frac{d}{dt} r_{\frac{1}{2}}(e^{(t \cdot a)}) \Big|_{t=0}$$

applied to a yields for any $h \in F^a$,

$$\mathrm{dr}_{\frac{1}{2}}(a)(h) = -h \cdot a. \tag{121}$$

Clearly, $\mathrm{dr}_{\frac{1}{2}}(a)$ is a skew-adjoint map. Therefore,

$$< \mathrm{dr}_{\frac{1}{2}}(a)(h_1), h_2 >= - < h_1 \times a, h_2 > \qquad \forall\, h_1, h_2 \in F^a, \tag{122}$$

saying that $\mathrm{dr}(a)$ defines a symplectic structure $-\omega^a$ on F^a (cf. section 1.3.5). Vice versa, any symplectic structure on F^a is of this form since any symplectic form ω on a plane $F \subset E$ is determined by

$$\omega = \lambda \cdot \omega^a$$

for some $\lambda \in \mathbb{R}$ and $a \in S^2$ perpendicular to F in E.

CHAPTER 3

Internal Variables of Singularity Free Vector Fields in a Euclidean Space

This chapter concludes the introductory part of our work. While the first chapter looked in detail at the quaternions as the algebraic and geometric basis of our constructions, the second chapter paid attention to the embedding of the three-sphere into the quaternions. The chapter to come will generalize the *reference point* $a \in S^2$ used so far to cause the 2-by-2-splitting of $\mathbb{H}$ as well as the 1-by-2-splitting of the Euclidean space E contained in $\mathbb{H}$. As we already noted, there are as many such splittings of $\mathbb{H}$ and E, respectively, as there are elements in S^2. Hence S^2 parameterizes those splittings. This gives rise to the idea that given an open subset O of E, say, we might look at a nowhere vanishing vector field X on O with values in E. Thus every field vector of the vector field causes a 2-by-2-splitting in the way described earlier. The following sections will deal with these structures:

We will start with a more general setting (cf. section 3.1) but will also look at gradient fields as the origin of our construction (cf. section 3.3). In particular, we will show that the vector field X defines a complex line bundle with a Hermitian structure as well as a characteristic $U(1)$-principal bundle (cf. sections 3.2 and 3.7). A natural connection form links the fibres of this principal bundle (cf. section 3.6).

3.1. The Complex Line Bundle $\mathbb{F}^{\mathbf{a}}$

In the previous chapters our reference vector $a \in S^2$ has always been arbitrarily chosen and fixed. To make our setting more flexible, in this chapter we introduce a smooth, singularity free vector field X defined on an open subset O of E whose principal part $\mathbf{a}$ will take over the role of the reference point a. More precisely, for each $x \in O$ the field vector $\mathbf{a}(x)$ will provide us with a reference point $\frac{\mathbf{a}(x)}{|\mathbf{a}(x)|} \in S^2$ introducing a splitting of E, of the quaternions $\mathbb{H}$ and so on.

In doing so, we will be able to carry over many of the results of the previous chapters to a more general setting. This requires some repetitions, but we will limit them and take the benefit of those necessary to get an even better understanding in linking the previous chapters with the one we are entering right now.

3.1.1. The Vector Field. Let us start with the definition of the vector field just mentioned:

O shall be an open subset not containing 0 in a three-dimensional, oriented Euclidean $\mathbb{R}$-vector space E with scalar product $<\ ,\ >$. On O we define a smooth, *singularity free* vector field $X : O \longrightarrow O \times E$ with principal part $\mathbf{a} : O \longrightarrow E$, say. Thus X is of the form $X = (\mathrm{id}\,, \mathbf{a})$. The positive function $|\mathbf{a}|$ is called the *field strength function* or just the field strength.

The *direction* of the field vector $\mathbf{a}(x)$ is pointwisely described by the map

$$\varepsilon_{\mathbf{a}} : O \longrightarrow S^2, \qquad x \longmapsto \varepsilon_{\mathbf{a}(x)} := \frac{\mathbf{a}(x)}{|\mathbf{a}(x)|}, \tag{123}$$

the *Gauss map*, also called the *directional field.* If no confusion arises, the index $\mathbf{a}$ in $\varepsilon_{\mathbf{a}}$ will be omitted. Associated with X there is a natural smooth one-form

$$\alpha_X : O \times E \longrightarrow \mathbb{R}, \qquad \alpha_X(x; v) =< \mathbf{a}(x), v > . \tag{124}$$

A vector field for which $\mathbf{a}$ is constant is called *trivial.* If $\varepsilon = \frac{\mathbf{a}}{|\mathbf{a}|}$ is constant map, any integral curve of the vector field X as well as of ε is a straight line and vice versa.

3.1.2. The Differential of the Vector Field. It is of interest to know whether a vector field is locally a gradient. To give a criterion let us look at the differential of the vector field. The results of this paragraph will be valuable to us in the later sections of this chapter. Clearly, the differential $d\mathbf{a}(x) : E \rightarrow E$ can be decomposed for any $x \in O$ into

$$d\mathbf{a}(x) = \mathcal{W}_x^{\mathbf{a}} + A_x^{\mathbf{a}} \tag{125}$$

where $\mathcal{W}_x^{\mathbf{a}}$ and $A_x^{\mathbf{a}}$ are self-adjoint and skew-adjoint, respectively. Let us have a look at $\mathcal{W}_x^{\mathbf{a}}$ and $A_x^{\mathbf{a}}$ separately. First we observe that the exterior derivative $d\alpha_X$ (cf. [**2**]) can be written as

$$d\alpha_X(x; v_0, v_1) = 2 < A_x^{\mathbf{a}}(v_0), v_1 > \qquad \forall\, x \in O \text{ and } \forall\, v_0, v_1 \in E.$$

Obviously,

$$2A_x^{\mathbf{a}}(v) = \operatorname{curl} \mathbf{a}(x) \times v \qquad \forall\, x \in O \text{ and } \forall\, v \in E. \tag{126}$$

Since any one-form θ and any two-form ω on O can be written as

$$\theta(x; v) =< \mathbf{b}(x), v > \qquad \text{and} \qquad \omega(x; v_0, v_1) =< \mathbf{d}_\omega(x), v_0 \times v_1 > \tag{127}$$

for all $v, v_0, v_1 \in E$ and some smooth functions $\mathbf{b}, \mathbf{d}_\omega : O \longrightarrow E$, respectively, we verify

$$\omega = d\theta \qquad \Longleftrightarrow \qquad \mathbf{d}_\omega = \operatorname{curl} \mathbf{b}. \tag{128}$$

Moreover, $A^{\mathbf{b}} = 0$ iff $\mathbf{b} = \operatorname{grad} g$ (locally), where g is a locally defined smooth real-valued map or, equivalently,

$$d\theta = 0 \qquad \Longleftrightarrow \qquad \mathbf{b} = \operatorname{grad} g \quad \text{(locally)}. \tag{129}$$

Focussing on the self-adjoint component $\mathcal{W}^{\mathbf{a}}$ in (125) one easily sees that

$$\operatorname{div} \mathbf{a} = \operatorname{tr} \mathcal{W}^{\mathbf{a}} = \operatorname{tr} W^{\mathbf{a}} \tag{130}$$

where

$$W^{\mathbf{a}} := d\mathbf{a}. \tag{131}$$

If $|a| = 1$ as in the case of $\varepsilon_{\mathbf{a}}$, then $W^{\varepsilon_{\mathbf{a}}}$ is called the *Weingarten map of the vector field* X.

3.1.3. Complex Lines. Given any $x \in O$, the orthogonal complement $F^{\mathbf{a}(x)}$ of $\mathbf{a}(x) \in E$ is a complex line in E in the same way as was shown for F^a in section 1.3.2. To make this more plausible, let us have a look at the construction in detail:

Let $\mathbb{C}^{\frac{\mathbf{a}(x)}{|\mathbf{a}(x)|}} \subset \mathbb{H}$ be the orthogonal complement of $F^{\mathbf{a}(x)}$. Hence

$$\mathbb{H} = \mathbb{C}^{\frac{\mathbf{a}(x)}{|\mathbf{a}(x)|}} \oplus F^{\mathbf{a}(x)}$$

is an *orthogonal splitting*. Its commutative subfield

$$\mathbb{C}^{\frac{\mathbf{a}(x)}{|\mathbf{a}(x)|}} = \mathbb{R} \cdot e \oplus \mathbb{R} \cdot \frac{\mathbf{a}(x)}{|\mathbf{a}(x)|} \tag{132}$$

is naturally isomorphic to $\mathbb{C}$ due to $\left(\frac{\mathbf{a}(x)}{|\mathbf{a}(x)|}\right)^2 = -e$ for all $x \in O$; here $|\cdot|$ again denotes the norm in $\mathbb{H}$ defined by $<\ ,\ >$. This is seen as follows: In analogy to section 1.2.1, let

$$i_x^{\mathbf{a}} : \mathbb{C} \longrightarrow \mathbb{C}^{\frac{\mathbf{a}(x)}{|\mathbf{a}(x)|}}, \qquad i_x^{\mathbf{a}}(1) := e, \quad i_x^{\mathbf{a}}(i) := \frac{\mathbf{a}(x)}{|\mathbf{a}(x)|}$$

be the isomorphism linking $\mathbb{C}$ and $\mathbb{C}^{\frac{\mathbf{a}(x)}{|\mathbf{a}(x)|}}$. Then the plane $F^{\mathbf{a}(x)}$ is a $\mathbb{C}^{\frac{\mathbf{a}(x)}{|\mathbf{a}(x)|}}$-linear space under the (right) multiplication of $\mathbb{H}$ and hence is a $\mathbb{C}$-linear space, i.e. a complex line.

The multiplicative group on the unit circle of $\mathbb{C}^{\frac{\mathbf{a}(x)}{|\mathbf{a}(x)|}}$ is denoted by $U^{\frac{\mathbf{a}(x)}{|\mathbf{a}(x)|}}(1)$. It is a subgroup of $SU(2) \subset \mathbb{H}$. We will frequently use the simplified symbols $\mathbb{C}^{\mathbf{a}}$ and $U^{\mathbf{a}}(1)$, respectively.

3.1.4. Vector Fields and their Complex Line Bundles. As we just saw, the vector field X determines a collection of complex lines. We will investigate the relation between a vector field and these complex lines more closely. At first we concentrate on the collection of all these real planes.

Any vector field $X = (\mathrm{id}, \mathbf{a})$ is equipped with the *topological subspace* of $O \times E$, namely with

$$\mathbb{F}^{\mathbf{a}} := \dot{\bigcup_{x \in O}} \mathbb{F}_x^{\mathbf{a}} = \bigcup_{x \in O} \{x\} \times \mathbb{F}^{\mathbf{a}(x)} \subset O \times E \tag{133}$$

if $\mathbb{F}_x^{\mathbf{a}}$ is identified with $\mathbb{F}_x^{\mathbf{a}} = \{x\} \times \mathbb{F}^{\mathbf{a}(x)}$ where

$$\mathbb{F}^{\mathbf{a}(x)} := \mathbf{a}(x)^{\perp_E} \subset E \qquad \forall x \in O$$

is a two-dimensional linear space, perpendicular to $\mathbf{a}(x)$. Since

$$\mathbb{F}^{\mathbf{a}(x)} = \mathbb{F}^{\varepsilon(x)} = T_{\varepsilon(x)} S^2$$

for any $x \in O$,

$$\mathbb{F}^{\mathbf{a}} = \varepsilon^* T S^2. \tag{134}$$

Therefore, $\mathbb{F}^{\mathbf{a}}$ is a *real rank two vector bundle*, the *pull-back* of TS^2 by ε.

Moreover, since TS^2 is a smooth complex line bundle with the Hopf bundle $\mathbb{P}_2$ (cf. section 2.3.4) as its bundle of normed complex frames, the pull-back $\mathbb{F}^{\mathbf{a}}$ of TS^2 by ε is a *smooth complex line bundle*, as well (cf. [**46**]). The fibres of $\mathbb{F}^{\mathbf{a}}$ are complex lines in the sense of section 3.1.3.

Let us describe the notion of *pull-back* (cf. [**46**]) somewhat closer for the particular example of ε^*TS^2 as a real vector bundle. The fibre $(\varepsilon^*TS^2)_x$ of ε^*TS^2 at any $x \in 0$ is by definition

$$(\varepsilon^*TS^2)_x := T_{\varepsilon(x)}S^2.$$

Since $\varepsilon(x) \perp T_{\varepsilon(x)}S^2$ in E for any $x \in O$, the fibre $(\varepsilon^*TS^2)_x$ at $x \in O$ is

$$\mathbb{F}^{\mathbf{a}}_x = T_{\varepsilon(x)}S^2 = (\varepsilon^*TS^2)_x.$$

The disjoint union of all fibres $T_{\varepsilon(x)}S^2$ formed over all $x \in O$ is by definition ε^*TS^2 and hence

$$\varepsilon^*TS^2 = \bigcup_{x\in O} \mathbb{F}^{\mathbf{a}}_x = \mathbb{F}^{\mathbf{a}}.$$

An atlas for ε^*TS^2 is determined by ε and an atlas $\{V_\alpha, \varphi_\alpha | \alpha \in I\}$ of TS^2 over S^2. Here I is an index set. Let F be the typical fibre of TS^2, a two-dimensional $\mathbb{R}$-linear space. Moreover, let

$$\varphi_{\alpha,y} : F \longrightarrow T_yS^2 \qquad \forall\, y \in V_\alpha$$

be the isomorphism to the fibre given by the trivialization

$$V_\alpha \times F \xrightarrow{\varphi_\alpha} TS^2|_{V_\alpha}$$

of TS^2 over V_α. Here $TS^2|_{V_\alpha}$ is the restriction of TS^2 to V_α. The isomorphism

$$\psi_{\alpha,x} : F \longrightarrow (\varepsilon^*TS^2)_x = \mathbb{F}^{\mathbf{a}}_x$$

is given by

$$\psi_{\alpha,x} := \varphi_{\alpha,\varepsilon(x)} \qquad \forall\, x \in V_\alpha.$$

Setting

$$U_\alpha = \varepsilon^{-1}(V_\alpha) \qquad \forall\, \alpha \in I$$

we obtain a covering $\{U_\alpha | \alpha \in I\}$ of O and the chart map

$$\psi_\alpha : U_\alpha \times F \longrightarrow \varepsilon^*TS^2$$

which is defined by setting

$$\psi_{\alpha,x} = \varphi_{\alpha,\varepsilon(x)} \qquad \forall\, x \in U_\alpha.$$

Thus the map $\psi^{-1}_{\alpha,x} \circ \psi_{\beta,x}$ for $x \in U_\alpha \cap U_\beta$ is the linear transformation of $(\varepsilon^*TS^2)_x = \mathbb{F}^{\mathbf{a}}_x$ into itself and hence is of the form

$$\psi^{-1}_{\alpha,x} \circ \psi_{\beta,x} = g_{\alpha\beta}(\varepsilon(x)) \qquad \forall\, x \in U_\alpha \cap U_\beta$$

where $g_{\alpha\beta}$ is the coordinate transformation of T_yS^2 into itself. Hence $\psi^{-1}_{\alpha,x} \circ \psi_{\beta,x}$ is smooth in x. We thus have a bundle atlas $\{(U_\alpha, \psi_\alpha) | \alpha \in I)\}$ on ε^*TS^2. The vector bundle ε^*TS^2 is called the pull-back of TS^2 by ε. By construction

$$\varepsilon^*TS^2 \subset O \times E$$

and

$$\mathbb{F}^{\mathbf{a}} = \varepsilon^*TS^2.$$

This immediately yields the *bundle map*

$$\widehat{\varepsilon} : \mathbb{F}^{\mathbf{a}} \longrightarrow TS^2, \qquad \widehat{\varepsilon}(k_x) = k_x \tag{135}$$

for all $k_x \in \mathbb{F}^{\mathbf{a}}$ and any $x \in O$. Thus $\widehat{\varepsilon}$ is fibrewisely the *identity*, although it does not need to be an embedding. The maps ε and $\widehat{\varepsilon}$ are linked by

$$\mathrm{pr}_{S^2} \circ \widehat{\varepsilon} = \varepsilon \circ \mathrm{pr}_O$$

where pr_{S^2} and pr_O are the projections from TS^2 and $O \times E$ onto S^2 and O, respectively.

We can complement the fibre $\mathbb{F}^{\mathbf{a}}_x$ by the real line $\mathbb{R} \cdot \varepsilon(x)$ to

$$\{x\} \times \left(\mathbb{F}^{\mathbf{a}(x)} + \mathbb{R} \cdot \varepsilon(x)\right) = \{x\} \times E \tag{136}$$

and, in consequence,

$$\mathbb{F}^{\mathbf{a}} \oplus \mathcal{L}^{\mathbf{a}} = O \times E \tag{137}$$

with a *trivial* real line bundle

$$\mathcal{L}^{\mathbf{a}} := \bigcup_{x \in O} \{x\} \times \mathbb{R} \cdot \varepsilon(x), \tag{138}$$

being the pull-back $\varepsilon^* \mathcal{N}_{S^2}$ of the (trivial) normal bundle $\mathcal{N}_{S^2}$ of S^2 by ε. Clearly, $\mathcal{L}^{\mathbf{a}} = \mathbb{R} \cdot \mathbf{a}$. Thus

$$\mathcal{L}^{\mathbf{a}} = \varepsilon^* \mathcal{N}_{S^2}$$

and hence

$$\mathbb{F}^{\mathbf{a}} \oplus \mathcal{L}^{\mathbf{a}} = \varepsilon^* TS^2 \oplus \varepsilon^* \mathcal{N}_{S^2}. \tag{139}$$

From the construction of the vector bundle $\mathbb{F}^{\mathbf{a}}$ we see that it does not characterize the field strength $|\mathbf{a}|$ of $\mathbf{a}$. In fact, $\mathbb{F}^{\mathbf{a}} = \mathbb{F}^{\varepsilon}$.

All of these constructions in connection with the pull-back (134) work for TS^2 as a complex line bundle (cf. section 2.3.6), as well. Hence $\mathbb{F}^{\mathbf{a}}$ is a complex line bundle. In summary, we therefore have:

THEOREM 12. *For any smooth vector field $X = (id, \mathbf{a})$ on an open set $O \subset E$ the topological subspace $\mathbb{F}^{\mathbf{a}}$ of $O \times E$ is a smooth real rank two vector bundle as well as a complex line bundle of the form*

$$\mathbb{F}^{\mathbf{a}} = \varepsilon^* TS^2 \tag{140}$$

and

$$\mathbb{F}^{\mathbf{a}} \oplus \mathcal{L}^{\mathbf{a}} = \varepsilon^* TS^2 \oplus \varepsilon^* \mathcal{N}_{S^2} = \varepsilon^*(S^2 \times E)$$

as real vector bundles. For any open chart (U, φ) of the vector bundle $\mathbb{F}^{\mathbf{a}}$ with typical fibre $F \subset E$ there is some fixed $b \in E$ transversal to F and a smooth bundle isomorphism

$$\Psi : O \times E \longrightarrow O \times E$$

respecting the splitting

$$O \times E = \mathbb{F}^{\mathbf{a}} \oplus \mathcal{L}$$

such that

$$\begin{array}{ccc} \mathbb{F}^{\mathbf{a}}|_U \oplus \mathcal{L}^{\mathbf{a}}|_U & \xrightarrow{\Psi} & \varphi(U) \times (F \oplus \mathbb{R} \cdot b) \\ \big\downarrow {\scriptstyle \mathrm{pr}_U} & & \big\downarrow {\scriptstyle \mathrm{pr}_1} \\ U & \xrightarrow{\varphi} & \varphi(U) \end{array}$$

commutes. Here φ is smooth and pr_1 projects to $\varphi(U)$.

From this we easily see:

COROLLARY 13. *The complex line bundle of a vector field X is trivial if its Gauss map ε is trivial.*

3.1.5. The Bundle of Complex Fields. As in the previous section the map $\mathbf{a} : O \longrightarrow E$ denotes the principal part of a smooth, singularity free vector field X on O. In a similar way as for $\mathbb{F}^{\mathbf{a}}$ we can use our vector field to define a bundle of complex fields. This is due to the fact that $\left(\frac{\mathbf{a}(x)}{|\mathbf{a}(x)|}\right)^2 = -e$.

Obviously, the topological subspaces

$$C^{\mathbf{a}} := \bigcup_{x \in O} \{x\} + \mathbb{R} \cdot e \quad \text{and} \quad \mathcal{L}^{\mathbf{a}} = \bigcup_{x \in O} \{x\} + \mathbb{R} \cdot \mathbf{a}(x) \tag{141}$$

of $O \times \mathbb{H}$ are real trivial line bundles (cf. (139)). Therefore, the Whitney sum of both, which is

$$\mathbb{C}^{\mathbf{a}} := \bigcup_{x \in O} \mathbb{C}^{\mathbf{a}}_x = \bigcup_{x \in O} \{x\} + \mathbb{C}^{\frac{\mathbf{a}(x)}{|\mathbf{a}(x)|}}, \tag{142}$$

is an $\mathbb{R}$-vector bundle. Clearly, the fibres $\mathbb{C}^{\mathbf{a}}_x$ are fields isomorphic to the field of complex numbers via the map $i^{\mathbf{a}}_x$. Hence $\mathbb{C}^{\mathbf{a}}$ is a *trivial* $\mathbb{C}$-vector bundle, as well, and, obviously,

$$O \times \mathbb{H} = \mathbb{C}^{\mathbf{a}} \oplus \mathbb{F}^{\mathbf{a}}$$

as vector bundles over O.

$\mathbb{C}^{\mathbf{a}}$ contains two natural subbundles, namely the group bundles $\dot{\mathbb{C}}^{\mathbf{a}}$ and $U^{\mathbf{a}}(1)$, whose fibres $\dot{\mathbb{C}}^{\mathbf{a}}_x$ and $U^{\mathbf{a}}_x(1)$ consist of the commutative groups $\mathbb{C}^{\mathbf{a}}_x \setminus \{0\}$ and $\{x\} \times U^{\frac{\mathbf{a}(x)}{|\mathbf{a}(x)|}}(1)$, respectively, for any $x \in O$.

3.2. Symplectic and Hermitian Structures on $\mathbb{F}^{\mathbf{a}}$

As we observed in the pervious section, the vector bundle $\mathbb{F}^{\mathbf{a}}$ of a vector field $\mathbf{a}$ admits a natural complex structure on each fibre; in fact, it is a complex line bundle. In the following we will see that it admits natural Hermitian structures, as well. A part of this structure is the fibrewisely given symplectic form which will prove to be much more than a technical tool only. In this section it will enable the bundle to determine the vector field uniquely, for example.

3.2.1. Symplectic Structure. Each fibre $\mathbb{F}^{\mathbf{a}}_x \subset \mathbb{F}^{\mathbf{a}}$ carries a natural *symplectic structure* $\omega^{\mathbf{a}}_x$ defined by

$$\omega^{\mathbf{a}}_x(h,k) := < h \times \mathbf{a}(x), k > = < h \cdot \mathbf{a}(x), k > \qquad \forall\, h,k \in F^{\mathbf{a}(x)} \tag{143}$$

(cf. section 1.3.5). We recall that the cross product $\times$ is identical with the product in $\mathbb{H}$ since h and $\mathbf{a}(x)$ are perpendicular elements in E. Each fibre $\mathbb{F}^{\mathbf{a}}_x$ is *oriented* by $\omega^{\mathbf{a}}_x$.

In fact, $\omega^{\mathbf{a}}_x(y,z) :=< y \times \mathbf{a}(x), z >$ is defined for all $x, y, z \in E$; but $\omega^{\mathbf{a}}$ is not a symplectic structure on O, it is a degenerate bilinear form on each fibre E of $TO = O \times E$. Hence we call it a *presymplectic structure* on O.

However, it is precisely this fact which allows us to reconstruct the principal part $\mathbf{a}$ of the vector field X. To show this recall that there is some $A(x) \in \operatorname{End} E$ such that

$$< A(x)(h), k >= \omega^{\mathbf{a}}_x(h,k) \qquad \forall\, h,k \in E$$

for any $x \in O$. The direction of $\mathbf{a}(x)$ is defined by $\ker A(x)$, and for any vector $u \in E$,

$$u \times \mathbf{a}(x) = A(x)(u)$$

holds true. Hence the sup-norm of A determined on $S^2 \subset E$ satisfies

$$\|A\|_{\text{sup}} = |\mathbf{a}(x)|.$$

Thus $A(x)$ determines $\mathbf{a}(x)$, and hence $\omega_x^{\mathbf{a}}$ does also.

3.2.2. Symplectic Structure, Volume Form and Divergence. The *orientation* on $\mathbb{F}_x^{\mathbf{a}}$ shall be specified by

$$(144) \qquad \omega_x^{\mathbf{a}} := -i_{\mathbf{a}(x)}\mu \qquad \text{or, equivalently,} \qquad \omega_x^{\varepsilon} := -i_{\varepsilon(x)}\mu$$

for any $x \in O$. (The minus sign is conventional.) Obviously,

$$\omega_x^{\varepsilon}(v_0, v_1) = -\omega^{S^2}(v_0, v_1) \qquad \forall v_0, v_1 \in \mathbb{F}_x^{\mathbf{a}} \text{ and } \forall x \in O$$

where ω^{S^2} is the volume form on S^2 determined by the *outward* directed unit normal and the volume form μ on E (cf. section 1.3.3). Clearly, ω^{S^2} is a symplectic structure on S^2. Comparing $\omega_x^{\mathbf{a}}$ with ω_x^{ε} we get on all of E,

$$\omega_x^{\mathbf{a}} = |\mathbf{a}(x)| \cdot \omega_x^{\varepsilon}.$$

Thus the pair $(\mathbb{F}^{\mathbf{a}}, \omega^{\mathbf{a}})$ determines X.

Since $< \varepsilon(x), \varepsilon(x) >= 1$ for all $x \in O$, the tangent map $T\varepsilon$ is the pair $(\varepsilon, D\varepsilon)$ where $D\varepsilon(x)(v) =: W_x^{\varepsilon}(v)$ is the Fréchet derivative, for which

$$(145) \qquad W_x^{\varepsilon}(v) \in \mathbb{F}_x^{\mathbf{a}} \qquad \forall v \in E$$

holds true (cf. (131)). In general, W_x^{ε} is not self-adjoint for all $x \in O$.

With these remarks we can now investigate the relation between the divergence of the vector field X and the form $\omega^{\mathbf{a}}$ trivially extended to all $TO = O \times E$ by setting

$$\omega^{\mathbf{a}}(x; u, \varepsilon) = \omega^{\mathbf{a}}(x; \varepsilon, \varepsilon) = 0$$

(cf. section 3.2.1). It is provided by the exterior derivative of $\omega^{\mathbf{a}}$ as

$$(146) \qquad d\omega^{\mathbf{a}} = \operatorname{div} X \cdot \mu$$

(cf. (130)). So $\omega^{\mathbf{a}}$ is closed iff the divergence vanishes. Due to (145), $\operatorname{div} \varepsilon = \operatorname{tr} W^{\varepsilon}$ so that the divergence of the directional field of X does not vanish, in general. Clearly, $\operatorname{div} X = 0$ means $\Delta f = 0$ if $\mathbf{a} = \operatorname{grad} f$.

3.2.3. A Hermitian Structure on $\mathbb{F}^{\mathbf{a}}$. In section 1.3 we defined a $\mathbb{C}$-linear structure on F^a for some $a \in S^2$, a construction we adapted to $F^{\mathbf{a}(x)}$ for each $x \in O$ in paragraph 3.1.3 and identified $\mathbb{F}^{\mathbf{a}}$ as a complex line bundle (cf. theorem 12). It is obvious that $\mathbb{F}^{\mathbf{a}} = \mathbb{F}^{\varepsilon}$ as complex line bundles for the directional vector field $\varepsilon(x) = \frac{\mathbf{a}(x)}{|\mathbf{a}(x)|}$ for all $x \in O$.

The scalar product $< \, , \, >$ on E together with ω^{ε} on $\mathbb{F}^{\varepsilon}$ causes a *Hermitian form* on $\mathbb{F}^{\varepsilon}$ defined for each $x \in O$ by

$$(147) \qquad \mathcal{H}^{\varepsilon}(k_1, k_2)_x \; := < k_1, k_2 > + \, \omega_x^{\varepsilon}(k_1, k_2) \cdot \varepsilon(x)$$

for each $k_1, k_2 \in \mathbb{F}^{\mathbf{a}} = \mathbb{F}^{\varepsilon}$ and any $x \in O$. Hence $\mathbb{F}^{\varepsilon}$ is a *Hermitian line bundle* determining ε uniquely. It determines $X = (\operatorname{id}, \mathbf{a})$ only up to the field strength. We easily see:

PROPOSITION 14. *A Hermitian line bundle $\mathbb{F}$ on O is of the form $\mathbb{F}^{\mathbf{a}}$ for some vector field $X = (id, \mathbf{a})$ iff $\mathbb{F}$ is embeddable into $O \times E$ as a real fibrewisely oriented vector bundle.*

Having a notion of a Hermitian line bundle at hand allows us to introduce the notion of *equivalent vector fields* in our setting:

DEFINITION 15. *Two vector fields* $(id, \mathbf{a}_s)$ *defined on* O *for* $s = 1, 2$ *are equivalent iff there is an isomorphism of Hermitian vector bundles*

$$\Psi : \mathbb{F}^{\varepsilon_1} \longrightarrow \mathbb{F}^{\varepsilon_2}.$$

With $\varepsilon_s = \frac{\mathbf{a}_s}{|\mathbf{a}_s|}$ *for* $s = 1, 2$, *the isomorphism* $\Psi|_{\mathbb{F}_x^{\varepsilon_1}}$ *is extended to all of* E *by setting* $\Psi(\varepsilon_1(x)) := \varepsilon_2(x)$.

As a first remark on the equivalence of vector fields let us point out the following:

LEMMA 16. *If* $\mathbf{a}_1$ *and* $\mathbf{a}_2$ *are homotopic maps, then the real vector bundles* $\mathbb{F}^{\mathbf{a}_1}$ *and* $\mathbb{F}^{\mathbf{a}_2}$ *are isomorphic and hence they are isomorphic as Hermitian vector bundles, as well.*

For a proof we refer to [**54**].

If we replace in the right-hand side of (147), the symplectic form ω_x^ε by $\omega_x^{\mathbf{a}}$ for each $x \in O$, then the resulting Hermitian form $\mathcal{H}^{\mathbf{a}}$ on the complex line bundle $\mathbb{F}^{\mathbf{a}}$ determines the vector field $X = (\mathrm{id}, \mathbf{a})$ since $\omega^{\mathbf{a}} \cdot \varepsilon$ determines the field strength (cf. section 3.2.1) and hence $\mathbf{a}$. Again ε is the principal part of the unit vector field caused by $X = (\mathrm{id}, \mathbf{a})$.

3.3. Gradient Fields

In this section we will look at the special case where X is a gradient field. This allows us to study the geometry of the complex line bundle $\mathbb{F}^{\mathbf{a}}$ in a more detailed and familiar way.

For general references for differential geometry of surfaces consult for example [**10**], [**9**] and [**34**].

3.3.1. Vector Field on O as a Gradient Field. Let X be a gradient field, i.e. $\mathbf{a} = \mathrm{grad}\, f$ for some smooth function $f : O \to \mathbb{R}$. The collection of singularities of f is a closed set which is assumed to be removed from the domain of X. Hence we may suppose that X is singularity free on O. Clearly, $\alpha_X = df$ (cf. (124)).

3.3.2. One-forms and Integrating Factors. Suppose for a given smooth map $\mathbf{a} : O \longrightarrow E$ that the form α_X admits an *integrating factor* g, meaning that $g \cdot \alpha_X = df$ for some smooth maps $g, f : O \to \mathbb{R}$. Clearly,

$$g \cdot \mathbf{a} = \mathrm{grad}\, f.$$

Moreover, if g is an integrating factor of $\mathbf{a}$, the function $g \cdot |\mathbf{a}|$ is an integrating factor of the directional field $(\mathrm{id}, \varepsilon_{\mathbf{a}})$ and $\mathcal{W}^{g \cdot \mathbf{a}}$, the self-adjoint part of $d(g \cdot \mathbf{a})$ (cf. (125)) determines the geometry of the level surfaces of f.

Note that in general, a one-form on $\mathbb{R}^3$ does not admit an integrating factor. An example is $\alpha := dx_1 + x_2 \cdot dx_3$, where x_1, x_2 and x_3 are the coordinate functions with respect to the standard basis in $\mathbb{R}^3$ (cf. [**79**]).

3.3.3. Level Surfaces and the Weingarten Map. We will now consider a *gradient field* $X = (\mathrm{id}, \mathbf{a})$ of a (smooth) function V called a potential. Clearly, $\mathbf{a} = \operatorname{grad} V$. A surface $S \subset E$ is called a *level surface* of the potential V if $V|_S = \text{const}$. Since grad V does not vanish at any $x \in O$, the kernel of dV has dimension two everywhere. Thus at each $x \in O$ there is a level surface passing through x. Let us call a level surface *maximal* if is not properly contained in any other one. If no confusion arises, all level surfaces are assumed to be maximal. The collection of all level surfaces foliates O. This means that O is decomposed into a disjoint union of (smooth) two-dimensional manifolds.

Each level surface S of V satisfies

$$TS = \mathbb{F}^{\mathbf{a}}|_S.$$

Here $\mathbb{F}^{\mathbf{a}}|_S = \bigcup_{x\in S}\{x\} \times F^{\mathbf{a}(x)}$. The relation between the principal part $\mathbf{a}$ of the vector field X and the geometry of level surfaces S in O is given by the directional field $\varepsilon_{\mathbf{a}}$ on O as we will see now: We expand ε and consider the unit normal bundle $\mathcal{N}_S$ where S shall be a level surface in E. Clearly, $\mathcal{N}_S(x) = \frac{\mathbf{a}(x)}{|\mathbf{a}(x)|}$ and $d\mathcal{N}_S(x)(h) \in T_xS$ for any $x \in O$ and any $h \in E$ since $< \mathcal{N}_S(x), \mathcal{N}_S(x) >= 1$. Clearly, $\mathcal{N}_S = \varepsilon|_S$. Thus the smooth map

$$W_x^\varepsilon : O \times E \longrightarrow T_xS,$$

called the *Weingarten map* in this case, given at each $x \in O$ by

$$W_x^\varepsilon(h) = \left\{ \begin{array}{l|l} 0 & \text{for } h = \lambda \cdot \mathbf{a}(x) \text{ with } \lambda \in \mathbb{R}, \\ d\mathcal{N}_S(x)(h) & \text{for } h \in T_xS, \end{array} \right.$$

satisfies $W^\varepsilon = \mathcal{W}^\varepsilon$ (cf. section 3.1.2).

An easily performed calculation shows

$$d\mathbf{a}(x)(h) = |\mathbf{a}(x)| \cdot W_x^\varepsilon(h) + d\ln|\mathbf{a}(x)|(h) \cdot \mathbf{a}(x) \tag{148}$$

for any $x \in O$ and any $h \in E$, a formula we will make frequent use of.

Taking the trace of both sides of (148) yields

$$\frac{1}{|\mathbf{a}(x)|} \cdot \operatorname{tr} d\mathbf{a}(x) = H_{\frac{1}{2}}^{\mathbf{a}}(x) + \operatorname{tr}\left(d\ln|\mathbf{a}(x)| \cdot \frac{\mathbf{a}(x)}{|\mathbf{a}(x)|}\right)$$

where

$$H_{\frac{1}{2}}^{\mathbf{a}}(x) := \operatorname{tr} W_x^\varepsilon$$

is twice *the mean curvature* at x of the level surface $S = V^{-1}(V(x))$ for any $x \in O$.

3.3.4. The Potential. Suppose $V(x) \neq 0$ and $H_{\frac{1}{2}}^{\mathbf{a}}(x) \neq 0$ for all $x \in O := \dot{E}$. On the level surface S the function V is constant and therefore

$$H_{\frac{1}{2}}^{\mathbf{a}}(x) = g(x) \cdot V(x) \qquad \forall x \in S$$

for some well defined smooth real-valued function g defined on $\dot{E}$. This is to say that the potential V can be expressed in terms of $H_{\frac{1}{2}}^{\mathbf{a}}$ and g by

$$V = g^{-1} \cdot H_{\frac{1}{2}}^{\mathbf{a}} \tag{149}$$

on all $\dot{E}$. This means that (149) expresses V in terms of the foliation of $\dot{E}$ into level surfaces and the function g varying along the field lines of X.

3.3.5. $SO(E)$-Invariant Potential. Suppose V on $\dot{E} = \backslash\{0\}$ is a potential invariant under $SO(E)$, i.e.

$$V(\ell(x)) = V(x) \qquad \forall x \in \dot{E} \text{ and } \forall \ell \in SO(E).$$

Thus each level surface S is a two-sphere S_r^2 of some radius r. By (149) we may set

$$V = \frac{2}{r} \cdot g^{-1} \text{ on } S = S_r^2,$$

where $g : \dot{E} \to \mathbb{R}$ is smooth and constant on S_r^2 for each r. Hence g varies only along the field lines of some vector field X and

$$V(x) = \frac{2 \cdot g^{-1}(x)}{|x|} \qquad \forall x \in \dot{E}.$$

The simplest case is if g^{-1} is constant with value $\frac{\overline{m}}{2}$, say. Then

$$V(x) = \frac{\overline{m}}{|x|} \qquad \forall x \in \dot{E}.$$

The principal part $\mathbf{a}$ of the gradient field of V is thus

$$\mathbf{a}(x) = -\frac{\overline{m}}{|x|^2} \cdot \frac{x}{|x|} \qquad \forall x \in \dot{E}.$$

For the general $SO(E)$-invariant potential V the principal part of the gradient field of V is at each $x \in \dot{E}$,

$$\mathbf{a}(x) = \frac{2 \cdot g^{-1}(x)}{|x|^2} \cdot \frac{x}{|x|} = \frac{2}{|x|} \cdot (\text{grad } g^{-1})(x).$$

3.4. Curvature Forms on Level Surfaces

In this section we will shortly repeat some widely used curvature forms and thereby fix our terminology.

3.4.1. Riemannian Metric. On any fixed (two-dimensional) level surface S of the potential f the scalar product $<\,,\,>$ on E determines a *Riemannian metric* g_S given by

$$g_S(x; v_x, w_x) :=< v_x, w_x >$$

for all $x \in S$ and all $v_x, w_x \in T_xS = F^{\mathbf{a}(x)}$ together with its *Levi-Civitá covariant derivative.* This derivative $\nabla^S_{v_x} Y$ of any given smooth vector field $Y : S \longrightarrow TS$ is defined by

$$\nabla^S_{v_x} Y := (dY(x)(v_x))^T = dY(x)(v_x) - < dY(x)(v_x), \mathcal{N}_S(x) > \cdot \mathcal{N}_S(x)$$

for any $v_x \in T_xS$ and any $x \in S$ where the upper index T means tangential component. Reformulated, the above equation reads

$$\nabla^S_{v_x} Y = dY(x)(v_x) + < Y(x), W^{\varepsilon}_x(v_x) > \cdot \mathcal{N}_S(x)$$

where ε is the unit normal vector field of X (cf. section 3.4.3).

3.4.2. Curvature. The *curvature* $R_S(V,W)U$ associated with three vector fields V,W,U of S is given by the *equation of Gauss*:

$$R_S(V,W)U := \nabla_V^S \nabla_W^S U - \nabla_W^S \nabla_V^S U - \nabla_{[V,W]}^S U.$$

Using this equation, the curvature R_S determined by the Levi-Civitá covariant derivative ∇^S is locally expressed by

$$\begin{aligned} R_S(v_x,w_x)u_x &= < W_x^\varepsilon(w_x),u_x > \cdot W_x^\varepsilon(v_x) \\ &\quad - < W_x^\varepsilon(v_x),u_x > \cdot W_x^\varepsilon(w_x) \end{aligned} \tag{150}$$

for all $v_x,w_x,u_x \in T_xS$ and any $x \in S$. To verify this formula it is enough to choose locally given vector fields V and Z on S with $[V,Z]=0$. This means that V and Z define a local coordinate system. Therefore,

$$\begin{aligned} \nabla_V^S \nabla_Z^S U &= d(\nabla_Z^S U)(V) + < \nabla_Z^S U, W^\varepsilon(V) > \cdot \mathcal{N}_S \\ &= d(dU(Z))(V) + d < U, W^\varepsilon(Z) > (V) \cdot \mathcal{N}_S \\ &\quad + < U, W^\varepsilon(Z) > \cdot\ W^\varepsilon(V) + < \nabla_Z^S U, W^\varepsilon(V) > \cdot \mathcal{N}_S \\ &= d(dU(Z))(V) + < \nabla_V^S U, W^\varepsilon(Z) > \cdot \mathcal{N}_S \\ &\quad + < U, \nabla_V^S(W^\varepsilon)(Z) > \cdot \mathcal{N}_S + < U, W^\varepsilon(Z) > \cdot\ W^\varepsilon(V) \\ &\quad + < \nabla_Z^S U, W^\varepsilon(V) > \cdot \mathcal{N}_S. \end{aligned}$$

Since the exterior differential of $dU : TS \longrightarrow E$ vanishes, we obtain

$$R_S(V,Z)U = < W^\varepsilon(Z), U > \cdot\ W^\varepsilon(V) - < W^\varepsilon(V), U > \cdot\ W^\varepsilon(Z) \tag{151}$$

as well as the *Codazzi-Mainardi equation*

$$\nabla_Y(W^\varepsilon)(Z) = \nabla_Z(W^\varepsilon)Y.$$

Since the right-hand side of (151) depends on $V(x)$, $Z(x)$ and $U(x)$ at any $x \in O$, the left-hand side does also. Thus (150) is established. In consequence, the *Riemannian curvature* defined as

$$\dot{R}(x; v_x, w_x : u_x, y_x) \quad := \quad < R_s(v_x,w_x)u_x, y_x >$$

is given by

$$\begin{aligned} \dot{R}(x; v_x, w_x : u_x, y_x) &= < W_x^\varepsilon(w_x),u_x > \cdot < W_x^\varepsilon(v_x), y_x > \\ &\quad - < W_x^\varepsilon(v_x),u_x > \cdot < W_x^\varepsilon(w_x), y_x > \end{aligned} \tag{152}$$

for any choice of $v_x,w_x,u_x,y_x \in T_xS$ and any $x \in S$.

3.4.3. Ricci Tensor and Scalar Curvature. The *Ricci tensor* Ric_x of the level surface S at $x \in S$ is determined by

$$\mathrm{Ric}_x(w_x,u_x) := \sum_{i=1}^{2} \dot{R}(x; e_i, w_x : u_x, e_i)$$

where e_1, e_2 is an orthonormal basis in T_xS. Hence

$$\mathrm{Ric}_x(u_x,w_x) = \mathrm{tr}\, W_x^\varepsilon \cdot < W_x^\varepsilon(w_x), u_x > - \ \mathrm{tr}(W_x^\varepsilon)^2 \cdot < u_x, w_x >$$

for all $x \in S$ and any choice of $w_x, u_x \in T_xS$.

The *scalar curvature* $R(x)$ of S at $x \in S$ is defined by contracting Ric_x, this is to say that

$$R(x) := (\mathrm{tr}\, W_x^\varepsilon)^2 - \mathrm{tr}(W^\varepsilon)_x^2.$$

Using the Cayley-Hamilton theorem, we get

$$(W_x^\varepsilon)^2 - \mathrm{tr}\, W^\varepsilon \cdot W_x^\varepsilon + \det W_x^\varepsilon \cdot \mathrm{id} = 0$$

which shows that

$$R(x) = 2 \cdot \kappa(x) \qquad \forall\, x \in S$$

where $\kappa(x) := \det \mathrm{tr}\, W_x^{\mathbf{a}}$ is the *Gaussian curvature.*

The $\mathbb{R}$-linear map $d\mathbf{a}(x)$ in (148) is an endomorphism of E. Its trace $\mathrm{tr}\; d\mathbf{a}(x)$ is computed at each $x \in O$ as follows:

$$\mathrm{tr}\; d\mathbf{a}(x) = |\mathbf{a}(x)| \cdot \mathrm{tr}\; W_x^\varepsilon + d\ln|\mathbf{a}(x)| \left(\frac{\mathrm{grad}\ |\mathbf{a}|}{|\mathrm{grad}\ |\mathbf{a}||}(x) \right) \cdot < \mathbf{a}(x), \frac{\mathrm{grad}\ |\mathbf{a}|}{|\mathrm{grad}\ |\mathbf{a}||}(x) >,$$

and thus

$$\mathrm{tr}\; d\mathbf{a}(x) = |\mathbf{a}(x)| \cdot H_{\frac{1}{2}}^{\mathbf{a}}(x) + d|\mathbf{a}(x)|(\mathbf{a}(x))$$

where $H_{\frac{1}{2}}^{\mathbf{a}}(x) = \mathrm{tr}\; W_x^\varepsilon$ (cf. section 3.4.2). In terms of the function V the above formula can be rewritten as

$$\mathrm{tr}\; d\mathbf{a}(x) = |\mathbf{a}(x)| \cdot H_{\frac{1}{2}}^{\mathbf{a}}(x) + dV(x)(\mathrm{grad}\ |\mathbf{a}|(x)).$$

Since $\mathrm{tr}\; d\mathbf{a}(x) = -\Delta V(x)$, we finally get

$$-\Delta V = |\mathbf{a}| \cdot H_{\frac{1}{2}}^{\mathbf{a}} + dV(\mathrm{grad}\ |\mathbf{a}|).$$

3.4.4. Symplectic Structure and Riemannian Curvature. The symplectic structure $\omega^{\mathbf{a}}$ is in fact related to the Riemannian curvature $\dot{R}$ of the level surface S of a gradient field $X = (\mathrm{id}\,, \mathbf{a})$ on O. To see this, let $v_x, w_x \in T_xS$ be an orthonormal basis. Keeping u_x, y_x fixed, without loss of generality, we may choose v_x and w_x to form an orthonormal eigenbasis of $W_x^{\mathbf{a}}$. Thus

$$\begin{aligned} \dot{R}(x; v_x, w_x : u_x, y_x) &= \kappa(x)(< w_x, u_x > \cdot < v_x, y_x > - < v_x, u_x > \cdot < w_x, y_x >) \\ &= -\kappa(x) \cdot < u_x \times y_x, v_x \times w_x >, \end{aligned}$$

due to (152) finally yielding for any $x \in S$,

$$\dot{R}(x; v_x, w_x : u_x, y_x) = \frac{\kappa(x)}{|\mathbf{a}(x)|} \cdot \omega^{\mathbf{a}}(x; u_x, y_x) \tag{153}$$

for any orthonormal basis $u_x, v_x \in T_xS$. Hence we have shown:

LEMMA 17. *By means of (153), the symplectic structure and the Gaussian curvature, both constructed out of X, together with the field strength determine the Riemannian curvature of any level surface of the locally given potential.*

3.5. Vector Fields Defined by Two-Forms

So far we have always started with the vector field generating the bundle $\mathbb{F}^{\mathbf{a}}$ with its symplectic structures. Now we will reverse the situation and investigate how a two-form on O will give rise to a vector field (and hence to a bundle of the type $\mathbb{F}^{\mathbf{a}}$).

3.5.1. Two-Form as a Starting Point. Instead of a vector field X on O we consider a non-vanishing presymplectic structure ω on O as our basic ingredient in this section. We will construct a vector field $X = (\mathrm{id}\,, \mathbf{a})$ such that

$$\omega^{\mathbf{a}} = \omega.$$

To do so we consider the fibred product $TO \times_O TO$ of $TO = O \times E$ which is as a set

$$TO \times_O TO := \bigcup_{x \in O} T_xO \times T_xO = O \times E \times E,$$

a subset of $TO \times TO$ and obviously a smooth manifold.

On $TO \times_O TO$ we define

$$\omega : TO \times_O TO \longrightarrow \mathbb{R}$$

assigning to each triple $(x, h, k) \in O \times E \times E$ the real $\omega(x; h, k)$. In addition, we take ω to be bilinear and skew-symmetric in $h, k \in E$ for any $x \in O$ so that ω is a two-form on O.

This two-form ω can be represented by the given scalar product as

$$\omega(x; h, k) =< A(x)(h), k > \qquad \forall\, x \in O \text{ and } h, k \in E,$$

where $A(x) \in \operatorname{End} E$ is a skew-adjoint map. Thus $A(x)$ is of the form

$$A(x)(h) = h \times \mathbf{a}_\omega(x) \qquad \forall\, x \in O \text{ and } h \in E$$

where $\mathbf{a}_\omega : O \longrightarrow E$ is a smooth map. Without loss of generality, we may assume that $\mathbf{a}_\omega(x) \neq 0$ for all $x \in O$. Obviously,

$$\omega(x; h, \mathbf{a}_\omega(x)) = 0 \qquad \forall\, x \in 0 \text{ and } \forall\, h \in E.$$

Therefore, we may state:

PROPOSITION 18. *Given a two-form ω on O with non-vanishing map $\mathbf{a}_\omega$ there is a unique non-vanishing vector field $X_\omega : O \longrightarrow O \times E$ defined by*

$$X_\omega(x) := (x, \mathbf{a}_\omega(x)) \qquad \forall\, x \in O$$

satisfying

$$\omega = \omega^{\mathbf{a}_\omega}.$$

3.5.2. Construction of a Gradient Field. In general X_ω is not a gradient field; for this it would need to satisfy $\operatorname{curl} X_\omega = 0$. To determine $\operatorname{curl} X_\omega$ (cf. (128)) we consider the one-form β_ω on O given by

$$\theta_\omega(x; h) :=< \mathbf{a}_\omega(x), h > \qquad \forall\, x \in O \text{ and } \forall\, h \in E.$$

Its exterior differential is

$$d\theta_\omega(x; h, k) =< d\mathbf{a}_\omega(x)(x; h), k > - < d\mathbf{a}_\omega(x)(x; k), h >$$

and by the definition of curl

$$d\theta_\omega(x; h, k) =< h \times \operatorname{curl} X_\omega, k >$$

for all $x \in O$ and any choice of $h, k \in E$. Thus X_ω is a gradient field iff

$$< d\mathbf{a}_\omega(x)(x; h), k >=< d\mathbf{a}_\omega(x)(x; k), h > \tag{154}$$

for all $x \in O$ and $h, k \in E$. In the context of proposition 18 above we may state:

COROLLARY 19. *X_ω is locally a gradient field on O with principal part $\mathbf{a}_\omega$ iff (154) holds true. The local potential can be determined by solving a boundary value problem given by (146).*

3.6. The Principal bundle $\dot{\mathbb{F}}^{\mathbf{a}}$ and its Natural Connection Form

As we will observe in this section, the complex line bundle $\mathbb{F}^{\mathbf{a}}$ admits a natural principal bundle $\dot{\mathbb{F}}^{\mathbf{a}}$ together with a natural connection form. This connection form on $\dot{\mathbb{F}}^{\mathbf{a}}$ will help us to classify isomorphism types of the complex line bundle $\mathbb{F}^{\mathbf{a}}$ in chapter 4. The purpose here is to define and study it in detail.

3.6.1. The Principal Bundle $\dot{\mathbb{F}}^{\mathbf{a}}$. The bundle $\dot{\mathbb{F}}^{\mathbf{a}}$ will be constructed out of $\mathbb{F}^{\mathbf{a}}$ by deleting the zero section. More precisely, we proceed as follows: Each tuple $(x, k) \in O \times E$ is also an element of $O \times \mathbb{H}$. A typical element in $O \times \mathbb{H}$ is of the form

$$(x, k + \mu \cdot e + \nu \cdot \varepsilon(x)) \in O \times (F^{\varepsilon(x)} \oplus \mathbb{C}^{\varepsilon(x)}) = O \times \mathbb{H}$$

where $\mu, \nu \in \mathbb{R}$. The bundle $O \times \mathbb{H}$ contains the trivial bundle $\mathbb{C}^\varepsilon$. This gives us a (right) action of the multiplicative group $\dot{\mathbb{C}}^\varepsilon_x := \mathbb{C}^\varepsilon_x \backslash \{0\}$ (cf. (132)) on $\dot{\mathbb{F}}^\varepsilon_x := \mathbb{F}^\varepsilon_x \backslash \{0\}$ for any $x \in O$. Let $\dot{\mathbb{F}}^\varepsilon \subset \mathbb{F}^\varepsilon$ consist of the complement of the *zero section* in $\mathbb{F}^\varepsilon$. The identification of $\mathbb{C}^{\varepsilon(x)}$ with $\mathbb{C}$ via $i^{\varepsilon(x)}$ (cf. section 1.2.1) hence causes a $\dot{\mathbb{C}}$-action on $\dot{\mathbb{F}}^\varepsilon$ given by

$$m_z(k_x) := k_x \cdot i^{\varepsilon(x)}(z) \qquad \forall\, k_x \in \dot{\mathbb{F}}^\varepsilon_x \text{ and } \forall z \in \dot{\mathbb{C}} \tag{155}$$

(cf. section 1.3.1). As can be easily verified, $\dot{\mathbb{F}}^\varepsilon$ is a $\dot{\mathbb{C}}^{\mathbf{a}}$- as well as a $\dot{\mathbb{C}}$-principal bundle.

This will lead us to a relation between the topology of O and the isomorphism type of the complex line bundle (cf. chapter 4). This relation will be provided by a connection form on $\dot{\mathbb{F}}^\varepsilon$ which we will study next.

3.6.2. Tangent Vectors to $\dot{\mathbb{F}}^{\mathbf{a}}$. In order to determine the form of a tangent vector $\dot{k}_x \in T_{k_x}\dot{\mathbb{F}}^{\mathbf{a}(x)}$ let $k_x(t)$ be a smoothly parameterized family of elements in $\dot{\mathbb{F}}^{\mathbf{a}}$. Then $k_x(t) =: (x(t), k(t)) \in O \times E$ and $\dot{k}_x(t) =: (\dot{x}(t), \dot{k}(t)) \in E \times E$. Projected to the basis O, the element $k_x(t)$ takes the value $\mathrm{pr}_O(k_x(t)) =: x(t)$. Thus an arbitrary tangent vector $\xi \in T_{k_x}\dot{\mathbb{F}}^{\mathbf{a}}$ is of the form

$$\xi = (x, k, v, \zeta)$$

where $k_x(0) = (x(0), k(0)) = (x, k)$, say, and $v := \dot{x}(0)$ as well as $\zeta := \dot{k}(0)$.

Clearly, any $\xi \in T_{k_x}\dot{\mathbb{F}}^{\mathbf{a}}_x$ can be expressed by

$$\xi = k_x \cdot z_x$$

for some $z_x \in \mathbb{C}^{\varepsilon(x)}$ where $z_x = \left(j^{\varepsilon(x)}\right)(z) \in \dot{\mathbb{C}}^\varepsilon_x$ and some $z \in \dot{\mathbb{C}}$ (cf. section 1.3.1). This means that

$$z_x = k^{-1} \cdot \zeta,$$

or, in quadruple notation,

$$k_x^{-1} \cdot \xi := (x, k, v, k^{-1} \cdot \zeta). \tag{156}$$

Hence any *vertical* vector of the form $(x, k, 0, \zeta)$ can be represented as

$$\xi = (x, k, 0, k \cdot z) \qquad \text{with } z_x \in \mathbb{C}^{\varepsilon(x)}.$$

In order to determine the vertical component of an *arbitrary* tangent vector we will study a natural connection form on $\dot{\mathbb{F}}^{\mathbf{a}}$ in the next paragraph.

3.6.3. A Natural Connection Form on $\dot{\mathbb{F}}^{\mathbf{a}}$. Our aim in this paragraph is to show that $\dot{\mathbb{F}}^{\mathbf{a}}$ admits a *natural connection form* $\dot{\alpha}^{\mathbf{a}}$ caused by the directional field ε of $\mathbf{a}$. This form will be derived by means of the $\dot{\mathbb{C}}^{\varepsilon}$-invariant bundle projection

$$\begin{array}{rcl} \mathrm{pr}_V : T\dot{\mathbb{F}}^{\mathbf{a}} & \longrightarrow & V\dot{\mathbb{F}}^{\mathbf{a}} \\ \xi = (x,k,v,\zeta) & \longmapsto & \xi^V := \left(x,k,0,k\cdot(k^{-1}\cdot\zeta)^{\mathbb{C}^{\varepsilon}}\right) \end{array}$$

onto the vertical vector ξ^V in the *vertical subbundle* $V\dot{\mathbb{F}}^{\mathbf{a}}$ of $T\dot{\mathbb{F}}^{\mathbf{a}}$. Here $(k^{-1}\cdot\zeta)^{\mathbb{C}^{\varepsilon(x)}}$ is the $\mathbb{C}^{\varepsilon(x)}$-component of $k^{-1}\cdot\zeta$ formed with respect to the splitting $\mathbb{H} = \mathbb{C}^{\varepsilon(x)} \oplus \mathbb{F}^{\mathbf{a}(x)}$ for any $x \in O$. This definition is motivated by equation (156). The fibrewise $\dot{\mathbb{C}}^{\varepsilon}$-invariance of pr_V is realized by the $\dot{\mathbb{C}}^{\varepsilon(x)}$-action on the fibres $\dot{\mathbb{F}}^{\mathbf{a}}_x = \dot{\mathbb{F}}^{\varepsilon}_x$ for all $x \in O$.

With $\mathbb{C}^{\varepsilon(x)} \oplus \mathbb{F}^{\varepsilon(x)} = \mathbb{H}$ and $\mathbb{C}^{\varepsilon}_x \oplus \mathbb{F}^{\varepsilon}_x = \{x\} \times \mathbb{H}$ any element $h \in \mathbb{H}$ can be split into the $\mathbb{C}^{\varepsilon(x)}$- and $\mathbb{F}^{\varepsilon(x)}$-component, i.e. into $h = h^{\mathbb{C}^{\varepsilon(x)}} + h^{\mathbb{F}^{\varepsilon(x)}}$. Hence any given tangent vector $\xi \in T_{k_x}\dot{\mathbb{F}}^{\varepsilon(x)}$ defines a complex number

$$z_x(k_x,\xi) := (k^{-1}\cdot\zeta)^{\mathbb{C}^{\varepsilon(x)}}$$

in $\mathbb{C}^{\varepsilon(x)}$ which we can rewrite as

$$z_x(k_x,\xi) = k^{-1}\cdot\zeta^V$$

with the vertical component

$$\zeta^V := k\cdot(k^{-1}\cdot\zeta)^{\mathbb{C}^{\varepsilon(x)}}$$

of ζ. Thus

$$z_x(k_x,\xi) = |k|^{-1}\cdot|\zeta^V|\cdot e^{\vartheta\cdot\varepsilon(x)}$$

for an appropriate ϑ. If $\dot{\xi} = 0$, we set $\vartheta = 0$.

Next let $\xi = (x,k,v,\zeta) \in T\dot{\mathbb{F}}^{\mathbf{a}}$ be an arbitrary tangent vector. Then we define the $\mathbb{C}$-valued connection form $\dot{\alpha}^{\mathbf{a}}$ as follows:

$$\begin{array}{rcl} \dot{\alpha}^{\mathbf{a}} : T\dot{\mathbb{F}}^{\mathbf{a}} & \longrightarrow & \mathbb{C} \\ \dot{\alpha}^{\mathbf{a}}(k_x,\xi) & := & z(k_x,\xi). \end{array}$$

Thus

$$\dot{\alpha}^{\mathbf{a}}(k_x,\xi) = \left(i^{\varepsilon(x)}\right)^{-1}\left((k^{-1}\cdot\zeta)^{\mathbb{C}^{\varepsilon(x)}}\right) = |k|^{-1}\cdot|\zeta^V|\cdot e^{\vartheta(x)\cdot i}. \tag{157}$$

It is straightforward to verify that $\dot{\alpha}^{\mathbf{a}}$ is indeed a *connection form.*

In fact, $\dot{\alpha}^{\mathbf{a}} = \dot{\alpha}^{\varepsilon}$, showing that $\dot{\alpha}^{\mathbf{a}}$ does *not* depend on the function $|\mathbf{a}|$, the field strength.

3.6.4. Horizontal Distribution. The *horizontal distribution*

$$\text{hor}^{\dot{\mathbb{F}}^{\mathbf{a}}} := \ker \dot{\alpha}^{\mathbf{a}} \subset T\mathbb{F}^{\mathbf{a}} \tag{158}$$

is related to TO by the tangent map $T\text{pr}_O : \text{hor}_{k_x}^{\dot{\mathbb{F}}^{\mathbf{a}}} \longrightarrow T_x O$, an isomorphism. Obviously,

$$\xi \in \text{hor}_{k_x}^{\dot{\mathbb{F}}^{\mathbf{a}}} \qquad \Longleftrightarrow \qquad z_x(k_x, \zeta) = 0.$$

Due to $< k, \varepsilon(x) >= 0$ for $k \in \dot{\mathbb{F}}^{\mathbf{a}}$, we see that $< \zeta, \varepsilon(x) >= - < W_x^\varepsilon(v), k >$ which shows that ζ has a non-vanishing component along $\varepsilon(x)$ (in general) and thus is decomposed into

$$\zeta = \zeta^V + \zeta^{\text{hor}}$$

where again

$$\zeta^V := k \cdot z_x(k, \xi) \quad \text{and} \quad \zeta^{\text{hor}} := - < W_x^\varepsilon(v), k > \cdot\, \varepsilon(x) \tag{159}$$

with $z_x(k, \zeta) \in \mathbb{C}^{\varepsilon(x)}$. Therefore,

$$\xi = \big(x, k, v, k \cdot z_x(k, \xi) - < W_x^\varepsilon(v), k > \cdot\, \varepsilon(x)\big)$$

holds true for any $k \in \dot{\mathbb{F}}_x^{\mathbf{a}}$. Evidently, $\dot{\alpha}^{\mathbf{a}}$ is the pull-back of the according connection form on S^2 inherited from the inclusion $S^2 \subset E$.

The Hermitian product $\mathcal{H}^\varepsilon$ (cf. section 3.6.2) yields the quadratic smooth function $\hat{\mathcal{H}}^\varepsilon : \dot{\mathbb{F}}^{\mathbf{a}} \longrightarrow \mathbb{C}^{\mathbf{a}}$ defined by

$$\hat{\mathcal{H}}^\varepsilon(k) := \mathcal{H}^\varepsilon(k, k) = |k|^2 \qquad \forall\, k \in \dot{\mathbb{F}}^{\mathbf{a}}. \tag{160}$$

$\hat{\mathcal{H}}^\varepsilon$ is called *covariantly constant* on $\dot{\mathbb{F}}^{\mathbf{a}}$ iff $d\hat{\mathcal{H}}^\varepsilon$ vanishes on $\text{hor}\,\dot{\mathbb{F}}^{\mathbf{a}} \subset T\dot{\mathbb{F}}^{\mathbf{a}}$, which is true for $\hat{\mathcal{H}}^\varepsilon$ (cf. [**76**]).

3.6.5. Curvature. The curvature $\dot{\Omega}^{\mathbf{a}}$ of the connection form $\dot{\alpha}^{\mathbf{a}}$ is the exterior covariant derivative of $\dot{\alpha}^{\mathbf{a}}$. In more detail $\dot{\Omega}^{\mathbf{a}}$ emanates from a general principle: The horizontal distribution $\dot{\mathbb{F}}^{\mathbf{a}}$ (cf. (158)) is a subbundle of $T\dot{\mathbb{F}}^{\mathbf{a}}$. Together with $V\dot{\mathbb{F}}$ (cf. section 3.6.3) we obtain the bundle splitting

$$T\dot{\mathbb{F}}^{\mathbf{a}} = V\dot{\mathbb{F}}^{\mathbf{a}} \oplus \text{hor}^{\dot{\mathbb{F}}^{\mathbf{a}}}.$$

The general construction of the exterior derivative is as follows. The projection $H^{\dot{\mathbb{F}}^{\mathbf{a}}}$, say, from $T\dot{\mathbb{F}}^{\mathbf{a}}$ to $\text{hor}^{\dot{\mathbb{F}}^{\mathbf{a}}}$ is a *bundle map*, more precisely a *bundle projection*. Given any differential n-form θ on $\dot{\mathbb{F}}^{\mathbf{a}}$ (with values in a linear space) we may form the *exterior derivative* $d\theta$ and set

$$d^{\text{hor}^{\dot{\mathbb{F}}^{\mathbf{a}}}}\theta := (H^{\dot{\mathbb{F}}^{\mathbf{a}}})^* d\theta$$

which applied to the vector fields $Y_0, \ldots, Y_n \in \Gamma T\dot{\mathbb{F}}^{\mathbf{a}}$ reads

$$d^{\text{hor}^{\dot{\mathbb{F}}^{\mathbf{a}}}}\theta(Y_0, \ldots, Y_n) = d\theta(H^{\dot{\mathbb{F}}^{\mathbf{a}}}(Y_0), \ldots, H^{\dot{\mathbb{F}}^{\mathbf{a}}}(Y_n)).$$

$d^{\text{hor}^{\dot{\mathbb{F}}^{\mathbf{a}}}}$ is called the *exterior covariant derivative* on $\dot{\mathbb{F}}^{\mathbf{a}}$ or simply a *connection* on $\dot{\mathbb{F}}^{\mathbf{a}}$ (cf. [**46**]). If $n = 1$ and $\theta = \dot{\alpha}^{\mathbf{a}}$, we have by definition

$$\dot{\Omega}^{\mathbf{a}} := d^{\text{hor}^{\dot{\mathbb{F}}^{\mathbf{a}}}}\dot{\alpha}^{\mathbf{a}}. \tag{161}$$

To determine the exterior covariant derivative of $\dot{\alpha}^{\mathbf{a}}$ in detail we use the splitting $\zeta = \zeta^V + \zeta^{\text{hor}}$ yielding

$$\zeta^V = \zeta + < W_x^\varepsilon(v), k > \cdot\, \varepsilon(x)$$

for $\xi = (x, k, v, \zeta)$ (cf. (159)). Thus

$$\dot{\alpha}^{\mathbf{a}}(k_x, \xi) := \left(i^{\varepsilon(x)}\right)^{-1}\left(k^{-1} \cdot \zeta + < W^{\varepsilon}_x(v), k > \cdot\, k^{-1} \cdot\, \varepsilon(x)\right)$$

and for the Fréchet derivative $D\dot{\alpha}^{\mathbf{a}}$ we compute

$$\begin{aligned}
&(D\dot{\alpha}^{\mathbf{a}}(k_x, \xi_1)(\xi_0^{\text{hor}})) \\
&= \; -k^{-1} \cdot \zeta_0^{\text{hor}} \cdot\, k^{-1} \cdot \zeta_1^{\text{hor}} + < D^2\varepsilon(x)(v_1, v_0), k > \cdot\, k^{-1} \cdot\, \varepsilon(x) \\
&\quad + < W^{\varepsilon}_x(v_1), k > \cdot\, k^{-1} \cdot \zeta_0^{\text{hor}} \cdot\, k^{-1} \cdot\, \varepsilon(x) - < W^{\varepsilon}_x(v_1), k > \cdot\, k^{-1} \cdot W^{\varepsilon}_x(v_0)
\end{aligned}$$

for $\xi_r = (x, k, v_r \zeta_r)$ with $r = 1, 2$. Using

$$\dot{\Omega}^{\mathbf{a}} = d^{\text{hor}\,\dot{\mathbb{F}}^{\mathbf{a}}}\dot{\alpha}^{\mathbf{a}}$$

and evaluating it on the vectors ξ_0 and ξ_1, i.e., applying it to the pair ξ_0, ξ_1, we get

$$\begin{aligned}
&i^{\varepsilon(x)}(\dot{\Omega}^{\mathbf{a}}(k_x; \xi_0, \xi_1)) \\
&= \; < W^{\varepsilon}_x(v_0), k > \cdot\, k^{-1} \cdot W^{\varepsilon}_x(v_1) - < W^{\varepsilon}_x(v_1), k > \cdot\, k^{-1} \cdot W^{\varepsilon}_x(v_0) \\
&= \; < W^{\varepsilon}_x(v_1), \frac{k}{|k|} > \cdot \frac{k}{|k|} \times W^{\varepsilon}_x(v_0) - < W^{\varepsilon}_x(v_0), \frac{k}{|k|} > \cdot \frac{k}{|k|} \times W^{\varepsilon}_x(v_1)
\end{aligned}$$

due to $k^{-1} = \frac{k}{|k|^2}$ (cf. (151)). The right-hand side is a real multiple of $\varepsilon(x) \in \dot{\mathbb{C}}^{\varepsilon(x)}$. This multiple hence is

$$< i^{\varepsilon(x)}(\dot{\Omega}^{\mathbf{a}}(k_x; \xi_0, \xi_1)), \varepsilon(x) > = < W^{\varepsilon}_x(v_0) \times W^{\varepsilon}_x(v_1), \varepsilon(x) >$$

showing that

$$\dot{\Omega}^{\mathbf{a}}(k_x; \xi_0, \xi_1) = \omega^{S^2}(W^{\varepsilon}_x(v_0), W^{\varepsilon}_x(v_1)) \cdot i \tag{162}$$

for any $k_x \in \dot{\mathbb{F}}^{\mathbf{a}}$ any $x \in O$ and any pair $\xi_0, \xi_1 \in T_{k_x}\dot{\mathbb{F}}^{\mathbf{a}}$.

Rewriting (162) we obtain the basic formulae

$$\dot{\Omega}^{\mathbf{a}} = \varepsilon^*\omega^{S^2} \cdot i$$

and hence

$$\dot{\Omega}^{\mathbf{a}} = -\kappa_{\varepsilon} \cdot \omega^{\varepsilon} \cdot i. \tag{163}$$

Thus $\dot{\Omega}^{\mathbf{a}}$ is purely imaginary and does *not* depend on the field strength $|\mathbf{a}|$ either. This two-form is closed, i.e.

$$d\dot{\Omega}^{\mathbf{a}} = 0 \tag{164}$$

since the exterior derivative commutes with pull-backs and $d\omega^{S^2} = 0$.

As a consequence of the $\dot{\mathbb{C}}$-equivariance of $\dot{\alpha}^{\mathbf{a}}$, the curvature $\dot{\Omega}^{\mathbf{a}}$ is a two-form on O rather than on $\dot{\mathbb{F}}^{\mathbf{a}}$.

(If $\mathbf{a} = \operatorname{grad} f$, the function $\kappa_{\varepsilon}|_S$ is the *Gaussian curvature* of any level surface S of f.)

The curvature two-form $\dot{\Omega}^{\mathbf{a}}$ defines a (smooth) vector field $X_{\dot{\Omega}^{\mathbf{a}}} := (\text{id}, \mathbf{c})$ on O whose principal part $\mathbf{c}$ is determined by

$$\dot{\Omega}^{\mathbf{a}}(x; v, w) = < \mathbf{c}(x), v \times w > \qquad \forall\, x \in O \text{ and } \forall\, v, w \in E \tag{165}$$

(cf. section 3.5.1). Given a smooth singularity free vector field on O we call $X_{\dot{\Omega}^{\mathbf{a}}}$ the *curvature field* of the complex line bundle $\mathbb{F}^{\mathbf{a}}$ on O. Since $\dot{\Omega}^{\mathbf{a}}$ is closed, according to (130) the curvature field $\mathbf{c}$ is *divergence free*: Due to (163)

$$\mathbf{c} = \kappa_\varepsilon \cdot \varepsilon, \tag{166}$$

which is equivalent to $\mathbf{a}$ on the common domain of $\mathbf{a}$ and $\mathbf{c}$.

Clearly, it is divergence free due to $d\Omega^{\mathbf{a}} = 0$. In fact, $\operatorname{div} \varepsilon = D(\operatorname{tr} W^\varepsilon|_{\mathbb{F}^{\mathbf{a}}})(\varepsilon)$ and

$$-\kappa_\varepsilon \operatorname{div} \varepsilon = d\kappa_\varepsilon(\varepsilon),$$

which again shows that the divergence of ε does not vanish, in general.

Hence the following holds true:

PROPOSITION 20. *Given the Hermitian line bundle $\mathbb{F}^{\mathbf{a}}$, for all $x \in O$ the curvature $\dot{\Omega}^{\mathbf{a}}$ is a closed two-form on O satisfying*

$$\dot{\Omega}^{\mathbf{a}} = \varepsilon^* \omega^{S^2} \cdot i = -\kappa_\varepsilon \cdot \omega^\varepsilon \cdot i \tag{167}$$

where the curvature field $\kappa_\varepsilon \cdot \varepsilon$ is divergence free. Hence $\dot{\Omega}^{\mathbf{a}} = 0$ iff $\kappa_\varepsilon = 0$.

If $\mathbf{a}$ is constant, $\kappa_{\mathbf{a}} = 0$ and hence $\dot{\Omega}^{\mathbf{a}} = 0$.

Note that due to the definition of κ_ε the directional field ε cannot be reconstructed from equation (167), in general. The constant vector field is a counterexample.

Since on any compact oriented two-dimensional submanifold of O the Gaussian curvature does not vanish everywhere, we have:

COROLLARY 21. *Let $\mathbf{a} = \operatorname{grad} f$. If $\dot{\Omega}^{\mathbf{a}} = 0$ none of the level surfaces of $\mathbf{a}$ is compact.*

The local form of the curvature is obviously expressed by:

PROPOSITION 22. *For any Hermitian bundle $\mathbb{F}^{\mathbf{a}}$ the curvature field $\kappa_\varepsilon \cdot \varepsilon$ can be represented on each contractible open set $U \subset O$ as*

$$(i) \qquad \kappa_\varepsilon \cdot \varepsilon = \operatorname{curl} \mathbf{b}_U$$

for some vector field $\mathbf{b}_U$ on U; this is to say $\dot{\Omega}^{\mathbf{a}} = d\theta^{\mathbf{b}}_U$ for some smooth one-form $\theta^{\mathbf{b}}_U$ on U; in other words, $\dot{\Omega}^{\mathbf{a}}$ is exact on U. Hence $\dot{\Omega}^{\mathbf{a}} = 0$ iff $\mathbf{b}_U$ is a gradient. For any additional contractible set V with $U \cap V \neq 0$ the difference $b_U - b_V$ satisfies

$$(ii) \qquad b_U - b_V = \operatorname{grad} g_{U,V}$$

for some smooth $\mathbb{R}$-valued function $g_{U,V}$ on $U \cap V$.

PROOF. Since U is contractible and $\dot{\Omega}^{\mathbf{a}}$ is closed as pointed out by (164), the *Poincaré Lemma* (cf. [**2**]) immediately yields $\dot{\Omega}^{\mathbf{a}} = d\theta^{\mathbf{b}_U}$ for some smooth one-form $\theta^{\mathbf{b}_U}$ on U. Thus (i) follows from equation (128). Claim (ii) is obvious due to (129). □

In contrast to the local statements in the proposition above, let us investigate under what condition the curvature $\dot{\Omega}^{\mathbf{a}}$ vanishes on all of the open set O. To this end we use *de Rham cohomology groups*. It will allow us to work globally. For any integer k the *de Rham cohomology group* on a smooth manifold M with or without boundary is defined by

$$_{\mathrm{deR}}H^k(M, \mathbb{R}) := \ker d^k / \operatorname{im} d^{k-1}$$

where

$$d^k : A^k(M, \mathbb{R}) \longrightarrow A^{k+1}(M, \mathbb{R})$$

is the *exterior differential* from the $\mathbb{R}$-linear space $A^k(M, \mathbb{R})$ of all smooth $\mathbb{R}$-valued k-forms to $A^{k+1}(M, \mathbb{R})$ (cf. [**2**] and [**46**]). In the sequel, if no confusion arises, we replace d^k by d, for simplicity. This differential exists for manifolds with or without boundary (cf. [**74**]) and $d^2 = 0$. A k-form β is called exact iff $\beta = d^{k-1}\theta$ for some $(k-1)$-form θ.

In globalizing claim *(ii)* of proposition 22, let us see now under what conditions $\dot{\Omega}^{\mathbf{a}} = 0$ holds true on all of O. From the definition of a de Rham cohomology group, the *Poincaré Lemma* again and (128) immediately yield the next statements:

PROPOSITION 23. *The cohomology class* $[\dot{\Omega}^{\mathbf{a}}]$ *of the curvature* $\dot{\Omega}^{\mathbf{a}}$ *vanishes if it is exact. Formally expressed this reads*

$$[\dot{\Omega}^{\mathbf{a}}] = 0 \iff \dot{\Omega}^{\mathbf{a}} = d\theta$$

for some smooth one-form θ *on* O*, unique up to the differential of a globally defined, smooth, real-valued function; formulated in terms of the curvature field, this means*

$$[\dot{\Omega}^{\mathbf{a}}] = 0 \iff \kappa_\varepsilon \cdot \varepsilon = curl\, \mathbf{b}$$

for some vector field on O *with principal part* $\mathbf{b}$*, unique up to a gradient. Hence* $\dot{\Omega}^{\mathbf{a}} = 0$ *implies*

$$\theta = d\phi$$

for some smooth function ϕ *on each contractible set* U *in* O*; the function* ϕ *is defined on all of* O *if* ${}_{\mathrm{deR}}H^1(O, \mathbb{R}) = 0$*. If* ${}_{\mathrm{deR}}H^2(O, \mathbb{R}) = 0$ *all curvatures are exact, i.e. all curvature fields are curls.*

3.7. The Characteristic Principal Bundle

The characteristic principal bundle of a vector field X on O is a principal bundle on O which characterizes the vector field pointwisely. This principal bundle, contained in $\dot{\mathbb{F}}^{\mathbf{a}}$, admits a natural connection form which in addition characterizes the vector field, as well. In this section we will present both and will relate them to $\dot{\mathbb{F}}^{\mathbf{a}}$ introduced in section 3.6.1 and its natural connection form $\dot{\alpha}^{\mathbf{a}}$ (cf. section 3.6.3).

3.7.1. Characteristic $U^a(1)$-Principal Bundles and a Natural Connection Form. The principal bundle $\dot{\mathbb{F}}^{\mathbf{a}}$ equipped with the covariantly constant Hermitian structure $\mathcal{H}^\varepsilon$ does not determine the vector field $X = (\mathrm{id}\,, \mathbf{a})$ entirely. It only determines ε, the directional field of X. However, there is a natural $U^a(1)$-principal bundle $\mathbb{P}^{\mathbf{a}} \subset \dot{\mathbb{F}}^{\mathbf{a}}$ whose fibres are circles the radii of which will be specified by a natural connection form determining the vector field X.

To understand this let us consider a $U(1)$-reduction $\mathbb{P}^{\mathbf{a}}$ of a $\dot{\mathbb{C}}$-principal bundle $\dot{\mathbb{F}}^{\mathbf{a}}$ emanating from a $\mathbb{C}^{\mathbf{a}}$-complex line bundle $\mathbb{F}^{\mathbf{a}} \subset O \times E$ over O. Since $U(1)$ is homotopy equivalent to $\dot{\mathbb{C}}$ this sort of *reduction* obviously exists. Furthermore, we assume a connection form α to be specified on the reduction $\mathbb{P}^{\mathbf{a}}$. It takes its values in $\mathbb{R} \cdot i$, the Lie algebra of $U(1)$. Our primary aim is to determine the general form of α. Our starting point is equation (157) in section 3.6.3 expressing the value of $\dot{\alpha}$ on some $\xi \in T\dot{\mathbb{F}}^{\mathbf{a}}$ in terms of ξ:

$$\dot{\alpha}(k_x, \xi) \quad = \quad \left(i^{\varepsilon(x)}\right)^{-1} \left((k^{-1} \cdot \zeta)^{\mathbb{C}^{\varepsilon(x)}}\right) = |k|^{-1} \cdot |\zeta^V| \cdot e^{\vartheta(x) \cdot i}.$$

The fibres $\mathbb{P}^{\mathbf{a}}_x$ of the given $U(1)$-reduction $\mathbb{P}^{\mathbf{a}}$ of $\dot{\mathbb{F}}^{\mathbf{a}}$ are circles centered at 0 in the respective fibres of $\mathbb{F}^{\mathbf{a}}$. Clearly, the multiplication in $\mathbb{H}$ (cf. (25)) by the imaginary unit is a rotation by $\frac{\pi}{2}$. Therefore, for any *vertical tangent* vector ξ in $T_{k_x}\mathbb{P}^{\mathbf{a}}_x = \mathbb{F}^{\mathbf{a}}_x$ and $\alpha(k_x;\xi) \in \mathbb{R} \cdot i$,

$$\xi = k_x \cdot \alpha(k_x;\xi)$$

(cf. (157)), or more explicitly expressed in terms of the multiplication in $\mathbb{H}$,

$$\xi = k_x \cdot i^{\varepsilon(x)}(\alpha(k_x;\xi)) \qquad \forall k_x \in \mathbb{P}^{\mathbf{a}}$$

(cf. section 1.3.1) requiring $i^{\varepsilon(x)}(\alpha(k_x;\xi))$ to be

$$i^{\varepsilon(x)}(\alpha(k_x;\xi)) = k_x^{-1} \cdot \zeta$$

provided $\xi = (x, k, v, \zeta)$.

Since for any $\xi \in T_{k_x}\mathbb{P}^{\mathbf{a}}$ the *vertical part* ξ^V is given by

$$\zeta^V = k \cdot \alpha(k_x;\xi) \qquad \forall k_x \in \mathbb{P}^{\mathbf{a}} \text{ and } \forall \xi \in T_{k_x}\mathbb{P}^{\mathbf{a}},$$

in analogy to the case of $\dot{\mathbb{F}}^{\mathbf{a}}$ we find that

$$i^{\varepsilon(x)}(\alpha(k_x;\xi)) = k_x^{-1} \cdot \zeta^V$$

for the fourth component ζ in quadruple notation and hence

$$k^{-1} \cdot \zeta^V = k_x^{-1} \cdot \zeta - < \zeta, \varepsilon(x) > \cdot\, k_x^{-1} \cdot \varepsilon(x)$$

again. Thus we finally obtain $\alpha(k_x;\xi) = (k^{-1} \cdot \zeta)^{\mathrm{Im}_{\dot{\mathbb{C}}}} \cdot\, i$ which turns out to be

$$\alpha(k_x;\xi) = - < k_x^{-1} \times \varepsilon(x), \zeta > \cdot\, i.$$

Therefore, due to $k_x^{-1} = -\frac{k_x}{|k_x|^2}$ we deduce for the given connection form α

$$\alpha(k_x;\xi) =< k_x \times \mathbf{b}_\alpha(x), \zeta > \cdot\, i \tag{168}$$

with

$$\mathbf{b}_\alpha(x) = \frac{1}{|k_x|^2} \cdot \varepsilon(x) \qquad \forall x \in O.$$

From this little analysis of the vertical component of a tangent vector $\xi \in T_{k_x}\mathbb{P}^{\mathbf{a}}$, we observe that the connection form α is determined by the reduction $\mathbb{P}^{\mathbf{a}}$. Hence we call α the *natural connection form* on $\mathbb{P}^{\mathbf{a}}$. Clearly, the vector field $X = (\mathrm{id}\,, \mathbf{b}_\alpha)$ with principal part $\mathbf{b}_\alpha$ is singularity free.

Since these steps can be reversed, we may state:

THEOREM 24. *Any connection form α on any given $U(1)$-reduction $\mathbb{P}^{\mathbf{a}} \subset \dot{\mathbb{F}}^{\mathbf{a}}$ of the $\dot{\mathbb{C}}^{\mathbf{a}}$-complex line bundle $\dot{\mathbb{F}}^{\mathbf{a}} \subset O \times E$ over O has the form*

$$\alpha(k_x;\xi) = (k^{-1} \cdot \zeta)^{\mathrm{Im}} \cdot i \qquad \textit{and} \qquad \forall\, k_x \in \mathbb{P}^{\mathbf{a}} \qquad \forall \xi \in T_{k_x}\mathbb{P}^{\mathbf{a}}$$

where $k_x = (x, k)$ and $\xi = (x, k, v, \zeta)$. This is equivalent to saying that there is a singularity free vector field $X = (id, \mathbf{b}_\alpha)$ on O satisfying

$$\mathbf{b}_\alpha(x) = \frac{1}{|k|^2} \cdot \varepsilon(x) \qquad \forall\, x \in O$$

and yielding a connection form α by (168).

From this theorem it is clear that by *breaking the $\dot{\mathbb{C}}$-symmetry* to $U(1)$ produces both a natural connection form and the field strength of X. As an immediate consequence of the above theorem we have:

COROLLARY 25. *Any singularity free vector field $X = (id, \mathbf{a})$ on an open set $O \subset E$ determines a $U(1)$-reduction $\mathbb{P}^{\mathbf{a}}$ of $\dot{\mathbb{F}}^{\mathbf{a}}$ whose connection form $\alpha^{\mathbf{a}}$ is*

$$(i) \qquad \alpha^{\mathbf{a}} = \iota^* \dot{\alpha}^{\mathbf{a}}$$

(cf. (157)) which is equivalent to

$$(ii) \qquad \alpha^{\mathbf{a}}(k_x; \xi) =< k \times \mathbf{a}(x), \zeta > \cdot\, i \qquad \forall\, k_x \in \mathbb{P}^{\mathbf{a}} \text{ and } \forall\, \xi \in T_{k_x}\mathbb{P}^{\mathbf{a}}$$

for any $x \in O$. Here $\iota : \mathbb{P}^{\mathbf{a}} \longrightarrow \mathbb{F}^{\mathbf{a}}$ is the inclusion map and ξ enters in quadruple form. Moreover, the field strength and the radii of the fibres are related by

$$|k_x| = (|\mathbf{a}(x)|)^{-\frac{1}{2}} \qquad \forall\, x \in O.$$

For $x \in O$, the form $\alpha^{\mathbf{a}}$ is equivalently expressed in terms of $\omega^{\mathbf{a}}$ as

$$\alpha^{\mathbf{a}}(k_x; \xi) = \omega^{\mathbf{a}}_x(k_x, \xi) \cdot i \qquad \forall\, k_x \in \mathbb{P}^{\mathbf{a}} \text{ and } \forall\, \xi \in T_{k_x}\mathbb{P}^{\mathbf{a}}.$$

With theorem 24 and corollary 25 the following corollary is obvious:

COROLLARY 26. *The collection of all $U(1)$-reductions of the $\dot{\mathbb{C}}$-principal bundles of all complex line bundles in $O \times E$ is parameterized by the collection of all singularity free smooth vector fields on O. In other words, there is a bijection between the collection of all $U(1)$-reductions of the $\dot{\mathbb{C}}$-principal bundles of all complex line bundles in $O \times E$ and the collection of all singularity free vector fields on O.*

The principal bundle $\mathbb{P}^{\mathbf{a}}$ determined by $X = (\mathrm{id}, \mathbf{a})$ is characteristic for $\mathbf{a}$ and thus called the *characteristic bundle* of X. Due to (140) in theorem 12

$$\mathbb{P}^{\varepsilon} = \varepsilon^* \mathbb{P}_2$$

holds true.

The following two examples illustrate this type of bundle:

- **i): Constant vector fields.** Let $X = (\mathrm{id}, \mathbf{a})$ be a *constant* vector field on E, where $|\mathbf{a}(x)| \neq 0$ for any $x \in O$. Clearly, the circles $\mathbb{P}^{\mathbf{a}}_{x_1}$ and $\mathbb{P}^{\mathbf{a}}_{x_2}$ are congruent for any two points in $x_1, x_2 \in E$. Hence $\mathbb{P}^{\mathbf{a}}|_{\mathrm{im}\,\beta}$ is a cylinder for each integral curve β of $\mathbf{a}$.
- **ii): The Coulomb field on $\dot{E}$.** Obviously, $\dot{E}$ is foliated into 2-spheres S^2_r centered at $0 \in E$ where r determines the radius. The map

  $$\varepsilon : \dot{E} \longrightarrow S^2$$

  assigning the unit vector $\frac{x}{|x|}$ to any $x \in \dot{E}$ yields the bundle $\varepsilon^* TS^2$ on $\dot{E}$; it contains all tangent planes to any S^2_r for all $r > 0$. This bundle, called $\mathbb{F}^C$ in the sequel, is a Hermitian line bundle with symplectic structure ω^{ε} on each fibre. Let

  $$\mathbb{P}^c \subset \dot{\mathbb{F}}^c$$

  be the $U(1)$-reduction with fibre $\mathbb{P}^c_x$, a circle of radius $|x|$ for any $x \in \dot{E}$. $\mathbb{P}^c$ is the characteristic principal bundle of the vector field $\mathbf{a}$ defined by

  $$|x| = \frac{1}{\sqrt{|\mathbf{a}(x)|}} \qquad \forall\, x \in E.$$

  This is to say that

  $$\mathbf{a}(x) = \frac{1}{|x|^2} \cdot \varepsilon(x) \qquad \forall\, x \in O;$$

it is the *Coulomb field* on $\dot{E}$ with potential $f(x) = -\frac{1}{|x|}$ for all $x \in \dot{E}$ (cf. also [**68**]).

3.7.2. Horizontal Vectors and Curvature on $\mathbb{P}^{\mathbf{a}}$. The horizontal space of $T\mathbb{P}^{\mathbf{a}}$ is given as the kernel of $\alpha^{\mathbb{P}^{\mathbf{a}}}$:

$$\mathrm{hor}_{k_x}^{\mathbb{P}^{\mathbf{a}}} := \ker \alpha^{\mathbb{P}^{\mathbf{a}}}(k_x; \ldots)$$

(cf. section 3.6.4). Since $\mathbb{F}^{\mathbf{a}} \subset O \times E$, we have $\mathbb{P}^{\mathbf{a}} \subset O \times E$. Due to equation (168) for any vector $\xi = (x, k, v, \zeta) \in T\mathbb{P}^{\mathbf{a}}$ and for all $x \in O$,

$$\mathrm{hor}_{k_x}^{\mathbb{P}^{\mathbf{a}}} = \{(x, k, v, \zeta) \mid \zeta \perp k \times \mathbf{a}(x)\} = \mathrm{hor}_{k_x}^{\dot{\mathbb{F}}^{\mathbf{a}}} \cap T_{k_x}\mathbb{P}^{\mathbf{a}}.$$

It is easily verified that the Hermitian form $\mathcal{H}^{\varepsilon}$ in (160) is covariantly constant on $\mathrm{hor}^{\mathbb{P}^{\mathbf{a}}}$ (cf. section 3.6.4).

By (i) in corollary 25 the curvature $\Omega^{\mathbf{a}}$ of $\alpha^{\mathbf{a}}$ is

$$\Omega^{\mathbf{a}} = \dot{\Omega}^{\mathbf{a}} \tag{169}$$

(cf. (163)).

Of course $\alpha^{\mathbf{a}}$ determines an exterior covariant derivative, as well. Let us call it $d^{\mathrm{hor}^{\mathbb{P}^{\mathbf{a}}}}$. We easily verify that

$$\Omega^{\mathbf{a}} = d^{\mathrm{hor}^{\mathbb{P}^{\mathbf{a}}}} \alpha^{\mathbf{a}}.$$

Following [**46**] both $d^{\mathrm{hor}^{\dot{\mathbb{F}}^{\mathbf{a}}}}$ and $d^{\mathrm{hor}^{\mathbb{P}^{\mathbf{a}}}}$ determine a covariant derivative on $\mathbb{F}^{\mathbf{a}}$ acting on smooth sections of $\mathbb{F}^{\mathbf{a}}$.

To present a special feature of the connection form α, for any fixed $x \in O$ we consider the curve $\sigma : \mathbb{R} \to \mathbb{P}_x^{\mathbf{a}}$ defined by

$$\sigma(t) = h \cdot e^{\nu \cdot t \cdot a} \qquad \forall\, t \in \mathbb{R} \tag{170}$$

for a given fixed frequency ν and some fixed $h \in \mathbb{P}_x^{\mathbf{a}}$. Thus $\alpha(\sigma; \dot{\sigma}) = \nu \cdot a$ and the phase factor in (170) is $e^{t \cdot \alpha(\sigma; \dot{\sigma})}$. The integral

$$\int_{t_0}^{t_1} e^{t \cdot \alpha(\sigma; \dot{\sigma})} dt$$

is thus the *phase difference*. Given an integral curve β_x of $X = (\mathrm{id}, \mathbf{a})$ with initial condition $\beta_x(t_0) = x$ there is a unique *horizontal lift* $\beta_x^{\mathrm{hor}} \subset \mathbb{P}^{\mathbf{a}}|_{\mathrm{im}\,\beta_x}$ of β_x with initial condition $\beta_x^{\mathrm{hor}}(t_0) = k \in \mathbb{P}^{\mathbf{a}}|_{\mathrm{im}\,\beta_x}$. Clearly, $\mathrm{pr}_O \beta_x^{\mathrm{hor}} = \beta_x$. A *periodic lift* θ of β_x is given by

$$\theta(t) := \beta_x^{\mathrm{hor}}(t) \cdot e^{\nu \cdot t \cdot \varepsilon(\beta_x(t))} \equiv \beta_x^{\mathrm{hor}}(t) \cdot e^{t \cdot \alpha(\theta(t); \dot{\theta}(t))} \qquad \forall\, t$$

due to the isomorphism of $\mathbb{C}^{\varepsilon}$ and $\mathbb{C}^{\mathbf{a}}$ and the operation defined in (170).
For $\nu = 1$ the integral

$$\chi(t_1) := \int_{t_0}^{t_1} e^{t \cdot \alpha(\theta(t); \dot{\theta}(t))} dt$$

is called the *helical phase* of β_x with initial condition $\beta_x(t_0) = x$; it neither depends on θ nor on $\dot{\theta}$ due to the $U(1)$-invariance of α. Thus α naturally yields a notion of phase of an integral curve with specified initial condition.

3.7.3. $\mathbb{F}^{\mathbf{a}}$ is Associated to $\mathbb{P}^{\mathbf{a}}$. As we saw in section (3.7.1) the characteristic bundle $\mathbb{P}^{\mathbf{a}}$ of $X = (\mathrm{id}\,, \mathbf{a})$ determines the vector field and hence $\mathbb{F}^{\mathbf{a}}$. However, there is a more direct way to construct $\mathbb{F}^{\mathbf{a}}$ out of $\mathbb{P}^{\mathbf{a}}$, namely to show that $\mathbb{F}^{\mathbf{a}}$ is isomorphic to the complex line bundle $\mathbb{P}^{\mathbf{a}} \times_{U(1)} F$, say, associated to $\mathbb{P}^{\mathbf{a}}$ (cf. [**46**]). For a short description of the construction of vector bundles associated to principal bundles we refer to section 2.3.6.

As mentioned there, the basic ingredients of the construction of $\mathbb{P}^{\mathbf{a}} \times_{U(1)} F$ are a one-dimensional $\mathbb{C}$-linear space F, its typical fibre and a representation.

To see that $\mathbb{F}^{\mathbf{a}}$ is isomorphic to $\mathbb{P}^{\mathbf{a}} \times_{U(1)} F$ for the inclusion $r : U(1) \hookrightarrow \mathbb{C}$, we observe that $\dot{\mathbb{F}}^{\mathbf{a}}$ can be deformed into $\dot{\mathbb{P}}^{\mathbf{a}}$ and hence that

$$\mathbb{P}^{\mathbf{a}} \times_{U(1)} F \cong \dot{\mathbb{F}}^{\mathbf{a}} \times_{\dot{\mathbb{C}}} \mathbb{C} \tag{171}$$

(in fact, $\mathbb{P}^{\mathbf{a}}$ is a deformation retract of $\mathbb{F}^{\mathbf{a}}$). For $\dot{\mathbb{F}}^{\mathbf{a}}$ the structural functions have values in $\dot{\mathbb{C}}$. Hence $\dot{\mathbb{F}}^{\mathbf{a}}$ is the frame bundle of $\dot{\mathbb{F}}^{\mathbf{a}}$ as well as of $\dot{\mathbb{F}}^{\mathbf{a}} \times_{\dot{\mathbb{C}}} F$. With the same techniques one verifies that $\mathbb{F}^{\mathbf{a}}$ is associated to $\dot{\mathbb{F}}^{\varepsilon}$ (cf. section 2.3.6). In conjunction with (90) and (134) we therefore state:

PROPOSITION 27. *The bundles $\mathbb{F}^{\mathbf{a}}$, $\mathbb{P}^{\mathbf{a}} \times_{U(1)} F$, $\dot{\mathbb{F}}^{\varepsilon} \times_{\dot{\mathbb{C}}} F$ and $\varepsilon^*(\mathbb{F}_2)$ are mutually isomorphic.*

3.8. Horizontal and Periodic Lifts of Integral Curves

The horizontal distribution will now be the basis of lifting integral curves of X to the $U(1)$-principal bundle $\mathbb{P}^{\mathbf{a}}$. We will restrict $\mathbb{P}^{\mathbf{a}}$ to an integral curve β of $X = (\mathrm{id}\,, \mathbf{a})$ and discuss geodesics of $\mathbb{P}^{\mathbf{a}}|_\beta$ for the special cases of the constant vector field and a Coulomb field, respectively.

3.8.1. Horizontal Lifts. Since, in general, $\Omega^{\mathbf{a}} \neq 0$, the horizontal distribution in $T\mathbb{P}^{\mathbf{a}}$ does not need to be integrable along level surfaces. However, $\Omega^{\mathbf{a}}$ vanishes along field lines and thus the horizontal distribution is integrable (cf. [**46**]) along these curves. Let us look at $\mathbb{P}^{\mathbf{a}}|_\beta$ where $\beta : I \longrightarrow O$ is a field line of the singularity free vector field X. Here I is an open interval. For reasons of simplicity let us assume that the initial condition of β is

$$\beta(0) = x \in O.$$

A *horizontal lift* of the velocity field $\dot{\beta}$ of β is a curve $\dot{\beta}^{\mathrm{hor}}$ in $Hor_\beta = \ker \alpha^{\mathbf{a}}$ which satisfies $T\mathrm{pr}^{\mathbf{a}} \dot{\beta}^{\mathrm{hor}} = \dot{\beta}$ and obeys an initial condition in $T\mathbb{P}^{\mathbf{a}}|_\beta$. Hence there is a unique curve β^{hor} passing through $v \in \mathbb{P}^{\mathbf{a}}_{\beta(t_0)}$, say, called *horizontal lift* of β. In the case of a constant vector field or in the case of the Coulomb field this is nothing else but a meridian on the cylinder and the cone $\mathbb{P}^{\mathbf{a}}|_\beta$ containing v, respectively.

In more detail, the horizontal lift $\dot{\beta}^{\mathrm{hor}}_v$ of $\dot{\beta}$ with initial condition $v \in \mathbb{P}^{\mathbf{a}}_x$ for a *constant* vector field $\mathbf{a}(x) = a$ for all $x \in O$ satisfies

$$T\mathrm{pr}^{\mathbf{a}} \dot{\beta}^{\mathrm{hor}}_v = \dot{\beta} = a.$$

To give an explicit description of β^{hor}_v we first determine the components of $\xi \in \mathrm{hor}^{\mathbb{P}^{\mathbf{a}}}_v$ more precisely. Since the ζ-component in $\xi = (x, v, h, \zeta)$ satisfies

$$\zeta^V := k \cdot z_x(k, \xi) \quad \text{and} \quad \zeta^{\mathrm{hor}} := - < W^\varepsilon_x(v), k > \cdot\, \varepsilon(x)$$

with $z_x(k,\zeta) \in \mathbb{C}^{\varepsilon(x)}$ (cf. (159)) we conclude

$$\xi = (x, v, h, 0)$$

due to $a = $ const., which implies $W_x^{\mathbf{a}} = 0$ and $d|a| = 0$. Hence

$$\dot{\beta}^{\mathrm{hor}}(t) = \Big(\beta(t), \beta^{\mathrm{hor}}(t), \dot{\beta}(t), 0\Big) \qquad \forall t.$$

This means that β^{hor} is a meridian of the cylinder $\mathbb{P}^{\mathbf{a}}|_\beta$; it passes through $v \in \mathbb{P}^{\mathbf{a}}|_\beta$. Let $\beta^{\mathrm{hor}}(0) = v$.

Obviously, β is a geodesic on $\mathbb{P}^{\mathbf{a}}|_\beta$ equipped with the metric g_{hor_β} induced by the scalar product $<\,,\,>$ on E.

For the initial value x of β let $\frac{\mathbf{a}(x)}{|\mathbf{a}(x)|}$ be denoted by a_1.

3.8.2. Lifts and Geodesics. A curve γ on $\mathbb{P}^{\mathbf{a}}|_\beta$ is called a *lift* of β through v iff it is of the form

$$\gamma(s) = \beta_v^{\mathrm{hor}}(s) \cdot e^{p \cdot s \cdot a_1} \tag{172}$$

for all s such that $\beta(s)$ is in a suitable neighborhood of O and a fixed real p.

Clearly, γ is a *horizontal lift* through v iff $\gamma = \beta_v^{\mathrm{hor}}$, i.e. iff $p = 0 \pmod{2\pi}$. In fact, any lift γ of β is a geodesic on $\mathbb{P}^{\mathbf{a}}|_\beta$. This means that γ has to satisfy

$$\nabla_{\frac{d}{ds}} \dot{\gamma} = \ddot{\gamma} - S(\dot{\gamma}, \dot{\gamma}) = 0.$$

Here ∇ and S are the covariant derivative of the metric on $\mathbb{P}^{\mathbf{a}}|_\beta$ defined by the scalar product $<\,,\,>$ and the second fundamental tensor of the cylinder $\mathbb{P}^{\mathbf{a}}|_\beta$, respectively. Hence γ is a geodesic iff $\ddot{\gamma}$ is perpendicular to $\mathbb{P}^{\mathbf{a}}|_\beta$. Due to the $U(1)$-symmetry of $\mathbb{P}^{\mathbf{a}}|_\beta$, a geodesic σ on $\mathbb{P}^{\mathbf{a}}|_\beta$ passing through v is of the form

$$\sigma(s) = \beta_v^{\mathrm{hor}}(\alpha \cdot s) \cdot e^{p \cdot \alpha \cdot s \cdot a_1}$$

or, reformulated as a curve in $O \times E$,

$$\sigma(s) = (\beta(\alpha \cdot s), v \cdot e^{p \cdot \alpha \cdot s \cdot a_1}).$$

Here p and α are positive reals. To verify that $\ddot{\sigma}(s)$ is perpendicular to $T_{\sigma(s)}\mathbb{P}^{\mathbf{a}}|_\beta$ in $E \times E$ we first compute $\dot{\sigma}(s)$ (omitting the variables in $\mathbb{P}^{\mathbf{a}}|_\beta$) as

$$\dot{\sigma}(s) = (\alpha \cdot a, v \cdot p \cdot \alpha \cdot a \cdot e^{p \cdot \alpha \cdot a \cdot a_1}).$$

Accordingly, $\ddot{\sigma}(s)$ (without the variables $\dot{\sigma}(s)$) reads

$$\ddot{\sigma}(s) = \Big(0, v \cdot (-p^2) \cdot \alpha^2 \cdot e^{p \cdot \alpha \cdot s \cdot a_1}\Big),$$

showing that $\ddot{\sigma}(s)$ is perpendicular to $T_{\sigma(s)}\mathbb{P}^{\mathbf{a}}|_\beta$ in $E \times E$. The parameter α determines the speed of the geodesic. Thus σ and β have accordant speeds if $\alpha = 1$. Obviously,

$$\dot{\gamma}(s) = \Big(\beta_v^{\mathrm{hor}}(s) \cdot p \cdot a + \dot{\beta}_v^{\mathrm{hor}}(s)\Big) \cdot e^{p \cdot s \cdot a_1}$$

and, in particular,

$$\dot{\gamma}(0) = p \cdot v \cdot a + \dot{\beta}_v^{\mathrm{hor}}(0).$$

3.8.3. Initial Velocity and Frequency. In the last formula of paragraph 3.8.2 there is hidden a relation between the initial velocity and a frequency. The F_x^a-component of the initial velocity $\dot{\gamma}(0)$ is $p \cdot |v|$. Hence the pair $(v, p \cdot v \cdot a)$ is an element of the phase space

$$F_x^a = T_v\mathbb{R} \cdot \frac{v}{|v|}.$$

We may identify $p \cdot |v|$ as the F_x^a-component of the momentum; we will simply call it the momentum $\mathbf{p}$. On the other hand, the geodesic γ winds around $\mathbb{P}^{\mathbf{a}}|_\beta$. The projection along the meridians of this motion to $\mathbb{P}_x^{\mathbf{a}}$ yields a circular motion on $\mathbb{P}_x^{\mathbf{a}}$. This motion is characterized by the phase, i.e. the angle $e^{\lambda \cdot a}$ measured from $\mathbb{R} \cdot \frac{v}{|v|}$ and the *frequency* ω, given by

$$\omega := \frac{2\pi}{T} \quad \text{and} \quad T \cdot p \cdot |v| = 2\pi |v|.$$

Thus the frequency ω is

$$\omega = p$$

and T is the period of revolution. Hence

$$\omega \in T\mathbb{R} \cdot \frac{v}{|v|} = F_x^a.$$

Therefore, F_x^a appears as a phase space in two ways, namely as consisting of space-momentum and as phase-frequency variables (cf. subsection 1.3.4). The geodesic motion sets up a correspondence q between these two sorts of variables. To define it more precisely we represent any $h \in F_x^a$ as

$$h = \lambda \cdot \frac{v}{|v|} + p \cdot v \cdot a$$

and define

$$q : F_x^a \longrightarrow F_x^a$$

by

$$q(h) := \lambda \cdot \frac{v}{|v|} + p \cdot \frac{v}{|v|} \cdot a.$$

Hence the relation between the momentum $\mathbf{p} \cdot |v|$ and the frequency ω is

$$\omega = \frac{1}{|v|} \cdot \mathbf{p} = p.$$

Let us point out that $\omega = p$ does not depend on the field strength $|a|$ of a; nor does it depend on the length of the curve β_x.

3.8.4. Geodesics of the Coulomb Field. For the Coulomb field $(\mathrm{id}, \mathbf{a})$ given explicitly by $X(x) = \left(x, -\frac{x}{|x|^3}\right)$ with $x \in O$ let a parametrization of the surface of revolution $\mathbb{P}^{\mathbf{a}}|_\beta$ be given in Clairaut coordinates via the map $\mathbf{x} : \mathcal{U} \to E$ defined by

$$\mathbf{x}(u, v) := -(3v - 2)^{\frac{1}{3}} \cdot r\left(e^{u \cdot \frac{a}{|a|}}\right) \cdot \left(v + \frac{a}{|a|}\right)$$

on an open set $\mathcal{U} \subset \mathbb{R}^2$. Here r is the representation of $U^{\frac{a}{|a|}}(1)$ onto $SO\left(F^{\frac{a}{|a|}}\right)$ for any $x \in O$. Then a geodesic γ on $\mathbb{P}^{\mathbf{a}}|_\beta$ takes the form

$$\gamma(s) = \mathbf{x}(u(s), w(s)) = -(3w(s) - 2)^{\frac{1}{3}} \cdot r\left(e^{u(s) \cdot \frac{a}{|a|}}\right) \cdot \left(v + \frac{a}{|a|}\right)$$

where the functions u and w are determined by

$$
\begin{aligned}
u(s) &= \sqrt{2}\cdot\arctan\left(\frac{s}{\sqrt{2}\,d}+\frac{c_1}{2\,d}\right)+c_2,\\
w(s) &= \pm\frac{1}{3}\left(\left(\frac{1}{\sqrt{2}}s+c_1\right)^2+d^2\right)^{\frac{3}{2}}+\frac{2}{3}
\end{aligned}
\tag{173}
$$

with s in an open interval $I\subset\mathbb{R}$ containing 1 (cf. [**68**]). Here c_1 and c_2 are integration constants determining the initial conditions. Since we are concerned with a forward movement along the channel $\mathbb{R}\cdot\frac{a}{|a|}$, only the positive sign in (173) is of interest. The constant d fixes the slope of the geodesic via

$$
\cos\vartheta=\frac{d}{\sqrt{\left(\frac{1}{\sqrt{2}}s+c_1\right)^2+d^2}}
$$

where ϑ is the constant angle between the geodesic γ, called periodic lift, again, and the parallels given in Clairaut coordinates (cf. [**68**]). This means that d vanishes precisely for a meridian. A periodic lift γ is a horizontal lift of β iff γ is a meridian.

Next we determine the parametrization of an integral curve β of X. It is given by

$$
\beta(t):=x_0\cdot|3\cdot t-c|^{\frac{1}{3}}
$$

for all t in a suitable small interval of $1\in\mathbb{R}$. Here x_0 is the initial condition and c a constant which is used to specify a particular initial condition. Choosing $c_1=2$ yields

$$
\beta(1)=x_0.
$$

Since the image of a horizontal lift of an integral curve β is a meridian of the cone (and vice versa) we obtain for $v_x\in\mathbb{P}^{\mathbf{a}}_{\beta(1)}$,

$$
\beta^{\text{hor}}(t)=v_x\cdot(3t-2)^{\frac{1}{3}}\qquad \frac{2}{3}\le t
$$

as the horizontal lift of β with

$$
\beta^{\text{hor}}(1)=v_x.
$$

While for the constant vector field, any periodic lift γ of β through v_x is uniquely determined by the $U^a(1)$-valued map

$$
s\mapsto e^{p\cdot s\cdot\frac{a}{|a|}},
$$

for the Coulomb field a periodic lift is characterized by

$$
s\mapsto e^{u(s)\cdot\frac{a}{|a|}}
$$

with $u(s)$ as in (173). These two maps here are called an elementary periodic function and an elementary Clairaut map, respectively. Therefore, we can state:

PROPOSITION 28. *Let $x=\beta(1)$. Under the hypothesis that a is a non-zero constant, there is a one-to-one correspondence between all elementary periodic $U^a(1)$-valued functions and all periodic lifts of β passing through a given $v\in\mathbb{P}^{\mathbf{a}}_x$. In case X is the Coulomb field there is a one-to-one correspondence between all periodic lifts passing through a given $v\in\mathbb{P}^{\mathbf{a}}_x$ and all elementary Clairaut maps.*

CHAPTER 4

Isomorphism Classes, Chern Classes and Homotopy Classes of Singularity Free Vector Fields in 3-Space

The aim of this chapter is to investigate the notion of equivalence of singularity free vector fields (as introduced in section 3.2.3) from a *topological* view point. In this context two such vector fields are called equivalent if their respective complex line bundles are isomorphic. Hence we will study *isomorphism classes* of the complex line bundles emanating from singularity free vector fields. (If no confusion arises all vector fields in this chapter are assumed to be smooth).

To be more precise, on the one hand we focus on the topological characterization of the *complex line bundles* associated with singularity free vector fields on M. In doing so we emphasize primarily on the topology of M. The particular topological properties of M are inherited from the extraction of the *singularities* of a vector field X' defined on a larger manifold M' with or without boundary. The topological space M consists of all of M' but the collection of singularities (a closed set in M'). M is assumed to be a smooth *manifold with or without boundary*, in the sequel just called a manifold for simplicity. By construction

$$X := X'|_M$$

is singularity free. The bundle $\mathbb{F}^{\mathbf{a}}$ trivializes on a particular covering which provides a characteristic cocycle. Hence we have *cohomological characteristics* under investigation, being identical for $\mathbb{F}^{\mathbf{a}}$ and $\mathbb{F}^{\varepsilon}$ (cf. section 3.1). Such characteristics are, for example, Chern or Chern–de Rham classes of $\mathbb{F}^{\varepsilon}$.

On the other hand, the quality of the unit vector field ε emanating from X reflects these topological properties $\mathbb{F}^{\mathbf{a}} = \mathbb{F}^{\varepsilon}$ equally well. Hence we are interested in the quality of the map

$$\varepsilon : M \longrightarrow S^2,$$

i.e. in characteristics such as its homotopy class. In particular, we will relate the Chern-class of $\mathbb{F}^{\mathbf{a}}$ to the *mapping degree* of a particular map produced out of ε.

We begin with a global analytic description of the collection of isomorphism classes of characteristic principal bundles (cf. section 3.7). This collection turns out to be an infinite dimensional manifold, provided M is compact.

4.1. Isomorphism Classes of Characteristic Principal Bundles of Vector Fields

Given a singularity free vector field $X = (\mathrm{id}\,, \mathbf{a})$ on a three-dimensional, *compact* smooth manifold M with boundary in E, the complex structure of $\mathbb{F}^{\mathbf{a}}$ is defined via the unit vector field $\varepsilon = \frac{\mathbf{a}}{|\mathbf{a}|}$ determined by X (cf. section 3.1.4). Thus it is enough to study the isomorphism classes of unit vector fields. In this section we will hence investigate the notion of isomorphism classes of the characteristic principal bundles of two *unit vector fields* on M (cf. section 3.7).

4.1.1. Isomorphisms of Characteristic Principal Bundles and Equivariant Maps. For any $\varepsilon \in C^\infty(M, S^2)$ the characteristic principal bundle $\mathbb{P}^\varepsilon$ is the pull-back of the *bundle of oriented, orthonormal frames* $\mathbb{P}_2$ on S^2 (cf. section 2.3.5). Thus

(174)
$$\begin{array}{ccc} \mathbb{P}^\varepsilon & \xrightarrow{\hat{\varepsilon}} & \mathbb{P}_2 \\ {\scriptstyle \mathrm{pr}_M}\downarrow & & \downarrow{\scriptstyle \mathrm{pr}_{S^2}} \\ M & \xrightarrow{\varepsilon} & S^2 \end{array}$$

commutes where $\hat{\varepsilon}$ is fibrewisely the identity (cf. (135)).

Since

(175)
$$S^2 = SO(E)/SO(F)$$

where $F \subset E$ is a fixed oriented two-dimensional subspace of E, the bundle $\mathbb{P}_2 \to S^2$ is naturally isomorphic to the fibration determined by

$$\mathrm{pr}_{S^2} : SO(E) \longrightarrow S^2$$

(cf. section 2.3.5). Hence $F = T_{y_0}S^2$ for a fixed *reference point* $y_0 \in S^2$. Here the metric on $SO(E)$ is defined on the Lie algebra $so(E)$ consisting of all *skew-adjoint transformations* on E. It assigns to any two $A, B \in SO(E)$ the real number $-\frac{1}{3}\operatorname{tr} AB$ (cf. section 2.3.5). Thus we will *identify* $\mathbb{P}_2$ with $SO(E)$ in the sequel. In these terms TS^2 is of the form

$$TS^2 = SO(E) \times_{SO(F)} F$$

and, moreover,

(176)
$$\begin{array}{ccc} \varepsilon^*(SO(E)) = \mathbb{P}^\varepsilon & \xrightarrow{\hat{\varepsilon}} & SO(E) \\ {\scriptstyle \mathrm{pr}_M}\downarrow & & \downarrow{\scriptstyle \mathrm{pr}_{S^2}} \\ M & \xrightarrow{\varepsilon} & S^2 \end{array}$$

where again $\hat{\varepsilon}$ is fibrewisely the identity as in (135). With these fundamentals at hand we proceed to study bundle isomorphisms of characteristic principal bundles of singularity free vector fields on M.

For any two $\varepsilon, \varepsilon' \in C^\infty(M, S^2)$ and a *smooth bundle isomorphism*

$$\Psi : \mathbb{P}^\varepsilon \longrightarrow \mathbb{P}^{\varepsilon'}$$

there is a unique *diffeomorphism*

$$\Psi_M : M \longrightarrow M$$

for which

(177)
$$\begin{array}{ccc} \mathbb{P}^\varepsilon & \xrightarrow{\Psi} & \mathbb{P}^{\varepsilon'} \\ {\scriptstyle \mathrm{pr}_M}\downarrow & & \downarrow{\scriptstyle \mathrm{pr}_M} \\ M & \xrightarrow{\Psi_M} & M \end{array}$$

commutes. Clearly,

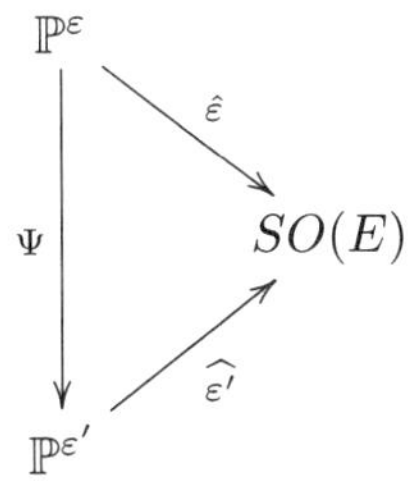

does *not* commute. However, there is a smooth map

$$\varphi_\Psi : \mathbb{P}^\varepsilon \longrightarrow SO(E)$$

such that

$$\hat{\varepsilon} \oslash \varphi_\Psi = \widehat{\varepsilon'} \circ \Psi \tag{178}$$

where the operation $\oslash$ on the left-hand side is defined by

$$\hat{\varepsilon}(k_x) \oslash \varphi_\Psi(k_x) \quad := \quad \varphi_\Psi(k_x)^{-1}(\hat{\varepsilon}(k_x)) \tag{179}$$

for all $k_x \in \mathbb{P}^\varepsilon$ and any $x \in M$. The inverse $\varphi_\Psi(k_x)^{-1}$ of $\varphi_\Psi(k_x)$ is formed in $SO(E)$ pointwisely. Inserting $k_x \in \mathbb{P}^\varepsilon_x$ into (178) yields

$$\hat{\varepsilon}(k_x) \oslash \varphi_\Psi(k_x) = \widehat{\varepsilon'}(\Psi(k_x)) \qquad \forall\, k_x \in \mathbb{P}^\varepsilon_x \text{ and } \forall\, x \in M \tag{180}$$

and therefore the smooth map

$$\hat{\varepsilon} \oslash \varphi_\Psi : \mathbb{P}^\varepsilon \quad \longrightarrow \quad \mathbb{P}^{\varepsilon'} \tag{181}$$

for which

$$\hat{\varepsilon}(k_x) \oslash \varphi_\Psi(k_x) \quad = \varphi_\Psi(k_x)^{-1}(\widehat{\varepsilon}(k_x)) = \widehat{\varepsilon'}(\Psi(k_x)) \tag{182}$$

holds true. Equation (182) implies the *uniqueness* of φ_Ψ.

The $U(1)$-action on $\mathbb{P}^\varepsilon$ (cf. section 1.3.1) corresponds to the $U(1)$-action on $\mathbb{P}_2$ (cf. section 2.3.5) and is related to the $SO(F)$-action on the bundle of oriented, orthonormal frames of S^2. More precisely, for each $x \in M$, each $k_x \in \mathbb{P}^\varepsilon_x$ and each $z \in U(1)$ there is a unique rotation $\ell_z \in SO(T_{\varepsilon(x)}S^2)$ with $\varepsilon(x)$ as the oriented axis of rotation and with an oriented rotation angle in $[0, 2\pi)$ such that

$$k_x \cdot z = \ell_z(\hat{\varepsilon}(k_x)) = \ell_z(k_x).$$

The multiplication $\cdot$ in $k_x \cdot z$ is the multiplication in $\mathbb{H}$ (cf. section 1.1.4). This holds accordingly for $\Psi(k_x)$ and unique $\ell'_z \in SO(T_{\varepsilon'(x)}S^2)$. Hence for all $z \in U(1)$ the equivariance of Ψ yields

$$\begin{aligned} \hat{\varepsilon}(k_x \cdot z) \oslash \varphi_\Psi(k_x \cdot z) \quad &= \quad \widehat{\varepsilon'}(\Psi(k_x \cdot z)) \\ &= \quad (\widehat{\varepsilon'}(\Psi(k_x))) \cdot z \\ &= \quad \ell'_z \circ \varphi_\Psi(k_x)^{-1}(\widehat{\varepsilon'}(k_x)) \end{aligned}$$

since Ψ is an isomorphism of principal bundles and, moreover, by (179)

$$\begin{aligned} \hat{\varepsilon}(k_x \cdot z) \oslash (\varphi_\Psi(k_x \cdot z)) \quad &= \quad \varphi_\Psi(k_x \cdot z)^{-1}(\widehat{\varepsilon}(k_x) \cdot z) \\ &= \quad \varphi_\Psi(k_x \cdot z)^{-1} \circ \ell_z(\hat{\varepsilon}(k_x)). \end{aligned}$$

Since $\hat{\varepsilon}$ is fibrewisely the identity we conclude for any $z \in U(1)$, any $k_x \in \mathbb{P}^\varepsilon$ and any $x \in M$ that

$$\varphi_\Psi(k_x \cdot z) = {\ell'_z}^{-1} \circ \varphi_\Psi(k_x) \circ \ell_z. \tag{183}$$

With this *reverse* conjugation with z equation (183) is rewritten as

$$({\ell'_z}^{-1} \circ \varphi_\Psi \circ \ell_z)(k_x) = \varphi_\Psi(k_x \cdot z) \qquad \forall\, k_x \in \mathbb{P}^\varepsilon \text{ and } \forall\, x \in M$$

which expresses equivariance in z. Clearly,

$$\varphi_\Psi(k_x)^{-1}(\hat{\varepsilon}(k_x)) \in T_{\varepsilon'(x)}S^2$$

for any $k_x \in \mathbb{P}^\varepsilon$ and any $x \in M$. Thus

$$\varphi_\Psi(k_x)(T_{\varepsilon(x)}S^2) = (T_{\varepsilon'(x)}S^2) \quad \text{ and } \quad \varphi_\Psi(k_x)(\varepsilon(x)) = \varepsilon'(x) \tag{184}$$

for any $x \in M$. The orthogonal map $\varphi_\Psi(k_x)$ depends on $x \in M$ only. If $\varepsilon(x) \neq \varepsilon'(x)$ the rotation axis of $\varphi_\Psi(k_x)$ is perpendicular to span $\{\varepsilon(x), \varepsilon'(x)\} \subset E$ and is oriented. Thus the rotation axis is given by $\mathbb{R} \cdot a(k_x)$ with $a(k_x) \in S^2$ satisfying

$$a(k_x \cdot z) = a(k_x) \qquad \forall\, z \in U(1).$$

If $\varepsilon(x) = \varepsilon'(x)$, then $\varphi(k_x)(\varepsilon(x)) = \varepsilon(x)$ with rotation axis $\mathbb{R} \cdot \varepsilon(x)$.

Therefore, each *isomorphism* Ψ between $\mathbb{P}^\varepsilon$ and $\mathbb{P}^{\varepsilon'}$ (if there is any) can be represented by a unique $SO(E)$-valued map φ_Ψ of $\mathbb{P}^\varepsilon$, *equivariant* with respect to the *inverse conjugation* by elements in $U(1)$ (cf. (183)). All vector and principal bundle isomorphisms under consideration are assumed to be smooth if not pointed out otherwise. By (180) and (183) the following is obvious:

LEMMA 29. *Given $\varepsilon, \varepsilon' \in C^\infty(M, S^2)$ and a $U(1)$-principal bundle isomorphism (if there is any at all)*

$$\mathbb{P}^\varepsilon \xrightarrow{\Psi} \mathbb{P}^{\varepsilon'},$$

the map

$$\varphi_\Psi : \mathbb{P}^\varepsilon \longrightarrow SO(E)$$

is a smooth $U(1)$-principal bundle homomorphism, equivariant under $U(1)$, satisfying equation (178).

Denoting by $\operatorname{Isom}(\mathbb{P}^\varepsilon, \mathbb{P}^{\varepsilon'})$ the collection of all smooth $U(1)$-principal bundle isomorphisms between $\mathbb{P}^\varepsilon$ and $\mathbb{P}^{\varepsilon'}$ (containing with each Ψ also its inverse) and by $C^\infty_{eq}(\mathbb{P}^\varepsilon, SO(E))$ the collection of all smooth maps from $\mathbb{P}^\varepsilon$ to $SO(E)$, equivariant under the reverse conjugation (cf. (183)), the map

$$\begin{array}{rcl} \tilde{\varepsilon} : \operatorname{Isom}(\mathbb{P}^\varepsilon, \mathbb{P}^{\varepsilon'}) & \longrightarrow & C^\infty_{eq}(\mathbb{P}^\varepsilon, SO(E)) \\ \Psi & \longmapsto & \varphi_\Psi \end{array}$$

is an *injection.* (Here it is assumed that $\operatorname{Isom}(\mathbb{P}^\varepsilon, \mathbb{P}^{\varepsilon'}) \neq \emptyset$). In case $\varepsilon = \varepsilon'$ we write $\operatorname{Aut} \mathbb{P}^\varepsilon$ instead of $\operatorname{Isom}(\mathbb{P}^\varepsilon, \mathbb{P}^\varepsilon)$.

The *gauge group* $\operatorname{Gau} \mathbb{P}^\varepsilon$ is the collection of all $\Psi \in \operatorname{Aut} \mathbb{P}^\varepsilon$ for which $\Psi_M = \operatorname{id}$. As shown in [**50**]

$$\operatorname{Gau} \mathbb{P}^\varepsilon \cong C^\infty(M, U(1)).$$

This isomorphism is canonical, hence we identify these two spaces. As is easily verified (178) is equivariant under $\operatorname{Gau} \mathbb{P}^\varepsilon$. In conclusion we state:

LEMMA 30. *For each* $\Psi \in \mathrm{Isom}(\mathbb{P}^{\varepsilon}, \mathbb{P}^{\varepsilon'})$ *and any* $\phi \in \mathrm{Gau}\,\mathbb{P}^{\varepsilon}$ *the injective map*

$$\tilde{\varepsilon} : \mathrm{Isom}(\mathbb{P}^{\varepsilon}, \mathbb{P}^{\varepsilon'}) \longrightarrow C^{\infty}_{eq}(\mathbb{P}^{\varepsilon}, SO(E)) \tag{185}$$

satisfies

$$\widehat{\varepsilon'} \oslash \tilde{\varepsilon}(\Psi \circ \phi)(k_x) = (\widehat{\varepsilon'} \circ \Psi)(\phi(k_x)) \qquad \forall\, k_x \in \mathbb{P}^{\varepsilon} \text{ and } \forall\, x \in M, \tag{186}$$

saying that $\tilde{\varepsilon}$ *is equivariant under* $\mathrm{Gau}\,\mathbb{P}^{\varepsilon}$ *and that*

$$\begin{array}{ccc} \mathbb{P}^{\varepsilon} & \xrightarrow{\widehat{\varepsilon'} \oslash \tilde{\varepsilon}(\Psi)} & \mathbb{P}, \\ {\scriptstyle \mathrm{pr}_M}\big\downarrow & & \big\downarrow{\scriptstyle \mathrm{pr}_{S^2}} \\ M & \xrightarrow{\varepsilon'} & S^2 \end{array} \tag{187}$$

commutes.

4.2. The Structure of Isomorphism Classes

In this section we will study in more detail the *isomorphism class*

$$\mathrm{Isom}\,\mathbb{P}^{\varepsilon} := \bigcup_{\varepsilon' \in C^{\infty}(M, S^2)} \mathrm{Isom}(\mathbb{P}^{\varepsilon}, \mathbb{P}^{\varepsilon'})$$

of ε, i.e. the collection of all *isomorphisms* of $\mathbb{P}^{\varepsilon}$ and relate it with the *homotopy class* $[\varepsilon]$ of ε.

4.2.1. Manifold Structure of $\mathrm{Isom}(\mathbb{P}^{\varepsilon})$**.** Let $\Gamma(\mathbb{P}^{\varepsilon} \times_{U(1)} SO(E))$ denote the collection of all smooth sections of M into $\mathbb{P}^{\varepsilon} \times_{U(1)} SO(E)$, the group bundle associated to $\mathbb{P}^{\varepsilon}$. Here $U(1)$ acts on $SO(E)$ by reverse conjugation, i.e. in the sense of (183). As shown in [**50**] there is a natural isomorphism

$$\gamma : C^{\infty}_{eq}(\mathbb{P}^{\varepsilon}, SO(E)) \longrightarrow \Gamma(\mathbb{P}^{\varepsilon} \times_{U(1)} SO(E)).$$

Under the pointwisely defined operations $\Gamma(\mathbb{P}^{\varepsilon} \times_{U(1)} SO(E))$ is a group. Moreover, under the C^{∞}-topology it is a smooth manifold with smooth group operations, a so-called *smooth differentiable group* (cf. section 6.2.5 and [**18**]). Pulling back this manifold structure to $C^{\infty}_{eq}(\mathbb{P}^{\varepsilon}, SO(E))$ turns this group into a smooth *infinite dimensional differentiable group*, as well.

The basic observation to determine the structure of $\mathrm{Isom}\,\mathbb{P}^{\varepsilon}$, the collection of all isomorphisms of $\mathbb{P}^{\varepsilon}$ into all $\mathbb{P}^{\varepsilon'}$ (if there is any), is therefore the following:

LEMMA 31. $C^{\infty}_{eq}(\mathbb{P}^{\varepsilon}, SO(E))$ *is a group under the pointwisely defined operations pulled back by* γ.

4.2.2. The Structure Theorem. With the preparations of the previous section we will now link $\mathrm{Isom}\,\mathbb{P}^{\varepsilon}$ with the homotopy class of ε. To do so, let $[[\varepsilon]] \subset C^{\infty}(M, S^2)$ be the collection of all smooth unit vector fields $\varepsilon' \in C^{\infty}(M, S^2)$ for which $\mathbb{P}^{\varepsilon}$ and $\mathbb{P}^{\varepsilon'}$ are *isomorphic* as principal $U(1)$-bundles. Again $C^{\infty}(M, S^2)$ is a smooth manifold (cf. [**18**]).

Since any two $\varepsilon, \varepsilon' \in C^{\infty}(M, S^2)$ are (smoothly) homotopic if they are joined by a smooth path, the homotopy class $[\varepsilon]$ is turned into a smooth infinite dimensional open submanifold of $C^{\infty}(M, S^2)$. Moreover, $C^{\infty}(M, SO(E))$ is a smooth differentiable manifold, too. Connected components are thus path-connected. Hence $[\varepsilon] \subset C^{\infty}(M, S)$ is the connected component of $\varepsilon \in C^{\infty}(M, S^2)$. In fact, by the usual techniques in infinite dimensional analysis and differential geometry (cf. [**18**]), the following holds true:

PROPOSITION 32. *The map*

$$\begin{aligned} \mathrm{pr}^*_{S^2} : C^\infty_{eq}(M, SO(E)) &\longrightarrow C^\infty(M, S^2) \\ \varphi &\longmapsto \mathrm{pr}_{S^2} \circ \varphi \end{aligned}$$

(cf. section 2.3.3) is a smooth surjection for which the quotient map

$$\overline{\mathrm{pr}^*_{S^2}} : C^*_{eq}(M, SO(E))/C^\infty(M, U(1)) \longrightarrow C^\infty(M, S^2)$$

is a smooth bijection. Here $U(1)$ acts on $SO(E)$ by reverse conjugation, i.e. in the sense of (183).

Approaching the theorem on the structure of $\mathrm{Isom}(\mathbb{P}^\varepsilon, \mathbb{P}^{\varepsilon'})$ (the structure theorem) we show that $C^\infty_{eq}(\mathbb{P}^\varepsilon, SO(E))$ is connected. To this end we observe that for all $\varphi \in C^\infty_{eq}(\mathbb{P}^\varepsilon, SO(E))$, any $h_x \in \mathbb{P}^\varepsilon$ and any $x \in M$, both equations in (184) hold true.

The orthogonal linear map $\varphi(h_x)$ has $\mathbb{R} \cdot a(h_x)$, say, as *oriented rotation axis* in E carried by $a(h_x) \in S^2$. As any element in $SO(E)$, the rotation $\varphi(h_x)$ has a uniquely determined rotation angle $\lambda(h_x)$ in $[0, 2\pi)$. Hence for any $x \in M$,

$$\varphi(h_x)(T_{\varepsilon(x)}S^2) = T_{\varepsilon'(x)}S^2.$$

Thus the value $\varphi(h_x)$ of the smooth map φ is of the form

$$\varphi(h_x)(x) = e^{\lambda(h_x)a(h_x)} \qquad \forall\, h_x \in \mathbb{P}^\varepsilon.$$

Clearly, the map

$$a : \mathbb{P}^\varepsilon \longrightarrow S^2$$

is a smooth function. Hence φ_t, given for each $h_x \in \mathbb{P}^\varepsilon$ by

$$\varphi_t(h_x) := e^{t\lambda(h_x)\cdot a(x)} \qquad \forall\, t \in [0,1] \text{ and } \forall\, x \in M, \tag{188}$$

is a smooth family of maps in $C^\infty_{eq}(\mathbb{P}^\varepsilon, SO(E))$, showing that $\varphi_t = e^{t\lambda\cdot a}$ is a homotopy with $\varphi_0 = \mathbf{id}$ and $\varphi_1 = \varphi$. Here $\mathbf{id}$ is the constant map with value id $\in SO(E)$; it is the unit element in the group $C^\infty_{eq}(\mathbb{P}^\varepsilon, SO(E))$. We hence have established:

PROPOSITION 33. *For each $\varepsilon \in C^\infty(M, S^2)$ the differentiable group $C^\infty_{eq}(\mathbb{P}^\varepsilon, SO(E))$ is path-connected and thus connected.*

The two above propositions in this section have the following implication:

PROPOSITION 34. *Let $\varepsilon \in C^\infty(M, S^2)$ be fixed. Then*

$$\tilde{\varepsilon} : \mathrm{Isom}(\mathbb{P}^\varepsilon) \longrightarrow C^\infty_{eq}(\mathbb{P}^\varepsilon, SO(E))$$

in (185) assigning to any $\Phi \in \mathrm{Isom}(\mathbb{P}^\varepsilon)$ the map φ_Φ is a bijection. $\mathrm{Isom}(\mathbb{P}^\varepsilon)$ *is a group if the group structure on $C^\infty_{eq}(\mathbb{P}^\varepsilon, SO(E))$ is pulled back by $\tilde{\varepsilon}$.*

PROOF. The range of $\widetilde{\varepsilon}$ is obviously a group. To prove the bijectivity of $\widetilde{\varepsilon}$, by lemma 30 it is enough to show surjectivity. With this goal in mind we observe the following: For a given $\varphi \in C^\infty_{eq}(\mathbb{P}^\varepsilon, SO(E))$ we set

$$\varepsilon' \circ \mathrm{pr}_M := \mathrm{pr}_{S^2} \circ \hat{\varepsilon} \oslash \varphi$$

which yields a well defined ε' due to the equivariance of φ. By proposition 33 the maps ε and ε' are homotopic. Thus there is a $U(1)$-equivariant isomorphism

$$\Phi^\varphi : \mathbb{P}^\varepsilon \longrightarrow \mathbb{P}^{\varepsilon'}$$

satisfying

$$\Phi^\varphi(k_x) := \widehat{\varepsilon'}^{-1}_{\varepsilon'(\Phi^\varphi(x))}(\hat{\varepsilon}(k_x) \oslash \varphi(k_x)) \qquad \forall\, k_x \in \mathbb{P}^\varepsilon \text{ and } \forall\, x \in M.$$

Here $(\widehat{\varepsilon'}_x)^{-1} := (\widehat{\varepsilon'}|_{\mathbb{P}^\varepsilon_x})^{-1}$ (cf. (135)) for all $x \in M$. Thus

$$\widehat{\varepsilon'} \circ \Phi^\varphi = \hat{\varepsilon} \oslash \varphi. \tag{189}$$

Clearly, $\mathrm{Gau}\,\mathbb{P}^\varepsilon$ operates on both $\mathrm{Isom}\,\mathbb{P}^\varepsilon$ and $C^\infty_{eq}(\mathbb{P}^\varepsilon, SO(E))$. The equivariance of $\tilde{\varepsilon}$ is obvious since the action of $\mathrm{Gau}\,\mathbb{P}^\varepsilon$ on $C^\infty_{eq}(M, SO(E))$ is the pointwisely formed reverse conjugation (cf. (183)). Hence the rest of the proposition is immediate. □

Therefore, we have the *isomorphism theorem*:

THEOREM 35. *For each $\varepsilon \in C^\infty(M, S^2)$ both* $\mathrm{Isom}(\mathbb{P}^\varepsilon)$ *and* $C^\infty_{eq}(\mathbb{P}^\varepsilon, SO(E))$ *are smooth differentiable groups diffeomorphic by the isomorphism $\tilde{\varepsilon}$, which is equivariant under* $\mathrm{Gau}\,\mathbb{P}^\varepsilon$.

Thus the above propositions in this section imply the following structure theorem on $\mathrm{Isom}(\mathbb{P}^\varepsilon)$:

THEOREM 36. *For each $\varepsilon \in C^\infty(M, S^2)$ the collection $[\varepsilon] \subset C^\infty(M, S^2)$ forms an open and closed submanifold of $C^\infty(M, S^2)$. Moreover,*

$$\mathrm{pr}_{S^2} : C^\infty_{eq}(\mathbb{P}^\varepsilon, SO(E)) \longrightarrow [\varepsilon]$$

is smooth, surjective and

$$\overline{\mathrm{pr}^*_{S^2}} : C^\infty_{eq}(\mathbb{P}^\varepsilon, SO(E))/C^\infty(M, U(1)) \longrightarrow [\varepsilon]$$

is a smooth bijection. Hence

$$\overline{\mathrm{pr}^*_{S^2}} \circ \tilde{\varepsilon} : \mathrm{Isom}\,\mathbb{P}^\varepsilon / \mathrm{Gau}\,\mathbb{P}^\varepsilon \longrightarrow [\varepsilon]$$

is a smooth bijection, as well.

By equations (187) and (188) we observe that $\varepsilon_t := \hat{\varepsilon} \oslash \varphi_t$ is a homotopy for which $\varepsilon_0 = \varepsilon$ and $\varepsilon_1 = \varepsilon'$ hold true. Using proposition 32 and theorem 36 we finally state:

COROLLARY 37. *The two smooth (infinite dimensional) manifolds* $\mathrm{Isom}(\mathbb{P}^\varepsilon)$ *and* $C^\infty_{eq}(\mathbb{P}^\varepsilon, SO(E))$ *are both connected. Moreover, the homotopy class $[\varepsilon] \subset C^\infty(M, S^2)$ of ε is a submanifold and*

$$\overline{\mathrm{pr}^*_{S^2}} : \mathrm{Isom}(\mathbb{P}^\varepsilon)/C^\infty(M, U(1)) \longrightarrow [\varepsilon]$$

is a smooth bijection.

The structure theorem implies that $\mathbb{P}^{\varepsilon_1}$ and $\mathbb{P}^{\varepsilon_2}$ belong to $\mathrm{Isom}(\mathbb{P}^{\varepsilon_1}, \mathbb{P}^{\varepsilon_2})$ if $\varepsilon_2 \in [\varepsilon_1]$. By (140) or by (171) we thus have:

COROLLARY 38. *For any $\varepsilon \in C^\infty(M, S^2)$ the collection $[[\varepsilon]]$ consisting of all $\varepsilon' \in C^\infty(M, S^2)$ for which $\mathbb{P}^\varepsilon$ and $\mathbb{P}^{\varepsilon'}$ are isomorphic is identical with the homotopy class $[\varepsilon]$ of ε, i.e. $[[\varepsilon]] = [\varepsilon]$. In other words, for all $\varepsilon_1, \varepsilon_2 \in C^\infty(M, S^2)$,*

$$\mathbb{P}^{\varepsilon_1} \cong \mathbb{P}^{\varepsilon_1} \textit{ iff } \quad \varepsilon_1 \simeq \varepsilon_2.$$

Therefore,

$$\mathbb{F}^{\varepsilon_1} \cong \mathbb{F}^{\varepsilon_1} \textit{ iff } \quad \varepsilon_1 \simeq \varepsilon_2.$$

4.2.3. Isomorphism Classes of Vector Bundles of Unit Vector Fields. Given a unit vector field $\varepsilon \in C^\infty(M, S^2)$ we will describe the collection $\operatorname{Isom} \mathbb{F}^\varepsilon$ of all complex line bundles $\mathbb{F}^{\varepsilon'}$ isomorphic to $\mathbb{F}^\varepsilon$ (cf. theorem 37). Here we have $\varepsilon' \in C^\infty(M, S^2)$. This description emanates from corollary 37; its technical basis is proposition 27.

More precisely, the isomorphism class $\operatorname{Isom} \mathbb{F}^\varepsilon$ shall consist of all complex vector bundle isomorphisms Φ (if there are any) satisfying

(190)
$$\begin{array}{ccc} \mathbb{F}^\varepsilon & \xrightarrow{\Phi} & \mathbb{F}^{\varepsilon'} \\ \downarrow & & \downarrow \\ M & \xrightarrow{\Phi_M} & M \end{array}$$

with $\varepsilon' \in C^\infty(M, S^2)$ where the vertical arrows are the natural projections. (Clearly, any two bundles in the same isomorphism class are isomorphic as Hermitian line bundles (cf. definition 15)). Hence $\widehat{\varepsilon'} \circ \Phi(k_x) \in T_{\Phi \circ \varepsilon(x)} S^2$ for any $k_x \in \mathbb{F}^\varepsilon_x$ and any $x \in M$. As in the case of $\mathbb{P}^\varepsilon$ we have

$$\widehat{\varepsilon'} \circ \Phi(k_x) = \varphi(k_x)^{-1}(\widehat{\varepsilon}(k_x)) \qquad \forall\, k_x \in \mathbb{F}^\varepsilon \text{ and } \forall\, x \in M$$

where $\varphi \in C^\infty_{eq}(\mathbb{P}^\varepsilon, SO(E))$, equivariant with respect to $SO(F)$ (cf. (179)). Again as in case of $\mathbb{P}^\varepsilon$ we find:

THEOREM 39. $\operatorname{Isom}(\mathbb{F}^\varepsilon) = \operatorname{Isom}(\mathbb{P}^\varepsilon)$ *and hence is a smooth differentiable group for which the natural map*

$$\operatorname{Isom}(\mathbb{F}^\varepsilon)/\operatorname{Gau} \mathbb{P}^\varepsilon \longrightarrow [\varepsilon]$$

is a smooth bijection.

4.3. Chern Classes

In this section we give a short overview over the description of a general complex line bundle $\mathbb{F}$ over $M \subset E$ in terms of Čech-cohomology.

4.3.1. General Cohomological Description of Complex Line Bundles. First we sketch the *cohomological description* of isomorphism classes of complex line bundles (cf. [**54**]). We begin by pointing out that a complex line bundle $\mathbb{F}$ over M as constructed in section 3.1.4 is uniquely determined by the *principal bundle*

$$\dot{\mathbb{F}} := \mathbb{F} \backslash \mathcal{O}(M)$$

with $\dot{\mathbb{C}} := \mathbb{C} \backslash \{0\}$ as its structure group. Here $\mathcal{O} : M \longrightarrow \mathbb{F}$ in the *zero section*. By (171) in section 3.7.3

$$\mathbb{F} = \dot{\mathbb{F}} \times_{\dot{\mathbb{C}}} \mathbb{C}.$$

Thus we may work with the principal bundle $\dot{\mathbb{F}}$ with structure group $\dot{\mathbb{C}}$.

Now we consider a *locally finite covering* $\mathcal{W} := \{U_\iota | \iota \in I\}$ of the manifold M consisting of open sets $U_\iota \subset M$ in M. This is to say that there is an index set I such that $M = \bigcup_{\iota \in I} U_\iota$ where locally finite means that each point $x \in M$ admits a neighborhood U_x, say, intersecting only finitely many members of $\mathcal{W}$ non-trivially.

Moreover, we require $\dot{\mathbb{F}}$ to *trivialize* on U_ι for each $\iota \in I$. The restriction of $\dot{\mathbb{F}}$ to U_ι is denoted by $\dot{\mathbb{F}}_\iota$.

For each pair $U_\sigma, U_\iota \in \mathcal{W}$ we set

$$U_{\sigma\iota} := U_\sigma \cap U_\iota \qquad \sigma, \iota \in I.$$

The *Čech cohomology class* of the bundle $\dot{\mathbb{F}}$ is determined as follows: For any U_σ, U_ι with $\sigma, \iota \in I$ and continuous sections

$$s_\iota : U_\iota \longrightarrow \dot{\mathbb{F}}_\iota \quad \text{and} \quad s_\sigma : U_\sigma \longrightarrow \dot{\mathbb{F}}_\sigma$$

there is a smooth function

$$\varphi_{\sigma\iota} : U_{\iota\sigma} \longrightarrow \dot{\mathbb{C}}$$

such that

$$s_\iota = s_\sigma \cdot \varphi_{\sigma\iota} \text{ on } U_{\iota\sigma}.$$

Each triple $\iota, \tau, \sigma \in I$ determines a continuous map, a *transition function*

$$\varphi_{\iota\tau} : U_{\iota,\tau} \longrightarrow \dot{\mathbb{C}},$$

satisfying the *cocycle condition*

$$\varphi_{\tau\iota} \cdot \varphi_{\iota\sigma} = \varphi_{\tau\sigma}.$$

The $\cdot$ denotes the pointwisely formed multiplication in $\mathbb{C}$. An equivalent formulation is

$$\varphi_{\tau\iota} \cdot \varphi_{\iota\sigma} \cdot \varphi_{\sigma\tau} = 1.$$

Vice versa, each system

$$\varphi := \{\varphi_{\iota\tau} | \iota, \tau \in I\} \tag{191}$$

of cocycles yields a principal bundle with $\dot{\mathbb{C}}$ as structure group, as shown, for example, in [**62**].

Let $Z^1(U_{\iota\tau})$ denote the collection of all continuous transition functions $\varphi_{\iota\tau}$ satisfying the *cocycle condition*. Two cocycles $\varphi_{\iota\tau}$ and $\varphi'_{\iota\tau}$ are called *equivalent* if for each $\iota \in I$ there is a continuous map

$$\psi_\iota : M \longrightarrow \dot{\mathbb{C}}$$

such that

$$\varphi'_{\iota\tau} = \psi_\iota^{-1} \cdot \varphi_{\iota\tau} \cdot \psi_\iota.$$

The collection of *equivalence classes of cocycles* shall be denoted by $H^1(\mathcal{W})$. It is a *commutative* group. For each element in this group, there is a complex line bundle determining the cocycle given (again cf. [**62**]).

If the covering $\mathcal{W}'$ over an index set I' is a *refinement* (cf. [**85**]) of the covering $\mathcal{W}$, then there is a *homomorphism*

$$r^{\mathcal{W}}_{\mathcal{W}'} : H^1(\mathcal{W}) \longrightarrow H^1(\mathcal{W}'),$$

called the *restriction homomorphism*, given by restricting the $\mathcal{W}$-cocycles to the respective open sets in $\mathcal{W}'$. In fact, this homomorphism is independent of the map from I' to I defining the refinement. Thus

$$\{H^1(\mathcal{W}) | \mathcal{W} \text{ locally finite trivializing open cover of } \dot{\mathbb{F}}\} \tag{192}$$

is a directed set. Its limit, called *direct limit* (cf. [**54**] and [**78**]) shall be denoted by

$$H^1(M). \tag{193}$$

Any element in $H^1(M)$ is a limit of a cocycle system. Thus we have the following results (cf. [**54**], [**63**] and [**78**]):

LEMMA 40. *The $\dot{\mathbb{C}}$-principal bundles $\dot{\mathbb{F}}$ and $\dot{\mathbb{F}}'$ over M are isomorphic iff their cocycle systems yield the same element in $H^1(M)$ and vice versa. Thus $H^1(M)$ is in a one-to-one correspondence to the isomorphism classes of $\dot{\mathbb{C}}$-principal bundles over M.*

Next we will reformulate $H^1(M)$ in terms of the *cohomology* with *coefficients* in a *presheaf* and in a *sheaf*, respectively. This step formalizes what we have done in this section and allows us to prepare a geometric description of the cohomology classes under consideration later. To do so we give a brief summary of some fundamentals in sheaf theory. (For a more detailed description we recommend one of the many texts on sheaf theory).

4.3.2. Review of the Very Elements of Sheaf Theory. In this section we will collect some details of *sheaf theory* in order to express the characteristics of a $\dot{\mathbb{C}}$-principal bundle in terms of sheaf cohomology of M. To this end we consider a *sheaf* (cf. for example [**54**] or [**78**]) $\mathcal{S}$ of *abelian groups*, a triple

$$\mathcal{S} = (S, \mathrm{pr}, M)$$

with the following three properties:

- **i):** S is a *topological space* and $\mathrm{pr} : S \longrightarrow M$ a *continuous surjective* map.
- **ii):** Each point $q \in S$ admits a neighborhood $N \subset S$ for which

  $$\mathrm{pr}|_N : N \longrightarrow \mathrm{pr}(N) \subset M$$

  is a *homeomorphism* and $\mathrm{pr}|_N(N)$ is a neighborhood of $\mathrm{pr}(q) \in M$.
- **iii):** $\mathrm{pr}^{-1}(x)$, the *stalk* over x, is an *abelian group* for any $x \in M$. The stalkwisely formed group operations depend continuously on the group elements in S.

Thus $\mathrm{pr}^{-1}(x) \subset S$ is a *discrete topological space*.

For each open set $U \subset M$ and the given sheaf $\mathcal{S}$ the collection of continuous sections from U into S is denoted by $\Gamma(U, \mathcal{S})$.

Sheaves can be constructed out of presheaves: A *presheaf* $\mathcal{G}$ of *abelian groups* over M associated with an open locally finite covering

$$\mathcal{U} := \{U_\iota | \iota \in I\}$$

of M consists of the following ingredients:

- **i):** To each U_ι there exists an abelian group S_{U_ι}, which is $\{0\}$ if $U_\iota = \emptyset$.
- **ii):** If $U_\iota \subset U_\tau$ for some pair $\iota, \tau \in I$ of indices there is a group homomorphism

  $$r^{U_\iota}_{U_\tau} : S_{U_\iota} \longrightarrow S_{U_\tau},$$

  called *restriction homomorphism*, satisfying the conditions

$$r^{U_\iota}_{U_\iota} = \mathrm{id} \qquad \forall\, \iota \in I \qquad \text{and} \qquad r^{U_\iota}_{U_\sigma} = r^{U_\iota}_{U_\tau} \circ r^{U_\tau}_{U_\sigma} \qquad \forall\, \iota, \tau, \sigma \in I.$$

This is to say that the presheaf $\mathcal{G}$ determines a *directed system* (cf. [**54**], [**63**] and [**78**]).

Therefore, each presheaf $\mathcal{G}$ defines a *sheaf* $\mathcal{S}$ by setting

$$\mathcal{S}_x := \varinjlim_{x \in U} S_U \tag{194}$$

for each fixed $x \in M$ formed with respect to the directed system determined by the presheaf $\mathcal{G}$. Here $\varinjlim$ denotes the *direct limit* (cf. [**54**], [**77**]). For each $f \in S_U$ and each $x \in M$ the direct limit yields an element $f_x \in \mathcal{S}_x$, called the *germ* of f at x.

The topology on the disjoint union $\mathcal{S} := \dot{\bigcup}_{x\in M} \mathcal{S}_x$ is prescribed by the following open basis: For each open set $U \subset M$ let

$$f_U := \{f_x | x \in U\};$$

then $\{f_U | U \in \mathcal{U}\}$ is a basis of open sets of $\mathcal{S}$. It is easily shown that $\mathcal{S}$ is indeed a sheaf.

For each covering $\mathcal{W}$ and any specified sheaf $\mathcal{S}$ the collection

$$\{\Gamma(U, \mathcal{S}) | U \in \mathcal{W}\}$$

forms a presheaf, the *canonical presheaf* $\mathcal{G}_\mathcal{S}$ of $\mathcal{S}$. Here again $\Gamma(U, \mathcal{S})$ is the collection of all *continuous* sections of U into $\mathcal{S}$.

A *homomorphism* h mapping a presheaf $\mathcal{G}$ into another one $\mathcal{G}'$ is a system

$$h_U : \mathcal{G}_U \longrightarrow \mathcal{G}'_{U'}$$

of *group homomorphisms* for which

$$r^{U'}_{V'} \circ h_U = h_V \circ r^U_V \qquad U \subset U' \text{ and } V \subset V'$$

where U, V vary in the covering belonging to $\mathcal{G}$, while U', V' vary in the covering belonging to $\mathcal{G}'$. This directed system of homomorphisms determines for each $x \in M$ a group homomorphism in the direct limit (194) (cf. [**54**], [**63**] or [**78**]). Thus a *homomorphism of presheaves* determines a *sheaf homomorphism*.

4.3.3. Čech Cohomology. In this subsection we will construct a cohomology group to a presheaf derived from a complex line bundle. Later we will relate this cohomology group to $H^1(M)$ defined in (193).

Consider a complex line bundle $\mathbb{F}$ associated with the principal bundle $\dot{\mathbb{F}}$ with structure group $\dot{\mathbb{C}}$, and a locally finite covering $\mathcal{U} = \{U_\iota | \iota \in I\}$ on which $\mathbb{F}$ trivializes. Let $C(U_\iota, \dot{\mathbb{C}})$ be the collection of all *continuous* functions of U_ι into $\dot{\mathbb{C}}$. The collection

$$\mathcal{G} := \{C(U_\iota, \dot{\mathbb{C}}) | \iota \in I\}$$

forms a presheaf. Moreover, there is a unique sheaf $\dot{\mathbf{C}}_c$, the *sheaf* of *germs of continuous* $\dot{\mathbb{C}}$*-valued functions* (cf. [**77**] and [**54**]) for which

$$\Gamma(U_\iota, \dot{\mathbf{C}}_c) = C(U_\iota, \dot{\mathbb{C}}) \qquad \forall \iota \in I.$$

Given a sheaf $\mathcal{S}$ of germs of G-valued continuous functions over M we will construct $H^q(M, \mathcal{S})$, the q-th cohomology group of M with coefficients in a sheaf $\mathcal{S}$. Here G is a topological group. This group is defined as follows: Let $\mathcal{G}_\mathcal{S}$ be the *canonical presheaf* of $\mathcal{S}$ belonging to a covering $\mathcal{W}$. A *q-cochain* f associates to any $(q+1)$-tuple $\iota_0, \ldots, \iota_q$ of indices in I an element

$$f(\iota_0, \ldots, \iota_q) \in S_{U_{\iota_0} \cap \ldots \cap U_{\iota_q}} = C(U_{\iota_0} \cap \ldots \cap U_{\iota_q}, G).$$

The collection of all *q-cochains* is denoted by $C^q(\mathcal{W}, G)$. The *coboundary* $\delta^q f$ is then defined by

$$\delta^q f(\iota_0, \ldots, \iota_{q+1}) := \sum_{k=0}^{q+1} r_W^{W_k} f(\iota_0, \ldots, \hat{\iota}_k, \ldots, \iota_{q+1}) \qquad \forall\, f \in C^q(\mathcal{W}, G);$$

here $W = U_{\iota_0} \cap \ldots \cap U_{\iota_{q+1}}$ and $W_k = U_{\iota_0} \cap \ldots \cap \hat{U}_{\iota_k} \ldots \cap U_{q+1}$ where ^ over ι_k and U_{ι_k} means to omit the ι_k-th element, respectively, the ι_k-th set. We set

$$H^q(\mathcal{U}, G) := \ker \delta^q / \operatorname{im} \delta^{q-1} \qquad \forall\, q > 0$$

and let

$$H^q(M, \mathcal{S}) := \varinjlim_{\mathcal{U}} H^q(\mathcal{U}, G)$$

where $\mathcal{U}$ varies in the directed system of locally finite coverings $\mathcal{U}$ of M. The group $H^q(M, \mathcal{S})$ is called the q-th *Čech-cohomology group* of M with coefficients in $\mathcal{S}$.

For the *canonical presheaf* $\mathcal{G}_{\mathcal{S}}$ of a sheaf $\mathcal{S}$ define

$$H^q(M, \mathcal{G}_{\mathcal{S}}) := H^q(M, \mathcal{S}) \qquad \forall\, q \geq 0$$

(cf. [**54**], [**77**]).

In particular, for the *sheaves* $\dot{\mathbf{C}}_{\mathbf{c}}$ and $\mathbf{C}_c$ of *germs* of *continuous* $\dot{\mathbb{C}}$-valued and $\mathbb{C}$-valued functions, respectively, we have

$$(195) \quad H^q(M, \mathcal{G}_{\dot{\mathbf{C}}_c}) = H^q(M, \dot{\mathbf{C}}_c) \quad \text{and} \quad H^q(M, \mathcal{G}_{\mathbf{C}_c}) = H^q(M, \mathbf{C}_c),$$

respectively, for any $q > 0$.

Thus we have a natural isomorphism

$$(196) \qquad H^1(M) \cong H^1(M, \dot{\mathbf{C}}_{\mathbf{c}})$$

(cf. lemma 40). This is to say that the first cohomology group

$$H^1(M, \dot{\mathbf{C}}_{\mathbf{c}})$$

classifies the collection of all *isomorphism classes* of complex line bundles over M. This is a basic statement for the following investigation. The goal is to replace $\dot{\mathbf{C}}_{\mathbf{c}}$ in (195) by $\mathbf{Z}$, the sheaf of germs of *constant functions* with values in $\mathbb{Z}$ (cf. [**54**]). To this end we need a few more technical instruments.

4.3.4. The First Chern Class of a Complex Line Bundle. Consider the *exact sequence* of sheaves

$$(197) \qquad 0 \longrightarrow \mathbf{Z} \xrightarrow{\iota_c} \mathbf{C}_c \xrightarrow{\exp} \dot{\mathbf{C}}_c \longrightarrow 0.$$

determined by the exact sequence of groups

$$0 \longrightarrow \mathbb{Z} \xrightarrow{\iota} \mathbb{C} \xrightarrow{\exp} \dot{\mathbb{C}} \longrightarrow 0$$

where ι is the inclusion. For each locally finite covering $\mathcal{W}$ this sequence yields the commuting diagram

$$\begin{array}{ccccccc}
0 \longrightarrow & C^1(\mathcal{W}, \mathbf{Z}) & \xrightarrow{\iota_*} & C^1(\mathcal{W}, \mathbf{C}_c) & \xrightarrow{\exp_*} & C^1(\mathcal{W}, \dot{\mathbf{C}}_c) & \longrightarrow \\
& \Big\downarrow{\scriptstyle \delta^1} & & \Big\downarrow{\scriptstyle \delta^1} & & \Big\downarrow{\scriptstyle \delta^1} & \\
0 \longrightarrow & C^2(\mathcal{W}, \mathbf{Z}) & \xrightarrow{\iota_*} & C^2(\mathcal{W}, \mathbf{C}_c) & \xrightarrow{\exp_*} & C^2(\mathcal{W}, \dot{\mathbf{C}}_c) & \longrightarrow
\end{array}$$

with *exact rows*. The maps in the rows are defined by composing the cochains with either ι_* or $\exp_*$ as specified in the above diagram. Hence we obtain an exact sequence of cohomology groups

$$(198)\quad \to H^1(M, \mathbf{C}_c) \longrightarrow H^1(M, \dot{\mathbf{C}}_c) \xrightarrow{\delta_c} H^2(M, \mathbf{Z}) \xrightarrow{(\iota_c)_*} H^2(M, \mathbf{C}_c) \to .$$

Here $(\iota_c)_*$ is the homomorphism determined by the inclusion ι_c, and δ_c is the *connecting homomorphism* defined as follows: We take any element $[\varphi] \in H^1(\mathcal{W}, \dot{\mathbf{C}}_c)$ and observe that there is some $\psi \in C^1(\mathcal{W}, \mathbf{C}_c)$ satisfying

$$\varphi = \exp_*(\psi),$$

due to the exactness of the rows. Since

$$\exp_*(\delta^1 \psi) = \delta^1 \exp_*(\psi) = \delta^1 \varphi = 0,$$

there is some element $\delta^1 \zeta$ with

$$\iota_*(\delta^1 \zeta) = \delta^1 \psi.$$

Due to $\delta^2 \circ \delta^1 = 0$ the cochain satisfies $\delta_b^2(\delta_b^1 \zeta) = 0$. This suggests to set

$$(199)\qquad \delta_c[\psi] := \delta^1 \zeta.$$

The definition of δ_c is independent of the particular elements representing cohomology classes.

Since the sheaf $\mathbf{C}_c$ is *fine*, i.e. all cohomology groups with coefficient in $\mathbf{C}_c$ vanish, we have the *isomorphism*

$$(200)\qquad H^1(M, \dot{\mathbf{C}}_c) \xrightarrow{\delta_c} H^2(M, \mathbf{Z})$$

due to (198) (cf. [**54**]). Instead of taking $\dot{\mathbb{C}}$ as structure group we can replace it by $U(1)$. The reason is that the quotient map

$$\dot{\mathbb{C}} \longrightarrow \dot{\mathbb{C}}/U(1) = \mathbb{R}$$

admits a *global section*, implying that the $\dot{\mathbb{C}}$-principal bundle can be reduced to $U(1)$-principal bundle (cf. [**46**]). This is to say that the natural inclusion of

$$\mathbf{U}(1)_c \xrightarrow{\iota} \dot{\mathbf{C}}$$

yields an *isomorphism*

$$H^q(M, \mathbf{U}(1)_c) \xrightarrow{\iota_*} H^q(M, \dot{\mathbf{C}}) \qquad \forall\, q.$$

Here $\mathbf{U}(1)_c$ denotes the sheaf of germs of all continuous $U(1)$-valued functions of M. The résumé of this little investigation is hence:

PROPOSITION 41. *The complex line bundles on M as well as the $\dot{\mathbb{C}}$-, respectively, $U(1)$-principal bundles are classified by $H^1(M, \dot{\mathbf{C}}_c)$ and $H^1(M, \mathbf{U}(1)_c)$. These two groups are naturally isomorphic.*

By (200) for the exact sequence

$$(201)\qquad 0 \longrightarrow \mathbf{Z} \xrightarrow{\iota} \mathbf{R}_c \xrightarrow{\exp} \mathbf{U}(1)_c \longrightarrow 0$$

of sheaves the connecting homomorphism

$$(202)\qquad H^1(M, \mathbf{U}(1)_c) \xrightarrow{\delta_c} H^2(M, \mathbf{Z})$$

is an *isomorphism*.

Since $H^1(M, \mathbf{U}(1)_c)$ classifies all $U(1)$-principal bundles, (202) thus shows that these bundles are classified by $H^2(M, \mathbf{Z})$, as well. This is to say that

$$H^1(M, \mathbf{U}(\dot{\mathbb{C}})_c) \xrightarrow{\delta_c \circ \iota} H^2(M, \mathbf{Z}) \tag{203}$$

is an isomorphism, as well.

The cohomology class in $H^1(M, \dot{\mathbf{C}})$ of a complex line bundle $\mathbb{F}$ (cf. (196)) can thus be viewed as a class in $H^2(M, \mathbf{Z})$. This class is called the *first Chern class* of $\mathbb{F}$ and is denoted here by

$$\check{c}_1(\mathbb{F});$$

it is *characteristic* for $\mathbb{F}$ and hence called a *characteristic class* for $\mathbb{F}$.

In fact, if $H^2(M, \mathbb{Z})$ denotes the *singular cohomology* group (cf. [**43**] or [**78**]) with coefficients in $\mathbb{Z}$, then

$$H^m(M, \mathbf{Z}) \cong H^m(M, \mathbb{Z}) \tag{204}$$

for all $m > 0$ as shown, for example, in [**35**], page 250. We will thus identify $H^2(M, \mathbf{Z})$ with $H^2(M, \mathbb{Z})$ and we may state:

PROPOSITION 42. *The collection of all characteristic classes of all complex line bundles is parameterized by $H^2(M, \mathbb{Z})$ and vice versa.*

4.3.5. The Naturality of Chern Classes. So far we have studied the cohomological characteristics of general complex line bundles. The specific quality of a complex line bundle of type $\mathbb{F}^{\mathbf{a}}$ emanating from a singularity free vector field $X = (\mathrm{id}, \mathbf{a})$ enters via the directional field ε. Equation (140) implies

$$\varepsilon^* \check{c}_1(TS^2) = \check{c}_1(\varepsilon^* TS^2) = \check{c}_1(\mathbb{F}^{\mathbf{a}}), \tag{205}$$

the *naturality of the (first) Chern classes* (cf. [**53**]).

Remark. *If $X' = (id, \mathbf{a}')$ is a vector field on an open set $M' \subset E$ with finitely many singularities forming a closed subset Q, then $X := X'|_M$ is singularity free on $M := M' \backslash Q$ (assumed to be a manifold). Let $\mathbf{a} := \mathbf{a}'|_M$. Thus the Chern class of $\mathbb{F}^{\mathbf{a}}$ on M encodes the effect of the singularities of X' on the topology of M.*

We will come back to this in section 4.4 in connection with the mapping degree.

The *classifying space* of a complex line bundle over any manifold M is the complex projective space $\mathbb{CP}^\infty$ of a separable complex Hilbert space with S^∞ as its *unit sphere.* Hence the universal $U(1)$-bundle for the collection of all complex line bundles on M is $S^\infty \longrightarrow \mathbb{CP}^\infty$, i.e. all complex bundles over M are pull-backs of this infinite dimensional $U(1)$-bundle (cf. [**39**]). Here is an analogy to the universal complex line bundle. Due to the naturality of Chern classes (205) a sort of "universal space" for all complex line bundles of singularity free vector fields on M is the two-sphere S^2 in E. The universal bundle is mimicked by the frame bundle $\mathbb{P}_2$ of the tangent bundle TS^2 (cf. theorem 12 and section 2.3.5). Clearly, S^2 is the complex projective space of the quaternions $\mathbb{H}$. The Chern class of any bundle of the form $\mathbb{F}^{\mathbf{a}}$ is hence a characteristic for $\varepsilon = \frac{\mathbf{a}}{|\mathbf{a}|}$.

The first Chern class of $\mathbb{F}^\varepsilon$ is encoded in $\mathrm{Isom}\, \mathbb{P}^\varepsilon / C^\infty(M, U(1))$ via the homotopy class $[\varepsilon]$ (cf. theorem 36). Let $[M, S^2]$ denote the collection of all homotopy classes of the smooth maps from M to S^2. To get a more direct link we may define

$$\check{c}_1 : C^\infty(M, S^2) \longrightarrow H^2(M, \mathbb{Z})$$

by assigning to each $\varepsilon \in C^\infty(M, S^2)$ the first Chern class $\check{c}_1(\varepsilon^* TS^2)$. By the isomorphism theorem 36 and by corollary 38 we have:

PROPOSITION 43. *The mapping*

$$\check{c}_1 : [M, S^2] \longrightarrow H^2(M, \mathbb{Z})$$

is injective and constant on homotopy classes.

4.3.6. The de Rham Isomorphism. In this subsection we will begin to reveal the connection of the Chern class of $\mathbb{F}^{\mathbf{a}}$ with the cohomology class of $\Omega^{\mathbf{a}}$, the curvature of $\mathbb{P}^{\mathbf{a}}$ (cf. section 3.6.5). For the manifold M with or without boundary the theorem of de Rham states

$$ {}_{\mathrm{deR}}H^q(M, \mathbb{R}) \cong H^q(M, \mathbf{R}) \tag{206}$$

where $\mathbf{R}$ denotes the sheaf of *germs* of *constant* $\mathbb{R}$-valued maps of M (cf. [**54**]). Moreover,

$$ {}_{\mathrm{deR}}H^q(M, \mathbb{R}) \cong H^q(M, \mathbf{R}) \cong H^q(M, \mathbb{R}) \tag{207}$$

where $H^q(M, \mathbb{R})$ denotes the *singular cohomology* with coefficients in $\mathbb{R}$ (cf. [**77**]).

If M is a manifold with boundary in E, it admits an open tubular neighborhood in E of any of its boundary components. This enlarged manifold in E shall be denoted by M'. Clearly, M is a deformation retract of M' (cf. [**77**], [**54**] and [**61**]) and $M' \subset E$ is open. Thus

$$ {}_{\mathrm{deR}}H^q(M', \mathbb{R}) \cong {}_{\mathrm{deR}}H^q(M, \mathbb{R})$$

holds true.

In analogy to the sheaf $\mathbf{Z}$ let us denote by $\mathbf{U}(1)$ the sheaf of *germs of constant* $U(1)$-valued maps of M. To relate $H^2(M, \mathbf{Z})$ with $H^2(M, \mathbf{R})$ we consider the exact sequence

$$0 \stackrel{\iota}{\longrightarrow} \mathbf{Z} \subset \mathbf{R} \stackrel{\exp}{\longrightarrow} \mathbf{U}(1) \longrightarrow 1$$

which causes a long exact sequence

$$\cdots \longrightarrow H^1(M, \mathbf{R}) \stackrel{\exp_*}{\longrightarrow} H^1(M, \mathbf{U}(1)) \stackrel{\delta_c}{\longrightarrow} H^2(M, \mathbf{Z}) \stackrel{\iota_*}{\longrightarrow} H^2(M, \mathbf{R}) \longrightarrow \cdots$$

of Čech cohomology groups. Here δ_c is the connecting homomorphism and ι_* is the homomorphism determined by the inclusion $\mathbf{Z} \subset \mathbf{R}$ of constant sheaves. We thus derive the following commuting diagram:

$$\begin{array}{ccccccc}
 & & & H^1(M, \mathbf{U}(1)) & & & \\
 & & & \downarrow{\scriptstyle \delta'_c} & & & \\
\longrightarrow & H^1(M, \mathbf{U}(1)_c) & \stackrel{\delta_c}{\longrightarrow} & H^2(M, \mathbf{Z}) & \stackrel{(\iota_c)_*}{\longrightarrow} & H^2(M, \mathbf{R}_c) & \longrightarrow \\
 & & {\scriptstyle \iota_* \circ \delta_c}\searrow & \downarrow{\scriptstyle \iota_*} & & & \\
 & & & {}_{\mathrm{deR}}H^2(M, \mathbb{R}) & & &
\end{array} \tag{208}$$

with an exact column and an exact row. In fact,

$$(\iota_c)_* = 0.$$

In the vertical sequence δ'_c denotes the connecting homomorphism.

By (204) we may replace $H^2(M, \mathbf{Z})$ by the singular cohomology group $H^2(M, \mathbb{Z})$.

An element in

$$\text{(209)} \qquad \operatorname{im} \iota_* \subset {}_{\mathrm{deR}}H^2(M, \mathbb{R})$$

is called an *integral de Rham class.*

In general, ι_* is not an isomorphism as we will see at the end of the next section.

Since the vertical sequence in (208) is *exact*, any $[\zeta] \in H^2(M, \mathbf{Z})$ with $\iota^*([\zeta]) = 0$ is of the form

$$[\zeta] = \delta'_c([\xi])$$

for $[\xi] \in H^1(M, \mathbf{U}(1))$.

As we have stated in proposition 20 in chapter 3, any complex line bundle of the form $\mathbb{F}^{\mathbf{a}}$ admits a closed curvature two-form $\dot{\Omega}^{\mathbf{a}}$. The cohomology class

$$[\dot{\Omega}^{\mathbf{a}}] \in {}_{\mathrm{deR}}H^2(M, \mathbb{R})$$

shall here be called the *Chern–de Rham class* of $\mathbb{F}^{\mathbf{a}}$; it shall be denoted by $c_1(\mathbb{F}^{\mathbf{a}})$.

We will show next that $[\dot{\Omega}^{\mathbf{a}}]$ is *integral* (as defined in (209)) for *any* smooth singularity free vector field $X = (\mathrm{id}\,, \mathbf{a})$. By proposition 20 we may be restricted to $\mathbf{a} = \varepsilon$.

By [**46**] the map

$$\hat{\varepsilon} : \varepsilon^*(TS^2) = \mathbb{F}^{\varepsilon} \longrightarrow TS^2,$$

which is fibrewisely the identity (cf. (135)), yields

$$\varepsilon^*[\omega^{S^2}] = [\dot{\Omega}^{\varepsilon}] = c_1(\mathbb{F}^{\varepsilon})$$

(cf. section 3.6.5). Here we used (167) and (169). This equation is valid in $H^2(M, \mathbb{R})$. On the other hand, by the naturality of the Chern classes (cf. (204)) this equation holds in $H^2(M, \mathbb{Z})$. Replacing M in (208) by S^2 and ι as well as δ_c by their analogies for the two-sphere ι^{S^2} and $\delta_c^{S^2}$, respectively, we conclude

$$\iota_*^{S^2} \circ \delta_c^{S^2}(\check{c}_1(TS^2)) = [\omega^{S^2}]$$

where $\iota_*^{S^2}$ is injective and, moreover,

$$\varepsilon^* \circ \iota_*^{S^2} \circ \delta_c^{S^2} = \iota_* \circ \delta_c \circ \varepsilon^*$$

showing

$$\varepsilon^*[\omega^{S^2}] = \iota_* \circ \delta_c \circ [\varepsilon^*(\check{c}_1(TS^2))].$$

Therefore, we may state:

PROPOSITION 44. *For any* $\varepsilon \in C^\infty(M, S^2)$,

$$c_1(\mathbb{F}^{\varepsilon}) = [\dot{\Omega}^{\varepsilon}] = \varepsilon^*[\omega^{S^2}] = \iota_* \circ \delta_c \circ \varepsilon^*(\check{c}_1[TS^2])$$

holds true. Thus $\dot{\Omega}^{\varepsilon}$ *is integral. If* $[\Omega^{\varepsilon}] = 0$, *then* $\mathbb{F}^{\varepsilon}$ *is trivial.*

4.3.7. A Two-Chart Trivialization and a Natural Čech Cocycle of $\mathbb{F}^{\mathbf{a}}$. The goal of this subsection is to demonstrate by an explicit construction of a Čech cocycle that the de Rham class $[\dot{\Omega}^{\mathbf{a}}] \in {}_{\mathrm{deR}}H^2(M, \mathbb{R})$, of the curvature $\dot{\Omega}^{\mathbf{a}}$ of the singularity free vector field $(\mathrm{id}, \mathbf{a})$, is the image of the Chern class of $\mathbb{F}^{\mathbf{a}}$ under $\iota_* \circ \delta_c$ and does not depend on the field strength (cf. (167) and (169)). Therefore, we make the following preparations: For any open set $V \subset S^2$ we have

$$(\varepsilon\;|_{\varepsilon^{-1}(V)})^*(TS^2|_V) = \mathbb{F}^{\mathbf{a}}|_{\varepsilon^{-1}(V)}$$

cf. (140)).

Thus the curvature forms $\dot{\Omega}^{\mathbf{a}}$ and Ω_{S^2} of $\mathbb{F}^{\mathbf{a}}|_{-1}(V)$ and of S^2, respectively, are related by

$$\varepsilon^*(\Omega_{S^2})|_{\varepsilon^{-1}(V)} = \dot{\Omega}^{\mathbf{a}}|_{\varepsilon^{-1}(V)},$$

which implies

$$(\varepsilon|_{\varepsilon^{-1}(V)})^*[\Omega_{S^2}] = [\dot{\Omega}^{\mathbf{a}}|_{\varepsilon^{-1}(V)}].$$

If $TS^2|_V$ is trivial, then $[\Omega^{S^2}|_V] = 0$ and hence

$$c_1(\mathbb{F}^{\mathbf{a}}|_{\varepsilon^{-1}(V)}) = 0.$$

In particular, if $y \in S^2$ and $V := S^2 \backslash \{y\}$ (being contractible), then $[\Omega_{S^2}|_V] = 0$ and we find that

$$[\dot{\Omega}^{\mathbf{a}}|_{\varepsilon^{-1}(V)}] = 0.$$

To link $c_1(\mathbb{F}^{\mathbf{a}})$ with the Čech cocycle of $\mathbb{F}^{\mathbf{a}}$ associated with an open trivializing cover of M, we construct the following *two-chart trivialization* of $\mathbb{F}^{\mathbf{a}}$ which is initiated by a cover of the two-sphere $S^2 \subset E$:

Due to (140) such a cover can be constructed based on (205). Let N, S be two antipodal points on S^2 and $F \subset E$ a two-dimensional subspace intersecting perpendicularly the line passing through N and S. Now F shall be equipped with an oriented, orthonormal coordinate system and hence F is naturally turned into a complex plane. As linear spaces $F = T_S S^2 = T_N S^2$.

Two open charts of S^2 containing S and N, respectively, are the contractible sets $U_S := S^2 \backslash \{N\}$ and $U_N := S^2 \backslash \{S\}$. The chart maps onto $T_S S^2$ and $T_N S^2$, respectively, are the *stereographic projections* from N and S onto F. These projections are linked to the oriented equator S^1 contained in the open set $U_{NS} := U_N \cap U_S$ by a smooth $U(1)$-valued map. The transition function φ_{NS} is a one-cocycle with cohomology class in the first Čech cohomology group $H^1(S^2, \mathbf{U}_c(1))$ (cf. proposition 41). This cocycle will be pulled back to M by ε as follows:

Setting

$$U_\nu := \varepsilon^{-1}(U_N) \text{ and } U_\sigma := \varepsilon^{-1}(U_S)$$

we obtain an open cover of M on which $\mathbb{F}^\varepsilon$ trivializes.

The restrictions $\mathbb{F}^{\mathbf{a}}|_{U_\nu}$ and $\mathbb{F}^{\mathbf{a}}|_{U_\sigma}$ of the bundle $\mathbb{F}^{\mathbf{a}}$ to U_ν and U_σ, respectively, are trivial. This is a consequence of (140) and of the fact that

$$TS^2|_{U_N} \text{ and } TS^2|_{U_S};$$

both are trivial. Moreover, the restrictions $\dot{\Omega}^{\mathbf{a}}|_{U_\nu}$ and $\dot{\Omega}^{\mathbf{a}}|_{U_\tau}$ of the curvature form $\dot{\Omega}^{\mathbf{a}}$ to U_ν and U_σ are pull-backs by ε^{-1} of the respective restrictions $\omega^{S^2}|_{U_N}$ and $\omega^{S^2}|_{U_S}$ of the curvature ω^{S^2} to U_N and U_S in S^2. We conclude

$$(\varepsilon|_{\varepsilon^{-1}(U_N)})^*[\omega^{S^2}] = [\dot{\Omega}^{\mathbf{a}}|_{\varepsilon^{-1}(U_N)}] = 0 \tag{210}$$

and

$$(\varepsilon|_{\varepsilon^{-1}(U_S)})^*[\omega^{S^2}] = [\dot{\Omega}^{\mathbf{a}}|_{\varepsilon^{-1}(U_S)}] = 0 \tag{211}$$

due to the fact that both U_N and U_S are contractible. The according Čech-cocycle $\varphi_{\nu\sigma}$ on $U_{\nu\sigma} := U_\nu \cap U_\sigma$ evaluated at $x \in U_{\nu\sigma}$ may be expressed as

$$\varphi_{\nu\sigma}(x) = e^{-2g_{\nu\sigma}(x)\cdot\varepsilon(x)} \in U^{\varepsilon(x)}(1) \qquad \forall\, x \in U_{\nu\sigma}. \tag{212}$$

Here any complex number $\lambda + \zeta \cdot i \in \mathbb{C}$ is replaced by $\lambda \cdot e + \zeta \cdot \varepsilon(x) \in \mathbb{C}^{\varepsilon(x)}$ for all $x \in U_{\nu\sigma}$ and $g_{\nu\sigma}$ is a smooth real-valued function defined on $U_{\nu\sigma}$. Thus the isomorphism class of $\mathbb{F}^{\mathbf{a}}$ is encoded in the exponent of (212) (cf. [**63**]).

The cover $\{U_\nu, U_\sigma\}$ on M is the *two-chart trivialization* of $\mathbb{F}^{\mathbf{a}}$ on M.

Note the following: The $U(1)$-principal bundle $\mathbb{P}^\varepsilon$ is turned into an $SO(2)$-principal bundle by replacing the multiplication (in $\mathbb{H}$) on the right-hand side of

$$h \cdot \varphi_{\nu\sigma}(x) = h \cdot e^{-2g_{\nu\sigma}(x)\cdot\varepsilon(x)} \qquad \forall\, h \in \mathbb{F}^{\varepsilon(x)} \text{ and } \forall\, x \in U_{\nu\sigma}$$

by

$$\tau_{e^{g_{\nu\sigma}(x)\cdot\varepsilon(x)}}(h)$$

due to $\mathbb{C}^{\varepsilon(x)} \subset \mathbb{H}$ (cf. (26)). Clearly, $\tau : SU(2) \longrightarrow SO(E)$ is the two-to-one map expressing the absorption of the factor 2 in the exponent of (212).

Using the two-chart trivialization of $\mathbb{F}^{\mathbf{a}}$ we will establish an explicitly formulated link between its Chern class and its Chern-de Rham class. Moreover, we will determine conditions on M and X such that $\iota_x \circ d_c$ in diagram (208) is injective if restricted to the collection of Chern classes emanating from singularity free vector fields on M.

To do so, we will at first study the according situation on S^2. The curvature of $S^2 \subset E$ is ω^{S^2}; its cohomology class $[\omega^{S^2}]$ is the first Chern-class of TS^2 (the map ι_* in (208) is injective for $M = S^2$). We will show that ω^{S^2} is *exact* on U_S and accordingly on U_N. Indeed, if X_1^S and X_2^S are orthogonal unit vector fields on U_S, then we have the one–form θ_S which assumes the value

$$\theta_S(x; v) :=< \nabla_v X_1^S(x), X_2^S(x) >$$

or any $x \in U_S$ and any $v \in T_x U_S$. Here ∇ is the covariant derivative of Levi-Civitá on S^2. The exterior derivative $d\theta_S$ of θ_S is hence

$$d\theta_S(x; V_0(x), V_1(x)) =< R(V_0(x), V_1(x))X_1^S(x), X_2^S(x) >$$

where V_0 and V_1 are any two vector fields on U_S and R denotes the curvature tensor of ∇. Since the left-hand side is of the form

$$< R(V_0(x), V_1(x))X_1^S(x), X_2^S(x) >= f(x) \cdot \omega^{S^2}(X_1(x), X_2(x))$$

for some real-valued smooth function f on U_S we conclude

$$< R(X_1^S(x), X_2^S(x))X_1^S(x), X_2^S(x) >= \kappa^{S^2}(x) \cdot \omega^{S^2}(X_1^S(x), X_2^S(x)).$$

Here $\kappa^{S^2}(x)$ is the Gaussian curvature which is identical to the one on U_S. Thus

$$\omega^{S^2} = d\theta_S$$

on U_S; in other words, ω^{S^2} is exact on U_S and accordingly on U_N.

To construct the unit vector fields X_1^S and X_2^S, say, on U_S we choose two constant vector fields $\hat{X}_1^S$ and $\hat{X}_2^S$ on $T_N S^2$ with orthogonal values. The stereographic projection

$$\mathrm{pr}_S : U_S \longrightarrow T_N S^2$$

preserves circles and angles. Hence

$$(T_N \mathrm{pr}_S)^{-1}(\hat{X}_1^S) \quad \text{and} \quad (T_N \mathrm{pr}_S)^{-1}(\hat{X}_2^S)$$

are nowhere vanishing and orthogonal vector fields on U_S. Normalizing $\hat{X}_1^S$ and $\hat{X}_2^S$ yields the two desired orthonormal fields X_1^S and X_2^S. Accordingly, we construct X_1^N and X_2^N, two orthonormal unit vector fields on U_N.

On U_{NS} we thus have

$$d\theta_S - d\theta_N \neq 0.$$

This is to say that $\theta_S - \theta_N$ is *not* exact since U_{NS} contains the equator $S^1 \subset S^2$ and ${}_{\mathrm{deR}}H^2(S^2, \mathbb{R}) \cong \mathbb{R}$.

The next step is to refine the covering $\{U_N, U_S\}$. There are finitely many open and geodesically convex sets in S^2 bounded by circles which are either in U_S or in U_N. Let $\mathcal{W} := \{U_\alpha | \alpha \in I\}$ be this covering with finite index set I. This covering is *simple* (cf. [**46**]), which means that each member in $\mathcal{W}$ and all non-void intersections of members in $\mathcal{W}$ are *contractible*. Hence all closed forms on U_α are exact for all $\alpha \in I$. On each $U_\alpha \in \mathcal{W}$, the cohomology class $[\omega_\alpha^{S^2}]$ of $\omega_\alpha^{S^2} := \omega^{S^2}|_{U_\alpha}$ vanishes, saying that

$$\omega_\alpha^{S^2} = d\theta_\alpha \qquad \forall\, \alpha \in I$$

for some one-form θ_α. Since $U_{\alpha\beta} := U_\alpha \cap U_\beta$ is contractible, as well.

$$\theta_\alpha - \theta_\beta = dg_{\alpha\beta}$$

for some $\mathbb{R}$-valued smooth function $g_{\alpha\beta}$ on $U_{\alpha\beta}$. On $U_\alpha \cap U_\beta \cap U_\sigma$ for $\alpha, \beta, \sigma \in I$ we thus find for the Fréchet derivative D on $U_{\alpha\beta}$,

$$D(g_{\alpha\beta} + g_{\beta\sigma} + g_{\sigma\alpha}) = 0$$

implying that the $\mathbb{R}$-valued function

$$Z_{\alpha\beta\sigma} := g_{\alpha\beta} + g_{\beta\sigma} + g_{\sigma\alpha} \tag{213}$$

is constant. For $S^2 \subset E$ the cohomology groups are

$$H^q(S^2, \mathbb{Z}) = \begin{cases} \mathbb{Z} & q = 2, \\ 0 & \text{otherwise} \end{cases}$$

and

$$H^q(S^2, \mathbb{R}) = \begin{cases} \mathbb{R} & q = 2, \\ 0 & \text{otherwise.} \end{cases}$$

Hence we have the inclusion

$$H^q(S^2, \mathbb{Z}) \subset H^q(S^2, \mathbb{R}).$$

Since on S^2 the homomorphism ι_* in (208) is an inclusion, i.e.

$$H^2(S^2, \mathbb{Z}) \subset H^2(S, \mathbb{R}) \cong_{\mathrm{deR}} H^2(S^2, \mathbb{R}),$$

$[\omega^{S^2}]$ is integral, i.e. (up to the isomorphism ι_*)

$$[\omega^{S^2}] \in H^2(S^2, \mathbb{Z})$$

and the map $Z_{\alpha\beta\sigma}$ (213) is $\mathbb{Z}$-valued (and hence identified with its value). The system of maps

$$\varphi_{\mu\nu} := e^{2\pi i g_{\mu\nu}} \qquad \mu, \nu \in I$$

satisfies the cocycle condition saying that

$$\varphi_{\alpha\beta} \cdot \varphi_{\beta\sigma} \cdot \varphi_{\sigma\alpha} = 1$$

holds true. Hence the system $\{\varphi_{\alpha\beta}\}$ where α, β varies in I defines a cohomology class $[\varphi_{\alpha\beta}]$ in $H^1(S^2, \mathbf{U}_c(1))$ for which

$$[\omega^{S^2}] = \iota_* \circ \delta_c[\varphi_{\alpha\beta}]. \tag{214}$$

Therefore, both the Chern–de Rham class $[\omega^{S^2}]$ and the Chern class $\delta_c[\varphi_{\alpha,\beta}]$ are identical and characterize the bundle TS^2 over S^2.

Now we pass to $\mathbb{F}^{\mathbf{a}}$, the complex line bundle of X on M. The isomorphism class of $\mathbb{F}^{\mathbf{a}}$ is $\varepsilon^*[\varphi_{\alpha,\beta}]$ constructed out of the simple covering $\mathcal{W}$ of M which is generated by the two–chart covering on M. By propositions 42 and 43 the map δ_c in diagram (208) is injective. Thus $\check{c}(\varepsilon)$ and $\varepsilon^*[\varphi_{\alpha\beta}]$ determine each other uniquely.

However, $\varepsilon^*[\omega^{S^2}] = [\dot{\Omega}^\varepsilon]$ in ${}_{\text{deR}}H^2(M, \mathbb{R})$ fails to determine $\delta_c(\varepsilon^*[\varphi_{\alpha\beta}])$ uniquely since ι_* in (208) is not injective, in general (as we will see below).

Let $H^1(M, \mathbf{U}(1)_c)_{\text{vect}} \subset H^1(M, \mathbf{U}(1_c))$ be the collection of all Čech classes emanating from all singularity free vector fields on M. Thus by corollary 38 we have the following proposition:

PROPOSITION 45. *The Chern class $\check{c}_1(\mathbb{F}^a)$ and the Chern–de Rham class $c_1(\mathbb{F}^{\mathbf{a}})$ of $\mathbb{F}^{\mathbf{a}}$ are given by*

$$\check{c}_1(\mathbb{F}^{\mathbf{a}}) = \varepsilon^*[\{\varphi_{\alpha\beta}\}] \quad \textit{and} \quad c_1(\mathbb{F}^{\mathbf{a}}) = [\varepsilon^*\omega^{S^2}] = [\dot{\Omega}^{\mathbf{a}}].$$

Thus $c_1(\mathbb{P}^{\mathbf{a}})$ can be reconstructed from $[\dot{\Omega}^{\mathbf{a}}]$. Moreover,

$$\iota_* \circ \delta_c : H^1(M, \mathbf{U}(1)_c) \longrightarrow {}_{\text{deR}}H^2(M, \mathbb{R})$$

is injective on $H^1(M, U(1)_c)_{\text{vect}}$ iff ι_* is injective.

In combination with corollary 38 we hence obtain:

PROPOSITION 46. *Let $[M, S^2]$ be the collection of all homotopy classes in $C^\infty(M, S^2)$. The map*

$$[M, S^2] \xrightarrow{c_1} {}_{\text{deR}}H^2(M, \mathbb{R})$$

assigning with any homotopy class $[\varepsilon]$ the Chern–de Rham class $\dot{\Omega}^\varepsilon$ is injective iff ι_* in (208) is injective.

In general, ι_* in (208) is not injective: If $M = SO(\mathbb{R}^3)$, then $H^2(SO(\mathbb{R}^3), \mathbb{R}) = 0$ while $H^1(SO(\mathbb{R}^3), \mathbb{Z}) = \mathbb{Z}_2$, being identical with the fundamental group. In fact, the injectivity of ι_* is related to torsion (cf. [**39**]). If $\delta_c^1 = 0$, then

$$H^2(M, \mathbb{Z}) \subset {}_{\text{deR}}H^2(M, \mathbb{R}). \tag{215}$$

Examples for the injectivity of ι_* in (208) shall be given next.

4.3.8. Examples. A class of examples for which ι_* in (208) is injective consists of manifolds which are homotopy equivalent to spherical complexes (cf. [**42**]):

i): Consider a vector field X' with two different singularities inside the closed unit ball B and vanishing on the boundary of B.

Deleting the boundary and the two singular points inside of B yields a manifold M together with a *singularity free* vector field $X = (\mathrm{id}\,, \mathbf{a})$. The first Chern–de Rham class $c_1(\mathbb{F}^{\mathbf{a}})$ of $\mathbb{F}^{\mathbf{a}}$ is in $H^2(M,\mathbb{Z})$. The unit vector field ε yields the map

$$H^2(\varepsilon) : H^2(S^2,\mathbb{Z}) \longrightarrow H^2(M,\mathbb{Z})$$

with

(216) $$H^2(\varepsilon)(c_1(TS^2)) = \varepsilon^*(c_1(TS^2)) = c_1(\mathbb{F}^{\mathbf{a}}).$$

Here $H^2(\varepsilon)$ denotes the homomorphism determined by ε. To see that $c_1(\mathbb{F}^{\mathbf{a}})$ determines the complex line bundle $\mathbb{F}^{\mathbf{a}} = \mathbb{F}^{\varepsilon}$ uniquely we compute the first and the second cohomology group of M with coefficients in $\mathbb{Z}$ and $\mathbb{R}$, respectively. To this end we place two spheres $S^2_{r_1}$ and $S^2_{r_2}$ into M with identical radius centered at the singular points. Without loss of generality, we may assume that these two-spheres touch each other in some point x_0, say. These two-spheres with $\{x_0\}$ in common form a spherical complex $S^2_{r_1} \vee S^2_{r_2}$, the *join* of $S^2_{r_1}$ and $S^2_{r_2}$ (cf. [**42**]).

The join of the two spheres is obtained by attaching the closed two-cell e_2 to $S^2_{r_2}$; it is done by mapping the boundary ∂e_2 of e_2 to $\{x_0\}$. This is to say that e_2 is turned into a two-sphere which may be identified with $S^2_{r_1}$, yielding the inclusion of $S^2_{r_1}$ into the join $S^2_{r_1} \vee S^2_{r_2}$.

The relevant second singular cohomology groups of the join $S^2_{r_1} \vee S^2_{r_2}$ are

$$H^2(S^2_{r_1} \vee S^2_{r_2},\mathbb{Z}) \cong H^2(S^2_{r_1},\mathbb{Z}) \oplus H^2(S^2_{r_1},\mathbb{Z}) \cong \mathbb{Z} \oplus \mathbb{Z}$$

and

$$H^2(S^2_{r_1} \vee S^2_{r_2},\mathbb{R}) \cong H^2(S^2_{r_1},\mathbb{R}) \oplus H^2(S^2_{r_1},\mathbb{R}) \cong \mathbb{R} \oplus \mathbb{R},$$

as shown in [**42**].

ii): Clearly, this example is generalized to M obtained from B_1, where *finitely* many points are extracted.

iii): If we extract *finitely* many closed discs from B_1, we obviously obtain some manifold M, which is homotopic to a spherical complex. Hence the map ι_* in (208) is injective, too.

4.3.9. Chern Numbers on S^2. The curvature $\dot{\Omega}^{\mathbf{a}}$ of any complex line bundle of type $\mathbb{F}^{\mathbf{a}}$ on M defines an integral cohomology class $[\dot{\Omega}^{\mathbf{a}}] \in \mathrm{deR}^2_H(M,\mathbb{R})$, the second de Rham cohomology group.

$H^2(S^2,\mathbb{Z})$ is naturally isomorphic to $\mathbb{Z}$ and is a subgroup of $H^2(S^2,\mathbb{R})$. The collection of isomorphism classes of complex line bundles on S^2 is in one-to-one correspondence to $\mathbb{Z}$ (cf. section 4.3.7 and [**62**]); each isomorphism class is given by a first Chern class and vice versa. The isomorphism is expressed by the natural map

$$H^2(S^2,\mathbb{R}) \longrightarrow \mathbb{R}, \qquad [\Omega] \longmapsto \frac{1}{4\pi} \cdot \int_{S^2} \Omega.$$

The integral $\frac{1}{4\pi} \cdot \int_{S^2} c_1(\mathbb{F}) \in \mathbb{Z}$ is called the *Chern number* $C_1(\mathbb{F})$ of a complex line bundle $\mathbb{F}$ on S^2; this number is 2 for $\mathbb{F} = TS^2$ (cf. [**63**] and [**46**]).

All Chern numbers of S^2 are realized by all Hopf bundles. Indeed, for each $n \in \mathbb{Z}\backslash\{0\}$ we consider the Hopf bundles $\mathbb{P}_n$ on S^2 as introduced in section 2.3.4. The transition map $S^2_\pm$ is the cocycle with respect to the two-chart trivialization. The integer n is the Chern number of $\mathbb{F}_n$ associated to $\mathbb{P}_n = SU(2)/\mathbf{C}^a_n$ (cf. sections 2.3.4, 2.3.6 and [**63**])

The complex line bundle $\mathbb{F}_n$ associated to $\mathbb{P}_n$ on S^2 (cf. section 2.3.4) is characterized by n. For $n = 2$ obviously

$$\mathbb{F}_2 = TS^2.$$

The Chern number of $\mathbb{F}_1$ associated to the Hopf fibration $\mathbb{P}_1$ is 1 (cf. [**46**]).

Since $C_1(\mathbb{L}_1) = 1$ and $C_1(TS^2) = 2$ (cf. [**46**]), we have, in particular,

$$c_1(\varepsilon^*(TS^2)) = 2 \cdot c_1(\varepsilon^*(\mathbb{L}_1)).$$

Therefore, we call $c_1(\varepsilon^* TS^2)$ *even.* Let $H^2(O, \mathbb{Z})_2 \subset H^2(O, \mathbb{Z})$ denote the collection of all *even* second de Rham cohomology classes. Hence the de Rham class $c_1(\varepsilon^* TS^2) = c_1(\mathbb{F}^{\mathbf{a}})$ is even. Clearly, $c_1(\varepsilon^* \mathbb{L}_1)$ might be even, as well, depending on ε (cf. section 4.4.2 below). Using lemma 55 we immediately obtain:

PROPOSITION 47. *For any fibrewisely oriented $\mathbb{F}$ embeddable in $O \times E$ the de Rham class $c_1(\mathbb{F})$ is even and is hence of the form $c_1(\mathbb{F}^{\mathbf{a}})$ for some $\mathbf{a} \in C^\infty_c(M, S^2)$.*

4.4. Mapping Degree and First Chern–de Rham Classes

So far we classified smooth vector fields such as $X = (\mathrm{id}\,, \mathbf{a})$ in terms of the isomorphism class of $\mathbb{F}^{\mathbf{a}}$ which is determined by $\mathbb{F}^\varepsilon$. The domain of X was a compact smooth manifold M with boundary (cf. section 4.1), embedded in the Euclidean space E. By equation (205) in section 4.3.5 the Chern class $\check{c}_1(\mathbb{F}^\varepsilon)$ of $\mathbb{F}^\varepsilon$ depends on $\varepsilon \in C^\infty(M, S^2)$. We classified the quality of ε partly by its homotopy class. Here we will use the mapping degree (as defined in [**48**]) to compute $\check{c}_1(\mathbb{F}^\varepsilon)$ directly out of $\check{c}_1(TS^2)$. We will observe that for oriented ∂-manifolds this degree is half the Chern number $C_1(\mathbb{F}^\varepsilon)$.

4.4.1. Mapping Degree.

In this section $M \subset \dot{E} := E\backslash\{0\}$ shall be a three-dimensional oriented compact manifold with boundary ∂M, i.e. a ∂-manifold, with non-empty interior O. On M let $X = (\mathrm{id}\,, \varepsilon)$ be a given smooth singularity free unit vector field. The orientation shall be determined by $\mu_{\dot{E}} := \mu|_{\dot{E}}$ (cf. section 1.1.9) restricted to M yielding μ_M.

The aim of this section is to relate the first Chern–de Rham class $c_1(\mathbb{F}^\varepsilon)$ with the one of $\mathbb{F}^c$, the complex line bundle of the Coulomb field X^c on $\dot{E}$ (cf. section 3.7). The principal part of X^c shall be denoted by $\mathbf{a}^c$. The key ingredient to establish this relation is the mapping degree. For this notion of *mapping degree* we refer to [**48**].

The *mapping degree* $\deg f$ of some map $f \in C^\infty(M, E)$ has the following property: Let $G_0 \subset E$ be a compact set with $f(\partial M) \subset G_0$ and $\tau \in C^\infty(E, \mathbb{R})$, a function with $\operatorname{supp} \tau \subset E\backslash G_0$. Then

$$\int_M \tau \circ f \cdot f^* \mu_{\dot{E}} = \sum_j \deg(f, M, G_j) \cdot \int_{G_j} \tau \cdot \mu_{\dot{E}} \tag{217}$$

holds true (cf. [**48**]). Here G_j is a *connected component* of $E\backslash G_0$ and $\deg(f, M, G_j)$ denotes the *mapping degree* of f with respect to any $x \in G_j$. The index j varies on all of the (finitely many) connected components of $E\backslash G_0$. It coincides with the notion of mapping degree in [**49**] for manifolds with boundaries and with the one in [**56**].

In order to apply the notion of mapping degree to ε which maps a three-dimensional ∂-manifold to S^2, we need to enlarge S^2 to a three-dimensional space. Here is how we do it. In the following we will consider the smooth map

$$\Xi_r : M \longrightarrow \dot{E} \qquad 0 < r \leq 1$$

defined by

$$\Xi_r(x, \lambda) := r \cdot (\varepsilon(x, \lambda), \lambda) \qquad \forall\, x \in S^2 \text{ and } \forall\, \lambda \in \dot{\mathbb{R}}.$$

(We will write just Ξ for Ξ_1.) Here $\dot{E} = S^2 \times \dot{\mathbb{R}} \equiv S^2 \times \dot{\mathbb{R}} \cdot \mathcal{N}_{S^2}$ where $\mathcal{N}_{S^2}$ is the (outward directed) unit normal field on S^2. To determine $\Xi^*\mu_E$ for $y = (x, \lambda)$ we split T_yM into

$$T_yM = \mathbb{F}^\varepsilon_y + \ker W^\varepsilon_y \qquad \forall\, y \in M$$

where W^ε_y is the Weingarten map of the vector field as defined in section 3.1.2, and observe that for any $v = w + u$ with $w \in \mathbb{F}^\varepsilon_y$ and $u \in \ker W^\varepsilon_y$,

$$T_y\Xi((w + u), \nu \cdot \mathcal{N}_{S^2}(\varepsilon(x))) = (d\varepsilon(x, \lambda)(w + u), \nu \cdot \mathcal{N}_{S^2}(\varepsilon(x)))$$

for any $\nu \in \mathbb{R}$. We will set

$$\mu_{S^2} \wedge d\lambda := \mathrm{pr}^*_{S^2}\mu_{S^2} \wedge \mathrm{pr}^*_{\mathbb{R}} d\lambda.$$

Moreover,

$$\mu_{\dot{E}} = \mathrm{pr}^*_{S^2}\mu_{S^2} \wedge \mathrm{pr}^*_{\mathbb{R}} d\lambda$$

(cf. [**46**]). Since $\mathbb{F}^\varepsilon_y = T_{\varepsilon(y)}S^2$ and $W^\varepsilon_y(v) \in \mathbb{F}^\varepsilon_y$ for all $y \in M$ and any $v \in T_yM$ we obtain for any choice of $v_1, v_2 \in T_yM$ and any $\nu \in \mathbb{R}$,

$$\Xi^*\mu_{\dot{E}}((x, \lambda); v_1, v_2, \nu) = \mathrm{pr}^*_{S^2}\mu_{S^2}\big(\varepsilon(x, \lambda); W^\varepsilon_{x,\lambda}(v_1), W^\varepsilon_{x,\lambda}(v_2)\big) \wedge \mathrm{pr}^*_{\mathbb{R}} d\lambda(\nu)$$

or

$$\Xi^*\mu_{\dot{E}}((x, \lambda); v_1, v_2, \nu) = \kappa(x, \lambda) \cdot \mu_{\dot{E}}(; \varepsilon(x, \lambda), v_1, v_2, \nu).$$

Here

$$\kappa(x, \lambda) = \det W^\varepsilon_{x,\lambda} \tag{218}$$

for any $x, \lambda \in M$. Therefore, we obtain the formula

$$\Xi^*\mu_{\dot{E}} = \varepsilon^*\mu_{S^2} \wedge d\lambda. \tag{219}$$

Due to $\mathbb{F}^\varepsilon = \varepsilon^*TS^2$ the naturality of the first Chern–de Rham class c_1 yields

$$c_1(\mathbb{F}^\varepsilon) = \varepsilon^* c_1(TS^2) = \varepsilon^*\mu_{S^2}.$$

Since $\mathbb{F}^\varepsilon = \Xi^*\mathbb{F}^c$ we observe that

$$\Xi^*\mu_{\dot{E}} = \kappa \cdot \mu_{\dot{E}}$$

as well as

$$c_1(\mathbb{F}^\varepsilon) = c_1(\Xi^*(\mathbb{F}^c)) \equiv \Xi^* c_1(\mathbb{F}^c),$$

where $\mathbb{F}^c$ is the complex line bundle of the *Coulomb field* on $\dot{E}$.

We will apply the transformation formula for integrals to the transformation Ξ_r. To this end we need some technical preparations. In order to specify G_0 we will consider M_η, an open *tubular neighborhood* (cf. [**18**] and [**46**]) of $M \subset E$ where η is the finite width of $M_\eta \backslash M$ measured along the outward directed unit normal field $\mathcal{N}_{S^2}$ of $M \subset E$. Now we set

$$G_0 := B_{r_1} \backslash M_\eta \cup \overset{\circ}{B}_{r_2}.$$

Here B_{r_1} is a closed ball in $\dot{E}$ with radius r_1 centered at $0 \in E$ such that

$$\overset{\circ}{B}_{r_1} \supset \overline{M_\eta}.$$

Moreover, $B_{r_2} \subset B_{r_1}$ is a concentric ball outside of $\overline{M_\eta}$. Hence

$$E \backslash G_0 = M_\eta \cup \overset{\circ}{B}_{r_2}.$$

Next we assume, without loss of generality, that

$$\Xi_r(\partial M) \subset G_0.$$

Finally, we choose τ in (217) such that

$$\operatorname{supp} \tau \subset M_\eta \text{ and } \tau|_M = 1.$$

Hence

$$\int_M \Xi_r^*(\mu_{S^2} \wedge d\lambda) = \deg(\Xi_r, \overset{\circ}{M}_1, M_\eta) \cdot \int_{M_\eta} \tau \cdot \mu_{\dot{E}}.$$

If η tends to zero, the mapping degree $\deg(\Xi_r, \overset{\circ}{M}, M_\eta) \in \mathbb{Z}$ does not change and we obtain

$$\int_M \Xi_r^*(\mu_{S^2} \wedge d\lambda) = \deg(\Xi_r, \overset{\circ}{M}, M_\eta) \cdot \int_M \mu_{\dot{E}}$$

or

$$\int_M \Xi_r^*(\mu_{S^2} \wedge d\lambda) = \deg \Xi_r \cdot \operatorname{vol}(M).$$

Here $\operatorname{vol}(M)$ denotes the volume of M in E and $\deg \Xi_r$ replaces $\deg(\Xi_r, \overset{\circ}{M}, M_\eta)$ (cf. (219)). Using (167) and (219) yields

$$\int_M c_1(\mathbb{F}^\varepsilon) \wedge d\lambda = \deg \Xi_r \cdot \operatorname{vol}(M).$$

Letting r tend to 1 the degrees $\deg \Xi_r$ and $\deg \Xi$ are identical and therefore we conclude

$$\int_M c_1(\mathbb{F}^\varepsilon) \wedge d\lambda = \deg \Xi \cdot \operatorname{vol}(M).$$

Next we consider $M \subset \dot{E}$, some x in the interior $\overset{\circ}{M}$ of M and choose open balls V_{ζ_1} and U_{ζ_1} as well as some non-negative $\tau \in C^\infty(M, \mathbb{R})$ with

$$\overset{\circ}{M} \supset V_{\zeta_1} \supset \operatorname{supp} \tau \supset V_{\zeta_2} \ni x \qquad \forall r > \zeta_1 > \zeta_2 > 0.$$

In addition, we require $\tau|_{V_{\zeta_2}} = 1$. Let $G_0 \subset \dot{E} \backslash V_{\zeta_1}$ be compact containing $M \backslash V_{\zeta_1}$ and let G_1 be the bounded component of $E \backslash G_0$ with $G_0 = M$. Thus

$$G_0 \backslash G_1 = V_{\zeta_1}.$$

Obviously,

$$\Xi_r(\partial M) \subset G_0$$

for r small enough. Then

$$\int_M \tau \circ \Xi_r \cdot \Xi_r^* \mu_{\dot{E}} = \deg \Xi_r \cdot \int_{\overline{V}_1} \tau \cdot \mu_{\dot{E}}$$

as well as

$$\int_{\operatorname{supp} \tau} \tau \circ \Xi_r \cdot \Xi_r^* \mu_{\dot{E}} = \deg \Xi_r \cdot \int_{\operatorname{supp} \tau} \tau \cdot \mu_{\dot{E}}.$$

Letting $\zeta_1 \longrightarrow \zeta_2$ we get

$$\int_{\overline{V}_{\zeta_1}} \Xi_r^* \mu_E = \int_{\overline{V}_{\zeta_1}} \det T\Xi_r \mu_E = \deg \Xi_r \cdot \int_{\overline{V}_{\zeta_1}} \mu_E.$$

Thus in the last equation $\det T\Xi_r$ and $\deg \Xi_r$ must have the same sign and the densities on both sides are identical (cf. [**7**]). Therefore, we conclude

$$\Xi_r^* \mu_E(x; \ldots, \ldots, \ldots) = \deg \Xi_r \cdot \mu_E(x; \ldots, \ldots, \ldots). \tag{220}$$

From the above and proposition 46 we immediately deduce:

THEOREM 48. *For any unit vector field with principal part ε on a compact connected ∂-manifold M the equations*

$$c_1(\mathbb{F}^\varepsilon) = \varepsilon^* c_1(TS^2) = \Xi^* c_1(\mathbb{F}^c) = \deg \Xi \cdot c_1(\mathbb{F}^c) \tag{221}$$

hold true. In addition, $\deg \Xi$ is injective on the collection $[M, S^2]$ of homotopy classes in $C^\infty(M, S^2)$ if ι_* is injective.

4.4.2. Examples of Vector Fields on Manifolds in $\dot{E}$. At first let us illustrate some of the results obtained above from a simple example on a spherical shell M centered at 0 in the Euclidean space E. The relation between TS^2 and $\mathbb{F}^{\mathbf{a}}$ on M is encoded in the (homotopy invariant) mapping degree $\deg \Xi$. Let $\varepsilon \in C^\infty(M, S^2)$ be a radial vector field on M. Since S^2 is a deformation retract of M for any complex line bundle $\mathbb{F}^{\mathbf{a}}$ on $\dot{E}$ due to (217) we may set for the field ε,

$$\deg \varepsilon := \frac{1}{4\pi} \cdot \int_{S^2} \varepsilon^* \omega^\varepsilon = \deg \Xi.$$

Clearly, $\delta_c \iota_*$ is injective (cf. proposition 45 and section 4.3.8).
Since $c_1(TS^2) = 2[\omega^{S^2}]$ (cf. [**46**] and (221)) we have the *Chern Number*

$$C_1(\mathbb{F}^\varepsilon) := \int_M \varepsilon^* c_1(TS^2) = 2 \cdot \frac{1}{4\pi} \cdot \int_{S^2} \varepsilon^* \omega^\varepsilon = 2 \cdot \deg \varepsilon. \tag{222}$$

Hence

$$C_1(\mathbb{F}^{\mathbf{a}}) = 2 \cdot \deg \varepsilon.$$

Since any integer can be realized as the mapping degree of a smooth map by a diffeomorphism from S^2 onto itself (cf. [**46**]), any even multiple of $\deg \varepsilon$ can be realized as a first Chern–de Rham class of a complex line bundle of a vector field on $\dot{E}$:

In fact, we can explicitly construct gradient fields on $\dot{E}$ whose directional fields have mapping degree n for any $n \in \mathbb{Z}$. To do so we will use the skew field of quaternions $\mathbb{H}$ to make use of higher powers. The primary tool is the map

$$g_n : \dot{\mathbb{H}} \longrightarrow \dot{\mathbb{H}}, \qquad \text{with} \qquad g_n(k) := k^n \qquad \forall\, k \in \dot{\mathbb{H}} := \mathbb{H} \backslash \{0\}$$

for a fixed integer $n \in \mathbb{Z}$. It is equivariant under the automorphism group of $\mathbb{H}$. As $\mathbb{H} = \mathbb{R} \cdot e \oplus E$, any $k \in \mathbb{H}$ is as usual of the form $k = \lambda e + \mu \cdot a$ with $a \in S^2 \subset \dot{E}$ and $\lambda, \mu \in \mathbb{R}$ and hence can be written as

$$k = |k| \cdot e^{\vartheta \cdot a} \qquad \vartheta \in \mathbb{R}.$$

Obviously, $e^{\vartheta \cdot a} \in S^3$ for any $a \in S^2 \subset E$. The map

$$\arg : \dot{\mathbb{H}} \longrightarrow \mathbb{R}$$

assigns with any $k = |k| \cdot e^{\vartheta \cdot a}$ the argument $\vartheta \in [0, 2\pi)$. This terminology is justified since $e^{\vartheta \cdot a} \in U^a(1)$. The composition

$$f_n := \arg \circ g_n, \qquad f_n(k) = n \cdot \vartheta,$$

will provide the right sort of gradients on a special Euclidean 3-space $E' \subset \mathbb{H}$ constructed as follows: Choosing a two-dimensional subspace $F \subset E$ let

$$E' := \mathbb{R} \cdot e \oplus F$$

where again e is the neutral element in $\mathbb{H}$. Thus for each unit vector $a \in F$, the unitary group $U^a(1)$ is in the unit sphere S'^2 in E'. The function f_n restricted to E' shall be called f_n again. In E' the gradient of f_n at $k \in E'$ is perpendicular to the level surface S_k passing through $k \neq 0$. Clearly, S_k is a cone in E' without its vertex $0 \in E'$. Hence it is tangent to the circle $|k| \cdot U^a(1)$ provided $k = \lambda \cdot e + \mu \cdot a$ for $a \in S^2$ and $\lambda, \mu \in \mathbb{R}$.

To compute grad f_n we hence can restrict to $U^a(1)$ as f_n is $SO(E')$-equivariant. Since

$$df_n(k)(h) = d \arg(k)(Dg_n(k)(h)) \qquad \forall\, h \in \dot{E}'$$

(where $\dot{E}' = E' \setminus \{0\}$), and, in particular, for $k \in U^a(1)$ and any $h = \zeta \cdot k \cdot a$ with $\zeta \in \mathbb{R}$,

$$Dg_n(k)(\zeta \cdot k \cdot a) = \zeta \cdot n \cdot k^{n-1} \cdot k \cdot a,$$

the differential $df_n(k)$ on $U^a(1)$ is of the form

$$\begin{aligned} df_n(k)(\zeta \cdot k \cdot a) &= <\text{grad } \arg(k), \zeta \cdot n \cdot k^{n-1} \cdot k \cdot a> \\ &= <n \cdot k^{-(n-1)} \cdot \text{grad } \arg(k), \zeta \cdot k \cdot a>. \end{aligned}$$

To determine grad arg at k we differentiate $e^{\arg(k) \cdot a} = k$ and find

$$\text{grad } \arg(k) = -a \cdot k^{-1}$$

for each $k \in U^a(1)$. Therefore,

$$\text{grad } f_n(k) = -n \cdot a \cdot k^{-n} \qquad \forall\, k \in \dot{E}'$$

with $k = |k| \cdot e^{\vartheta \cdot a}$ and any $a \in S^2$. Thus for any $k \in \dot{E}'$ the directional field ε_n of grad f_n is

$$\varepsilon_n := -|k|^n a \cdot k^{-n}$$

and hence $\varepsilon_n|_{S'^n} : S'^2 \longrightarrow S'^2$ has mapping degree

$$\deg \varepsilon_n = -n \qquad \forall\, n \in \mathbb{Z}$$

on S'^2 according to [**46**].

Identifying E' with E we thus have the following:

THEOREM 49. *Let M be a spherical shell centered at 0 containing S^2 in its interior. For any even integer $2 \cdot n$ there is a smooth map $f_n : M \longrightarrow \mathbb{R}$ such that*

$$C_1(\mathbb{F}^{\mathbf{a}}) = 2 \cdot n$$

for $\mathbf{a} := \operatorname{grad} f_n$. Any other vector field on M is equivalent to $\operatorname{grad} f_m$ for some $m \in \mathbb{Z}$ since $H^2(M, \mathbb{Z}) = H^2(S^2, \mathbb{Z}) = \mathbb{Z}$.

4.4.3. Applications. Here we consider the setting of a $U(1)$-gauge theory (cf. [**63**]) for the Maxwell equations for the vacuum in conjunction with complex line bundles of the type $\mathbb{F}^{\mathbf{a}}$ and compare it with the above formalism.

In electrodynamics in a vacuum we are given a $U(1)$-principal bundle $\mathbb{P}_{\text{Maxw}}$ on an open set O' in Minkowski space $\mathbb{R}^{1,3}$ which can be identified with $\mathbb{H}$ (cf. section 5.7). $\mathbb{R} \cdot e$ is the time axis and $\mathbb{R}^3$ is orthogonal to e. The metric η shall be diagonally represented and negative definite on $\mathbb{R}^3$. The $\mathbb{R} \cdot i$-valued curvature Ω on $\mathbb{R}^{1,3}$ is assumed to be exact, i.e.

$$\Omega = d\theta \tag{223}$$

for some one-form θ on $\mathbb{R}^{1,3}$. Both Ω and θ shall be pulled back to $s(O') \subset \mathbb{P}_{\text{Maxw}}$ by the inverse of a section $s : O' \longrightarrow \mathbb{P}_{\text{Maxw}}$ and equivariantly extended to all of $\mathbb{P}_{\text{Maxw}}$, yielding an equivariant field $\mathcal{F}$ and a gauge potential $\mathcal{A}$. These two forms on $\mathbb{P}_{\text{Maxw}}$ are related by

$$\mathcal{F} = d\mathcal{A}$$

(cf. [**62**] and [**63**]). The real forms A and F, given by $\mathcal{F} = i \cdot F$ and $\mathcal{A} = i \cdot A$, are related accordingly. The transition functions $\varphi_{\iota\nu}$ of $\mathbb{P}_{\text{Maxw}}$ are of the form

$$\varphi_{\iota\nu}(x) = e^{-i\Lambda_{\iota\nu}(x)}$$

for real-valued maps $\Lambda_{\iota\nu}$, which yields

$$\varphi_{\iota\nu}^{-1} \cdot d\varphi_{\iota\nu} = i \cdot d\Lambda_{\iota\nu} \quad \text{and} \quad \mathcal{A}_\nu = \mathcal{A}_\iota - i \cdot d\Lambda_{\iota\nu}.$$

In matrix form F (regarded as a two-form on $\mathbb{R}^{1,3}$) we see that

$$(F_{\alpha\beta}) = \begin{pmatrix} 0 & -E^1 & -E^2 & -E^3 \\ E^1 & 0 & B^3 & B^2 \\ E^2 & -B^3 & 0 & B^1 \\ E^3 & -B^2 & -B^1 & 0 \end{pmatrix}. \tag{224}$$

The electric and the magnetic fields in $\mathbb{R}^3$ have E^1, E^2 and E^3, respectively, B^1, B^2 and B^3 as components (cf. [**63**]). Both F and its Hodge dual $^*F$ in Minkowski space satisfy (by construction)

$$dF = 0 \quad \text{and} \quad d\,^*F = 0,$$

respectively. These equations are equivalent with Maxwell's equations for the vacuum (cf. [**63**]) again.

Let us turn to our setting using bundles of the type $\mathbb{F}^{\mathbf{a}}$ as studied in the previous sections. Let $O := O' \cap \mathbb{R}^3$. From (224) we see that the spatial part of F is the two-form B, the magnetic field which is exact due to (223). Hence the cohomology class $[B] \in H^2(O, \mathbb{R})$ vanishes. Thus B may be interpreted as the curvature of a *trivial complex line bundle* which obviously may be embedded into $O \times \mathbb{R}^3$. The curvature field of B is therefore of the form $\mathbf{c} = \kappa_\varepsilon \cdot \varepsilon$ with $\varepsilon \in C^\infty(O, S^2)$ homotopic to a point (cf. (166)) and $\kappa_\varepsilon = \det W^\varepsilon$.

On the other hand, a magnetic field B on an open set $O \subset \mathbb{R}^3$ is the curvature of a complex line bundle $\mathbb{F}$ (cf. [**76**] or [**63**]). However, $\mathbb{F} = \mathbb{F}^{\mathbf{a}}$ iff the curvature field is of the form (166).

4.5. Hodge-Morrey Decomposition

In Hodge theory (cf. [**74**]) closed n-forms on compact manifolds are decomposed into an *exact*, a *coexact* and a *harmonic component.* The harmonic component determines uniquely the cohomology class of a given closed n-form and vice versa. Thus there is a one-to-one correspondence between the $\mathbb{R}$-linear space of harmonic forms and the linear space of n-th cohomology classes (cf. [**85**]).

The analogue on a ∂-manifold is the Hodge-Morrey decomposition. Of course on the ∂-manifold an appropriate boundary condition has to be specified (cf. [**74**]).

Therefore, it is interesting to know under which circumstances $\dot{\Omega}^{\mathbf{a}}$ is harmonic for a singularity free vector field $X = (\mathrm{id}\,, \mathbf{a})$ on M where M shall be a ∂-manifold.

4.5.1. The Hodge $*$-Operator. The oriented Euclidean space $E, <\,,\,>$ naturally admits a star operator

$$* : \wedge E^* \longrightarrow \wedge E^*$$

on the exterior algebra $\wedge E^*$ of the dual E^* of E. To define it let μ be the Euclidean volume form specified in section 1.1.9. Then

$$* : \mathbb{R} \longrightarrow \wedge^3 E$$

assigns to any real λ the volume form $\lambda \cdot \mu$ and $*(\lambda \cdot \mu) = \lambda$ for any $\lambda \in \mathbb{R}$. Moreover, let $\ell_a :=< a, \ldots >$ for any $a \in E$. The Hodge $*$-*operator* on $\wedge E^*$ restricted to $\wedge^1 E^*$, i.e.

$$* : \wedge^1 E^* \longrightarrow \wedge^2 E^*$$

maps ℓ_a into

$$*\ell_a(v, w) :=< a, v \times w > \qquad \forall\, v, w \in E.$$

In addition,

$$* : \wedge^2 E^* \longrightarrow \wedge^1 E^*$$

is given by $*\alpha(x, v) =< a, \ldots >$ where $\alpha(x; v, w) =< a, v \times w >$.

4.5.2. The Hodge $*$-Operation for Forms on M. Let $M \subset E$ be an oriented compact ∂-manifold of dimension three endowed with the Riemannian metric defined by $<,>$. Moreover, let X be a (smooth) vector field on M with principal part $\mathbf{a}_X$. The one-form α_X given by

$$\alpha_X(Y) =< \mathbf{a}_X, \mathbf{a}_Y > \qquad \forall Y \in \Gamma TM \tag{225}$$

is exact if X is a gradient field, since $d\alpha_X(Y, Z) =< \mathrm{curl}\, X, \mathbf{a}_Y \times \mathbf{a}_Z >$ (cf. section 3.1.2). Here $\mathbf{a}_Y$ and $\mathbf{a}_Z$ are the principal parts of Y and Z in ΓTM, respectively. Hence α_X does not reflect topological properties encoded in terms of algebraic topological ingredients.

Therefore, we concentrate on the two-form ω_X defined by

$$\omega_X(Y,Z) = < \mathbf{a}_X, \mathbf{a}_Y \times \mathbf{a}_Z > \tag{226}$$

for any pair $Y, Z \in \Gamma TM$ of vector fields on M. It is a symplectic form on each fiber of $\mathbb{F}^{\mathbf{a}_X}$. Clearly, ω_X is exact on M iff $a_X = \operatorname{curl} c$ for some smooth vector field $(\mathrm{id}, c) \in \Gamma TM$. This two-form determines ω_X uniquely. Indeed, if $\omega_X = \omega_{X'}$, we observe that

$$< \mathbf{a}_X, \mathbf{a}_Y \times \mathbf{a}_Z > = < \mathbf{a}_{X'}, \mathbf{a}_Y \times \mathbf{a}_Z > .$$

Since any vector $v \in E$ is of the form $v = u_1 \times u_2$ for some $u_1, u_2 \in E$ we conclude that $\mathbf{a}_X = \mathbf{a}_{X'}$.

Thc (unique) Hodge-Morrey decomposition of ω_X reads

$$\omega_X = d\alpha^X + \delta\beta^X + \gamma^X \tag{227}$$

complemented with *boundary conditions* to be specified later. Here α^X is a one-form, β^X a three-form and δ is the co-differential on M given by

$$\delta\alpha = (-1)^{3\cdot k+4} * d * \alpha \tag{228}$$

for any k-form. $*$ denotes the Hodge $*$-operator as defined above (depending on the Riemannian metric on M inherited by $< \, , \, >$). Finally, γ^X is a *harmonic two-form*, i.e.

$$d\gamma^X = \delta\gamma^X = 0 \tag{229}$$

with either vanishing tangential or normal components as possible *boundary conditions.*

According to the (unique) *Friederichs decomposition* (cf. [**74**]) this harmonic part γ^X splits further into

$$\gamma^X = d\eta^X + \varphi^X \quad \text{with} \quad \varphi^X = \delta\psi^X. \tag{230}$$

Thus

$$\omega_X = d\alpha + \partial\beta^X + d\eta^X + \delta\psi^X.$$

Since $d\gamma^X = 0$, the harmonic form γ^X determines a cohomology class $[\gamma^X] = [\delta\psi^X]$. Therefore, it is natural to ask under what conditions $\dot{\Omega}^{\mathbf{a}_X}$ is harmonic, i.e.

$$\delta\dot{\Omega}^{\mathbf{a}_X} = 0. \tag{231}$$

To answer this question is the goal for the rest of this section.

To do so, we investigate under what conditions a given two-form Ω on M is harmonic. One of our technical tools is the formula

$$*\beta(x; v_2, v_3) = \beta(x; v_1) \tag{232}$$

for a one-form β, any orthonormal base $v_1, v_2, v_3 \in T_pM$ and any $p \in M$. In our case the linear map β is replaced by

$$\alpha_c = < c, \ldots >$$

where $c : M \longrightarrow E$ is a smooth map satisfying

$$\Omega(x; v_1, v_2) = < c(x), v_1 \times v_2 > \qquad \forall\, v_1, v_2 \in T_xM \text{ and } \forall\, x \in M. \tag{233}$$

Thus

$$(*\alpha_c)(x; v_1, v_2) = \alpha_c(x; v_1 \times v_2) \qquad \forall\, v_1, v_2 \in T_xM \text{ and } \forall\, x \in M \tag{234}$$

saying that

$$(*\alpha_c)(x; v_1, v_2) = \Omega(x; v_1 \times v_2).$$

Thus

$$\alpha_c = *\Omega$$

due to the identity

$$** = (-1)^{k\cdot(n-k)}\mathrm{id}$$

for any k-form on M. Here, $n = 3$. Therefore,

$$d\alpha_c = d * \Omega,$$

implying

(235) $$\omega_{\mathrm{curl}\ c} = d * \Omega.$$

In turn we find:

LEMMA 50. *For a given two-form Ω on M*

(236) $$\delta\Omega = \alpha_{curl\, c}$$

and therefore

(237) $$\delta\Omega = 0 \quad \textit{iff} \quad curl\, c = 0 \quad \textit{saying} \quad c = grad\, f \quad \textit{(locally)}.$$

Let us recall the notion of the curvature field $\kappa \cdot \varepsilon$ of the curvature $\dot{\Omega}^{\mathfrak{a}\chi}$ in (166). Here ε is the principal part of the unit vector field of X. An immediate consequence from lemma 50 for the curvature two-form $\dot{\Omega}^{\mathfrak{a}\chi}$ and its curvature field is stated in:

COROLLARY 51. *The following local statements are equivalent:*

i): $\delta\dot{\Omega}^{\mathfrak{a}\chi} = 0$.

ii): $\kappa \cdot \varepsilon = grad\, f$ *(for some locally defined smooth $\mathbb{R}$-valued map).*

iii): $\mathbb{F}^{\mathfrak{a}\chi}$ *is integrable where the leaves are level surfaces locally defined smooth $\mathbb{R}$-valued functions f satisfying*

$$\kappa = df(\varepsilon).$$

4.5.3. The Curvature as a Harmonic Form. As motivated by lemma 50 and corollary 51 we now want to find conditions for $\delta\dot{\Omega}^{\mathfrak{a}\chi} = 0$. Among technical tools concerning the curvature field $\kappa \cdot \varepsilon$ we focus on the integrability of $\mathbb{F}^{\mathfrak{a}\chi}$. To do so we first compute $d\alpha_{\kappa\varepsilon}$. As it is easily seen

(238) $$\begin{aligned} d\alpha_{\kappa\varepsilon}(X', Y') &= d\kappa(X')\cdot < \varepsilon, Y' > - d\kappa(Y')\cdot < \varepsilon, X' > \\ &\quad + \kappa\cdot < d\varepsilon(X'), Y' > - \kappa\cdot < d\varepsilon(Y'), X' > \end{aligned}$$

for all $X', Y' \in \Gamma TM$ or, equivalently,

(239) $$\begin{aligned} d\alpha_{\kappa\varepsilon}(X', Y') &= d\kappa(X')\cdot < \varepsilon, Y' > - d\kappa(Y')\cdot < \varepsilon, X' > \\ &\quad + \kappa\cdot < W^{\varepsilon}(X'), Y' > - \kappa\cdot < W(Y'), X' > \end{aligned}$$

for the Weingarten map W^{ε} (cf. (131) and section 3.1.2). Moreover,

(240) $$d\alpha_{\kappa\varepsilon}(X', Y') = < \mathrm{curl}\ \kappa \cdot \varepsilon, X' \times Y' > .$$

By a well-known identity for the cross product (cf. [**43**], page 189), the curl of $\kappa \cdot \varepsilon$ in equation (240) is split as follows:

LEMMA 52.

(241) $$curl\, \kappa\varepsilon = -\varepsilon \times grad\, \kappa + \kappa \cdot curl\, \varepsilon.$$

From this equation we see that:

PROPOSITION 53.

$$\textit{curl}\, \kappa\varepsilon = 0 \qquad \textit{iff} \qquad \varepsilon \times \textit{grad}\, \kappa = \kappa \cdot \textit{curl}\, \varepsilon. \tag{242}$$

Since

$$\varepsilon \times \operatorname{curl} \kappa \cdot \varepsilon = -\varepsilon \times (\varepsilon \times \operatorname{grad} \kappa) + \kappa \cdot \varepsilon \times \operatorname{curl} \varepsilon \tag{243}$$

we obtain

$$\varepsilon \times \operatorname{curl} \kappa \cdot \varepsilon = - < \varepsilon, \operatorname{grad} \kappa > \cdot \varepsilon + \operatorname{grad} \kappa + \kappa \cdot \varepsilon \times \operatorname{curl} \varepsilon. \tag{244}$$

Moreover,

$$\begin{aligned} d\alpha_\varepsilon(X', Y') &= < W^\varepsilon(X'), Y' > - < W^\varepsilon(Y'), X' > \\ &= < \operatorname{curl} \varepsilon, X' \times Y' > \end{aligned} \tag{245}$$

for all $X', Y' \in \Gamma TM$ which implies for $Y' = (\mathrm{id}, \varepsilon)$

$$\varepsilon \times \operatorname{curl} \varepsilon = -W^\varepsilon(\varepsilon) \qquad \text{and} \qquad \widetilde{W^\varepsilon}(\varepsilon) = 0 \tag{246}$$

due to $\operatorname{im} W^\varepsilon \subset \mathbb{F}^{\mathbf{a}\chi}$. Here $\widetilde{W^\varepsilon}$ is the adjoint of W^ε. Thus we may state:

LEMMA 54.

$$\textit{grad}\, \kappa = d\kappa(\varepsilon) \cdot \varepsilon + \kappa \cdot \varepsilon \times \textit{curl}\, \varepsilon - \varepsilon \times \textit{curl}\, \kappa \cdot \varepsilon \tag{247}$$

or, equivalently,

$$\textit{grad}\, \kappa = d\kappa(\varepsilon) \cdot \varepsilon + \kappa \cdot W^\varepsilon(\varepsilon) - \varepsilon \times \textit{curl}\, \kappa \cdot \varepsilon. \tag{248}$$

To analyze $\operatorname{curl} \kappa \cdot \varepsilon$ we determine its ε- and the $\mathbb{F}^{\mathbf{a}\chi}$-component, called the vertical and the tangential component, respectively. At first let us exhibit the tangential component $\operatorname{curl}^T \varepsilon$ of $\operatorname{curl} \varepsilon$. From (246) we see that

$$-\varepsilon \times \operatorname{curl}^T \varepsilon = W^\varepsilon(\varepsilon)$$

rewritten in terms of the multiplication in $\mathbb{H}$ as

$$-\varepsilon \cdot \operatorname{curl}^T \varepsilon = W^\varepsilon(\varepsilon).$$

Thus

$$\operatorname{curl}^T \varepsilon = \varepsilon \times W^\varepsilon(\varepsilon). \tag{249}$$

The vertical component $\operatorname{curl}^\perp \varepsilon$ of $\operatorname{curl} \varepsilon$ follows from (245) at any $x \in M$,

$$\operatorname{curl}^\perp_x \varepsilon = < (W^\varepsilon_x - \widetilde{W^\varepsilon_x})(e_1), e_2 > \cdot \varepsilon(x)$$

where $e_1, e_2 \in \mathbb{F}^{\mathbf{a}\chi}$ is an orthogonal basis. Obviously, $\operatorname{curl}^\perp_x \varepsilon$ does not depend on this particular basis. Therefore, we may assume, without loss of generality, that

$$(W^\varepsilon_x - \widetilde{W^\varepsilon_x})(e_1) = \vartheta(x) \cdot e_2, \tag{250}$$

for all $x \in M$ and for some $\vartheta(x) \in \mathbb{R}$, yielding

$$\operatorname{curl}^\perp_x \varepsilon = \vartheta(x) \cdot \varepsilon(x) \qquad \forall\, x \in M \tag{251}$$

and thus $\operatorname{curl}^\perp_x \varepsilon(x)$ splits into

$$\operatorname{curl}_x \varepsilon = \varepsilon \cdot W^\varepsilon_x(\varepsilon(x)) + \vartheta(x) \cdot \varepsilon(x). \tag{252}$$

Obviously,

$$\vartheta = 0 \;\text{ iff }\; \mathbb{F}^{\mathbf{a}\chi} \;\text{ is integrable.}$$

To prepare further integrability conditions complemented with a requirement on $(\text{curl}\,\kappa\varepsilon)^T$ to guarantee the harmoniticity of $\delta\dot{\Omega}^{\mathbf{a}\mathbf{x}}$, we need a few more techniques. The vertical and tangential decompositions of curl $\kappa\cdot\varepsilon$ according to (241) and (246) are

$$\text{curl}^{\,T}\kappa\cdot\varepsilon = -\varepsilon\times(\text{grad}^{\,T}\kappa - W^\varepsilon(\varepsilon)) \tag{253}$$

and

$$\text{curl}^{\,\perp}\kappa\cdot\varepsilon = \vartheta d\kappa(\varepsilon)\cdot\varepsilon, \tag{254}$$

respectively. This shows, in particular, for the tangential component of grad that

$$\text{grad}^{\,T}\kappa = W^\varepsilon(\varepsilon) + \varepsilon\times\text{curl}^{\,T}\kappa\varepsilon.$$

Let $e_1, \varepsilon(x)\times e_1$ in the fibre of $\mathbb{F}^{\mathbf{a}\mathbf{x}}$ at x be an orthonormal basis where $\varepsilon(x)\times e_1$ is perpendicular to curl $\kappa\varepsilon$. Then the constituents in (243) have the following components along e_1 and $e_2 := \varepsilon(x)\times e_2$, respectively:

$$\text{grad}_x^{\,e_1}\kappa = \kappa\cdot(W^\varepsilon(\varepsilon(x)))^{e_1}$$

and

$$\varepsilon(x)\times\text{curl}_x^{\,e_1}\kappa\varepsilon = \text{grad}^{\,e_2}\kappa - \kappa(x)\cdot(W_x^\varepsilon(\varepsilon(x)))^{e_2}.$$

From (246) we observe:

LEMMA 55. *$W^\varepsilon(\varepsilon) = 0$ iff curl $\varepsilon = f\cdot\varepsilon$ for some $f\in C^\infty(M,\mathbb{R})$.*

Obviously,

$$\begin{aligned} < d\varepsilon(X'), Y' > - < d\varepsilon(Y'), X' > &= - < \varepsilon, [X', Y'] > \\ &= < W^\varepsilon(X'), Y' > - < W^\varepsilon(Y'), X' > \end{aligned} \tag{255}$$

for all $X', Y' \in \Gamma\mathbb{F}^{\mathbf{a}\mathbf{x}}$ and

$$d\alpha_{\kappa\varepsilon}(X', Y') = -\kappa\cdot < \varepsilon, [X', Y'] > \qquad \forall\, X', Y' \in \mathbb{F}^{\mathbf{a}\mathbf{x}}. \tag{256}$$

Hence the following implication holds true:

$$d\alpha_{\kappa\cdot\varepsilon} = 0 \ \text{ along } \mathbb{F}^{\mathbf{a}\mathbf{x}} \qquad \Longleftrightarrow \qquad \mathbb{F}^{\mathbf{a}\mathbf{x}} \text{ is integrable.}$$

If, on the other hand, $Y' \in \Gamma\mathbb{F}^{\mathbf{a}\mathbf{x}}$ and $X' = \varepsilon$,

$$\begin{aligned} d\alpha_{\kappa\varepsilon}(\varepsilon, Y') &= -d\kappa(Y') + \kappa\cdot < d\varepsilon(\varepsilon), Y' > \\ &= -d\kappa(Y') + \kappa\cdot < W^\varepsilon(\varepsilon), Y' > . \end{aligned} \tag{257}$$

A criterion for the integrability of $\mathbb{F}^{\mathbf{a}\mathbf{x}}$ is furthermore observed from (255), namely:

LEMMA 56. *The following conditions are equivalent:*

- **i):** $\mathbb{F}^{\mathbf{a}}$ *is integrable.*
- **ii):** $W^\varepsilon|_{\mathbb{F}^{\mathbf{a}\mathbf{x}}}$ *is self-adjoint.*
- **iii):** $d\alpha_{\kappa\cdot\varepsilon}(X', Y') = 0 \qquad \forall\, X', Y' \in \Gamma\mathbb{F}^{\mathbf{a}}$.
- **iv):** $(\textit{curl}\,\kappa\cdot\varepsilon)^\varepsilon = 0$.

Condition ii) in lemma 56 allows the following reformulation:

LEMMA 57. *The following statements are equivalent:*

- **i):** *curl $\kappa\varepsilon = \varphi\cdot\varepsilon$ for some $\varphi\in C^\infty(M,\mathbb{R})$.*
- **ii):** $d\kappa(Y') = < \textit{grad}\,\kappa, Y' > = < \kappa\cdot W^\varepsilon(\varepsilon), Y' > \qquad \forall\, Y' \in \Gamma\mathbb{F}^{\mathbf{a}\mathbf{x}}$.

PROOF. i) yields

$$< \text{curl}\, \kappa\varepsilon, X' \times \varepsilon >= 0 \qquad \forall X' \in \Gamma\mathbb{F}^{\mathbf{a}_X}$$

implying ii) by using (238) and by choosing $X' \in \Gamma\mathbb{F}^{\mathbf{a}_X}$. On the other hand, ii) requires by (248) that

$$< \text{curl}\, \kappa\varepsilon, X' \times \varepsilon >= \varphi d\kappa(X') - \varphi \cdot \kappa < W^{\varepsilon}(\varepsilon), X' >= 0 \qquad \forall X' \in \Gamma\mathbb{F}^{\mathbf{a}_X}$$

and hence

$$\text{curl}\, \kappa\varepsilon \perp X' \times \varphi \cdot \varepsilon$$

to hold, which implies i). □

Thus we may conclude due to (238) and (256):

THEOREM 58. $\delta\dot{\Omega}^{\mathbf{a}_X} = 0$ *holds true iff both of the following two conditions are satisfied:*

- **i):** $\mathbb{F}^{\varepsilon}$ *is integrable.*
  This is equivalent to $(curl\, \kappa\varepsilon)^{\perp} = 0$.
- **ii):** $< grad\, \kappa, Y' >= \kappa \cdot < W^{\varepsilon}(\varepsilon), Y' > \qquad \forall Y' \in \Gamma\mathbb{F}^{\mathbf{a}_X}$.
  This is equivalent to$(curl\, \kappa\varepsilon)^{T} = 0$.

Thus if $X = \text{curl}\, Z$ on M, the bundle $\mathbb{F}^{\mathbf{a}_X}$ is not integrable. This is the case if X is source-free and fulfills the boundary condition

$$\int_M g(X, L)_{\mu} = 0 \qquad \forall L \in \kappa_D^{\gamma}(M)$$

(cf. [**74**]). In summary the above lemmata 55, 56, 57 and proposition 58 show the following:

LEMMA 59. *If* $\delta\dot{\Omega}^{\mathbf{a}_X} = 0$, *then*

$$\kappa \cdot < W^{\varepsilon}(Z), Y >= \kappa \cdot < W^{\varepsilon}(Y), Z > \qquad \forall Y, Z \in \Gamma TS_{\kappa}$$

for each level surface S_{κ} *of* κ.

To understand

$$\kappa = df(\varepsilon) =< \text{grad}\, f, \varepsilon > \tag{258}$$

in the context of $\delta\dot{\Omega}^{\mathbf{a}_X} = 0$, i.e. for $\kappa \cdot \varepsilon = \text{grad}\, f$, we first observe from (238) that

$$d\alpha_{\kappa\varepsilon}(X', \varepsilon) = d\kappa(X') - d\kappa(\varepsilon) \cdot < \varepsilon, X' > - \kappa \cdot < W^{\varepsilon}(\varepsilon), X' > \qquad \forall X' \in \Gamma TM.$$

Therefore, we obtain in the case of $\delta\dot{\Omega}^{\mathbf{a}_X} = 0$,

$$\text{grad}\, \kappa = d\kappa(\varepsilon) \cdot \varepsilon + \kappa \cdot W^{\varepsilon}(\varepsilon) \tag{259}$$

or, equivalently,

$$\text{grad}\, \kappa = d\kappa(\varepsilon) \cdot \varepsilon + W^{\varepsilon}(\text{grad}\, f).$$

Denoting by $H(f)$ the Hessian for f and inserting (258) and

$$d\kappa(\varepsilon) = H(f)(\varepsilon, \varepsilon)$$

(obtained from (258)) into (259), we finally relate κ to f by

$$\text{grad}\, \kappa = H(f)(\varepsilon, \varepsilon) \cdot \varepsilon + df(\varepsilon) \cdot W^{\varepsilon}(\varepsilon). \tag{260}$$

Hence

$$d\kappa(Y) = df(\varepsilon) \cdot < W^{\varepsilon}(\varepsilon), Y > \qquad \forall Y \in \Gamma\mathbb{F}^{\mathbf{a}_X}$$

and

$$H(f)(\varepsilon,\varepsilon)\cdot <\varepsilon, Y>= -df(\varepsilon)\cdot <W^{\varepsilon}(\varepsilon), Y> \qquad \forall Y \in \Gamma\mathbb{F}^{\kappa\cdot\varepsilon}. \tag{261}$$

Finally, we look at the self-adjointness of W^{ε} within the context of $\delta\dot{\Omega}^{\mathbf{a}\mathsf{X}} = 0$. From $d\alpha_{\varepsilon}(Y,Z)$ expressed for any pair $Y, Z \in \Gamma TM$ as

$$< \mathrm{curl}\,\varepsilon, \mathbf{a}_Y \times \mathbf{a}_Z >=< W^{\varepsilon}(Y), Z- < W^{\varepsilon}(Z), Y > \tag{262}$$

and (151) we see that

PROPOSITION 60. *The following conditions are equivalent:*

- **i):** *W^{ε} is self-adjoint.*
- **ii):** *$curl\,\varepsilon = 0$.*

If, moreover, $\delta\dot{\Omega}^{\mathbf{a}\mathsf{X}} = 0$, then the following are equivalent:

- **a):** *W^{ε} is self-adjoint.*
- **b):** *The leaves of $\mathbb{F}^{\mathbf{a}\mathsf{X}}$ are level surfaces.*

CHAPTER 5

Heisenberg Algebras, Heisenberg Groups, Minkowski Metrics, Jordan Algebras and $\mathrm{SL}(2, \mathbb{C})$

The main goal in this chapter is to provide us with the fundamental geometric structure one meets, for example, in information transmission, signal analysis, optics, quantization, quantum field theory and the symplectic group of the plane as well in its Lie algebra. The structure mentioned is that of a Heisenberg group and a Heisenberg algebra (also called Heisenberg Lie algebra), respectively. Concerning these objects in finite dimensions we concentrate on the $\mathbb{R}$-linear space

$$E = \mathbb{R} \cdot \frac{a}{|a|} \oplus F^a$$

(cf. (22)) which naturally carries the structure of a Heisenberg group and that of its Lie algebra. These two geometric objects are closely related to the quaternions and the Minkowski geometry in four dimensions.

5.1. Natural Symplectic Structure on a Plane in 3-Space

In this section we study the symplectic plane in more detail and shed some light on it by presenting the very fundamentals of the symplectic group of this plane embedded in the Euclidean 3-space E. Such symplectic planes are found in the $(2+1)$-splitting of E caused by a fixed non-vanishing vector $a \in E$. In the next section of this chapter we will link these splittings to the notion of the Heisenberg group.

5.1.1. Symplectic Form. Let us start by refreshing the setting given in the first chapter, also adding some more details. Given a non-vanishing vector a in the oriented Euclidean space $(E, < \, , \, >, \, \mu_E)$, the orthogonal complement F^a of $a \in \dot{E} = E \backslash \{0\}$ carries the natural symplectic structure ω^a, defined by

$$\omega^a(h_1, h_2) := < h_1 \times a, h_2 > \tag{263}$$

for all $h_1, h_2 \in E$ (cf. sections 1.3.5 and 3.2.2). If ω^a is restricted to F^a, it is obviously non-degenerate; it is closed, as well, since it does not depend on an additional variable in F^a. This is to say that ω^a is a (constant) symplectic form on F^a.

The vector $a \in \dot{E}$ causes an oriented splitting

$$E = \mathbb{R} \cdot a \oplus F^a.$$

In fact, for all $h_1, h_2 \in E$,

$$\omega^a(h_1, h_2) = < h_1 \cdot a, h_2 > = - < a, h_1 \cdot h_2 > \tag{264}$$

where $\cdot$ denotes the product in $\mathbb{H}$, or differently formulated,

$$\omega^a(h_1, h_2) = -\mu_E(a, h_1, h_2) = -|a| \cdot \mu_E\left(\frac{a}{|a|}, h_1, h_2\right) \tag{265}$$

for any choice of $h_1, h_2 \in E$. In other words,

$$\omega^a = |a| \cdot \mu_{F^a}$$

where

$$\mu_{F^a}(h_1, h_2) := \mu_E\left(h_1, \frac{a}{|a|}, h_2\right) \qquad \forall\, h_1, h_2 \in F^a$$

(cf. section 1.3.3). Hence ω^a determines the orientation on F^a which is reverse to $\mu_E(a, \ldots, \ldots)$.

Clearly, given ω^a, the volume form μ_E is determined by (265) for all $a \in \dot{E}$.

Conversely, given a non-degenerate two-form ω on a plane F in E, it is of the form $\omega = \omega^a$ for some $a \in \dot{E}$ perpendicular to F. To show this, let $J_r \in \operatorname{End} F$ be the skew linear map for which

$$< J_r(h_1), h_2 >= \omega(h_1, h_2) \qquad \forall\, h_1, h_2 \in E. \tag{266}$$

Extending J_r to all of E by setting $J_r|_{F^\perp} = 0$, yields a skew-adjoint map J_r on E for which

$$J_r(h_1) = h_1 \times a \qquad \forall\, h_1 \in F \tag{267}$$

for some vector $a \in \dot{E}$ perpendicular to F. Whence $\omega = \omega^a$. Here the index r in J_r stands for "right" referring to the right multiplication with a in (267) (cf. section 3.2.1).

Any two forms ω^a and ω^b on F where both a and b in E are transversal to F differ by a real multiple $\lambda = \frac{1}{|a|^2} < b, a >$.

Fixing a unit vector $v \in F^a$, the splitting

$$F^a = \mathbb{R} \cdot v + \mathbb{R} \cdot v \times a$$

defines an orthogonal coordinate system. Hence if

$$h_s = q_s \cdot v + p_s \cdot v \times a, \quad s = 1, 2,$$

the value of ω^a on them is

$$\omega^a(h_1, h_2) = |a| \cdot (q_1 \cdot p_2 - p_1 \cdot q_2) \tag{268}$$

and, in particular,

$$\omega^a(q \cdot v, p \cdot v \times a) = |a| \cdot q \cdot p. \tag{269}$$

Let $a \in S^2$. The symplectic structure ω^a is directly related to a natural Hermitian structure on F^a determined by a unit vector $v \in F^a$ (cf. section 3.2.3). Since F^a is a $\mathbb{C}$-linear space, the choice of v turns F^a into a complex field as mentioned in section 1.2.1. The real axis and the imaginary one are $\mathbb{R} \cdot v$ and $\mathbb{R} \cdot v \times a$, respectively. This field is isometrically isomorphic to $\mathbb{C}$ via

$$i^v_{F^a} : F^a \longrightarrow \mathbb{C},$$

say; it sends v into 1 and $v \times a$ into i. The $v \times a$-component of some $h \in F^a$ is called the *imaginary part* of h. The natural Hermitian structure on F^a is then determined by

$$\mathcal{H}^a(h_1, h_2) := i^v_{F^a}(h_1) \cdot \overline{i^v_{F^a}(h_2)} \qquad \forall\, h_1, h_2 \in F^a. \tag{270}$$

Thus

$$\omega^a(h_1, h_2) = -\operatorname{Im}\left(\mathcal{H}^a(h_1, h_2)\right). \tag{271}$$

Here Im denotes the imaginary part of the complex number $\mathcal{H}^a(h_1, h_2)$. The minus sign is caused by the right multiplication with a in (264).

In contrast to mathematical usage, in physics literature the bar in (270) is over the first factor of the product. We will follow this tradition in chapters 9 and 10, where the physical approach is in focus.

5.1.2. Symplectic Basis. The matrix $M(J_r)$ of J_r is easily computed with respect to a *symplectic basis* $\{e_1, e_2\}$ of F^a, which satisfies per definition

$$\omega^a(e_1, e_2) = 1. \tag{272}$$

This basis need not be orthogonal, as we will see below. If in addition to (272) $e_2 := e_1 \cdot a$, we have

$$\begin{array}{ll} < e_1 \cdot a, e_2 >= \omega^a(e_1, e_2) = 1, & < e_2 \cdot a, e_1 >= \omega^a(e_2, e_1) = -1, \\ < e_1 \cdot a, e_1 >= \omega^a(e_1, e_1) = 0, & < e_2 \cdot a, e_2 >= \omega^a(e_1, e_2) = 0 \end{array} \tag{273}$$

showing that the matrix $M(J_r)$ of J_r in (266) is written as

$$M(J_r) \;=\; \begin{pmatrix} 0 & -1 \\ 1 & 0 \end{pmatrix} \tag{274}$$

formed with respect to the orthogonal, symplectic basis $\{e_1, e_2\}$. The inverse of J_r is J with matrix

$$M(J) = \begin{pmatrix} 0 & 1 \\ -1 & 0 \end{pmatrix}. \tag{275}$$

Hence, both J_r and J are orthogonal maps with oriented rotation angles $-\frac{\pi}{2}$ and $\frac{\pi}{2}$, respectively. Clearly,

$$\omega^a(h_1, h_2) = -< a \times h_1, h_2 >= -< J(h_1), h_2 >$$

for all $h_1, h_2 \in F^a$.

To investigate a general (i.e. not necessary orthogonal) symplectic basis $\{e_1, e_2\}$ in further detail we write

$$e_2 = \alpha \cdot e_1 + \beta \cdot e_1 \cdot a.$$

The coefficients α and β are determined by

$$\begin{aligned} \omega^a(e_1, e_2) = 1 \quad &= \quad < e_1 \cdot a, e_2 >= \beta \cdot |e_1|^2 \cdot |a|^2, \\ < e_2, e_1 > \quad &= \quad \alpha \cdot |e_1|^2, \\ |e_2|^2 \quad &= \quad \alpha^2 |e_1|^2 + \beta^2 \cdot |a|^2 \cdot |e_1|^2. \end{aligned}$$

Thus given e_1, the second vector e_2 of the symplectic basis has to satisfy

$$e_2 = \frac{< e_2, e_1 >}{|e_1|^2} \cdot e_1 + \frac{1}{|a|^2 \cdot |e_1|^2} \cdot e_1 \cdot a, \tag{276}$$

and its norm is computed from

$$|e_2|^2 = \frac{< e_1, e_2 >^2}{|e_1|^2} + \frac{1}{|a|^2 \cdot |e_1|^2}.$$

This shows that e_2 is determined by e_1 and the prescribed value of $< e_1, e_2 >$. In case $< e_1, e_2 >= 0$, the vectors e_1 and e_2 satisfy

$$|a| = \frac{1}{|e_1| \cdot |e_2|}. \tag{277}$$

Hence the collection of all symplectic basis forms a four parameter family $\mathbb{R}^4$. In fact, if e_1 varies in $\mathbb{R}^2$, the second vector e_2 runs through $\mathbb{R}^2$, as well, as seen from the factor $< e_2, e_1 >$ in (276). If, however, e_1 varies on S^1, the second vector runs through $\mathbb{R}^2$, yielding a *three parameter family.* Hence this collection forms a topological space homeomorphic to $S^1 \times \mathbb{R}^2$ (cf. [**5**]).

5.1.3. The Symplectic Group. Let us consider the group $\mathrm{Sp}(F^a)$ of all linear maps A of F^a preserving ω^a, i.e.

$$\omega^a(A(h_1), A(h_2)) = \omega^a(h_1, h_2) \qquad \forall\, h_1, h_2 \in F^a. \tag{278}$$

$\mathrm{Sp}(F^a)$ is called the *symplectic group* of F^a. This subgroup of $\mathrm{GL}(F)$ of all invertible linear maps of F^a will be a basic ingredient in quantum mechanics (cf. chapter 9). Since $A \in \mathrm{Sp}(F^a)$ is characterized by

$$\omega(x, y) = \omega(A(x), A(y)) = \det A \cdot \omega(x, y) \qquad \forall\, x, y \in F^a,$$

it has to satisfy $\det A = 1$. By (278) the symplectic map A is determined by

$$< J_r(h_1), h_2 >=< J_r \circ A(h_1), A(h_2) >=< \tilde{A} \circ J_r \circ A(h_1), h_2 >$$

and hence satisfies

$$J_r = \tilde{A} \circ J_r \circ A \qquad \text{or} \qquad J_r \circ A^{-1} = \tilde{A} \circ J_r \tag{279}$$

where $\tilde{A}$ is the adjoint of A formed with respect to the scalar product $< \, , \, >$. Since $J_r^{-1} = J$ the inverse of A satisfies

$$A^{-1} = J \circ \tilde{A} \circ J^{-1}.$$

Clearly, J_r is an element of $\mathrm{so}(F^a)$, the Lie algebra of $\mathrm{SO}(F^a)$. By (279) we deduce that the center of $\mathrm{Sp}(F^a)$ is $\{\mathrm{id}\, , -\mathrm{id}\, \}$.

Obviously, any $A \in \mathrm{Sp}(F^a)$ sends a symplectic basis into another one. Therefore, $\mathrm{Sp}(F^a)$ is homeomorphic to $S^1 \times \mathbb{R}^2$ (cf. [**5**]).

Since F^a is two-dimensional, $\det A = 1$ for all $A \in \mathrm{Sp}(F^a)$ is the characteristic property of A (by (278)). Hence

$$\mathrm{Sp}(F^a) = \mathrm{SL}(F^a) \tag{280}$$

for any $a \in S^2$. Here $\mathrm{SL}(F^a)$, the special linear group, is the collection of all elements in $\mathrm{GL}(F^a)$ with determinant 1. The Lie algebra $\mathrm{sp}(F^a)$ of $\mathrm{Sp}(F^a)$, i.e. the tangent space $T_{\mathrm{id}}\, \mathrm{Sp}(F^a)$ to $\mathrm{Sp}(F^a)$ at id, hence consists of all *trace-free endomorphisms* of F^a, as is easily verified.

5.1.4. Symplectic Transvections. Among the transformation in $\mathrm{Sp}(F^a)$ is a class of maps of a very special kind, the so-called *symplectic transvections.* These maps will play a fundamental role in quantum mechanics (cf. chapter 9), as well. In this section $a \in \dot{E}$.

Given any $\alpha \in \mathbb{R}$ and any vector $v \in F^a$, the linear transformation $A_{\alpha,v} \in \mathrm{Sp}(F^a)$ shall be defined by

$$A_{\alpha,v}(h) := h - \alpha \cdot \omega^a(h, v) \cdot v \qquad \forall\, h \in F^a. \tag{281}$$

As is easily checked, v is an eigenvector with the double eigenvalue 1. In the basis $\{v, v \cdot \frac{a}{|a|}\}$ the matrix $M(A_{\alpha,v})$ of $A_{\alpha,v}$ is of the form

$$M(A_{\alpha,v}) = \begin{pmatrix} 1 & \alpha \\ 0 & 1 \end{pmatrix}.$$

Thus $A_{-\alpha,v}$, having

$$M(A_{-\alpha,v}) = \begin{pmatrix} 1 & -\alpha \\ 0 & 1 \end{pmatrix}$$

as its matrix, is the inverse of A_α.

Obviously,

$$\begin{pmatrix} 1 & \alpha \\ 0 & 1 \end{pmatrix} = \exp_{\mathrm{GL}(F)} \alpha \cdot \begin{pmatrix} 0 & 1 \\ 0 & 0 \end{pmatrix} \tag{282}$$

and

$$\begin{pmatrix} 1 & 0 \\ \alpha & 1 \end{pmatrix} = \exp_{\mathrm{GL}(F)} \alpha \cdot \begin{pmatrix} 0 & 0 \\ 1 & 0 \end{pmatrix}$$

where $\exp_{\mathrm{GL}(F)}$ is the exponential of the Lie group $\mathrm{GL}(F)$.

Let $v \in F^a$ be a given unit vector. For any $\alpha \in \mathbb{R}\backslash\{0\}$,

$$A_{\alpha,v}(h) = h \qquad \forall\, h \in \mathbb{R} \cdot v.$$

To understand the meaning of α we consider $\nu \cdot v \cdot \frac{a}{|a|}$, which is perpendicular to the axis $\mathbb{R} \cdot \frac{a}{|a|}$, and evaluate $A_{\alpha,v}$ to obtain

$$\begin{aligned} A_{\alpha,v}\left(\nu \cdot v \cdot \frac{a}{|a|}\right) &= \nu \cdot v \cdot \frac{a}{|a|} - \alpha \cdot < \nu \cdot v \cdot \frac{a}{|a|} \times a, v > \cdot v \\ &= \nu \cdot (v \times \frac{a}{|a|} + \alpha \cdot |a| \cdot v). \end{aligned} \tag{283}$$

Since $\omega^{\frac{a}{|a|}}$ is the metric volume associated to the orientation ω^a of F^a, the oriented angle β between v and $A_{\alpha,v}(v \cdot \frac{a}{|a|})$ satisfies

$$\cos\beta = \frac{< v \times \frac{a}{|a|} + \alpha \cdot \frac{a}{|a|} \cdot v, v >}{|v \times \frac{a}{|a|} + \alpha \cdot |a| \cdot v|} = \frac{\alpha|a|}{|v \times \frac{a}{|a|} + \alpha \cdot |a| \cdot v|}$$

and

$$\sin\beta = \frac{< v \times \frac{a}{|a|}, v \times \frac{a}{|a|} + \alpha \cdot |a| \cdot v >}{|v \times \frac{a}{|a|} + \alpha \cdot |a| \cdot v|} = \frac{1}{|v \times \frac{a}{|a|} + \alpha \cdot |a| \cdot v|}$$

and hence

$$\tan\beta = \frac{1}{\alpha \cdot |a|}. \tag{284}$$

Thus α determines the *slope* of the line $A_{\alpha,v}(\mathbb{R}\cdot v\cdot\frac{a}{|a|})$ with respect to $\mathbb{R}\cdot v$. Moreover, for any $\nu\in\mathbb{R}$ and each fixed $\lambda\in\mathbb{R}$,

$$A_{\alpha,v}\left(\nu\cdot v\cdot\frac{a}{|a|}+\lambda v\right)=\lambda\cdot v+\nu\left(v\times\frac{a}{|a|}+\alpha\cdot|\frac{a}{|a|}|\cdot v\right),$$

which is parallel to (283). In particular, for coordinates x_1 and x_2 with respect to the basis $v, v\cdot\frac{a}{|a|}$,

$$A_{\alpha,v\cdot\frac{a}{|a|}}(x_1\cdot v)=x_1\cdot\left(v+\alpha\cdot|a|\cdot v\cdot\frac{a}{|a|}\right)\qquad\forall x_1\in\mathbb{R}$$

and

$$A_{\alpha,v}\left(x_2\cdot v\cdot\frac{a}{|a|}\right)=x_2\cdot\left(v\times\frac{a}{|a|}+\alpha\cdot|a|\cdot v\right)\qquad\forall x_2\in\mathbb{R}.$$

Hence fixing $v\in F^a$ and $\alpha\in\mathbb{R}$,

$$\left\{\lambda\cdot v+A_{\alpha,v}\left(\mathbb{R}\cdot v\cdot\frac{a}{|a|}\right)\mid\lambda\in\mathbb{R}\right\}$$

and

$$\left\{\lambda\cdot v\cdot\frac{a}{|a|}+A_{\alpha,v\cdot\frac{a}{|a|}}(\mathbb{R}\cdot v)\mid\lambda\in\mathbb{R}\right\}$$

form a (possibly degenerate) *grid* of the plane F^a .

These grids play a role in the encoding procedure of MRI (cf. [**72**]).

As a conclusion we will note an important fact concerning

$$\{A_{\alpha,v}|\alpha\in\mathbb{R}, v\in F^a\},$$

which is easily verified:

LEMMA 61. *Given any $h_1,h_2\in F^a$, there is one or a product of two symplectic transvections which maps h_1 into h_2; hence* $\mathrm{Sp}(F^a)$ *operates transitively and irreducibly on F^a.*

Here irreducible means that the only invariant subspaces of F^a are $\{0\}$ and $\{F^a\}$.

Let us point out that $\mathrm{Sp}(F^a)$ is generated by the linear automorphisms of F^a with matrices

$$\begin{pmatrix}0&-1\\1&0\end{pmatrix},\begin{pmatrix}1&0\\t&1\end{pmatrix}\text{ and }\begin{pmatrix}t&0\\0&t^{-1}\end{pmatrix}$$

for all $t\in\mathbb{R}$ (cf. [**37**]).

5.2. The Notion of a Heisenberg Algebra

5.2.1. Lie Brackets. Given a non-vanishing vector a in the oriented Euclidean space E, we form the oriented splitting (cf. section 5.1)

$$E=\mathbb{R}\cdot\frac{a}{|a|}\oplus F^a. \tag{285}$$

The plane F^a is equipped with the symplectic structure ω^a defined in (263). This two-form now yields a natural Heisenberg algebra structure on E, i.e. it turns E into a Heisenberg algebra, called $\mathfrak{h}^a$, as we will see below. The bracket will be derived from the multiplication in $\mathbb{H}$. In fact, E carries several Lie algebra structures such as su(2) and so(3), for example, which are directly related to the geometry of the Euclidean space E as a whole. From a geometric view point, su(2) describes the geometry of E within $\mathbb{H}$ while so(3) emanates directly from E itself. The Heisenberg algebra will describe the geometry of the splitting (285) within E.

Moreover, the Lie algebra structures on su(2) and so(3) are linked in a natural way and are defined from the commutator on $\mathbb{H}$ introduced in section 1.1.7 and equation (7), respectively. They yield two different Lie brackets on the splitting (285); this is to say that for

$$k_s = \lambda_s \cdot \frac{a}{|a|} + h_s \qquad \text{with } \lambda_s \in \mathbb{R} \text{ and } h_s \in F^a \text{ for } s = 1, 2$$

the respective commutators in su(2) and so(3) mentioned take the form

$$[k_1, k_2]_{\mathrm{su}(2)} = \lambda_1 \cdot \left[\frac{a}{|a|}, h_2\right] + \lambda_2 \left[h_1, \frac{a}{|a|}\right] + [h_1, h_2]_{\mathrm{so}(3)}$$

and

$$[h_1, h_2]_{\mathrm{so}(3)} = h_1 \times h_2.$$

Thus by the definition of the multiplication in $\mathbb{H}$ (cf. (7) and (8)),

$$[h_1, h_2]_{\mathrm{su}(2)} = 2 \cdot h_1 \times h_2 = 2 \cdot [h_1, h_2]_{\mathrm{so}(3)}.$$

Therefore, the negative value of the a-components of these expressions are

$$- < a, [k_1, k_2]_{\mathrm{su}(2)} > \cdot \frac{a}{|a|} = -2 \cdot < a, h_1 \times h_2 > \cdot \frac{a}{|a|} = 2 \cdot \omega^a(h_1, h_2) \cdot \frac{a}{|a|}$$

and

$$- < a, [h_1, h_2]_{\mathrm{so}(3)} > \cdot \frac{a}{|a|} = - < a, h_1 \times h_2 > \cdot \frac{a}{|a|} = \omega^a(h_1, h_2) \cdot \frac{a}{|a|}.$$

These observations suggest to consider the following two types of brackets on E (differing by a factor 2) expressed in terms of ω^a on the splitting (285) as:

$$[k_1, k_2]_1^a := 2 \cdot \omega^a(h_1, h_2) \cdot \frac{a}{|a|} \tag{286}$$

and

$$[k_1, k_2]_2^a := \omega^a(h_1, h_2) \cdot \frac{a}{|a|}. \tag{287}$$

Both turn the splitting (285) into a Lie algebra: Given $k_3 = \lambda_3 \cdot \frac{a}{|a|} + h_3$ for any choice of $\lambda_3 \in \mathbb{R}$ and $h_3 \in F^a$, then

$$[[k_1, k_2]_1^a, k_3]_1^a = 2 \cdot \omega^a(h_1, h_2) \cdot \omega^a\left(\frac{a}{|a|}, h_3\right) = 0$$

and

$$[[k_1, k_2]_2^a, k_3]_2^a = \omega^a(h_1, h_2) \cdot \omega^a\left(\frac{a}{|a|}, h_3\right) = 0$$

hold true. Therefore, these brackets satisfy Jacobi's identity and thus both brackets $[\ ,\ ]_1^a$ and $[\ ,\ ]_2^a$ define Lie algebra structures on E.

In the sequel E equipped with the bracket (287) is called the *Heisenberg algebra* $\mathfrak{h}^a$ determined by (the non-vanishing) $a \in E$. The bracket will be denoted by $[..,..]_{\mathfrak{h}^a}$. Clearly,

$$[k_1, k_2]_{\mathfrak{h}^a} = - < k_1 \cdot k_2, a > \cdot \frac{a}{|a|} = - < h_1 \times h_2, a > \cdot \frac{a}{|a|} \tag{288}$$

and hence for $h_1, h_2 \in F^a$

$$[h_1, h_2]_{\mathfrak{h}^a} = -h_1 \times h_2 \tag{289}$$

(cf. (287)). Thus the Lie bracket $[h_1, h_2]_{\mathfrak{h}^a}$ in $\mathfrak{h}^a$ is — up to the sign — identical with the Lie bracket in so(3) $= E$. (The minus sign in (288) is caused by the orientation ω^a on F^a in opposition to $\omega^{-a} = \bar{\omega}^a$, introduced in section 1.3.5. Hence the minus sign in (288) vanishes if we had used the left multiplication by a as in $\bar{\omega}^a$.) In addition,

$$\left[\frac{a}{|a|}, k\right] = 0 \qquad \forall\, k \in \mathfrak{h}^a$$

holds true, revealing $\mathbb{R} \cdot \frac{a}{|a|}$ as the *center* of $\mathfrak{h}^a$, i.e. the collection of all elements commuting with any other element in $\mathfrak{h}^a$. Here $\cdot$ is the multiplication in $\mathbb{H}$.

The reason for this choice of the bracket is hidden in the Euclidean geometry of the oriented linear space E. The cross product is a Lie product on E fitting in this geometry naturally, it is the Lie product of SO(3). Moreover, the cross product and the two-form ω^a emanate directly out of the product in $\mathbb{H}$ and the axis $\mathbb{R} \cdot a$, as expressed by (288).

Given $a \in E\backslash\{0\}$ and a symplectic structure ω^a on a two-dimensional linear subspace $F \subset E$ transversal to $\mathbb{R} \cdot a$, the two-form ω^a extends to μ_E by using (265). Thus the Heisenberg algebra

$$\mathfrak{h}^a = \mathbb{R} \cdot a \oplus F$$

yields an oriented splitting of E as introduced in section 5.1.1.

As an example we take the Heisenberg algebra of an inner automorphism τ_k of $\mathbb{H}$ with $k \in \mathbb{H}$ (cf. (26)). Decomposing $k \in \mathbb{H}$ into $k = \lambda \cdot e + u$ with $\lambda \in \mathbb{R}$ and $u \in E$, the unit vector $\frac{u}{|u|}$ and its orthogonal complement $F^{\frac{u}{|u|}}$ determine the axis of rotation of τ_k and its plane of rotation, respectively (cf. section 1.4.4). The skew bilinear form $\omega^{\frac{u}{|u|}}$ as defined in section 1.3.5 or else in equation (264) turns $F^{\frac{u}{|u|}}$ into a symplectic plane. Moreover,

$$\mathfrak{h}^{\frac{u}{|u|}} := \mathbb{R} \cdot \frac{u}{|u|} + F^{\frac{u}{|u|}}$$

is a Heisenberg algebra with Lie product as in (287). This algebra is called the *(normalized) Heisenberg algebra of the inner automorphism* τ_k. It determines the plane of rotation and the rotation axis, but not the angle of rotation. Obviously, the Heisenberg algebras of $\tau_{t\cdot h}$ and τ_k are the same for any non-vanishing real number t. On the other hand,

$$\mathfrak{h}^u := \mathbb{R} \cdot u + F^u$$

with ω^u on F^u as in section 1.3.5 yields a Heisenberg algebra of τ_u, as well. Since we may assume $|k| = 1$, the rotation angle ϑ_k of τ_k is entirely determined by $|u|^2$ due to $\lambda^2 + |u|^2 = 1$ (cf. (32)). Moreover, u encodes the direction of the rotation with respect to the orientation ω^u on F^u. Thus ω^u contains this information of τ_k. This is to say $\mathfrak{h}^u$ fully determines the rotation τ_k. We therefore call $\mathfrak{h}^u$ *the Heisenberg algebra of the inner automorphism* τ_k provided $k = \lambda \cdot e + u$ with $|k| = 1$.

Therefore, the collection of all *normalized* Heisenberg algebras of inner automorphisms is parameterized by S^2. Moreover, $S^3/\mathbb{Z}_2$ parameterizes the collection of all inner automorphisms of $\mathbb{H}$, as we saw in section 1.4.3 in chapter 1.

5.3. Heisenberg Group and its Lie Algebra

5.3.1. Definition of the Heisenberg Group. With $a \in \dot{E}$ let us consider the splitting

$$E = \mathbb{R} \cdot \frac{a}{|a|} \oplus F^a$$

together with the symplectic form ω^a on F^a. The linear space E shall now be endowed with a multiplication: Given two vectors $\lambda_1 \cdot \frac{a}{|a|} + h_1$ and $\lambda_1 \cdot \frac{a}{|a|} + h_2$ in E, the multiplication is given by

$$\begin{aligned} &\left(\lambda_1 \cdot \frac{a}{|a|} + h_1\right) \cdot \left(\lambda_2 \cdot \frac{a}{|a|} + h_2\right) \\ &:= \left(\lambda_1 + \lambda_2 + \frac{1}{2}\omega^a(h_1, h_2)\right) \cdot \frac{a}{|a|} + h_1 + h_2. \end{aligned} \tag{290}$$

E together with this multiplication is called a *Heisenberg group* and will be denoted by $\mathcal{H}^a$.

The relation (271) allows us to reformulate this multiplication in $\mathbb{R} \cdot \frac{a}{|a|} \oplus F^a$ in terms of the complex line F^a (cf. section 1.3.1): By choosing some unit vector $v \in F^a$ this plane turns into a field *isometrically isomorphic* to $\mathbb{C}$ via $i^v_{F^a}$ as mentioned above. Hence by (271) the expression of the multiplication can be reformulated as

$$\left(\lambda_1 \cdot \frac{a}{|a|} + h_1\right) \cdot \left(\lambda_2 \cdot \frac{a}{|a|} + h_2\right) = \left(\lambda_1 + \lambda_2 - \frac{1}{2}\operatorname{Im}\mathcal{H}^a(h_1, h_2)\right) \cdot \frac{a}{|a|} + h_1 + h_2$$

for any choice of $\lambda_1, \lambda_2 \in \mathbb{R}$ and any $h_1, h_2 \in F^a$.

5.3.2. Reduced Heisenberg Group. Instead of looking at the direct splitting $\mathbb{R} \cdot \frac{a}{|a|} \oplus F^a$ we may consider the three-dimensional cylinder

$$\mathcal{H}^a_{\text{red}} := U^a(1) \oplus F^a \tag{291}$$

in $\mathbb{H}$. It is a smooth manifold with $\mathbb{Z}$ as its fundamental group. The law of multiplication shall be the following: The product of two pairs $(z_1 + h_1)$ and $(z_2 + h_2)$ is defined by

$$(z_1 + h_1) \cdot (z_2 + h_2) := z_1 \cdot z_2 \cdot e^{\frac{1}{2}\omega^a(h_1,h_2) \cdot \frac{a}{|a|}} + h_1 + h_2 \tag{292}$$

or, if we write $z_s = e^{t_s \cdot \frac{a}{|a|}}$ for $s = 1, 2$,

$$\left(e^{t_1 \cdot \frac{a}{|a|}} + h_1\right) \cdot \left(e^{t_2 \cdot \frac{a}{|a|}} + h_2\right) = e^{(t_1 + t_2) \cdot \frac{a}{|a|}} \cdot e^{\frac{1}{2}\omega^a(h_1,h_2) \cdot \frac{a}{|a|}} + h_1 + h_2$$

for any two $z_1, z_2 \in U^a(1)$ and any pair $h_1, h_2 \in F^a$. The inverse element of $z + h$ in $\mathcal{H}^a_{\text{red}}$ is $z^{-1} - h$, which is identical to $\bar{z} - h$. Note that in physics literature one frequently uses $(z_1 \cdot z_2)^{-1}$ instead of $z_1 \cdot z_2$ in equation (292) (cf. [**50**]). We call $\mathcal{H}^a_{\text{red}}$ the *reduced Heisenberg group* (cf. [**37**]).

The center of $\mathcal{H}^a_{\text{red}}$ is $U^a(1)$, while the center of $\mathcal{H}^a$ is $\mathbb{R} \cdot \frac{a}{|a|}$.

5.3.3. The Lie Algebras of $\mathcal{H}^a$ and $\mathcal{H}^a_{\text{red}}$. For $a \in \dot{E}$ let us concentrate on $\mathcal{H}^a_{\text{red}}$. At first we study the *conjugation*. The conjugation of $z_1 + h_1$ by $z + h$ is given by

$$\begin{aligned} \operatorname{con}_{z+h}(z_1 + h_1) &:= (z+h)\cdot(z_1+h_1)\cdot(z+h)^{-1} \\ &= (z+h)\cdot(z_1+h_1)\cdot(\bar{z}-h) \\ &= \left(z\cdot z_1\cdot e^{\frac{1}{2}\omega^a(h,h_1)\cdot\frac{a}{|a|}} + h + h_1\right)\cdot(\bar{z}-h) \\ &= z_1\cdot e^{\frac{1}{2}\omega^a(h,h_1)\cdot\frac{a}{|a|}}\cdot e^{\frac{1}{2}\omega^a(h+h_1,-h)\cdot\frac{a}{|a|}} + h_1 \end{aligned}$$

and hence

$$\operatorname{con}_{z+h}(z_1 + h_1) = z_1\cdot e^{\omega^a(h,h_1)\cdot\frac{a}{|a|}} + h_1. \tag{293}$$

The *adjoint representation* Ad^a applied to $z+h$ is defined to be $T_e\operatorname{con}_{z+h}$. Thus for $e^{\vartheta\cdot t\cdot\frac{a}{|a|}} + h_1(t)$ with $h_1(0) = 0$ and $\dot{h}_1(0) = k$,

$$\begin{aligned} \operatorname{Ad}^a_{z+h}\left(\vartheta\cdot\frac{a}{|a|} + k\right) &= \frac{d}{dt}\left(\operatorname{con}_{z+h}(t\cdot\vartheta\cdot\frac{a}{|a|} + k)\right)\Big|_{t=0} \\ &= \frac{d}{dt}\left(e^{\vartheta\cdot t\cdot\frac{a}{|a|}}\cdot e^{\omega^a(h,h_1(t))\cdot\frac{a}{|a|}} + h_1(t)\right)\Big|_{t=0} \\ &= (\vartheta + \omega^a(h,k))\cdot\frac{a}{|a|} + k. \end{aligned} \tag{294}$$

Thus $\operatorname{Ad}^a_{z+h}$ does not depend on z. In particular,

$$\operatorname{Ad}^a_{z+h}\left(\vartheta\cdot\frac{a}{|a|}\right) = \vartheta\cdot\frac{a}{|a|} \qquad \forall\,\vartheta\in\mathbb{R} \tag{295}$$

and

$$\operatorname{Ad}^a_{z+h}(k) = \omega^a(h,k)\cdot\frac{a}{|a|} + k \qquad \forall\,k\in F^a. \tag{296}$$

From this we get the Lie bracket of $T_{\frac{a}{|a|}}\mathcal{H}^a_{\text{red}}$ defined (for Lie groups in general) by:

$$\left[\vartheta_1\cdot\frac{a}{|a|} + h_1, \vartheta_2\cdot\frac{a}{|a|} + h_2\right]_{\mathfrak{h}^a} := \frac{d}{dt}\operatorname{Ad}_{e^{\vartheta_1\cdot t\cdot\frac{a}{|a|}} + t\cdot h_1}\left(\vartheta_2\cdot\frac{a}{|a|} + h_2\right)\Big|_{t=0}$$

yielding

$$\left[\vartheta_1\cdot\frac{a}{|a|} + h_1, \vartheta_2\cdot\frac{a}{|a|} + h_2\right]_{\mathfrak{h}^a} = \omega^a(h_1,h_2)\cdot\frac{a}{|a|} \tag{297}$$

for any choices $\vartheta_1, \vartheta_2 \in \mathbb{R}$ and $h_1, h_2 \in F^a$. In other words:

LEMMA 62. *The Lie algebra of $\mathcal{H}^a_{\text{red}}$ and thus of $\mathcal{H}^a$ is the Heisenberg algebra $\mathfrak{h}^a$.*

Remark. *Let $F \subset E$ be a $2n$-subspace where $\dim E = 2n+1$. Given a symplectic form ω on F the multiplication (290) extends to $F + \mathbb{R}\cdot a$ verbatim for some fixed $a \subset E$ transversal to F. The resulting Heisenberg group is referred to as $\mathcal{H}_{2n}$. Its Lie algebra is denoted by $\mathfrak{h}_{2n}$ and is called a $2n+1$-dimensional Heisenberg algebra.*

5.3.4. Exponential Map. To determine the *exponential map* $\exp_{\mathfrak{h}^a}$ from $\mathfrak{h}^a$ to $\mathcal{H}^a$ we use the Campbell-Hausdorff series. Here it reduces to

$$(\lambda_1 \cdot \frac{a}{|a|} + h_1) + (\lambda_2 \cdot \frac{a}{|a|} + h_2) - \frac{1}{2} \cdot [\lambda_1 \cdot \frac{a}{|a|} + h_1, \lambda_2 \cdot \frac{a}{|a|} + h_2] + \text{iterated brackets}$$
$$= (\lambda_1 \cdot \frac{a}{|a|} + h_1) + (\lambda_2 \cdot \frac{a}{|a|} + h_2) + \frac{1}{2} \cdot \omega^a(h_1, h_2) \cdot \frac{a}{|a|}$$

due to (297) since the iterated brackets vanish. The (uniquely determined) exponential map

$$\exp_{\mathfrak{h}^a} : \mathfrak{h}^a \longrightarrow \mathcal{H}^a$$

thus satisfies

$$\exp_{\mathfrak{h}^a}(\lambda_1 \cdot \frac{a}{|a|} + h_1) \cdot \exp_{\mathfrak{h}^a}(\lambda_2 \cdot \frac{a}{|a|} + h_2)$$
$$= \exp_{\mathfrak{h}^a}\left((\lambda_1 \cdot \frac{a}{|a|} + h_1) + (\lambda_2 \cdot \frac{a}{|a|} + h_2) + \frac{1}{2} \cdot \omega^a(h_1, h_2) \cdot \frac{a}{|a|}\right).$$

From this we see that

$$\exp_{\mathfrak{h}^a} = \mathrm{id} \tag{298}$$

(cf. [**37**]).

5.3.5. Adjoint and Coadjoint Orbits. The *adjoint action* of $\mathcal{H}^a$ on $\mathfrak{h}^a$ is defined by

$$\mathrm{Ad}^a : \mathcal{H}^a \times \mathfrak{h}^a \longrightarrow \mathfrak{h}^a$$

given by (296). From equation (294) we conclude immediately:

PROPOSITION 63. *The adjoint orbit through $\vartheta \cdot \frac{a}{|a|}$ is $\{\vartheta \cdot \frac{a}{|a|} \mid \vartheta \in \mathbb{R}\}$. Otherwise each adjoint orbit of $\mathcal{H}^a$ is a straight line parallel to $\mathbb{R} \cdot \frac{a}{|a|}$. The collection of all adjoint orbits through elements in $F^a \setminus \{0\}$ is the collection of all straight lines parallel to $\mathbb{R} \cdot \frac{a}{|a|}$ and hence is in a one-to-one correspondence with F^a.*

To find the *coadjoint action* the adjoint operation Ad^a shall be expressed in more detail:

As

$$\mathrm{Ad}^a_{z+h}\left(\vartheta' \cdot \frac{a}{|a|} + k\right) = (\vartheta' + \omega^a(h,k)) \cdot \frac{a}{|a|} + k$$

for any $z + h \in \mathcal{H}^a$ and any $\vartheta' \cdot \frac{a}{|a|} + k \in \mathfrak{h}^a$ the coadjoint action

$$\mathrm{Ad}^{a^*} : \mathcal{H}^a \times \mathfrak{h}^{a^*} \longrightarrow \mathfrak{h}^{a^*}$$

sends each $< \vartheta \cdot \frac{a}{|a|} + h_1, .. > \in \mathfrak{h}^{a^*}$ applied to $\vartheta' \cdot \frac{a}{|a|} + k \in \mathfrak{h}^a$ into

$$\mathrm{Ad}^{a^*}_{(z+h)^{-1}}\left(< \vartheta \cdot \frac{a}{|a|} + h_1, \dots >\right)\left(\vartheta' \cdot \frac{a}{|a|} + k\right)$$
$$=< \vartheta \cdot \frac{a}{|a|} + h_1, \mathrm{Ad}_{\bar{z}-h}\left(\vartheta' \cdot \frac{a}{|a|} + k\right) >$$
$$=< \vartheta \cdot \frac{a}{|a|} + h_1, (\vartheta' - \omega^a(h,k)) \cdot \frac{a}{|a|} + k >$$

or, reformulated for $|a| = 1$, into

$$\begin{aligned}
&\mathrm{Ad}^{a*}_{(z+h)^{-1}}(< \vartheta \cdot a + h_1, \dots >)(\vartheta' \cdot a + k) \\
= \quad & < \vartheta \cdot a + h_1, \mathrm{Ad}^{a}_{\bar{z}-h}(\vartheta' \cdot a + k) > \\
= \quad & < \vartheta \cdot a + h_1, (\vartheta' - \omega^a(h,k)) \cdot \ a + k > \\
= \quad & < \vartheta \cdot a + h_1, \vartheta' \cdot a + k > + < \vartheta \cdot a + h_1, - < h \times a, k > \cdot\, a + k > \\
= \quad & < \vartheta \cdot a + h_1, \vartheta' \cdot a + k > - < \vartheta \cdot a + h_1, a > \cdot < h \times a, k > \\
& \qquad\qquad\qquad\qquad\qquad\qquad\qquad\qquad + < \vartheta \cdot a + h_1, k > \\
= \quad & < \vartheta \cdot a + h_1, \vartheta' \cdot a + k > - \vartheta < h \times a, k > + < h_1, k > \\
= \quad & < \vartheta \cdot a + h_1, \vartheta' \cdot a + k > - \vartheta < h \times a, \vartheta' \cdot a + k > + < h_1, \vartheta' \cdot a + k > \\
= \quad & < \vartheta \cdot a + 2h_1 - \vartheta \cdot h \times a, \vartheta' \cdot a + k > .
\end{aligned}$$

Identifying $\mathfrak{h}^{a^*}$ with $\mathfrak{h}^a$ by means of $<,>$, the *coadjoint orbit* (cf. [**50**], [**60**] and [**71**]) is

$$\vartheta \cdot \frac{a}{|a|} + F^a;$$

it carries a natural symplectic structure ω_ϑ defined as follows: For each

$$k = \vartheta \cdot \frac{a}{|a|} + h' \text{ with } h' \in F^a$$

we set

$$\begin{aligned}
&\omega_\vartheta\left(\vartheta \cdot \frac{a}{|a|} + h_1, \vartheta \cdot \frac{a}{|a|} + h_2\right) \\
&:= < k, \left[\vartheta \cdot \frac{a}{|a|} + h_1, \vartheta \cdot \frac{a}{|a|} + h_2\right] > . \qquad (299)
\end{aligned}$$

Here $h_1, h_2 \in F^a$. Therefore, the right-hand side is converted into

$$\omega_\vartheta\left(\vartheta \cdot \frac{a}{|a|} + h_1, \vartheta \cdot \frac{a}{|a|}, h_2\right) = \vartheta \cdot \omega^a(h_1, h_2)$$

due to (287) and hence ω_ϑ is of the form

$$\omega_\vartheta = \vartheta \cdot \omega^a = \omega^{\vartheta \cdot a}. \qquad (300)$$

In case $\vartheta = 0$, the form ω_ϑ on $\{h\}$ for any $h \in F^a$ vanishes identically.

The parametrization of coadjoint orbits by means of the points in $\mathbb{R} \cdot \frac{a}{|a|}$ causes a natural family of Heisenberg algebras

$$\mathfrak{h}^\vartheta_\omega := (\mathbb{R} - \vartheta) \cdot \frac{a}{|a|} + F^a \qquad \forall \vartheta \in \mathbb{R}$$

(as underlying linear spaces) with

$$\omega^\vartheta = \omega + \vartheta \cdot \omega^{\frac{a}{|a|}}$$

as a symplectic form on F^a. Here ω is a fixed symplectic form on F^a. Clearly, F^a in $\mathfrak{h}^0$ carries ω as symplectic form. We thus may state:

PROPOSITION 64. *The coadjoint orbit of Ad^{a^*} passing through $< \vartheta \cdot \frac{a}{|a|} + h_1, \dots >$ for $\vartheta \neq 0$ is*

$$\vartheta \cdot \frac{a}{|a|} \oplus F^a. \qquad (301)$$

Moreover, if $\vartheta = 0$ for any $k \in F^a$, the set $\{h\}$ is a coadjoint orbit, as well. The symplectic form on $\vartheta \cdot \frac{a}{|a|} + F^a$ is $\omega^{\vartheta \cdot a}$ for $\vartheta \neq 0$ for all $h_1, h_2 \in F^a$.

The underlying vector space of this coadjoint orbit (a real affine space) is a complex line and a symplectic manifold, of course.

The above family of Heisenberg algebras plays a key role, for example, in MRI (cf. [**72**] and [**68**]).

5.3.6. Automorphisms. Next let $a \in E$. Any $l \in \mathrm{Sp}(F^a)$ uniquely extends to the Heisenberg group $\mathcal{H}^a_{\text{red}}$ by setting the extension of l on the center of $U^{\frac{a}{|a|}}(1)$ to be the identity. Let this extension of l to $\mathcal{H}^a_{\text{red}}$ be called l again. Obviously, l is an automorphism of $\mathcal{H}^a_{\text{red}}$ meaning that

$$l\left((z_1 + k_1) \cdot (z_2 + k_2)\right) = l(z_1 + k_1) \cdot l(z_2 + k_2)$$

for all $z_1, z_2 \in U^{\frac{a}{|a|}}(1)$ and any $k_1, k_2 \in F^a$. These sorts of automorphisms play a fundamental role in chapter 9 on quantization. An according statement holds for $\mathcal{H}^a$.

5.4. $\mathcal{H}^a_{\text{red}}$ as a Semi-direct Product

5.4.1. Abelian Subgroups of $\mathcal{H}^a_{\text{red}}$. Let $a \in S^2$. The group structure of the Heisenberg group $\mathcal{H}^a_{\text{red}}$ obviously does *not* rely on a coordinate system in the symplectic plane F^a. However, if F^a is equipped with a coordinate system with axes $\mathbb{R} \cdot w$ and $\mathbb{R} \cdot w \cdot a$ where $w \in S^2$, the plane F^a turns into a *phase space* (cf. section 1.3.4). We observe

$$F^a = w \cdot \mathbb{C}^a. \tag{302}$$

(The $\cdot$ in (302) denotes the multiplication in $\mathbb{H}$.) This geometric structure causes an algebraic decomposition of $\mathcal{H}^a_{\text{red}}$. Indeed, any $w \in S^2$ yields the following two *abelian subgroups* of $\mathcal{H}^a_{\text{red}}$, namely the groups N^a and K^a which are

$$N^a := U^a(1) + \mathbb{R} \cdot w \cdot a \tag{303}$$

caused by the purely imaginary number $a \in \mathbb{C}^a$ and

$$K^a := \mathbb{R} \cdot w. \tag{304}$$

For any two $z_1 + \mu_1 \cdot w \cdot a$ and $z_2 + \mu_2 \cdot w \cdot a$ in N^a the product is given by

$$(z_1 + \mu_1 \cdot w \cdot a) \cdot (z_2 + \mu_2 \cdot w \cdot a) := z_1 \cdot z_2 + (\mu_1 + \mu_2) \cdot w \cdot a. \tag{305}$$

The product in K^a is

$$\lambda_1 \cdot w \cdot \lambda_2 \cdot w := (\lambda_1 + \lambda_2) \cdot w,$$

i.e. we use the structure of $\mathcal{H}^a_{\text{red}}$ to define the multiplication and addition for any two vectors $\lambda_1 \cdot w, \lambda_2 \cdot w$ in K^a.

5.4.2. Semidirect Product. Let us observe that N^a and K^a define a coordinate system on F^a with $\mathbb{R} \cdot w$ and $\mathbb{R} \cdot w \cdot a$ as axes. Any element in $\mathcal{H}^a_{\text{red}}$ can uniquely be written as a product (in $\mathcal{H}^a_{\text{red}}$) of an element in N^a and an element in K^a, as it is easily verified, and therefore

$$\mathcal{H}^a_{\text{red}} = N^a + K^a \tag{306}$$

(cf. (303) and (304)). Moreover, N^a is a normal subgroup of $\mathcal{H}^a_{\text{red}}$.

K^a operates on N^a by conjugation in $\mathcal{H}^a_{\text{red}}$, i.e.

$$(\lambda \cdot w) \cdot (z + \mu \cdot w \cdot a) \cdot (-\lambda \cdot w) = z \cdot e^{\lambda \cdot \mu \cdot a} + \mu \cdot w \cdot a \tag{307}$$

for any $\lambda \cdot w \in K^a$, any $z \in U^a(1)$ and any $\mu \in \mathbb{R}$. The orbit through $z + \mu \cdot w \cdot a$ is the circle $U^a(1) + \lambda \cdot w$.

$N^a + K^a$ in equation (306) has a semidirect product structure. The multiplication of this semidirect product is defined by

$$\begin{aligned} &(z_1 + \mu_1 \cdot w \cdot a + \lambda_1 \cdot w) \boxed{\mathrm{s}} (z_2 + \mu_2 \cdot w \cdot a + \lambda_2 \cdot w) \\ &:= \lambda_1 \cdot w + \lambda_2 \cdot w + (\lambda_2 \cdot w) \cdot (z_1 + \mu_1 \cdot w \cdot a) \cdot (-\lambda_2 \cdot w) \cdot (z_2 + \mu_2 \cdot w \cdot a) \end{aligned} \tag{308}$$

for all $z_1, z_2 \in U^a(1)$ and all $\mu_1, \mu_2, \lambda_1, \lambda_2 \in \mathbb{R}$, which rewritten in terms of (307) reads

$$\begin{aligned} &(z_1 + \mu_1 \cdot w \cdot a + \lambda_1 \cdot w) \boxed{\mathrm{s}} (z_2 + \mu_2 \cdot w \cdot a + \lambda_2 \cdot w) \\ &= \lambda_1 \cdot w + \lambda_2 \cdot w + (z_1 \cdot e^{\lambda_2 \cdot \mu_1 \cdot a} + \mu_1 \cdot w \cdot a) \cdot (z_2 + \mu_2 \cdot w \cdot a) \\ &= z_1 \cdot z_2 \cdot e^{\lambda_2 \cdot \mu_1 \cdot a} + \lambda_1 \cdot w + \lambda_2 \cdot w + \mu_1 \cdot w \cdot a + \mu_2 \cdot w \cdot a. \end{aligned}$$

The Lie group with $\mathcal{H}^a_{\text{red}}$ as the underlying topological space and $\boxed{\mathrm{s}}$ as the product shall be denoted by $\mathcal{H}^a_N$.

On the other hand, in the reduced Heisenberg group $\mathcal{H}^a_{\text{red}}$,

$$\begin{aligned} &(z_1 + \mu_1 \cdot w \cdot a + \lambda_1 \cdot w) \cdot (z_2 + \mu_2 \cdot w \cdot a + \lambda_2 \cdot w) \\ &= z_1 \cdot z_2 \cdot e^{\frac{1}{2}(\lambda_1 \cdot \mu_2 - \mu_1 \cdot \lambda_2) \cdot a} + (\lambda_1 + \lambda_2) \cdot w + (\mu_1 + \mu_2) \cdot w \cdot a \end{aligned} \tag{309}$$

holds true. This equation motivates a natural map

$$A : \mathcal{H}^a_{\text{red}} \longrightarrow \mathcal{H}^a_N = N^a \boxed{\mathrm{s}} K^a \tag{310}$$

assigning to any $z + h = z + \mu \cdot w \cdot a + \lambda \cdot w$ the value

$$A(z + \mu \cdot w \cdot a + \lambda \cdot w) := z \cdot e^{\mu \cdot \lambda \cdot a} + h.$$

With respect to the multiplication $\boxed{\mathrm{s}}$, A is a group isomorphism: This means that for any $z_1, z_2 \in U^a(1)$ and any $h_1, h_2 \in F^a$,

$$A((z_1 + h_1) \cdot (z_2 + h_2)) = A(z_1 + h_1) \boxed{\mathrm{s}} A(z_2 + h_2) \tag{311}$$

where $z_s + h_s$ is decomposed as $h_s = z_s + \mu_s \cdot w \cdot a + \lambda_s \cdot w$ for $s = 1, 2$. Indeed, the exponents involved in (310) and in turn (311) satisfy

$$\frac{1}{2}(\mu_2 \cdot \lambda_1 - \mu_1 \cdot \lambda_2) + \frac{1}{2}(\mu_1 + \mu_2) \cdot (\lambda_1 + \lambda_2) = \frac{1}{2}(\mu_1 \cdot \lambda_1 + \mu_2 \cdot \lambda_2) + \mu_2 \cdot \lambda_1$$

which shows that A is an isomorphism. Due to this we call $N^a \boxed{\mathrm{s}} K^a$ the *semidirect product decomposition* of $\mathcal{H}^a_{\text{red}}$ and write

$$\mathcal{H}^a_{\text{red}} = N^a \boxed{\mathrm{s}} K^a. \tag{312}$$

Note that the group structure on $\mathcal{H}^a_{\text{red}}$ does not involve a coordinate system while $N^a \boxed{\mathrm{s}} K^a$ does.

5.5. A Heisenberg Algebra Structure on sp(F)

5.5.1. sp(F) as a Heisenberg Algebra. Given any plane F in the oriented Euclidean space E equipped with a symplectic form ω there are associated a Heisenberg algebra (cf. section 5.2) on one hand and on the other hand a natural non-commutative field isomorphic to the skew field of quaternions causing this structure, as shown previously (cf. sections 1.1.5 and 5.2.1).

We will show next that sp(F) also carries natural Heisenberg algebra structures, again inherited from a skew field isomorphic to the quaternions.

At first we will construct a natural splitting of sp(F) and a symplectic structure on a two-dimensional subspace. The basis of this construction is End F, the four-dimensional $\mathbb{R}$-linear space of all ($\mathbb{R}$-linear) endomorphisms of F. The trace

$$\operatorname{tr} : \operatorname{End} F \longrightarrow \mathbb{R}$$

(being independent of a particular symplectic form on F) allows the splitting

$$\operatorname{End} F = \mathbb{R} \cdot \operatorname{id}_F \oplus \operatorname{sp}(F) \tag{313}$$

where sp(F) is the $\mathbb{R}$-linear space of all *traceless endomorphisms* of F. In fact, this splitting is obviously orthogonal with respect to the scalar product $< \, , \, >_{\mathbb{H}}$, say, for which the identity id_F is a unit vector and which is defined on sp(F) by

$$< A, B >_{\mathbb{H}} := \frac{1}{2} \operatorname{tr} A \circ \widetilde{B} \qquad \forall\, A, B \in \operatorname{sp}(F) \tag{314}$$

where $\widetilde{B}$ is the adjoint of B (cf. (1)). Moreover, let $\mu_{\operatorname{sp}(F)}$ be a constant *metric volume form* on sp(F) of the scalar product given by (314) (a normed determinant function in [**43**]).

Let $J \in \operatorname{sp}(F)$ be a solution of the equation

$$J^2 = -\operatorname{id}_F \tag{315}$$

(hence taking the role of the imaginary unit in $\operatorname{sp}(F) \subset \operatorname{End} F$). Both J_r and $(J_r)^{-1}$ in (267) satisfies (315). Given any solution J obviously $L \circ J \circ L^{-1}$ is another one, for any $L \in \operatorname{GL}(F)$. Clearly,

$$\operatorname{tr} J = 0$$

since the two eigenvalues of J are purely imaginary and conjugate to each other. Given the *orthogonal complement* Σ of J in sp(F) the orthogonal splitting

$$\operatorname{sp}(F) = \mathbb{R} \cdot J \oplus \Sigma$$

yields

$$\operatorname{End} F = \mathbb{R} \cdot \operatorname{id} + \mathbb{R} \cdot J + \Sigma. \tag{316}$$

Let J have matrix $\begin{pmatrix} 0 & 1 \\ -1 & 0 \end{pmatrix}$ as in (275.) The matrix of an element in Σ has the form $\begin{pmatrix} a & b \\ b & -a \end{pmatrix}$, obviously a *symmetric matrix* with vanishing trace. These matrices are formed with respect to an orthonormal basis in F. Hence

$$\Sigma = \operatorname{span} \left\{ a \cdot \begin{pmatrix} 1 & 0 \\ 0 & -1 \end{pmatrix}, c \cdot \begin{pmatrix} 0 & 1 \\ 1 & 0 \end{pmatrix} \middle| \, a, c \in \mathbb{R} \right\}.$$

If A and B in Σ have matrices $\begin{pmatrix} 1 & 0 \\ 0 & -1 \end{pmatrix}$ and $\begin{pmatrix} 0 & 1 \\ 1 & 0 \end{pmatrix}$, respectively, an explicit calculation shows

$$A \circ J = B \qquad J \circ B = A \qquad A \circ B = J \qquad \text{and} \qquad [A, B] = 2 \cdot J.$$

Here $[A, B]$ is the *commutator* of A and B. (Thus $\mathrm{sp}(F)$ is not closed under composition.) In addition, the linear two-dimensional space Σ inherits a *symplectic structure* ω_Σ defined by

$$\omega_\Sigma(A, B) := < \frac{1}{2} \cdot [A, B], J >_{\mathbb{H}} \qquad \forall\, A, B \in \mathrm{sp}(F). \tag{317}$$

Obviously,

$$< [A, J], J >_{\mathbb{H}} = < [J, A], J >_{\mathbb{H}} = 0$$

for any $A \in \Sigma$. Moreover,

$$[A, B] \in \mathbb{R} \cdot J \quad \text{if} \quad A, B \in \Sigma$$

implies

$$\mu_{\mathrm{sp}(F)}(A, B, J) = \alpha \cdot \omega_\Sigma(A, B) \qquad \forall\, A, B \in \Sigma$$

for some non-vanishing constant $\alpha \in \mathbb{R}$. Therefore, if $\| A \| = \| B \| = 1$

$$\mu_{\mathrm{sp}(F)}(A, B, J) = \omega_\Sigma(A, B) \qquad \forall\, A, B \in \Sigma. \tag{318}$$

This shows that $[A, B]$ satisfies

$$\mu_{\mathrm{sp}(F)}(A, B, J) = < \frac{1}{2} \cdot [A, B], J >_{\mathbb{H}} \tag{319}$$

identifying $\frac{1}{2} \cdot [A, B]$ as the *cross product* of $A, B \in \Sigma$ in $\mathrm{sp}(F)$. Equation (319) extends to any triplet of elements in $\mathrm{sp}(F)$. Hence $\mathrm{sp}(F)$ admits a natural cross product and the following is easily verified:

LEMMA 65. *The symplectic structure on Σ is uniquely extended to a metric volume form $\mu_{\mathrm{sp}(F)}$ on all of* $\mathrm{sp}(F)$ *by setting*

$$\mu_{\mathrm{sp}(F)}(\lambda \cdot J, A, B) := \lambda \cdot \omega_\Sigma(A, B) \qquad \forall\, A, B \in \mathrm{sp}(F) \text{ and } \forall\, \lambda \in \mathbb{R}.$$

With respect to $<,>_{\mathbb{H}}$ and the volume form $\mu_{\mathrm{sp}(F)}$ half of the Lie product of $\mathrm{sp}(F)$*, i.e. half of the commutator is a cross product on* $\mathrm{sp}(F)$.

The *Heisenberg algebra structure* on $\mathrm{sp}(F)$ is hence given by the Lie product

$$[A, B]_{\mathrm{sp}(F)} := < \frac{1}{2} \cdot [A, B], J >_{\mathbb{H}} \qquad \forall\, A, B \in \Sigma \tag{320}$$

(cf. (317) and (288)). The resulting Heisenberg algebra is denoted by $\mathrm{sp}(F)^H$. Clearly, J corresponds to the unit vector $a \in E$ in the construction of Heisenberg algebras in section 5.2. In fact, there are as many Heisenberg algebra structures on $\mathrm{sp}(F)$ as there are solutions to (315). Hence the collection of all Heisenberg algebra structures on $\mathrm{sp}(F)$ given by all solutions to (315) is parameterized by the unit sphere in $\mathrm{End}\, F$.

Turning back to the splitting of $\mathrm{End}\, F$ in (313), the following is obvious from lemma 65, the definition of quaternions and subsection 1.1.9 in the first chapter:

PROPOSITION 66. *Given the scalar product* $< \, , \, >_{\mathbb{H}}$ *and one of its normed determinant functions on* $\operatorname{End} F$*, the linear space* $\operatorname{End} F = \mathbb{R} \cdot id_F \oplus \operatorname{sp}(F)$ *is uniquely turned into a skew field isomorphic to the quaternions* $\mathbb{H}$ *such that the commutator of the multiplication restricted to* $\operatorname{sp}(F)$ *coincides with the Lie product in* $\operatorname{sp}(F)$ *(being twice the cross product) and causes the Lie product of the Heisenberg algebra* $\operatorname{sp}(F)^H$ *given by (320).*

5.5.2. Natural Isomorphisms of F onto Σ. The plane F allows many natural isomorphisms onto Σ. Indeed, given $q_0' \in F$ and the unit vector $A \in \Sigma$ with matrix $\begin{pmatrix} 1 & 0 \\ 0 & -1 \end{pmatrix}$, we define a linear map

$$j_\Sigma : F \longrightarrow \Sigma$$

on F by setting

$$\frac{j_\Sigma(q_0')}{\lambda} := A \qquad \text{and} \qquad \frac{j_\Sigma(J(q_0'))}{\lambda} := A \times J$$

supplemented by

$$\omega\left(\frac{q_0'}{\lambda}, \frac{J(q_0')}{\lambda}\right) = 1$$

for some real $\lambda > 0$. We observe

$$\omega\left(\frac{q_0'}{\lambda}, \frac{J(q_0')}{\lambda}\right) = \omega_\Sigma(A, A \times J) = 1$$

to obtain

$$\omega(q_0, J(q_0)) = \omega_\Sigma(A, A \times J) = 1 \tag{321}$$

provided $q_0 := \frac{q_0'}{\lambda}$. This is to say that j_Σ is a *symplectic* map and an *isometry* with respect to the scalar product $< \, , \, >_F$ given by

$$< A_1, A_2 >_F := < j_\Sigma(A_1), j_\Sigma(A_2) > \qquad \forall\, A_1, A_2 \in \Sigma$$

due to (314).

Clearly, if ω is not specified on F, equation (321) can be used to implement a symplectic structure by choosing any point in $F \backslash \{0\}$ as q_0.

$\operatorname{End} F$ operates on $j_\Sigma(F) = \Sigma$ by right multiplication. In addition,

$$\Sigma \circ \operatorname{End} F = \operatorname{End} \Sigma$$

holds true.

5.5.3. $\operatorname{Sp}(F)$ as a Reduced Heisenberg Group. There is a natural homeomorphism between $\operatorname{Sp}(F)$ and $S^1 \times \mathbb{R}^2$ as mentioned in section 5.1.3. This suggests that $\operatorname{Sp}(F)$ carries the structure of a reduced Heisenberg group with the *maximal compact subgroup* $U(1)$ of $\operatorname{Sp}(F)$ as its center.

To show the Heisenberg group structure of $\operatorname{Sp}(F)$ we consider the collection Sym^+ of all elements in $\operatorname{Sp}(F)$ with *symmetric, positive definite matrix* and uniquely decompose any $S \in \operatorname{Sp}(F)$ into

$$S = R \cdot e^{\lambda \cdot J}$$

with $\lambda \in \mathbb{R}$ and $R \in \mathrm{Sym}^+$. This is called the Cartan decomposition (cf. [**31**] and [**33**]). The map R is easily obtained by writing $R := (\tilde{S} \circ S)^{-\frac{1}{2}}$ ($\tilde{S}$ being the adjoint of S) and calculating the unitary map $U(\lambda) := e^{\lambda \cdot J}$ from the equation

$$S = (\tilde{S} \circ S)^{-\frac{1}{2}} \circ U.$$

Here $\lambda \in \mathbb{R}$. The collection Sym^+ has the following property: Let

$$\log S := \int_{-\infty}^{0} (\lambda \cdot \mathrm{id} - S)^{-1} - (\lambda - 1) \cdot \mathrm{id} \; d\lambda.$$

Since S is symmetric and positive no eigenvalue is less than or equal to zero. Developing both terms under the integral into power series yields the value of the integral (cf. [**83**] and [**32**]). We therefore may write

$$R = \exp \log R \qquad \forall\, R \in \mathrm{Sym}^+$$

with $\log R \in \mathrm{sp}(F)$, since $S \in \mathrm{Sp}(F)$. Hence any $S \in \mathrm{Sp}(F)$ can be written as

$$S = e^{\lambda J} \cdot e^{A} \qquad \text{for } A \in \mathrm{sp}(F)$$

with $A := \frac{1}{2} \log \tilde{S} \circ S$ in the linear space $\mathrm{sp}(F) \cap \mathrm{Sym}(F)$ where Sym is the collection of all symmetric elements in $\mathrm{End}\, F$ (cf. [**32**]). However, $\mathrm{sp}(F) \cap \mathrm{Sym}(F) = \Sigma$.

In fact, the exponential map

$$\exp : \mathrm{sp}(F) \longrightarrow \mathrm{Sp}(F)$$

yields the diffeomorphism

$$\exp|_{\Sigma} : \Sigma \longrightarrow \mathrm{Sym}^+$$

(cf. [**32**]). We transfer the addition of Σ to Sym^+ by $\exp|_{\Sigma}$. Hence we may write

$$\mathrm{Sp}(F) = U(1) \cdot \mathrm{Sym}^+ .$$

Therefore, the set $\mathrm{Sp}(F)$ inherits the structure of a Heisenberg group; the resulting group is denoted by $\mathrm{Sp}^H(F)$. Its Lie algebra is $\mathrm{sp}(F)^H$. Clearly, the group $\mathrm{Sp}^H(F)$ is not isomorphic to the group $\mathrm{Sp}(F)$, nor is the Heisenberg algebra $\mathrm{sp}(F)^H$ (cf. section 5.5.1) isomorphic to $\mathrm{sp}(F)$, of course.

5.6. The Spin Group and the Skew Field of Quaternions are Determined by Only One Heisenberg Group

5.6.1. su(2) **Determines $\mathfrak{h}^a$.** As we know from section 1.1.8 the three-sphere $S^3 \subset \mathbb{H}$ carries a group structure. In fact, $S^3 \subset \mathbb{H}$ determines $\mathbb{H}$ just by letting

$$k_1 \cdot k_2 = |k_1| \cdot |k_2| \cdot \frac{k_1}{|k_1|} \cdot \frac{k_2}{|k_2|} \qquad \forall\, k_1, k_2 \in \mathbb{H}.$$

On the other hand, given $a \in S^2$ the Heisenberg algebra

$$\mathfrak{h}^a = \mathbb{R} \cdot a \oplus F^a$$

inherits its bracket from $\mathbb{H}$ as can be seen from (288).

Therefore, the Lie algebra structure of $\mathfrak{h}^a$ is entirely determined by the spin group $\mathrm{SU}(2) \equiv S^3$. Hence the simply connected Lie group $\mathcal{H}^a$, i.e. the Heisenberg group, is entirely determined by SU(2). This is also true for the reduced Heisenberg group.

5.6.2. $\mathcal{H}^a$ **Determines** $\mathrm{SU}(2)$. To demonstrate the converse (cf. [**68**]), namely that given any $a \in S^2$ the Heisenberg group $\mathcal{H}^a$ determines $\mathrm{SU}(2)$ we will proceed as follows: Given S^3 in the *linear space* $\mathbb{H} = \mathbb{R} \cdot e \oplus E$ let $k_1, k_2 \in S^3$. Hence there are a_1, $a_2 \in S^2$ such that

$$k_s = e \cdot \cos t_s + a_s \cdot \sin t_s \qquad s = 1, 2$$

for some reals t_1 and t_2. Looking at the group law on S^3,

$$k_1 \cdot k_2 = e \cdot (\cos t_1 \cdot \cos t_2 + a_1 \cdot \sin t_2 + a_2 \sin t_1) + a_1 \cdot a_2 \cdot \sin t_1 \cdot \sin t_2$$

shows that we have to recover each of the terms on the right-hand side out of the multiplication in one Heisenberg group $\mathcal{H}^{a_1}$, say. The summation is performed in the linear space $\mathbb{H}$. Since

$$a_1 \cdot a_2 = a_1 \times a_2 - < a_1, a_2 > \cdot\, e$$

has to hold true, we need to investigate both $a_1 \times a_2$ and $< a_1, a_2 > \cdot\, e$.

The product $a_1 \cdot a_2$ emanates from the Heisenberg algebra. In fact, $a_1 \times a_2$ is in $F^{a_1} \cap F^{a_2}$. In case $a_1 = -a_2$, the product is $a_1 \cdot a_2 = |a|^2 \cdot e$ and we are done. In case $a_1 \notin \mathbb{R} \cdot a_2$, the component of a_1 in F^{a_1} is $a_1 - < a_1, a_2 > \cdot a_2$ and thus

$$< a_1 \times a_2, u >= -\omega^{a_1}(a_2, u) \qquad \forall\, u \in \mathbb{R} \cdot a_1 \times a_2$$

yielding

$$< a_1 \times a_2, \frac{u}{|u|} >= -\omega^{a_1}\left((a_2)_{F^{a_1}}, \frac{u}{|u|}\right) \quad \text{for } u \in \dot{\mathbb{R}} \cdot a_1 \times a_2$$

and in turn

$$a_1 \times a_2 = -\omega^{a_1}\left((a_2)_{F^{a_1}}, \frac{u}{|u|}\right) \cdot \frac{u}{|u|}.$$

Here $(a_2)_{F^{a_1}}$ denotes the F^{a_1}-component of a_2. Thus $a_1 \times a_2$ can be computed out of the Heisenberg group $\mathcal{H}^{a_1}$ and the elements a_1 and a_2. Hence $a_1 \cdot a_2$ emanates from data in $\mathcal{H}^{a_1}$.

Since $< a_1, a_2 >= \cos\psi$ and $|a_1 \times a_2| = \sin\psi$ for some $\psi \in \mathbb{R}$, we find

$$< a_1, a_2 >= \left(1 - \omega^{a_1}\left(\frac{u}{|u|}, a_2\right)^2\right)^{\frac{1}{2}}.$$

In fact, only one Heisenberg group is needed to determine the group structure of $\mathrm{SU}(2)$. Having established the group law on S^3 the multiplication on $\mathbb{H}$ is performed in the obvious way.

We have thus proved (cf. [**68**]):

THEOREM 67. *Given the linear space of* $\mathbb{H}$ *both the multiplication on* $\mathbb{H}$ *(and hence the structure of the skew field) and the multiplication on* S^3 *(and hence on* $\mathrm{SU}(2)$*) are determined by the Heisenberg group* $\mathcal{H}^a$ *for a fixed a in* S^2 *and vice versa.*

5.6.3. The Action of the Heisenberg Group on the Field of Quaternions. Since $\mathcal{H}^a$ for $a \in S^2$ determines S^3 one might expect that $\mathcal{H}^a$ acts on all of $\mathbb{H}$. In fact, this is indeed so.

Let $a \in S^2 \subset S^3$. Then $\mathbb{H}$ is split into $\mathbb{H} = \mathbb{C}^a \oplus F^a$ (cf. section 1.2.3). Each quaternion is hence of the form $z + h$ with $z \in \mathbb{C}^a$ and $h \in F^a$. The action $\underline{\psi}$ shall be given by

$$\begin{aligned} \underline{\psi} : \mathbb{C}^a \oplus F^a \times \mathcal{H}^a \quad &\longrightarrow \quad \mathbb{C}^a \oplus F^a \\ \underline{\psi}(z + h, e^{s\cdot a} + k) \quad &:= \quad z \cdot e^{s\cdot a} \cdot e^{\frac{1}{2}\omega^a(h,k)} + h + k \end{aligned}$$

for any $z \in \mathbb{C}^a$, any pair $h, k \in F^a$ and $s \in \mathbb{R}$. If $z \in U^a(1) \subset \mathbb{C}^a$, one sees that $\underline{\psi}(z + h, e^{s\cdot a} + k) \in \mathcal{H}^a$. Therefore, given any $\nu, s, r \in \mathbb{R}$ and any two $h, k \in F$,

$$\underline{\psi}(r \cdot e^{\nu \cdot t \cdot a} + h, e^{s\cdot a} + k) = r \cdot \left(e^{\nu\cdot t\cdot a} + \frac{h}{r}\right) \cdot \left(e^{-s\cdot t\cdot a} + k\right)$$

where the elements $e^{\nu\cdot t\cdot a} + \frac{h}{r}$ and $e^{s\cdot t\cdot a} + \frac{k}{r}$ of $\mathcal{H}^a$ are multiplied in $\mathcal{H}^a$.

5.7. Scalar Products and Minkowski Metrics on the Heisenberg Algebra

In this section we will show that the Minkowski geometry in *four* dimensions is entirely determined by the *three-dimensional* one which itself emanates from the Heisenberg algebra structure on $\mathfrak{h}^a$ with $a \in S^2$ (cf. (67)). In doing so will observe how tightly algebraic structures intertwine with symplectic and Minkowski geometry. At first, we prepare an algebraic support for this situation in the next section.

5.7.1. Heisenberg Algebra and Jordan Algebra. The Minkowski metric on $\mathbb{H}$ introduced in (11) in section 1.1.10 is based on the *Jordan product* $\odot$ given by

$$k_1 \odot k_2 = \frac{1}{2}(k_1 \cdot k_2 + k_2 \cdot k_1)$$

(cf. section 1.1.14). The following lemma expresses a remarkable fact, which is easily derived from (14):

LEMMA 68. *The linear subspace* $\mathfrak{h}^a_e := \mathbb{R} \cdot e \oplus F^a$ *of* $\mathbb{H}$ *is closed under* $\odot$.

Hence alternatively to (13) the Jordan product $\odot$ on $\mathfrak{h}^a_e$ can be introduced directly by (14). Thus $\mathfrak{h}^a_e$ is a natural Jordan algebra and thus admits a natural Minkowski metric $g_{\mathfrak{h}^a_e}$, defined by (11). Hence $g_{\mathfrak{h}^a_e}$ is a *canonical Minkowski metric* on $\mathfrak{h}^a_e$.

To understand the role of $\odot$ we consider a unit vector $u \in F^a$ and form span (e, u). This two-dimensional linear space is a field isomorphic to $\mathbb{C}$ with u as imaginary unit, this is to say $u \cdot u = -e$. The elements in span $(e, u_1) = \mathbb{C}^{u_1}$ and span $(e, u_2) = \mathbb{C}^{u_2}$ for two unit vectors u_1, u_2 in F^a cannot be multiplied together in the sense of complex numbers. However, $\odot$ unifies these multiplications.

5.7.2. Minkowski Metrics on $\mathbb{H}$ and on $\hbar^a$. In this subsection we will show that any *oriented* Minkowski space carries the structure of the skew field of quaternions. As shown in section 1.1.10 the $\mathbb{R}$-linear space $\mathbb{H}$ carries a natural Minkowski metric due to

$$\mathbb{H} = \mathbb{R} \cdot e \oplus E.$$

Given $h_j := \lambda_j \cdot e + u_j \in \mathbb{H}$ for $j = 1, 2$, the multiplication in $\mathbb{H}$ defines a natural Minkowski metric, again denoted by g_M, through

$$g_M(h_1, h_2) := < h_1 \cdot h_2, e > = \lambda_1 \cdot \lambda_2 - < u_1, u_2 >$$

(cf. (10)). Moreover, the commutator in $\mathbb{H}$ yields the cross product on E and hence a metric volume form on E, which extends naturally to an oricntation (in fact, to a metric volume form) on $\mathbb{H}$ by using the above splitting.

We will show next that on the other hand any oriented *four-dimensional* Minkowski space M with *Minkowski metric* g with signature $+ - --$ and a metric volume form μ (the orientation of M) admits a multiplication, turning M into a skew field isomorphic to the quaternions. To do so we first observe that any vector $e \in M$ for which $\mathbb{R} \cdot e$ is light-like, saying that

$$g(\lambda e, \lambda e) > 0 \qquad \forall\, \lambda \in \mathbb{R},$$

determines an orthogonal complement E *in* M on which g is negative definite, i.e. $<,> := g_M|_{E\times E}$ is a scalar product. Moreover, $\mu(e, \ldots, \ldots, \ldots)$ is an orientation on E and in fact a metric volume form. Together with the scalar product $<\, ,\, >$ it determines a vector product on E and M is split orthogonally into

$$M = \mathbb{R} \cdot e \oplus E.$$

This spitting allows us to introduce the quaternion multiplication: Since the mentioned splitting of M is g orthogonal we set

$$k_1 \cdot k_2 := g(k_1, k_2) \cdot e + \lambda_1 \cdot u_2 + \lambda_1 \cdot u_1 + \frac{1}{2} \cdot u_1 \times u_2$$

where $k_s = \lambda_s \cdot e + u_s \in M$ for $s = 1, 2$. Obviously, $g = g_M$. We therefore have:

PROPOSITION 69. *The skew field $\mathbb{H}$ is an oriented Minkowski space with natural metric g_M. Vice versa any oriented Minkowski space with metric g carries the structure of the quaternions; moreover, $g = g_M$.*

The next goal is to show that the four-dimensional Minkowski space $\mathbb{H}$ with its metric g_M is determined by a natural Minkowski metric on a Heisenberg algebra. To this end the metric g_M on a given skew field $\mathbb{H}$ shall naturally be transferred to the given Heisenberg algebra

$$\hbar^a = \mathbb{R} \cdot a \oplus F^a$$

with $a \in S^2$ and $F^a := a^{\perp_E}$, say. To do so we need the Heisenberg algebra

$$\hbar^a_e := \mathbb{R} \cdot e + F^a$$

with brackets

$$[\lambda_1 e + u_1, \lambda_2 e + u_2]_{\hbar^a_e} = \omega^a(u_1, u_2) \cdot e$$

for all $\lambda_i e + u_i \in \hbar^a_e, i = 1, 2$. The linear map

$$l_a : \mathbb{R} \cdot e \oplus F^a \quad \longrightarrow \quad \mathbb{R} \cdot a \oplus F^a$$

defined by

$$l_a(\lambda \cdot e + u) \quad := \quad \lambda \cdot a + u \cdot a \qquad \forall\, \lambda \in \mathbb{R} \text{ and } \forall\, u \in F^a \tag{322}$$

is an orthogonal, symplectic $\mathbb{R}$-linear isomorphism with respect to $< \, , \, >$ and ω^a, respectively. In fact, since F^a is closed under the multiplication by a, obviously,

$$(\mathbb{R} \cdot e + F^a) \cdot a = \mathbb{R} \cdot a \oplus F^a$$

holds true for each fixed $a \in S^2$. Moreover,

$$< l_a(\lambda_1 \cdot e + u_1), l_a(\lambda_2 \cdot e + u_2) >= \lambda_1 \cdot \lambda_2 + < u_1, u_2 > \tag{323}$$

for any choice of $\lambda_1, \lambda_2 \in \mathbb{R}$ and $u_1, u_2 \in F^a$, and hence ℓ_a is a Euclidean isometry.

We introduce the natural Minkowski metric $g_{\mathfrak{h}^a}$ on $\mathfrak{h}^a = \mathbb{R} \cdot a \oplus F^a$, given for any $k_s = \lambda_s \cdot a + u_s$ with $u_s \in F^a$ and $s = 1, 2$ by

$$g_{\mathfrak{h}^a}(k_1, k_2) := g_M(l_a^{-1} \cdot k_1, l_a^{-1} \cdot k_2) = \lambda_1 \cdot \lambda_2 - < u_1, u_2 > . \tag{324}$$

Since $k_s \cdot a^{-1} = \lambda_s \cdot e + u_s \cdot a^{-1}$ for $s = 1, 2$. Thus g_M determines $g_{\mathfrak{h}^a}$. Let us illustrate this a little in an algebraic fashion. Given $k = \lambda \cdot e + u$ in $\mathbb{H}$ with $u \in E$,

$$k^2 \cdot a = (\lambda^2 - |u|^2) \cdot a + 2\lambda \cdot u \cdot a$$

for any $a \in S^2$, yielding the following lemma:

LEMMA 70. *For any $a \in S^2$, the Minkowski metric $g_{\mathfrak{h}^a}$ on the Heisenberg algebra $\mathfrak{h}^a$ follows from the a-component of $k^2 \cdot a$ or more explicitly from*

$$< k^2 \cdot a, a >= g_{\mathfrak{h}^a}(k, k) \qquad \forall\, k \in \mathbb{H} \tag{325}$$

and the symplectic form satisfies

$$< k^2 \cdot a, v >= 2\lambda \cdot \omega^a(u, v) \qquad \forall\, v \in F^a$$

for all $k = \lambda \cdot e + u$ in $\mathbb{H}$ with $u \in E$. Obviously,

$$\omega^a(u_1, u_2) = g_{\mathfrak{h}^a}([u_1, u_2]_{\mathfrak{h}^a}, a) \quad \forall\, u_1, u_2 \in F^a. \tag{326}$$

The light cone of g_M on $\mathbb{H}$ is a three-dimensional cone centered at $0 \in \mathbb{H}$. Its intersection with $\mathbb{R} \cdot e \oplus F^a$ is a two-dimensional cone centered at $0 \in \mathbb{R} \cdot e \oplus F^a$ with $\mathbb{R} \cdot e$ as axis of rotation. One Heisenberg algebra $\mathfrak{h}^a$ determines $\mathbb{H}$ together with g_M by theorem 67 and $g_M|_{\mathfrak{h}^a} = g_{\mathfrak{h}^a}$. We therefore obtained:

COROLLARY 71. *For any $a \in S^2$ the Heisenberg algebra $\mathfrak{h}^a$ carries a natural Minkowski metric $g_{\mathfrak{h}^a}$ and hence determines an oriented four-dimensional Minkowski space carrying the structure of the skew field of quaternions; moreover, $g_M|_{\mathfrak{h}^a} = g_{\mathfrak{h}^a}$.*

5.8. Symplectic Group, Special Linear Groups and Lorentz group

In this section we will observe from a group theoretic point of view that the algebraic structures studied in the previous sections are all emanating from the component $O_0(1, 3)$ of the identity in the Lorentz group of g_M.

5.8.1. $O_0(1,2)$. The relations between ω^a and $g_{\hbar^a}$ mentioned in (326) in the previous section can further be illustrated *group theoretically* as a relation between

$$\mathrm{Sp}(F^a) = \mathrm{SL}(F^a)$$

and the component $O_0(1,2)$ of id in the group $O(1,2)$ of all $g_{\hbar^a}$-preserving linear isomorphisms of F^a. To demonstrate this we consider the Lie algebra $\mathrm{sl}(F^a)$ of $\mathrm{SL}(F^a) \subset \mathrm{End}\, F^a$ consisting of all *trace-free endomorphisms.* $\mathrm{sl}(F^a)$ is a three-dimensional $\mathbb{R}$-linear subspace of $\mathrm{End}\, F^a$. It is generated by the elements J, B_1, B_2 with the matrices

$$J = \begin{pmatrix} 0 & -1 \\ 1 & 0 \end{pmatrix}, \qquad B_1 = \begin{pmatrix} 1 & 0 \\ 0 & -1 \end{pmatrix}, \qquad B_2 = \begin{pmatrix} 0 & 1 \\ 1 & 0 \end{pmatrix},$$

formed with respect to an orthonormal basis $\{e_1, e_2\} \subset F^a$ for which $\{a, e_1, e_2\}$ is a $g_{\hbar^a}$-orthonormal basis of $\hbar^a$ (cf. section 5.5.1). Let $R : \mathrm{sl}(F^a) \longrightarrow \hbar^a$ be the linear map for which

$$R(J) = a, \qquad R(B_1) = e_1, \qquad R(B_2) = e_2.$$

Obviously,

$$g_{\hbar^a}(R(B'), R(B'')) = \frac{1}{2} \operatorname{tr} B' \circ B'' \qquad \forall\, B', B'' \in O_0(1,2).$$

Since $A \in \mathrm{Sp}(F^a)$ is trace free we conclude

$$A \circ B \circ A^{-1} \in \mathrm{sl}(F^a) \qquad \forall\, B \in \mathrm{sl}(F^a).$$

Thus the adjoint representation

$$\tau_{\mathrm{Sp}} : \mathrm{Sp}(F^a) \longrightarrow O_0(1,2) \tag{327}$$

on $\mathrm{Sp}(F^a)$ mapping any $B \in \mathrm{sl}(F^a)$ into

$$\tau_{\mathrm{Sp}}(A)(B) := A \circ B \circ A^{-1}$$

turns any $A \in \mathrm{Sp}(F^a)$ into a Lorentz transformation on $\mathrm{sl}(F^a)$. The map τ_{Sp} is a surjection with $\{\mathrm{id}\,, -\mathrm{id}\,\}$ as its kernel. Hence

$$O_0(1,2) = \mathrm{Sp}(F^a)/\{\mathrm{id}\,, -\mathrm{id}\,\} \tag{328}$$

(cf. [**5**], page 321).

5.8.2. A Fundamental Diagram. Turning back to $\mathbb{H}$ with the Minkowski metric g_M we relate the groups $\mathrm{SL}(2,\mathbb{C}), \mathrm{Sp}(F), O_0(1,3)$ and $O_0(1,2)$ to each other.

We have the diagram

$$\begin{array}{ccc} \mathrm{SL}(2,\mathbb{C}) & \supset & \mathrm{Sp}(F) \\ \Phi_L \big\downarrow & & \big\downarrow \tau_{Sp} \\ O_0(1,3) & \supset & O_0(1,2) \end{array} \tag{329}$$

where the vertical arrows are the natural $2:1$ covering maps given by conjugations (cf. section 1.4.4). Including the maximal compact subgroups, i.e. implementing diagram (39) in section 1.4.4 into diagram (329) yields the fundamental diagram

(330)
$$\begin{array}{ccccccc}
\mathrm{SL}(2,\mathbb{C}) & & & \supset & & & \mathrm{Sp}(F^a) \\
 & \rotatebox{-45}{$\supset$} & & & & \rotatebox{-135}{$\supset$} & \\
\Big\downarrow \Phi_L & & \mathrm{SU}(2) & \supset & U^a(1) & & \Big\downarrow \tau_{\mathrm{Sp}} \\
 & & \tau \downarrow & & \downarrow \tau & & \\
 & & \mathrm{SO}(E) & \supset & \mathrm{SO}(F^a) & & \\
 & \rotatebox{45}{$\supset$} & & & & \rotatebox{135}{$\supset$} & \\
O_0(1,3) & & & \supset & & & O_0(1,2)
\end{array}$$

encoding the various geometric aspects treated so far in the $3+1$-splitting of $\mathbb{H}$ and the $2+1$-splitting of E group theoretically. In fact, Φ and τ are the restrictions of Φ_L and τ_{Sp} in (329).

Thus $O_0(1,3)$ determines the fundamental diagram entirely. Hence it is the oriented Minkowski structure on $\mathbb{H}$ which determines all the algebraic structures of the groups appearing in the fundamental diagram. This Minkowski metric and $\mathbb{H}$, however, are both obtained from $\mathrm{Sp}(F^a)$, a natural Heisenberg group (cf. theorem 67 and subsection 5.5.2). Group theoretically formulated diagram (330) is the reason for the appearance of the Heisenberg group in so many different looking algebraic structures.

CHAPTER 6

The Heisenberg Group and Natural C^*-Algebras of a Vector Field in 3-Space

Throughout this chapter $M \subset E$ is either an open set or an oriented ∂-manifold (manifold with boundary) (cf. [**2**]) in the oriented three-dimensional Euclidean space E. Occasionally M will be assumed to be compact.

The aim of this chapter is to show that to any singularity free vector field $X = (\mathrm{id}, \mathbf{a})$ on M there corresponds a type of Heisenberg group and natural C^*-algebras determining the field. However, all these objects will be infinite dimensional ones. Nevertheless, it will allow us to handle X algebraically. These algebraic objects will form the basis for the Weyl quantization of X, presented in the last chapter.

The first goal will be to construct an *infinite dimensional Heisenberg group* $\mathcal{H}_X^\infty$ out of a given singularity free vector field $X = (\mathrm{id}, \mathbf{a})$ on M. This group will turn out to be characteristic for the field X in the sense that $\mathcal{H}_{X_1}^\infty$ is isomorphic to $\mathcal{H}_{X_2}^\infty$ iff X_1 and X_2 are equivalent and have the same field strength. The link between the vector field on M and $\mathcal{H}_X^\infty$ to be constructed is a natural bundle, the Heisenberg group bundle $\mathbf{H}^{\mathbf{a}}$. The construction mentioned above will require

$$\sup \mid \mathbf{a}(x) \mid < \infty \text{ for all } x \in M.$$

This restriction shall hold throughout this chapter.

One part of the space underlying $\mathcal{H}_X^\infty$ is a *space of sections* of the Heisenberg group bundle $\mathbf{H}^{\mathbf{a}}$ of X. More precisely, the construction of $\mathcal{H}_X^\infty$ is based on $\Gamma\mathbf{H}^{\mathbf{a}}$, the collection of all *Schwartz sections* of the natural Heisenberg group bundle $\mathbf{H}^{\mathbf{a}}$ of X. Its L_2-completion is identical to the space of *square integrable sections* (cf. [**18**] or [**66**]). The bundle $\mathbf{H}^{\mathbf{a}}$ is the sum $\mathbb{F}^{\mathbf{a}} \oplus \mathcal{L}$ (cf. section 3.1.4). The underlying linear space of $\mathcal{H}_X^\infty$ is $\Gamma\mathbf{H}^{\mathbf{a}} \oplus \mathbb{R} \cdot e$ where $e \in \mathbb{H}$ is the unit element of the skew field of quaternions (any fibre of $\mathbf{H}^{\mathbf{a}}$ is equipped with the according degenerate symplectic structure).

A word on the space of all Schwartz sections: Any finite dimensional vector bundle V on a finite dimensional manifold M admits another vector bundle N on M with typical fibre G such that $V \oplus N$ is trivial, i.e. smoothly isomorphic to $G \times M$, say. Any section of G can be identified with a G-valued function. This applies, in particular, to any section of $V \subset G$. A section of V that corresponds to a G-valued *Schwartz function* (cf. [**38**]) is called a *Schwartz section*. Hence ΓV is naturally identified with a subspace of the collection of all G-valued Schwartz functions. The L_2-completion of this subspace hence determines the L_2-completion of ΓV (cf. [**66**]).

In the second part of this chapter the group $\mathcal{H}_X^\infty$ will be enlarged to the discrete group algebra of which a natural completion is the C^**-group algebra* of $\mathcal{H}_X^\infty$

(cf. [**13**]). An analogue construction, based on $\Gamma\mathbf{H}^{\mathbf{a}}$, will yield a natural *Weyl algebra* $W\Gamma\mathbf{H}^{\mathbf{a}}$, a C^*-algebra, as well as a natural *Poisson algebra* $P^{\mathbf{a}}$; both $*$-algebras are characteristic of X (cf. [**14**]).

The C^*-algebras determined by $\mathcal{H}_X^\infty$ and $\Gamma\mathbf{H}^{\mathbf{a}}$ will be basic ingredients in the construction of a *quantization scheme* of X in chapter 10.

At the beginning of this chapter we start with the construction of the bundles $\mathbf{H}^{\mathbf{a}}$ as well as $\mathbf{h}^{\mathbf{a}}$, the Heisenberg Lie algebra bundle, and continue in the second part with the demonstration that $\mathcal{H}_X^\infty$ is *characteristic* for X.

6.1. The Heisenberg Group Bundle of a Vector Field

6.1.1. The Heisenberg Algebra Bundle of a Vector field. Any given *singularity free* vector field $X = (\mathrm{id}\,, \mathbf{a})$ on M defines a *complex line bundle* $\mathbb{F}^{\mathbf{a}}$ equipped with a fibrewisely defined *symplectic structure* $\omega^{\mathbf{a}}$ (cf. sections 3.1.4 and 3.2.1).

Following [**68**] we define the *Heisenberg algebra bundle*

$$\mathbf{h}^{\mathbf{a}} := \bigcup_{x\in M} \{x\} \times \mathbb{F}_x^{\mathbf{a}} \oplus \mathbb{R}\cdot \frac{\mathbf{a}(x)}{|\mathbf{a}(x)|},$$

regarded as a topological subspace of $M \times \mathbb{H}$, with $\omega^{\mathbf{a}(x)} = \omega_x^{\mathbf{a}}$ (cf. sections 1.3.5 and 3.2.1) on each fibre $\mathbb{F}_x^{\mathbf{a}}$ in

$$\mathbf{h}_x^{\mathbf{a}} := \mathbb{F}_x^{\mathbf{a}} + \mathbb{R}\cdot \frac{\mathbf{a}(x)}{|\mathbf{a}(x)|}.$$

For any $x \in M$ the symplectic form $\omega^{\mathbf{a}(x)}$ on $\mathbb{F}_x^{\mathbf{a}}$ is trivially extended to a two-form defined on all of the fibre $\mathbf{h}_x^{\mathbf{a}}$; the extension is still denoted by $\omega^{\mathbf{a}(x)}$. Expressed more precisely, this means that $\omega^{\mathbf{a}(x)}(\mathbf{a}(x), h) = 0$ for any h in $\mathbb{F}_x^{\mathbf{a}}$. This two-form on $\mathbf{h}_x^{\mathbf{a}}$ shall be called the *presymplectic structure* on $\mathbf{h}_x^{\mathbf{a}}$ or the *symplectically orthogonal extension* of $\omega^{\mathbf{a}(x)}$.

As is easily seen,

$$\mathbf{h}^{\mathbf{a}} = \mathbb{F}^{\mathbf{a}} \oplus \mathcal{L}^{\mathbf{a}} \tag{331}$$

as a vector bundle. Here

$$\mathcal{L}^{\mathbf{a}} = \bigcup_{x\in M} \{x\} \times \mathbb{R}\cdot \frac{\mathbf{a}(x)}{|\mathbf{a}(x)|}$$

(cf. section 3.1.4). Thus by (137) and the equations following

$$\mathbf{h}^{\mathbf{a}} = \varepsilon^* TS \oplus \varepsilon^* \mathcal{N} \tag{332}$$

where $\mathcal{N}$ is the *normal bundle* of the *unit sphere* in E and $\varepsilon(x) = \frac{\mathbf{a}(x)}{|\mathbf{a}(x)|}$ for any $x \in M$. This is to say that $\mathbf{h}^{\mathbf{a}}$ naturally emanates from $TS^2 \oplus \mathcal{N}$ as a pull-back by means of ε and hence is a vector bundle.

The Lie bracket on each fibre $\mathbf{h}_x^{\mathbf{a}}$ is

$$\left[h_1 + \lambda_1 \cdot \frac{\mathbf{a}(x)}{|\mathbf{a}(x)|}, h_2 + \lambda_2 \cdot \frac{\mathbf{a}(x)}{|\mathbf{a}(x)|}\right] := \omega^{\mathbf{a}(\mathbf{x})}(h_1, h_2)\cdot \frac{\mathbf{a}(x)}{|\mathbf{a}(x)|} \tag{333}$$

for all $h_1, h_2 \in \mathbb{F}^{\mathbf{a}}_x$ and for all $\lambda_1, \lambda_2 \in \mathbb{R}$. Here x varies in M. Clearly, by construction

$$\omega^{\mathbf{a}(x)}(h_1, h_2) \cdot \frac{\mathbf{a}(x)}{|\mathbf{a}(x)|} = \omega^{\mathbf{a}(x)}(h_1, h_2) \cdot \varepsilon(x)$$

for all variables (cf. section 3.2.2 and equation (287)). The fibrewisely formed symplectic form satisfies

$$\omega^{\mathbf{a}}(s_1, s_2) = \mu_E(s_1, \mathbf{a}, s_2)$$

for all sections $s_1, s_2 \in \Gamma\mathbb{F}^{\mathbf{a}}$ (cf. section 5.3). The trivial bundle $\mathcal{L} = \varepsilon^*\mathcal{N}$ is the bundle of centers of $\mathbf{h}^{\mathbf{a}}$.

Due to $\mathbb{F}^{\mathbf{a}} \subset \mathbf{h}^{\mathbf{a}}$, the bundle $\mathbf{h}^{\mathbf{a}}$ characterizes the vector field uniquely since the *field strength* $|\mathbf{a}|$ is encoded in the brackets of the fibres (cf. section 3.1.4). In other words, the vector field can be reconstructed from $\mathbf{h}^{\mathbf{a}}$.

The bundle $\mathbf{h}^{\mathbf{a}}$ is *not trivial* as a Heisenberg algebra bundle. Otherwise there would be a plane F in E, a vector $w \in E$ transversal to the plane and a smooth map

$$\Phi : \mathbf{h}^{\mathbf{a}} \longrightarrow M \times (F \oplus \mathbb{R} \cdot w),$$

fibrewisely a Heisenberg algebra isomorphism, which, in particular, yields the line bundle isomorphism

$$\Phi|_{\mathbb{F}^{\mathbf{a}}} : \mathbb{F}^{\mathbf{a}} \longrightarrow M \times F.$$

This is not true in general unless $(\mathrm{id}, \varepsilon)$ is trivial (cf. theorem 13 in section 3.1).

Since the topological nature of $\mathbb{F}^{\mathbf{a}}$ is determined by the *first Chern class*, $\mathbf{h}^{\mathbf{a}}$ is characterized topologically by this class, as well (cf. sections 4.3.4 and 4.3.6).

The vector field $X = (\mathrm{id}, \mathbf{a})$ is entirely characterized by the *characteristic bundle* $\mathbb{P}^{\mathbf{a}}$ (cf. section 3.7). In fact, $\mathbf{h}^{\mathbf{a}}$ is associated with $\mathbb{P}^{\mathbf{a}}$ as can be seen as follows: $\mathbb{F}^{\mathbf{a}}$ is a $U(1)$-principal bundle. The representation of $U(1)$ on E is given by identifying $U(1)$ with $U^a(1)$ in the obvious way for some fixed $a \in S^2$. Then $U^a(1)$ acts on $F = a^{\perp_E}$ (cf. section 1.2.4) and trivially on $\mathbb{R} \cdot a$.

Therefore,

$$\mathbb{P}^{\mathbf{a}} \times_{U(1)} F = \mathbb{F}^{\mathbf{a}}$$

and

$$\mathbb{P}^{\mathbf{a}} \times_{U(1)} (F \oplus \mathbb{R} \cdot a) = \mathbf{h}^{\mathbf{a}}$$

as a real vector bundle.

For any $x \in M$ the bracket on $\mathbf{h}^{\mathbf{a}}_x$ is inherited from $\mathbb{P}^{\mathbf{a}}_x$ since

$$\omega^{\mathbf{a}}(x; h_1, h_2) = |\mathbb{P}^{\mathbf{a}}_x|^{-2} \cdot \omega^{\varepsilon}(x; h_1, h_2) \qquad \forall x \in M \text{ and } h_1, h_2 \in \mathbb{F}_x.$$

Here $|\mathbb{P}^{\mathbf{a}}_x|$ is the diameter of the circle $\mathbb{P}^{\mathbf{a}}_x$ (cf. theorem 24). In fact,

$$\mathbb{P}^{\varepsilon} = \varepsilon^* \mathbb{P}_2$$

where $\mathbb{P}_2$ is the bundle of orthonormal frames of TS^2, a Hopf bundle on S^2 (cf. section 2.3.4) with 2 as the first Chern number (cf. section 4.3.9).

Therefore, we observe that the characteristic principal bundle $\mathbb{P}^{\mathbf{a}}$ of X uniquely determines the Heisenberg algebra bundle $\mathbf{h}^{\mathbf{a}}$.

This Heisenberg algebra bundle admits the bundle $\mathbb{S}p^{\mathbf{a}}$, say, of *symplectic frames* with the symplectic group $\mathrm{Sp}(F)$ as the structure group. This group acts on F and hence on $F \oplus \mathbb{R} \cdot a$ where it operates trivially, i.e. by the identity, on $\mathbb{R} \cdot a$. Hence we may obtain $\mathbf{h}^{\mathbf{a}}$ as an associated vector bundle:

$$\mathbf{h}^{\mathbf{a}} = \mathbb{S}p^{\mathbf{a}} \times_{\mathrm{Sp}(F)} (F \oplus \mathbb{R} \cdot a)$$

We will see later that the converse is true, as well.

6.1.2. The Heisenberg Group Bundle. The Heisenberg algebra bundle $\mathbf{h}^{\mathbf{a}}$ of the vector field $X = (\mathrm{id}\,, \mathbf{a})$ determines a *Heisenberg group bundle* $\mathbf{H}^{\mathbf{a}}$ with identical fibres as $\mathbf{h}^{\mathbf{a}}$, saying that $\mathbf{H}^{\mathbf{a}} = \mathbb{F}^{\mathbf{a}} \oplus \mathcal{L}$ as vector bundles. The fibrewise *group operations*, however, are

$$(h_1 + \lambda_1 \varepsilon(x)) \cdot (h_2 + \lambda_2 \varepsilon(x)) = h_1 + h_2 + \left(\lambda_1 + \lambda_2 + \frac{1}{2}\omega^{\mathbf{a}(x)}(h_1, h_2)\right) \cdot \varepsilon(x)$$

for all $\lambda_1, \lambda_2 \in \mathbb{R}$ and any $h_1, h_2 \in \mathbb{F}^{\mathbf{a}}_x$ and any $x \in M$ (cf. section 5.3.1).

The *reduced Heisenberg group bundle* $\mathcal{H}^{\mathbf{a}}_{\mathrm{red}}$ is topologically given as a subbundle

$$\mathbb{T}^{\mathbf{a}} \oplus \mathbb{F}^{\mathbf{a}} \subset \mathbb{C}^{\mathbf{a}} \oplus \mathbb{F}^{\mathbf{a}}$$

on M where

$$\mathbb{T}^{\mathbf{a}} \subset \mathbb{C}^{\mathbf{a}}$$

is the bundle of unitary groups over M (cf. section 5.3.3). The underlying set of this bundle is the disjoint union

$$\mathbb{T}^{\mathbf{a}} = \dot{\bigcup}_{x \in M} U^{\mathbf{a}(x)}(1).$$

The topology (and hence the C^∞-manifold structure) is inherited from $\mathbb{C}^{\mathbf{a}}$. The group operation on any fibre of $\mathbf{H}^{\mathbf{a}}_{\mathrm{red}}$ is defined by

$$(z_1 + h_1) \cdot (z_1 + h_2) := z_1 \cdot z_2 \cdot e^{\frac{1}{2}\omega^{\mathbf{a}(x)}(h_1, h_2) \cdot \varepsilon(x)} + h_1 + h_2$$

for all $z_1, z_2 \in \mathbb{C}^{\mathbf{a}}$, for all $h_1, h_2 \in \mathbb{F}^{\mathbf{a}}_x$ and all $x \in M$. As we will see in section 6.3, the bundle $\mathbf{H}^{\mathbf{a}}_{\mathrm{red}}$ determines the vector field uniquely, as well. In the sequel the emphasis will be mainly on $\mathbf{H}^{\mathbf{a}}$.

6.1.3. The Use of the Heisenberg Algebra Bundle. The Heisenberg algebra bundle $\mathbf{h}^{\mathbf{a}}$ as well as the Heisenberg group bundle $\mathbf{H}^{\mathbf{a}}$ reflect the *global nature* of a singularity free vector field $X = (\mathrm{id}\,, \mathbf{a})$ on M. *Locally* any singularity free vector field is trivial. This fact corresponds with the local triviality of $\mathbf{h}^{\mathbf{a}}$ and $\mathbf{H}^{\mathbf{a}}$. The global nature of the vector field is manifested in the sensitivity to the topology of the two bundles $\mathbf{h}^{\mathbf{a}}$ and $\mathbf{H}^{\mathbf{a}}$ which emanate from the topological characterization of the complex line bundle $\mathbb{F}^{\mathbf{a}}$ (cf. section 4.3.4) contained in both of them.

A quantization of X as mentioned in the introduction of this chapter will thus involve the global nature of X. Based on $\mathbf{h}^{\mathbf{a}}$ in chapter 10 we will construct what is called the *Weyl quantization* of a singularity free vector field on M. To this end we will next construct *infinite dimensional*, algebraic objects such as a Heisenberg algebra, a Heisenberg group and Weyl algebras of X.

6.2. Infinite Dimensional Heisenberg Algebras and Infinite Dimensional Heisenberg Groups of Vector Fields

In this section we introduce an infinite dimensional Heisenberg group and its infinite dimensional Heisenberg algebra, its Lie algebra, both constructed from the spaces of Schwartz sections $\Gamma\mathbf{H^a}$ and $\Gamma\mathbf{h^a}$ of $\mathbf{H^a}$ and $\mathbf{h^a}$, respectively. These infinite dimensional objects will turn out to be characteristic for the singularity free vector field $X = (\mathrm{id}\,, \mathbf{a})$ on M.

6.2.1. The Infinite Dimensional Heisenberg Algebra. The basic ingredients of the infinite dimensional Heisenberg algebra are $\mathbf{h^a}$, the Heisenberg algebra bundle, and $\Gamma\mathbf{h^a}$, the collection of all Schwartz sections of $\mathbf{h^a}$. An element in $\Gamma\mathbf{h^a} = \Gamma\mathbb{F}^{\mathbf{a}} \oplus \Gamma\mathcal{L}$ (cf. (331)) is denoted by $\mathbf{s}$. It is of the form

$$\mathbf{s} = s + \lambda \cdot \varepsilon$$

with $s \in \Gamma\mathbb{F}^{\mathbf{a}}$, the collection of all Schwartz sections of $\mathbb{F}^{\mathbf{a}}$ and $\lambda \in \mathcal{S}(M, \mathbb{R})$, the collection of all real-valued *Schwartz functions* of M endowed with the C^∞-topology if necessary (cf. [**38**]). Here $\Gamma\mathbb{F}^{\mathbf{a}}$ is a $\mathcal{S}(M, \mathbb{R})$-module. The $\mathbb{R}$-linear space $\Gamma\mathbf{h^a} = \Gamma\mathbb{F}^{\mathbf{a}} \oplus \Gamma\mathcal{L}$ (cf. (332)) admits a $\mathcal{S}(M, \mathbb{R})$-module structure, as well.

$$\hbar_X^\infty := \Gamma\mathbf{h^a} \oplus \mathbb{R} \cdot e \tag{334}$$

will be turned into an *infinite dimensional* Heisenberg algebra. Here $e \in \mathbb{H}$ is the unit element in the skew field of quaternions (cf. section 1.1.1). Elements in $\Gamma\mathbf{h^a} = \Gamma\mathbf{H^a}$ (being identical as spaces of sections) shall be denoted by $\mathbf{s}$.

The *presymplectic structure* on $\Gamma\mathbf{h^a}$ is defined to be

$$\omega^{\Gamma\mathbf{h^a}}(\mathbf{s}_1, \mathbf{s}_2) := \int_M \omega^{\mathbf{a}}(\mathbf{s}_1, \mathbf{s}_2) \cdot \mu_M \qquad \forall\, \mathbf{s}_1, \mathbf{s}_2 \in \Gamma\mathbf{h^a} \tag{335}$$

where μ_M denotes the Riemannian volume form of M, which yields the bracket

$$[\mathbf{s}_1 + \lambda_1 \cdot e, \mathbf{s}_2 + \lambda_2 \cdot e] := \omega^{\Gamma\mathbf{h^a}}(\mathbf{s}_1, \mathbf{s}_2) \cdot e \qquad \forall\, \mathbf{s}_1, \mathbf{s}_2 \in \Gamma\mathbf{h^a} \tag{336}$$

on $\hbar_X^\infty$. Obviously, $\omega^{\Gamma\mathbf{h^a}}$ vanishes identically on $\Gamma\mathcal{L} \subset \Gamma\mathbf{h^a}$ and $\omega^{\Gamma\mathbf{h^a}}$ extends trivially and symplectically orthogonal to $\mathbb{R} \cdot e$ to become a *presymplectic structure* on all of $\hbar_X^\infty$. The center is $\Gamma\mathcal{L} + \mathbb{R} \cdot e$ (cf. (331)).

Let us replace $\omega^{\Gamma\mathbf{h^a}}$ on $\hbar_X^\infty$ by the more simple symbol ω_X^∞ if no confusion arises.

6.2.2. The Infinite Dimensional Heisenberg Group. The *infinite dimensional* Heisenberg group

$$\mathcal{H}_X^\infty := \Gamma\mathbf{H^a} \oplus \mathbb{R} \cdot e \tag{337}$$

is endowed with the *group operation*

$$(\mathbf{s}_1 + \lambda_1 \cdot e) \cdot (\mathbf{s}_2 + \lambda_2 \cdot e) := \mathbf{s}_1 + \mathbf{s}_2 + \frac{1}{2} \cdot \omega_X^\infty(\mathbf{s}_1, \mathbf{s}_2) + (\lambda_1 + \lambda_2) \cdot e \tag{338}$$

for all choices of $\mathbf{s}_1, \mathbf{s}_2 \in \Gamma\mathbf{h^a}$ and all $\lambda_1, \lambda_2 \in \mathbb{R}$.

The *commutator* on $\mathcal{H}_X^\infty$ yields the *Lie bracket* of $\hbar_X^\infty$.

Of course we could as well pass on to the reduced (infinite dimensional) Heisenberg group

$$\mathcal{H}_{X,\mathrm{red}}^\infty := \Gamma\mathbf{h^a} + U(1) \cdot e.$$

The group structure is defined in the obvious way using the exponential map on $\mathbb{R}$.

Let us point out that $\Gamma\mathbf{H}^{\mathbf{a}} = \Gamma\mathbf{H}^{\mathbf{a}}_{|\mathbf{a}|}$ as linear spaces, though they differ in the (degenerate) presymplectic structure.

Moreover, the L_2-scalar product

$$\ll \mathbf{s}_1, \mathbf{s}_2 \gg := \int_M < \mathbf{s}_1, \mathbf{s}_2 > \mu_M \tag{339}$$

turns $\Gamma\mathbf{H}^{\mathbf{a}}$ (regarded as an $\mathbb{R}$-*linear space*) into a pre-Hilbert space. Here any element in $\mathcal{H}_X^\infty$ is of the form

$$\mathbf{s} + \eta \cdot e = s + \lambda \cdot \varepsilon + \eta \cdot e$$

for some $s \in \Gamma\mathbb{F}^{\mathbf{a}}$, some $\lambda \in \mathcal{S}(M, \mathbb{R})$ and some constant $\eta \in \mathbb{R}$.

The scalar product $\ll \, , \, \gg$ on $\Gamma\mathbf{h}^{\mathbf{a}}$ orthogonally combined with $< \, , \, >$ on $\mathbb{R} \cdot e$ yields a scalar product on the underlying linear space of $\mathcal{H}_X^\infty$. This new scalar product is again denoted by $\ll \, , \, \gg$.

6.2.3. The Lie Algebra of $\mathcal{H}_X^\infty$. We will verify here that the Lie algebra

$$\operatorname{Lie} \mathcal{H}_X^\infty := T_e \mathcal{H}_X^\infty$$

of $\mathcal{H}_X^\infty$ is indeed $\mathfrak{h}_X^\infty$. To determine the right-hand side of the above definition we thus have to use differentiation along smooth curves. To this end we first have to specify a topology on $\mathcal{H}_X^\infty$: First we observe that the underlying linear spaces of $\mathcal{H}_X^\infty$ and $\mathfrak{h}_X^\infty$ are identical. This linear space shall be equipped with the C^∞-topology (cf. [**18**]). It is hence a locally convex space but in general not normed. The differential calculus on spaces of sections and functions adopted here is the one in [**18**] and [**51**]. They differ in the respective techniques, however, they agree for the C^∞-calculus as shown in [**57**]. To outline this very technical calculus at this time would be beyond the scope of this book. However, we will present the notion of a tangent vector on an intuitive level and leave it to the interested reader to verify the details.

The curve to determine any tangent vector to $\mathcal{H}_X^\infty$ can be chosen to be an affine vector, passing through a given vector (a section or a real-valued function) in the underlying linear space. It is hence smooth and we immediately deduce that

$$\operatorname{Lie} \mathcal{H}_X^\infty = \mathfrak{h}_X^\infty,$$

as linear spaces. Here "Lie" stands for the "Lie algebra of".

To compute the Lie bracket we proceed according to the method which yielded (296), where $\omega^{\mathbf{a}}$ has to be replaced by $\omega^{\Gamma\mathbf{h}^{\mathbf{a}}}$ in section 6.3. The differentiation of arguments of $\omega^{\Gamma\mathbf{h}^{\mathbf{a}}}$ with respect to the parameter t (under the integral) is justified by the calculus presented in [**18**] or [**51**], as well. The reason is that $\omega^{\mathbf{a}}$ is a continuous bilinear map and that the volume element in (335) is independent of t.

Thus we obtain for the Lie bracket

$$[\mathbf{s}_1 + \lambda \cdot e, \mathbf{s}_2 + \lambda \cdot e] = \omega_X^\infty(\mathbf{s}_1, \mathbf{s}_2) \cdot e$$

in $\operatorname{Lie} \mathcal{H}_X^\infty$ as is easily verified. This bracket holds true accordingly for $\mathcal{H}_{X,\mathrm{red}}^\infty$. Thus we may state:

LEMMA 72. *For each singularity free vector field X on M,*

$$\operatorname{Lie} \mathcal{H}_X^\infty = \mathfrak{h}_X^\infty \qquad \textit{and} \qquad \operatorname{Lie} \mathcal{H}_{X,\mathrm{red}}^\infty = \mathfrak{h}_{X,\mathrm{red}}^\infty.$$

6.2.4. The Exponential Map. The exponential map

$$\exp_x : \mathbf{h}^{\mathbf{a}}_x \longrightarrow \mathbf{H}^{\mathbf{a}}_x,$$

a map which is the *identity* (cf. [**37**]) for any x in M, yields the exponential map

$$\exp^X : \hbar^\infty_X \longrightarrow \mathcal{H}^\infty_X.$$

It is defined by

$$\begin{aligned}(\exp^X(\mathbf{s} + \lambda \cdot e))(x) &:= \exp_x(\mathbf{s}(x)) + \exp \lambda \cdot e \\ &= \mathbf{s}(\mathbf{x}) + \lambda \cdot \mathbf{e}\end{aligned}$$

for all $x \in M$ and all $\lambda \in \mathbb{R}$. By the so-called Ω lemma (cf. [**18**], [**51**] or [**66**]) this map is smooth. Since $\exp_x$ is the identity, for any $x \in M$,

$$\exp^X = \mathrm{id}$$

and thus is a diffeomorphism.

Finally, let us point out that $\Gamma\mathbf{H}^{\mathbf{a}}$ has the structure of an *infinite dimensional Heisenberg group*, as well; the operation is determined by

$$(s_1 + \lambda_1 \varepsilon) \cdot (s_2 + \lambda_2 \varepsilon) := s_1 + s_2 + \frac{1}{2}\omega^{\Gamma\mathbf{H}^{\mathbf{a}}}(s_1, s_2) \cdot \varepsilon$$

for all $s_1, s_2 \in \Gamma\mathbb{F}^{\mathbf{a}}$ and for all $\lambda_1, \lambda_2 \in \mathcal{S}(M, \mathbb{R})$. An according *Heisenberg algebra structure* is given for $\Gamma\mathbf{h}^a$.

6.2.5. $\mathcal{H}^\infty_X$ and $\hbar^\infty_X$ as Infinitesimal Objects. In this subsection we will reveal the infinitesimal character of the infinite dimensional Heisenberg group $\mathcal{H}^\infty_X$ and its Lie algebra. This view will rely on vector fields along ε. The bundle map $\hat{\varepsilon} : \mathbb{F}^{\mathbf{a}} \longrightarrow TS^2$ (cf. (135)) will cause a relation between $\Gamma\mathbb{F}^\varepsilon$ and ΓTS^2 by means of $\hat{\varepsilon}$ (cf. section 3.1.4). The reason is that $\hat{\varepsilon}$ is fibrewisely a linear isomorphism. More precisely, diagram (135) immediately yields the map

$$\hat{\varepsilon}^* : \Gamma TS^2 \longrightarrow \Gamma\mathbb{F}^{\mathbf{a}} \tag{340}$$

which sends each $s' \in \Gamma TS^2$ into

$$\hat{\varepsilon}^*(s') := \hat{\varepsilon}^{-1} \circ s' \circ \varepsilon.$$

The short hand $\hat{\varepsilon}^{-1} \circ s' \circ \varepsilon$ is defined by the map

$$\hat{\varepsilon}^{-1} \circ s' \circ \varepsilon(x) := \hat{\varepsilon}_x^{-1}(s'(\varepsilon(x))) \qquad \forall x \in M.$$

Here $\hat{\varepsilon}_x^{-1}$ denotes the fibrewisely formed inverse of $\hat{\varepsilon}_x := \hat{\varepsilon}|_{\mathbb{F}^{\mathbf{a}}_x}$ for any $x \in M$. The image of $\hat{\varepsilon}^*$ depends highly on the quality of ε. In general, it is neither *injective* nor *surjective*. Its kernel is generated by those vector fields s of S^2 vanishing on $\varepsilon(M)$. If there is a point $y \notin \overline{\varepsilon(M)} \subset S^2$, then there is a vector field $s \in \Gamma TS^2$ vanishing on $\varepsilon(M)$ but not on y. The map $\hat{\varepsilon}^*$ is obviously injective if ε is surjective.

Consequently, we have the following obvious facts:

LEMMA 73. *Given a singularity free vector field $X = (id, \mathbf{a})$ on M, the following holds true:*

- **i):** *$\hat{\varepsilon}^*$ is injective if ε is surjective.*
- **ii):** *If an open neighborhood of $y \in S^2$ is outside of $\overline{\varepsilon(M)}$, then ε fails to be surjective.*

For the classification of the maps in $C^\infty(M, S^2)$ by means of the mapping degree see section 4.4.

In order to show the infinitesimal character of both $\mathcal{H}_X^\infty$ and $\hbar_X^\infty$, the next technical goal is to relate $\Gamma\mathbb{F}^\varepsilon$ with the tangent space $T_\varepsilon C^\infty(M, S^2)$.

Therefore, let us consider some vector field $s'' \in \Gamma TS^2$ on S^2 in order to relate $\Gamma\mathbb{F}^{\mathbf{a}}$ with the geometry of S^2 and the map ε. The diagram

$$\begin{array}{ccc} & & TS^2 \\ & \overset{s''\circ\varepsilon}{\nearrow} & \uparrow s'' \\ M & \xrightarrow{\ \varepsilon\ } & S^2 \end{array} \tag{341}$$

commutes. Thus $s'' \circ \varepsilon$ is a tangent vector to $\varepsilon \in C^\infty(M, S^2)$ in the following sense: Let $\exp_{S^2}$ be the exponential map from TS^2 to S^2; then there is a smooth curve in $C^\infty(M, S^2)$ (carrying the C^∞-topology) (cf. [**18**]) passing through ε such that the derivative with respect to time is $s'' \circ \varepsilon$ as we will demonstrate now.

To do this we will need a C^∞-*calculus* on $C^\infty(M, S^2)$ again. As above, we refer to [**18**], [**51**] or [**40**] (which all produce the same C^∞-classes of maps (cf. [**57**])). In such a calculus in general it is rather easy to get the derivatives formally. However, to verify this rigorously, one of the formalisms referred to is needed. In fact, we will be in this situation: In the sequel we will obtain some TS^2-valued fields by means of ordinary calculus in finite dimensional linear spaces. To show that these fields are in fact derivatives of a curve c one of the formalisms is needed. We will omit the technicalities in verifying this, however, we will give some slight hints of details.

Turning back to $s'' \circ \varepsilon$ we will observe first that the family of maps

$$\sigma(t) := \exp_{S^2} \circ (ts'' \circ \varepsilon) \qquad \forall\, t \in (-1, 1)$$

in $C^\infty(M, S^2)$ yields the smooth curve

$$\sigma_x : (-1, 1) \longrightarrow M \qquad \forall\, x \in M$$

on M mapping any $t \in (-1, 1)$ into $\exp_{S^2} ts''(\varepsilon(x)) \in M$. Differentiation at $t = 0$ yields

$$\frac{d}{dt}(\sigma(t)(x))\bigg|_{t=0} = s'' \circ \varepsilon(x) \qquad \forall\, x \in M.$$

Hence $s'' \circ \varepsilon$ is a smooth vector field on M along ε. Here a smooth map $s : M \to TS^2$ is called a *vector field along* ε iff $\mathrm{pr}_{S^2} \circ s = \varepsilon$. Therefore, the collection of all smooth vector fields along ε is called the *tangent space* of $C^\infty(M, S^2)$ at ε; it is denoted by $T_\varepsilon C^\infty(M, S^2)$ (cf. [**18**]). We sketch a motivation of this definition as follows:

Let

$$\sigma : (-1, 1) \longrightarrow C^\infty(M, S^2)$$

be any smooth curve in the sense of [**18**]. Moreover, for each fixed $x \in M$ the point evaluation

$$\sigma : (-1, 1) \longrightarrow S^2, \qquad t \mapsto \sigma(t)(x)$$

is smooth in the ordinary sense. Assuming $\sigma(0) = \varepsilon$ the differentiation with respect to t yields the tangent vector

$$\frac{d}{dt}\sigma_x(t)\bigg|_{t=0} \in T_{\varepsilon(x)} S^2.$$

Indeed, for any x,

$$\frac{d}{dt}\sigma_x(t)\bigg|_{t=0} = \frac{d}{dt}\sigma(t)\bigg|_{t=0}(x).$$

Setting

$$\dot{\sigma}(0) := \frac{d}{dt}\sigma(t)\bigg|_{t=0}$$

we find the smooth map $\dot{\sigma}$, a vector field along ε, for which

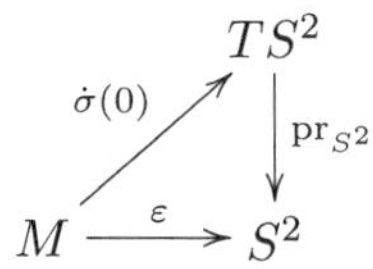

commutes. Here pr_{S^2} is the natural projection. On the other hand, given some vector field $s'' : M \longrightarrow TS^2$ along ε, the map $\sigma(t) \in C^\infty(M, S^2)$ is defined by

$$\sigma(t)(x) := \exp_{S^2} \circ (t \cdot s''(x)) \qquad \forall\, x \in M \text{ and } \quad \forall\, t \in \mathbb{R}.$$

Obviously,

$$\dot{\sigma}(0) = s''$$

showing that s'' is a tangent vector to ε. Thus the following is immediate:

PROPOSITION 74. *In the sequence*

$$\Gamma TS^2 \xrightarrow{\varepsilon^*} T_\varepsilon C^\infty(M, S^2) \xrightarrow{\hat{\varepsilon}_*} \Gamma\mathbb{F}^{\mathbf{a}} \tag{342}$$

the map $\hat{\varepsilon}_*$ is an injection. Its image $\mathrm{im}\,\hat{\varepsilon}_*$ *consists of all sections in $\Gamma\mathbb{F}^{\mathbf{a}}$ which are constant along $\varepsilon^{-1}(y)$ for all $y \in S^2$. (The lower star means the fibrewisely formed composition with $(\hat{\varepsilon})^{-1}$ from the left.)*

The surjectivity of $\hat{\varepsilon}_*$ in (342) in general fails to be true. This is rather easily seen. A vector field along ε is constant along $\varepsilon^{-1}(y)$ for any $y \in S^2$. On the other hand, $\Gamma\mathbb{F}^{\mathbf{a}}$ separates points in the sense that given any two different points $x_1, x_2 \in \varepsilon^{-1}(y)$ (if there are any), then there is some $s \in \Gamma\mathbb{F}^{\mathbf{a}}$ such that

$$s(x_1) \neq s(x_2).$$

Thus $\hat{\varepsilon} \circ s \notin T_\varepsilon C^\infty(M, S^2)$.

Now we also have:

LEMMA 75. *For each singularity free vector field $X = (id, \mathbf{a})$ on M the map $\hat{\varepsilon}_*$ factors as follows: Let M_ε be the topological space obtained by identifying $\varepsilon^{-1}(y)$ to one point for each $y \in S^2$; the set M_ε is equipped with the finest topology for which the projection map $\mathrm{pr}_\varepsilon : M \longrightarrow M_\varepsilon$ is continuous. Thus ε factors into*

$$\varepsilon = \varepsilon' \circ \mathrm{pr}_\varepsilon \tag{343}$$

with the continuous map $\varepsilon' : M_\varepsilon \longrightarrow S^2$ defined by equation (343). Thus ε' is injective and, moreover,

$$\mathbb{F}^{\mathbf{a}} = \varepsilon^* TS^2 = \mathrm{pr}_\varepsilon^*(\varepsilon'^*(TS^2)). \tag{344}$$

(The upper $*$ is the pull-back.) Therefore,

$$\begin{array}{ccc} T_\varepsilon C^\infty(M,S^2) & \xrightarrow{\hat{\varepsilon}_*} & \Gamma\mathbb{F}^{\mathbf{a}} \\ & \searrow^{(\hat{\varepsilon}')_*} & \uparrow {\scriptstyle (\widehat{\mathrm{pr}_\varepsilon})_*} \\ & & \Gamma\varepsilon'^* TS^2 \end{array}$$

commutes. Hence $(\hat{\varepsilon}')_*$ is an isomorphism. $\hat{\varepsilon}'$ and $\widehat{\mathrm{pr}_\varepsilon}$ are defined according to $\hat{\varepsilon}$. A complement of $\mathrm{im}\,\hat{\varepsilon}_*$ contains the sections which are not constant on fibres of pr_ε.

Therefore, the topological nature of ε^*TS^2 (as expressed, for example, by the Chern class of $\mathbb{F}^{\mathbf{a}}$) is influenced by ε' (cf. section 4.4) and by the fibres of $\mathrm{pr}_\varepsilon^{-1}(\mathrm{pr}_\varepsilon(x))$ for any $x \in M$ as expressed in (344).

Now we will be able to observe the infinitesimal character of $\mathcal{H}_X^\infty$. The infinite dimensional Heisenberg group $\mathcal{H}_X^\infty$ contains a natural subgroup. The underlying linear space is

$$\mathcal{H}_{T_\varepsilon}^\infty := \hat{\varepsilon}^*(T_\varepsilon C^\infty(M,S^2)) \oplus \Gamma\mathcal{L} \oplus \mathbb{R}\cdot e.$$

Hence

$$\mathcal{H}_{T_\varepsilon}^\infty \subset \mathcal{H}_X^\infty.$$

The group operations are inherited from those of $\mathcal{H}_X^\infty$, i.e. given by (338). Its Lie algebra is

$$\hbar_{T_\varepsilon}^\infty := \hat{\varepsilon}^*(T_\varepsilon C^\infty(M,S^2)) \oplus \Gamma\mathcal{L} \oplus \mathbb{R}\cdot e.$$

Hence

$$\hbar_{T_\varepsilon}^\infty \subset \hbar_X^\infty$$

with the bracket inherited from $\hbar_X^\infty$, i.e. given by (336).

Clearly, $\mathcal{H}_{T_\varepsilon}^\infty$ and $\hbar_{T_\varepsilon}^\infty$ are of an *infinitesimal* nature since they are built from $T_\varepsilon C^\infty(M,S^2)$ and $\mathbb{R}\cdot e$. Thus both $\mathcal{H}_X^\infty$ and $\hbar_X^\infty$ reflect *infinitesimal properties* of ε within $C^\infty(M,S^2)$ as well as topological properties inherited from pr_ε (cf. lemma 75).

Obviously,

$$\mathcal{H}_X^\infty / \mathcal{H}_{T_\varepsilon}^\infty = \Gamma\mathbb{F}^{\mathbf{a}} / \mathrm{im}\,\hat{\varepsilon}_* = \hbar_X^\infty / \hbar_{T_\varepsilon}^\infty.$$

Therefore, both $\mathcal{H}_{T_\varepsilon}^\infty$ and $\hbar_{T_\varepsilon}^\infty$ are entirely determined by ε and hence encode algebraically topological properties of $\varepsilon^*TS^2 = \mathbb{F}^{\mathbf{a}}$.

6.3. Maps Determined by Homomorphisms

So far we have constructed an infinite dimensional Heisenberg group, respectively, Heisenberg algebra for each singularity free vector field $X = (\mathrm{id}\,, \mathbf{a})$ on $M \subset E$. They both reflect global analytic and global topological properties of X. Immediate questions arising in this context are:

- **i):** To what extent do $\mathcal{H}_X^\infty$ and $\hbar_X^\infty$ determine X?
- **ii):** To what extent do they determine $\mathbf{H}^{\mathbf{a}}$, $\mathbf{h}^{\mathbf{a}}$ and $\mathbb{F}^{\mathbf{a}}$?

The answers to these questions will make apparent in which way the vector field admits a reformulation in algebraic terms, namely in terms of C^*-algebras. These will be introduced below in sections 6.4 and 6.5. They will play a major role in the quantization procedure of X in the last chapter.

But first we will answer question (ii). Since X is uniquely determined by the complex line bundle $\mathbb{F}^{\mathbf{a}}$ and $\omega^{\mathbf{a}}$, the answer to (ii) is almost immediate and allows us to derive the answer to question (i). Here is the setting we will start with.

Let $\mathbf{a}$ and $\mathbf{a}'$ be the principal parts of two singularity free vector fields $X = (\mathrm{id}\,, \mathbf{a})$ and $X' = (\mathrm{id}\,, \mathbf{a}')$, respectively, on M and

$$\Psi : \mathbb{F}^{\mathbf{a}'} \longrightarrow \mathbb{F}^{\mathbf{a}}$$

an *isomorphism* of *complex line bundles.* (By definition such an isomorphism is smooth). This is to say that the vector fields X and X' are equivalent, meaning that the bundles $\mathbb{F}^{\varepsilon}$ and $\mathbb{F}^{\varepsilon'}$ of the respective unit vector fields $X = (\mathrm{id}\,, \varepsilon)$ and $X = (\mathrm{id}\,, \varepsilon')$ are isomorphic as complex line bundles (cf. section 3.2.3). The diffeomorphism on M defined by Ψ shall be denoted by Ψ_M (cf. section 4.1.1). Thus we have the commuting diagram

$$\begin{array}{ccc} \mathbb{F}^{\mathbf{a}'} & \xrightarrow{\Psi} & \mathbb{F}^{\mathbf{a}} \\ {\scriptstyle \mathrm{pr}_M}\downarrow & & \downarrow{\scriptstyle \mathrm{pr}_M} \\ M & \xrightarrow{\Psi_M} & M \end{array} .$$

Setting

$$\Phi(s') := \Psi \circ s' \circ \Psi_M^{-1} \qquad \forall\, s' \in \Gamma\mathbb{F}^{\mathbf{a}'} \tag{345}$$

yields a $\mathcal{S}(M, \mathbb{C})$*-module isomorphism*

$$\Phi : \Gamma\mathbb{F}^{\mathbf{a}'} \longrightarrow \Gamma\mathbb{F}^{\mathbf{a}}.$$

Recall that $\mathcal{S}(M, \mathbb{C})$ denotes the collection of all Schwartz maps from M to $\mathbb{C}$. We write

$$\Phi = \Psi_*.$$

On the other hand, let a $\mathcal{S}(M, \mathbb{C})$-module isomorphism

$$\Phi : \Gamma\mathbb{F}^{\mathbf{a}'} \longrightarrow \Gamma\mathbb{F}^{\mathbf{a}}$$

be given. We will show that then there is a complex line bundle isomorphism

$$\Psi : \mathbb{F}^{\mathbf{a}'} \rightarrow \mathbb{F}^{\mathbf{a}}$$

for which

$$\Phi = \Psi_*. \tag{346}$$

Indeed, given $h \in \mathbb{F}^{\mathbf{a}}_x$, we choose some section $s \in \Gamma\mathbb{F}^{\mathbf{a}}$ with

$$s(x) = h$$

and set

$$\Psi_x(h) := \Phi(s)(x).$$

Remark. *$\Psi(h)$ depends only on $s(x)$ and not on s itself.*

PROOF. To show this we let

$$\Psi_x(h) = \Phi(s_1)(x)$$

with

$$s_1(x) = h$$

for some other $s_1 \in \Gamma\mathbb{F}^{\mathbf{a}}$. Due to the $\mathcal{S}(M,\mathbb{C})$-module structure of $\Gamma\mathbb{F}^{\mathbf{a}}$,

$$s_1(y) = s(y) \cdot f(y) \qquad \forall\, y \in M$$

for some $f \in \mathcal{S}(M,\mathbb{C})$ with $f(x) = 1$. Consequently, we have for $x \in M$,

$$\Phi(s_1)(x) = \Phi(s \cdot f)(x) = (\Phi(s) \cdot f)(x) = \Phi(s)(x) \cdot f(x) = \Phi(s)(x)$$

since Φ is a $\mathcal{S}(M,\mathbb{C})$-module isomorphism. Hence Ψ_x does not depend on the particular section s chosen. □

In verifying that $\Psi : \mathbb{F}^{\mathbf{a}'} \longrightarrow \mathbb{F}^{\mathbf{a}}$ is injective we let

$$\Psi(h) = 0$$

for $h \in \mathbb{F}^{\mathbf{a}}_x$ and represent h as

$$h = s(x)$$

for some $s \in \Gamma\mathbb{F}^{\mathbf{a}}$. Since Φ is an $\mathcal{S}(M,\mathbb{C})$-module isomorphism, there is some $s \in \Gamma\mathbb{F}^{\mathbf{a}}$ such that $\Phi(s) \neq 0$. Hence $\Phi(s)(x) \neq 0$. Due to

$$h(x) = (s \cdot f)(x) = s(x) \cdot f(x) \qquad \forall\, x \in M$$

for some $f \in \mathcal{S}(M,\mathbb{C})$,

$$\Psi(h) = 0 = \Phi(s)(x) \cdot f(x),$$

implying $f(x) = 0$ and, therefore, $h = 0$. In fact, Ψ is smooth, hence is a bundle isomorphism and in turn yields a diffeomorphism Ψ_M on M. Obviously, (345) holds true. Therefore, we have:

LEMMA 76. *Any $\mathcal{S}(M,\mathbb{C})$-module isomorphism of $\Phi : \Gamma\mathbb{F}^{\mathbf{a}'} \longrightarrow \Gamma\mathbb{F}^{\mathbf{a}}$, is of the form*

$$\Phi = \Psi_*. \tag{347}$$

The fibrewisely given symplectic form $\omega^{\mathbf{a}}$ on $\mathbb{F}^{\mathbf{a}}$ determines the bracket on $\mathbf{h}^{\mathbf{a}}$ with values in the bundle $\mathbb{R} \cdot \varepsilon$ of centers. Since $\mathbb{F}^{\mathbf{a}} = \mathbf{h}^{\mathbf{a}}/\mathbb{R} \cdot \varepsilon$, the Heisenberg algebra bundle determines the line bundle $\mathbb{F}^{\mathbf{a}}$ with the form $\omega^{\mathbf{a}}$ and vice versa. Obviously, a corresponding statement is valid for $\mathbf{H}^{\mathbf{a}}$.

Accordingly, $\hbar^{\infty}_{\varepsilon}$ uniquely determines $\Gamma\mathbb{F}^{\mathbf{a}}$ and vice versa.

Next let us assume that

$$\Phi^{\hbar^{\infty}} : \hbar^{\infty}_{\varepsilon_1} \longrightarrow \hbar^{\infty}_{\varepsilon_2}$$

is a *Heisenberg algebra isomorphism*. Since $\Phi^{\hbar^{\infty}}$ maps center into center it uniquely determines an isomorphism from $\Gamma\mathbb{F}^{\mathbf{a}_1}$ into $\Gamma\mathbb{F}^{\mathbf{a}_2}$ due to

$$\Gamma\mathbb{F}^{\mathbf{a}_r} = \hbar^{\infty}_{\varepsilon_r} / \,\mathrm{center}_r \qquad \forall\, r = 1,2.$$

In addition, let us suppose that $\Phi^{\hbar^{\infty}}$ is a $\mathcal{S}(M,\mathbb{C})$-*module isomorphism*, denoted by $\Phi^{\hbar^{\infty}}$, too. Obviously, $\Phi^{\hbar^{\infty}}$ is of the form Ψ_* by lemma 76 and $\omega^{\mathbf{a}_1}$ is the pull-back of $\omega^{\mathbf{a}_2}$ by Ψ. This is easily verified by using sections in $\Gamma\mathbb{F}^{\mathbf{a}}_r$ with compact support in local charts of $\mathbb{F}^{\mathbf{a}}_r$ for $r = 1,2$, implying that X_1 and X_2 have identical field strength functions (cf. section 3.1.1). Hence we have completed the answers to the questions above. In summary:

THEOREM 77. *The following equivalences hold:*

- **i):** $X_1 \cong X_2$.
- **ii):** $\Gamma\mathbb{F}^{\varepsilon_1} \cong \Gamma\mathbb{F}^{\varepsilon_2}$ *as $\mathcal{S}(M,\mathbb{C})$-modules.*
- **iii):** $\hbar^{\infty}_{\varepsilon_1} \cong \hbar^{\infty}_{\varepsilon_2}$ *as $\mathcal{S}(M,\mathbb{C})$-modules.*

Moreover, $\mathfrak{h}^{\infty}_{\mathbf{a}_1} \cong \mathfrak{h}^{\infty}_{\mathbf{a}_1}$ *as* $\mathcal{S}(M,\mathbb{C})$*-modules of Heisenberg algebras iff* $X_1 \cong X_2$ *and have identical field strength functions. This statement holds accordingly for the respective Heisenberg groups.*

6.4. Group Algebras of Infinite Dimensional Heisenberg Groups

Algebraic quantum field theory as formulated in [**13**], [**14**] and [**15**] relies on the notion of a C^*-algebra and thus provides the basis for the quantization of a vector field X on M in terms of the C^*-group algebra of $\mathcal{H}^{\infty}_X$, as formulated, for example, in the last chapter. In this context it is interesting to understand how close the C^*-group algebra of the vector field is related to the field itself. In fact, this C^*-group algebra determines the vector field uniquely. To show this, is the main goal of this section. Hence we begin with the introduction of the notion of a *general group algebra*.

6.4.1. Group Algebras. Given a group G, the underlying set of the *group algebra* $\mathcal{F}_{ae}(G,\mathbb{C})_\star$ of G consists of the collection $\mathcal{F}_{ae}(G,\mathbb{C})$ of all $\mathbb{C}$-valued functions of G with finite support. Thus an element in $\mathcal{F}_{ae}(G,\mathbb{C})$ does not vanish at most on *finitely many* points. The collection $\mathcal{F}_{ae}(G,\mathbb{C})$ is obviously a $\mathbb{C}$-linear space under the *pointwisely defined* operations. The *multiplication* in $\mathcal{F}_{ae}(G,\mathbb{C})_\star$, however, is the *convolution* given for all $f,g \in \mathcal{F}_{ae}(G,\mathbb{C})$ by

$$f \star g(s) := \sum_{s' \in G} f(s') \cdot g(s'^{-1} \cdot s) \qquad \forall\, s \in G. \tag{348}$$

The operation in G is written multiplicatively.

The *natural inclusion*

$$j : G \longrightarrow \mathcal{F}_{ae}(G,\mathbb{C})_\star \tag{349}$$

assigning the Dirac map δ_s to each $s \in G$ is a group homomorphism, i.e.

$$j(s \cdot s'^{-1}) := \delta_{s \cdot s'^{-1}} = \delta_s \star \delta_{s'^{-1}} \qquad \forall\, s, s' \in G \tag{350}$$

holds true, as can be seen directly form (348). (Recall that

$$\delta_s(s') = \begin{cases} 1 & \text{if } s = s', \\ 0 & \text{otherwise} \end{cases}$$

for any $s \in G$.) Clearly, the (multiplicative) unit element $\mathbb{1}$ in G is mapped into $\delta_{\mathbb{1}}$ which is the *unit element* in $\mathcal{F}_{ae}(G,\mathbb{C})_\star$. The *constant function* with value *zero* is the *zero element.*

The group algebra admits a natural *involution* $*$ assigning to δ_s the element

$$*\delta_s := \delta_{s^{-1}} \tag{351}$$

for any $s \in G$. Since $f \in \mathcal{F}_{ae}(G,\mathbb{C})_\star$ is (uniquely) decomposed into

$$f = \sum_{s' \in G} f(s') \cdot \delta_{s'}, \tag{352}$$

the map $*$ extends linearly to all of $\mathcal{F}_{ae}(G,\mathbb{C})$. Since, moreover, $* \circ * = \mathrm{id}$ this map is indeed an involution.

Together with this involution and the multiplication with *complex scalars* $\mathcal{F}_{ae}(G,\mathbb{C})_\star$ is a $*$-algebra.

For any group homomorphism

$$\Psi : G_1 \to G_2$$

the natural map

$$\Psi_* : \mathcal{F}_{ae}(G_1, \mathbb{C})_\star \longrightarrow \mathcal{F}_{ae}(G_2, \mathbb{C})_\star$$

defined by

$$\Psi_*(f)(s) := \sum_{s'} f(s')\delta_{\Psi(s')}(s) \tag{353}$$

for all $f \in \mathcal{F}_{ae}(G_1, \mathbb{C})_\star$ and all $s \in G_1$ is a (covariant) $*$-*algebra homomorphism*, as is easily verified. Clearly, $* \circ \Psi_* = \Psi_* \circ *$.

Now we replace G by the infinite dimensional Heisenberg group $\mathcal{H}_X^\infty = \Gamma \mathbf{H}^\mathbf{a} + \mathbb{R} \cdot e$ of a singularity free vector field $X = (\mathrm{id}, \mathbf{a})$ on M. The $*$-algebra $\mathcal{F}_{ae}(\mathcal{H}_X^\infty, \mathbb{C})_\star$ is called the *Heisenberg group algebra* of X.

Given the Heisenberg group algebra $\mathcal{F}_{ae}(\mathcal{H}^\infty, \mathbb{C})_\star$, the normal subgroup

$$C := \Gamma(\mathbb{R} \cdot \varepsilon) \oplus \mathbb{R} \cdot e \tag{354}$$

of $\mathcal{H}_X^\infty$ is the center and hence

$$\mathcal{H}_X^\infty / C = \Gamma \mathbb{F}^\mathbf{a}$$

is a *commutative group* under the addition. Clearly, we may rewrite C in the form

$$C = \mathcal{S}(\mathbb{R}, \mathbb{R}) \cdot \varepsilon + \mathbb{R} \cdot e$$

where $\mathcal{S}(\mathbb{R}, \mathbb{R})$ is the collection of all $\mathbb{R}$-valued Schwartz function of $\mathbb{R}$.

The exact sequence of groups

$$0 \longrightarrow C \stackrel{\iota}{\longrightarrow} \mathcal{H}_X^\infty \stackrel{\mathrm{pr}}{\longrightarrow} \Gamma \mathbb{F}^\mathbf{a} \longrightarrow 0$$

with ι and pr the inclusion and the projection, respectively, yields the commuting diagram

$$\begin{array}{ccccccccc}
0 & \longrightarrow & \mathcal{F}_{ae}(C, \mathbb{C})_\star & \stackrel{\iota_*}{\longrightarrow} & \mathcal{F}_{ae}(\mathcal{H}_X^\infty, \mathbb{C})_\star & \stackrel{\mathrm{pr}_*}{\longrightarrow} & \mathcal{F}_{ae}(\Gamma \mathbb{F}^\mathbf{a}, \mathbb{C})_\star & \longrightarrow & 0 \\
 & & \uparrow j & & \uparrow j & & \uparrow j & & \\
1 & \longrightarrow & C & \stackrel{\iota}{\longrightarrow} & \mathcal{H}_X^\infty & \stackrel{\mathrm{pr}}{\longrightarrow} & \Gamma \mathbb{F}^\mathbf{a} & \longrightarrow & 0
\end{array} \tag{355}$$

of groups and $*$-algebra homomorphisms. The maps in the upper row are linear extensions of the respective ones in the lower row.

The top row is an exact sequence of *linear* maps. However, it is not exact as a sequence of $*$-*homomorphisms*. This is easily verified by looking at

$$\mathrm{pr}_* \circ \iota_*(\delta_c) = \delta_{\mathrm{pr}(c)} = \delta_0 \qquad \forall c \in C.$$

Clearly, δ_0 is not the zero map in general, it is the *unit element*; the function δ_0 is of the form $j(0) = \delta_0$. Hence $\mathcal{F}_{ae}(C, \mathbb{C})_\star$ is not an ideal of $\mathcal{F}_{ae}(\mathcal{H}_X^\infty, \mathbb{C})_\star$. Obviously, the splitting

$$\mathcal{F}_{ae}(\mathcal{H}_X^\infty, \mathbb{C}) = \mathcal{F}_{ae}(C, \mathbb{C}) \oplus \mathcal{F}_{ae}(\Gamma \mathbb{F}^\mathbf{a}, \mathbb{C}) \tag{356}$$

holds true only as $\mathbb{C}$-linear spaces.

To understand the upper row in (355) in more detail we first determine $\ker \mathrm{pr}_*$. For any $g = \sum_r z^r \delta_{s_r+c_r}$ where $s_r \in \Gamma\mathbb{F}^{\mathbf{a}}$ and $j(c_r)$ in the center C_{ae} of $\mathcal{F}_{ae}(\mathcal{H}_X^\infty, \mathbb{C})_\star$ we set

$$\mathrm{pr}_* \left(\sum z^r \delta_{s_r+c_r}\right) = \sum_r z^r \cdot \delta_{s_r}.$$

Hence

$$\ker \mathrm{pr}_* = \text{ ideal generated by } \left\{\sum_r z^r \delta_{s_r+c_r} \middle| \sum z^r = 0\right\}, \tag{357}$$

implying

$$j(C) \not\subset \ker \mathrm{pr}_*.$$

Next let us determine the center C_{ae} of $\mathcal{F}_{ae}(\mathcal{H}_X^\infty, \mathbb{C})_\star$. To this end we observe that

$$j(C) \subset C_{ae}.$$

Clearly, $\mathcal{F}_{ae}(C, \mathbb{C})_\star \subset C_{ae}$. To obtain a more detailed form of C_{ae} we notice that any $g \in C_{ae}$ yields

$$g \star \delta_s = \delta_s \star g \qquad \forall\, s \in \Gamma\mathbb{F}^{\mathbf{a}}$$

and in turn

$$g = \delta_s \star g \star \delta_{-s} \qquad \forall\, s \in \Gamma\mathbb{F}^{\mathbf{a}}.$$

Due to (352) we hence verify for any element

$$\sum_r z_r \delta_{s_r+c_r} = \sum_r z^r \cdot \delta_{s_r} \star \delta_{c_r} = \sum_r z^r \delta_s \star \delta_{s_r} \star \delta_{-s} \star \delta_{c_r} \qquad \forall\, s \in \Gamma\mathbb{F}^{\mathbf{a}}$$

in C_{ae}. Here $s_r \in \Gamma\mathbb{F}^{\mathbf{a}}, c_r \in C$ and $z_r \in \mathbb{C}$. Since $\delta_{c_r} \in C_{ae}$ we finally obtain for any r,

$$\delta_{s_r} \star \delta_{c_r} \quad = \quad \delta_s \star \delta_{s_r} \star \delta_{-s} \star \delta_{c_r}$$

and thus

$$\delta_{s_r} \quad = \quad \delta_s \star \delta_{s_r} \star \delta_{-s}.$$

From

$$s_r = s \cdot s_r \cdot (-s) = s_r + 2\omega^\infty(s, s_r) \cdot e$$

holding for all $s \in \Gamma\mathbb{F}^{\mathbf{a}}$ and any index r we conclude

$$\omega^\infty(s, s_r) = 0$$

saying that

$$s_r = 0 \qquad \forall\, r.$$

The center C_{ae} of $\mathcal{F}_{ae}(\mathcal{H}_X^\infty, \mathbb{C})_\star$ is thus of the form

$$C_{ae} = \mathcal{F}_{ae}(C, \mathbb{C})_\star, \tag{358}$$

the group $*$-algebra of C. This clarifies the non-exactness of the upper row in (355) of $*$-homomorphisms.

6.4.2. Fourier Transformation, Character Groups and Group Algebras. In order to study (in section 6.4.4) the effect of homomorphisms of $\mathcal{H}_{X_1}^\infty$ and $\mathcal{H}_{X_2}^\infty$ to their respective $*$-group algebras and vice versa, some techniques have to be built up (cf. section 6.4.1). They will involve Fourier transforms and character groups (cf. [**58**] and [**81**]). This is the main topic of this section.

In this subsection G will be replaced by the commutative group $\Gamma\mathbb{F}^{\mathbf{a}}$ again which carries the *discrete topology*. The *Fourier transform* $\mathbf{F}$ of $f \in \mathcal{F}_{ae}(\Gamma\mathbb{F}^{\mathbf{a}}, \mathbb{C})$ is given by

$$\mathbf{F}(f)(s) := \sum_{s'} f(s') \cdot e^{2\pi i \ll s', s \gg} \tag{359}$$

for all $s, s' \in \Gamma\mathbb{F}^{\mathbf{a}}$ and for all $f \in \mathcal{F}_{ae}(\Gamma\mathbb{F}^{\mathbf{a}}, \mathbb{C})$. Here $\ll \ , \ \gg$ (cf. (339)) is the scalar product on $\Gamma\mathbf{H}^{\mathbf{a}}$. Clearly, $\mathbf{F}(f)$ is in the linear space $\mathcal{F}(\Gamma\mathbb{F}^{\mathbf{a}}, \mathbb{C})$ of all $\mathbb{C}$-valued function on $\Gamma\mathbb{F}^{\mathbf{a}}$, but is in general not in $\mathcal{F}_{ae}(\Gamma\mathbb{F}^{\mathbf{a}}, \mathbb{C})$ as is easily seen by considering

$$\mathbf{F}(\delta_{s''})(s) = e^{2\pi i \ll s, s'' \gg} \qquad \forall\, s, s'' \in \Gamma\mathbb{F}^{\mathbf{a}}. \tag{360}$$

The operations in $\Gamma\mathbb{F}^{\mathbf{a}}$ are defined *pointwisely.* The multiplication $\cdot$ in the *linear* space $\mathcal{F}_{ae}(\Gamma\mathbb{F}^{\mathbf{a}}, \mathbb{C})$ is defined pointwisely and yields the $\mathbb{C}$-algebra $\mathcal{F}(\Gamma\mathbb{F}^{\mathbf{a}}, \mathbb{C})_{\bullet}$. One verifies by a straightforward calculation that the Fourier transform $\mathbf{F}$ is $\mathbb{C}$-linear and satisfies

$$\mathbf{F}(f \star g) = \mathbf{F}(f) \cdot \mathbf{F}(g) \qquad \forall\, f, g \in \mathcal{F}(\Gamma\mathbb{F}^{\mathbf{a}}, \mathbb{C})_{\star}. \tag{361}$$

Thus the Fourier transform $\mathbf{F}$ is a *homomorphism*

$$\mathbf{F} : \mathcal{F}_{ae}(\Gamma\mathbb{F}^{\mathbf{a}}, \mathbb{C})_{\star} \longrightarrow \mathcal{F}(\Gamma\mathbb{F}^{\mathbf{a}}, \mathbb{C})_{\bullet}$$

converting the convolution into the pointwise multiplication.

To investigate $\mathbf{F}$ further we look at the Hermitian form

$$< f | g >_{L_2} := \int_{\mathcal{H}_X^\infty} f(s) \cdot \bar{g}(s) \qquad \forall\, s \in \Gamma\mathbb{F}^{\mathbf{a}} \tag{362}$$

on $\mathcal{F}_{ae}(\Gamma\mathbb{F}^{\mathbf{a}}, \mathbb{C})$ computed with respect to the *counting measure.* The resulting L_2-norm is denoted by $\| \ \|_{L_2}$.

Since any $f \in \mathcal{F}_{ae}(\Gamma\mathbb{F}^{\mathbf{a}}, \mathbb{C})$ is bounded, the linear space $\mathcal{F}_{ae}(\Gamma\mathbb{F}^{\mathbf{a}}, \mathbb{C})$ admits the *sup-norm* given by

$$\| f \|_{\sup} := \sup_{s \in \mathcal{H}_X^\infty} |f(s)|. \tag{363}$$

Clearly,

$$\| f \|_{L_2} \leq N \cdot \| f \|_{\sup}$$

where N is the number of points on which f does not vanish. Of course, $\| \ \|_{\sup}$ is defined accordingly on $\Gamma\mathbb{F}^{\mathbf{a}}$ and on $\Gamma\mathbf{H}^{\mathbf{a}}$. The $\| \ \|_{\sup}$-topology is finer than the topology of pointwise convergence. From (359) we immediately deduce Parseval's formula

$$\| f \|_{L_2} = \| \mathbf{F}(f) \|_{L_2} \qquad \forall\, f \in \mathcal{F}_{ae}(\Gamma\mathbb{F}^{\mathbf{a}}, \mathbb{C}), \tag{364}$$

implying

$$\mathbf{F}(f) = 0 \qquad \Longleftrightarrow \qquad f = 0$$

for $f \in \mathcal{F}_{ae}(\Gamma\mathbb{F}^{\mathbf{a}}, \mathbb{C})$. Hence

$$\mathbf{F} : \mathcal{F}_{ae}(\Gamma\mathbb{F}^{\mathbf{a}}, \mathbb{C})_\star \longrightarrow \operatorname{im}\mathbf{F} \subset \mathcal{F}(\Gamma\mathbb{F}^{\mathbf{a}}, \mathbb{C})_\bullet \tag{365}$$

is a natural algebra *isomorphism*. The reason for (365) to hold is that $\mathbf{F}$ is injective on the image $j(\Gamma\mathbb{F}^{\mathbf{a}})$ of the natural inclusion j of $\Gamma\mathbb{F}^{\mathbf{a}}$ into $\mathcal{F}_{ae}(\Gamma\mathbb{F}^{\mathbf{a}}, \mathbb{C})_\star$ (cf. (349)).

For each $s \in \Gamma\mathbb{F}^{\mathbf{a}}$ the group homomorphism

$$\chi_s : \Gamma\mathbb{F}^{\mathbf{a}} \longrightarrow U(1)$$

defined by

$$\chi_s := \mathbf{F}(\delta_s) \qquad \forall\, s \in \Gamma\mathbb{F}^{\mathbf{a}} \tag{366}$$

is called a *character* of $\Gamma\mathbb{F}^{\mathbf{a}}$; it is continuous.

More generally, any $U(1)$-valued homomorphism of any *commutative* topological group G the collection $\chi(G)$ of all continuous characters of G forms a group under the pointwisely defined operation. It is called the *character group* of G.

6.4.3. Homomorphisms and Characters. Next we will study relations between continuous $\mathbb{C}$-valued homomorphisms of $\mathcal{F}_{ae}(\Gamma\mathbb{F}^{\mathbf{a}}, \mathbb{C})_\star$, endowed with the topology of *pointwise convergence*, and the character group $\chi(\mathbb{F}^{\mathbf{a}})$ of the group $\Gamma\mathbb{F}^{\mathbf{a}}$ carrying the *discrete topology*. Both the pointwisely defined linear operations and the convolution are continuous, as is easily seen.

We observe first that

$$j : \Gamma\mathbb{F}^{\mathbf{a}} \longrightarrow \mathcal{F}_{ae}(\Gamma\mathbb{F}^{\mathbf{a}}, \mathbb{C})_\star$$

(cf. (349)) is obviously continuous. Second, for any $h \in \operatorname{Hom}\mathcal{F}_{ae}(\Gamma\mathbb{F}^{\mathbf{a}}, \mathbb{C})_\star$, the collection of all $\mathbb{C}$-*valued algebra homomorphisms* sending the unit element to $1 \in \mathbb{C}$, the composition

$$h \circ j : \Gamma\mathbb{F}^{\mathbf{a}} \longrightarrow \mathbb{C}$$

is a *continuous* group homomorphism and

$$h \circ j(s) \neq 0 \qquad \forall\, s \in \Gamma\mathbb{F}^{\mathbf{a}}.$$

To see this recall that the zero section $\mathcal{O}$ in $\Gamma\mathbb{F}^{\mathbf{a}}$ is the neutral element in $\Gamma\mathbb{F}^{\mathbf{a}}$. Hence $\delta_{\mathcal{O}}$ is the (two-sided) unit element with respect to the convolution. Therefore,

$$h \circ j(s - s) = h(\delta_s) \cdot h(\delta_{-s}) = h(\delta_{\mathcal{O}}) = 1$$

showing that $h(\delta_s) \neq 0$ for all $s \in \Gamma\mathbb{F}^{\mathbf{a}}$. Therefore,

$$h \circ j(\Gamma\mathbb{F}^{\mathbf{a}}) \subset \dot{\mathbb{C}}.$$

Since h is uniquely determined by the value $h(\delta_s)$ for any s, there is a map

$$\begin{aligned} \chi_h : \Gamma\mathbb{F}^{\mathbf{a}} &\longrightarrow \dot{\mathbb{C}} \\ s &\longrightarrow h(\delta_s). \end{aligned}$$

It satisfies

$$h(\delta_{s_1} \star \delta_{s_2}) = h(\delta_{s_1+s_2}) = \chi_h(s_1 + s_2)$$

as well as

$$h(\delta_{s_1} \star \delta_{s_2}) = h(\delta_{s_1}) \cdot h(\delta_{s_2}) = \chi_h(s_1) \cdot \chi_h(s_2)$$

for any choice of $s_1, s_2 \in \Gamma\mathbb{F}^{\mathbf{a}}$. Thus χ_h is a *group homomorphism* of the additive group $\Gamma\mathbb{F}^{\mathbf{a}}$. In fact, χ_h is $U(1)$-valued. Indeed, the topology on $\mathcal{F}_{ae}(\Gamma\mathbb{F}^{\mathbf{a}}, \mathbb{C})_\star$

determined by the sup-norm $\| \|_{\text{sup}}$ (cf. (363)) is finer than the topology of pointwise convergence. Therefore,

$$|h(\delta_s)| \leq c \qquad \forall\, s \in \Gamma\mathbb{F}^{\mathbf{a}}$$

has to hold for some positive constant c. In fact, $c \geq 1$ since $h(\delta_{\mathcal{O}}) = 1$. Because $h(j(\Gamma\mathbb{F}^{\mathbf{a}})) \subset \dot{\mathbb{C}}$ is a subgroup and $\chi_h(s) \cdot \chi_h(-s) = 1$ holds, we conclude

$$h(j(\Gamma\mathbb{F}^{\mathbf{a}})) \subset U(1) \qquad \forall\, h \in \operatorname{Hom}\mathcal{F}_{ae}(\Gamma\mathbb{F}^{\mathbf{a}}, \mathbb{C})_{\star}.$$

This is to say that

$$h|_{j(\Gamma\mathbb{F}^{\mathbf{a}})} \in \chi(\Gamma\mathbb{F}^{\mathbf{a}}) \qquad \forall\, h \in \operatorname{Hom}\mathcal{F}_{ae}(\Gamma\mathbb{F}^{\mathbf{a}}, \mathbb{C})_{\star}.$$

Vice versa, any character $\xi \in \chi(\Gamma\mathbb{F}^{\mathbf{a}})$ extends to a homomorphism h of $\mathcal{F}_{ae}(\Gamma\mathbb{F}^{\mathbf{a}}, \mathbb{C})_{\star}$ by setting

$$h(f) := \sum_r z^r \cdot \xi(s_r) \qquad \forall\, f \in \mathcal{F}_{ae}(\Gamma\mathbb{F}^{\mathbf{a}}, \mathbb{C})_{\star}$$

where s_r varies in $\Gamma\mathbb{F}^{\mathbf{a}}$ and f is decomposed in

$$f = \sum_r z^r \cdot \delta_{s_r} \qquad \forall\, z_r \in \mathbb{C};$$

in particular, this means

$$h(\delta_s) = \xi(s) \quad \forall\, s \in \Gamma\mathbb{F}^{\mathbf{a}}.$$

Hence

$$\begin{aligned} j^* : \operatorname{Hom}\mathcal{F}_{ae}(\Gamma\mathbb{F}^{\mathbf{a}}, \mathbb{C})_{\star} &\longrightarrow \chi(\Gamma\mathbb{F}^{\mathbf{a}}) \\ h &\longmapsto h \circ j \end{aligned} \tag{367}$$

is a bijection. Pulling back the group operations of $\chi(\Gamma\mathbb{F}^{\mathbf{a}})$ to $\operatorname{Hom}\mathcal{F}_{ae}(\Gamma\mathbb{F}^{\mathbf{a}}, \mathbb{C})_{\star}$ by j^* turns this set into a group, as well. Equipping both domain and range of j with the *topology of pointwise convergence* turns both spaces into topological groups. Since $\Gamma\mathbb{F}^{\mathbf{a}}$ is discrete, its character group $\chi(\Gamma\mathbb{F}^{\mathbf{a}})$ is *compact* (cf. [**8**]) and j^* is continuous. We thus have:

PROPOSITION 78. $j^* : \operatorname{Hom}\mathcal{F}_{ae}(\Gamma\mathbb{F}^{\mathbf{a}}, \mathbb{C})_{\star} \longrightarrow \chi(\Gamma\mathbb{F}^{\mathbf{a}})$ *is a continuous isomorphism of compact groups, if both range and domain of j are equipped with the topology of pointwise convergence.*

We are now in the realm of *Pontryagin's duality* theory (cf. [**8**]) for *locally compact* groups. An important feature of this theory is that a *discrete* group admits a *compact* character group, and vice versa, any compact group has a discrete character group. Formulated in our setting this implies that

$$\begin{aligned} \check{j} : \Gamma\mathbb{F}^{\mathbf{a}} &\longrightarrow \chi(\chi(\Gamma\mathbb{F}^{\mathbf{a}})) \\ s &\longmapsto (\xi \longmapsto \xi(s)) \end{aligned} \tag{368}$$

is a continuous isomorphism if $\chi(\chi(\Gamma\mathbb{F}^{\mathbf{a}}))$ is equipped with the topology of compact convergence, again. Thus $\Gamma\mathbb{F}^{\mathbf{a}}$ is a *reflexive group* (cf. [**8**]).
As a résumé we state:

Proposition 79. *The maps in the commuting diagram*

$$\begin{array}{ccc} \chi(\chi(\Gamma\mathbb{F}^{\mathbf{a}})) & \xrightarrow{(j^*)^*} & \chi(\operatorname{Hom}\mathcal{F}_{ae}(\Gamma\mathbb{F}^{\mathbf{a}},\mathbb{C})_\star) \\ \Big\uparrow{\scriptstyle \check{\jmath}} & & \Big\uparrow{\scriptstyle \check{\jmath}} \\ \Gamma\mathbb{F}^{\mathbf{a}} & \xrightarrow{\quad id \quad} & \Gamma\mathbb{F}^{\mathbf{a}} \end{array}$$

are all continuous isomorphisms of discrete groups. Here

$$(j^*)^*(\eta)(h) := \eta(j^*(h))$$

for any $\eta \in \chi(\Gamma\mathbb{F}^{\mathbf{a}})$ *and any* $h \in \operatorname{Hom}\mathcal{F}_{ae}(\Gamma\mathbb{F}^{\mathbf{a}},\mathbb{C})_\star$*; in addition,*

$$\check{\jmath}(s)(h) := h(\delta_s) \qquad \forall\, s \in \Gamma\mathbb{F}^{\mathbf{a}} \text{ and } \forall\, h \in \operatorname{Hom}\mathcal{F}_{ae}(\Gamma\mathbb{F}^{\mathbf{a}},\mathbb{C})_\star.$$

We close this section by relating $\operatorname{Hom}\operatorname{im}\mathbf{F}$ with $\chi(\Gamma\mathbb{F}^{\mathbf{a}})$; by $\operatorname{Hom}\operatorname{im}\mathbf{F}$ we denote the collection of all $\mathbb{C}$-valued algebra homomorphisms of $\operatorname{im}\mathbf{F} \subset \mathcal{F}_{ae}(\Gamma\mathbb{F}^{\mathbf{a}},\mathbb{C})_\bullet$, endowed with the *topology of pointwise convergence*. At first we consider

$$\mathbf{F}^* : \operatorname{Hom}\operatorname{im}\mathbf{F} \longrightarrow \operatorname{Hom}\mathcal{F}_{ae}(\Gamma\mathbb{F}^{\mathbf{a}},\mathbb{C})_\star$$

mapping any $h \in \operatorname{Hom}\operatorname{im}\mathbf{F}$ into $h \circ \mathbf{F}$. Here the upper star means composition from the right. This map is a continuous isomorphism. Pulling the group operations on the range back to the domain turns $\operatorname{Hom}\operatorname{im}\mathbf{F}$ into a topological group. In analogy to j we define

$$\hat{\jmath} : \Gamma\mathbb{F}^{\mathbf{a}} \longrightarrow \operatorname{im}\mathbf{F}$$

by

$$\hat{\jmath}(s)(\mathbf{F}(\delta_{s'})) := \check{\jmath}(s)(\mathbf{F}(\delta_{s'})) \qquad \forall\, s, s' \in \Gamma\mathbb{F}^{\mathbf{a}}.$$

Clearly, $\hat{\jmath}$ is a continuous injection for which

$$\hat{\jmath}(s)(\mathbf{F}(\delta_{s'})) = e^{2\pi i \ll s', s \gg}$$

holds true for any pair $s, s' \in \Gamma\mathbb{F}^{\mathbf{a}}$ (cf. (366)). Due to proposition 78,

$$\begin{array}{ccc} \operatorname{Hom}\operatorname{im}\mathbf{F} & \xrightarrow{\qquad \mathbf{F}^* \qquad} & \operatorname{Hom}\mathcal{F}_{ae}(\Gamma\mathbb{F}^{\mathbf{a}},\mathbb{C})_\star \\ & {\scriptstyle (\hat{\jmath})^*} \searrow \qquad \nearrow {\scriptstyle (j^*)^{-1}} & \\ & \chi(\Gamma\mathbb{F}^{\mathbf{a}}) & \end{array} \tag{369}$$

is a commuting diagram of continuous isomorphisms of *compact* groups.

Let us point out that the map

$$(\check{\jmath})_{\mathbb{F}^{\mathbf{a}}} : \Gamma\mathbb{F}^{\mathbf{a}} \longrightarrow \chi(\operatorname{Hom}\operatorname{im}\mathbf{F}) \tag{370}$$

defined by

$$(\check{\jmath})_{\mathbf{F}}(s)(h') := \mathbf{F}^*(h')(\delta_s) \qquad \forall\, s \in \Gamma\mathbb{F}^{\mathbf{a}} \text{ and } \forall\, h' \in \operatorname{Hom}\operatorname{im}\mathbf{F}$$

is a (continuous) isomorphism of *discrete* groups. Clearly,

$$\mathbf{F}^*(h')(\delta_s) = h'(e^{2\pi i \ll s, \ldots \gg})$$

for all variables.

Due to (356) all these results hold true accordingly if $\Gamma\mathbb{F}^{\mathbf{a}}$ is replaced by the $\mathbb{R}$-linear space $\Gamma\mathbf{H}^{\mathbf{a}}$, of course.

6.4.4. Bundle Homomorphism and $*$-Algebra Homomorphism. In section 6.3 we studied the relation between singularity free vector fields and *bundle isomorphisms* of complex line bundles associated with them. Here we study such isomorphisms in relation with *isomorphisms* of *$*$-algebras* of singularity free vector fields. Towards the end of this subsection we will combine both views.

We begin with a $\mathcal{S}(M,\mathbb{C})$-*module isomorphism*

$$\Phi : \Gamma\mathbb{F}^{\mathbf{a}_1} \longrightarrow \Gamma\mathbb{F}^{\mathbf{a}_2}$$

where $X_1 = (\mathrm{id}\,, \mathbf{a}_1)$ and $X_2 = (\mathrm{id}\,, \mathbf{a}_2)$ are two singularity free vector fields on M. This map causes a $*$-algebra isomorphism

$$\Phi_{ae} : \mathcal{F}_{ae}(\Gamma\mathbb{F}^{\mathbf{a}_1}, \mathbb{C})_\star \longrightarrow \mathcal{F}_{ae}(\Gamma\mathbb{F}^{\mathbf{a}_2}, \mathbb{C})_\star$$

by setting

$$(371) \qquad \Phi_{ae}(g) := \sum_r z^r \delta_{\Phi(s_r)} \qquad \forall\, g \in \mathcal{F}_{ae}(\Gamma\mathbb{F}^{\mathbf{a}_1}, \mathbb{C})_\star.$$

Here $g = \sum_r z^r \delta_{s_r}$. Since Φ is a $\mathcal{S}(M,\mathbb{C})$-module isomorphism we conclude for any $s \in \Gamma\mathbb{F}^{\mathbf{a}_1}$,

$$(372) \qquad \Phi_{ae}(\delta_{f\cdot s}) = \delta_{f\cdot\Phi(s_r)} \qquad \forall\, f \in \mathcal{S}(M,\mathbb{C}).$$

Clearly, Φ_{ae} as defined by (371) is continuous.

On the other hand, let

$$\Phi_{ae} : \mathcal{F}_{ae}(\Gamma\mathbb{F}^{\mathbf{a}_1}, \mathbb{C})_\star \longrightarrow \mathcal{F}_{ae}(\Gamma\mathbb{F}^{\mathbf{a}_2}, \mathbb{C})_\star$$

be a continuous $*$-algebra isomorphism. We will construct some $\mathcal{S}(M,\mathbb{C})$-module isomorphism

$$\Phi : \Gamma\mathbb{F}^{\mathbf{a}_1} \longrightarrow \Gamma\mathbb{F}^{\mathbf{a}_2}$$

such that

$$\Phi_{ae} = \Phi_*.$$

Here

$$\Phi_*(\delta_s) = \Phi(s) \qquad \forall\, s \in \Gamma\mathbb{F}^{\mathbf{a}}.$$

To this end we will consider j^* in (367) again which maps any $h \in \mathrm{Hom}\,\mathcal{F}_{ae}(\Gamma\mathbb{F}^{\mathbf{a}}, \mathbb{C})_\star$ into $h \circ j$. In more detail the group structure on $\mathrm{Hom}\,\mathcal{F}_{ae}(\Gamma\mathbb{F}^{\mathbf{a}}, \mathbb{C})_\star$ pulled back by j^* mentioned in the previous section, is given as follows: Let $\xi_1, \xi_2 \in \chi(\Gamma\mathbb{F}^{\mathbf{a}})$. The product $\xi_1 \cdot \xi_2$ in $\chi(\Gamma\mathbb{F}^{\mathbf{a}})$ is pointwisely given by

$$\xi_1 \cdot \xi_2(s) = \xi_1(s) \cdot \xi_2(s) \qquad \forall\, s \in \Gamma\mathbb{F}^{\mathbf{a}_1}.$$

Therefore, given any two $h_1, h_2 \in \mathrm{Hom}\,\mathcal{F}_{ae}(\Gamma\mathbb{F}^{\mathbf{a}_1}, \mathbb{C})_\star$ we consider

$$j_1^*(h_1) \cdot j_1^*(h_2) \in \chi(\Gamma\mathbb{F}^{\mathbf{a}}).$$

Since j_1 is a bijection as shown in proposition 78 in the previous section there is a unique $h_3 \in \mathrm{Hom}\,\mathcal{F}_{ae}(\Gamma\mathbb{F}^{\mathbf{a}_1}, \mathbb{C})_\star$ such that

$$j_1(h_3) = j_1(h_1) \cdot j_1(h_2)$$

which yields

$$h_1 \cdot h_2 = h_3$$

in $\mathrm{Hom}\,\mathcal{F}_{ae}(\Gamma\mathbb{F}^{\mathbf{a}}, \mathbb{C})_\star$. Now we pass from $\mathbf{a}$ to $\mathbf{a}_1$ and $\mathbf{a}_2$, respectively, and replace j by j_1 and j_2, respectively. Obviously, the description of the group structure holds accordingly for $\operatorname{Hom}\mathcal{F}_{ae}(\Gamma\mathbb{F}^{\mathbf{a}_1},\mathbb{C})_\star$ and $\operatorname{Hom}\mathcal{F}_{ae}(\Gamma\mathbb{F}^{\mathbf{a}_2},\mathbb{C})_\star$. In analogy, we take $h_1', h_2' \in \operatorname{Hom}\mathcal{F}_{ae}(\Gamma\mathbb{F}^{\mathbf{a}_2},\mathbb{C})_\star$ and study

$$h_1' \cdot h_2' \circ \Phi_{ae} = h_3' \circ \Phi_{ae}.$$

Clearly,

$$h_1' \cdot h_2' \circ \Phi_{ae}(j_2(s')) = (h_1' \cdot h_2') \circ (\Phi_{ae}(\delta_{s'})) \qquad \forall\, s' \in \Gamma\mathbb{F}^{\mathbf{a}_2}.$$

Thus

$$h_1' \cdot h_2' \circ \Phi_{ae}(j_2(s')) = h_1'(\Phi_{ae}(\delta_{s'})) \cdot h_2'(\Phi_{ae}(\delta_{s'}))$$

holds true, which shows that

$$\begin{array}{rcl}\Phi_{ae}^* : \operatorname{Hom}\mathcal{F}_{ae}(\Gamma\mathbb{F}^{\mathbf{a}_2},\mathbb{C})_\star & \longrightarrow & \operatorname{Hom}\mathcal{F}_{ae}(\Gamma\mathbb{F}^{\mathbf{a}_1},\mathbb{C})_\star \\ h' & \longmapsto & h' \circ \Phi_{ae}\end{array}$$

is a group homomorphism. Hence we consider the diagram

$$\begin{array}{ccc}\chi(\operatorname{Hom}\mathcal{F}_{ae}(\Gamma\mathbb{F}^{\mathbf{a}_1},\mathbb{C})_\star) & \xrightarrow{\Phi_{ae}^{**}} & \chi(\operatorname{Hom}\mathcal{F}_{ae}(\Gamma\mathbb{F}^{\mathbf{a}_2},\mathbb{C})_\star) \\ {\scriptstyle j_1^{**}}\Big\uparrow & & \Big\uparrow{\scriptstyle j_2^{**}} \\ \chi(\chi(\Gamma\mathbb{F}^{\mathbf{a}_1})) & & \chi(\chi(\Gamma\mathbb{F}^{\mathbf{a}_2})) \\ {\scriptstyle \check{j}_1}\Big\uparrow & & \Big\uparrow{\scriptstyle \check{j}_2} \\ \Gamma\mathbb{F}^{\mathbf{a}_1} & \overset{\Phi}{\dashrightarrow} & \Gamma\mathbb{F}^{\mathbf{a}_2}\end{array}$$

where we have set

$$\Phi := \check{j}_2^{-1} \circ j_2^{**\,-1} \circ \Phi_{ae}^{**} \circ j_1^{**} \circ \check{j}_1,$$

defined the maps j and $\check{j}$ as in proposition 79 and indexed them by 1 and 2 accordingly to the fields X_1 and X_2, respectively. All spaces in the diagram carry the discrete topology.

To understand Φ in more detail we consider for all variables

$$(j_1^{**} \circ \check{j}_1(s))(h) = \check{j}_1(s)(j_1^*(h)) = \check{j}_1(s)(h \circ j_1) = h(\delta_s)$$

for any section s in $\Gamma\mathbb{F}^{\mathbf{a}_1}$ and any homomorphism h in $\operatorname{Hom}\mathcal{F}_{ae}(\Gamma\mathbb{F}^{\mathbf{a}_1},\mathbb{C})_\star$. Letting

$$\check{j}_{\mathcal{F}_1}(\delta_s)(h) := h(\delta_s)$$

for all elements h in $\operatorname{Hom}\mathcal{F}_{ae}(\Gamma\mathbb{F}^{\mathbf{a}_1},\mathbb{C})_\star$ (saying that $\check{j}_{\mathcal{F}_1}$ is the *point evaluation*) we find

$$j_1^{**} \circ \check{j}_1(s) = \check{j}_{\mathcal{F}_1}(\delta_s) \qquad \forall\, s \in \Gamma\mathbb{F}^{\mathbf{a}_1}.$$

Therefore,

$$\begin{aligned}\Phi_{ae}^{**}(j_1^{**} \circ \check{j}_1)(s)(h) &= \check{j}_{\mathcal{F}}(\delta_s)(\Phi_{ae}^*(h)) \\ &= \check{j}_{\mathcal{F}_1}(\delta_s)(h \circ \Phi_{ae}) \\ &= h \circ \Phi_{ae}(\delta_s) \\ &= \check{j}_{\mathcal{F}_2}(\Phi_{ae}(\delta_s))(h)\end{aligned}$$

for any $h \in \operatorname{Hom}\mathcal{F}_{ae}(\Gamma\mathbb{F}^{\mathbf{a}_1},\mathbb{C})_\star$. Here $\check{j}_{\mathcal{F}_2}$ is defined accordingly to $\check{j}_{\mathcal{F}_1}$. Thus we have

$$\Phi_{ae}^{**} \circ \check{j}_{\mathcal{F}_1} \circ \delta_s = \check{j}_{\mathcal{F}_2} \circ \Phi_{ae}(\delta_s) \qquad \forall\, s \in \Gamma\mathbb{F}^{\mathbf{a}_1}.$$

To relate this with Φ from above we observe

$$j_2^{**} \circ \check{j}_2 \circ \Phi(s) = \check{j}_{\mathcal{F}_1} \circ \Phi_{ae}(\delta_s)$$

and, therefore,

$$\check{\jmath}_{\mathcal{F}_2}(\delta_{\Phi(s)}) = \check{\jmath}_{\mathcal{F}_2}(\Phi(s))$$

which immediately yields

$$\Phi_{ae} = \Phi_*.$$

Taking (372) and theorem 77 into account, we hence have:

PROPOSITION 80. *Given two singularity free vectors $X = (id, \mathbf{a}_r)$ for $r = 1, 2$ on M and a continuous $*$-isomorphism*

$$\Phi_{ae} : \mathcal{F}_{ae}(\Gamma\mathbb{F}^{\mathbf{a}_1}, \mathbb{C})_\star \longrightarrow \mathcal{F}_{ae}(\Gamma\mathbb{F}^{\mathbf{a}_2}, \mathbb{C})_\star$$

there is a (continuous) map

$$\Phi : \Gamma\mathbb{F}^{\mathbf{a}_1} \longrightarrow \Gamma\mathbb{F}^{\mathbf{a}_2}$$

such that

$$\Phi_{ae} = \Phi_* \tag{373}$$

and vice versa. If Φ satisfies

$$\Phi_{ae}(\delta_{f\cdot s}) = \delta_{f\cdot\Phi(s)} \qquad \forall\, s \in \Gamma\mathbb{F}^{\mathbf{a}_1} \text{ and } \forall\, f \in \mathcal{S}(M, \mathbb{C}), \tag{374}$$

Φ is a $\mathcal{S}(M, \mathbb{C})$-bundle isomorphism. Hence X_1 and X_2 are equivalent iff there is a $*$-isomorphism Φ_{ae} yielding a $\mathcal{S}(M, \mathbb{C})$-module isomorphism Φ.

6.4.5. $*$-Isomorphism and Heisenberg Groups of Singularity Free Vector Fields. We will show in this subsection that $\mathcal{F}_{ae}(\mathcal{H}_X^\infty, \mathbb{C})_\star$ determines the singularity free vector field X on M. To this end let

$$\Phi_{ae} : \mathcal{F}_{ae}(\mathcal{H}_{X_1}^\infty, \mathbb{C})_\star \longrightarrow \mathcal{F}_{ae}(\mathcal{H}_{X_2}^\infty, \mathbb{C})_\star$$

be a $*$-*isomorphism* for the two singularity free vector fields $X_1 = (\mathrm{id}, \mathbf{a}_1)$ and $X_2 = (\mathrm{id}, \mathbf{a}_2)$. We will show in the sequel that

$$\Phi_{ae} \circ j_1(\Gamma\mathbb{F}^{\mathbf{a}_1}) = j_2(\Gamma\mathbb{F}^{\mathbf{a}_2}).$$

To do so we consider the group homomorphism

$$\begin{aligned} \mathrm{pr}_1 : \mathcal{H}_{X_1}^\infty &\longrightarrow \Gamma\mathbb{F}^{\mathbf{a}_1} \\ s + f\varepsilon_1 + \lambda e &\longmapsto s. \end{aligned} \tag{375}$$

This projection pr_1 in (375) yields the $*$-homomorphism

$$\mathrm{pr}_*^1 : \mathcal{F}_{ae}(\mathcal{H}_{X_1}^\infty, \mathbb{C})_\star \longrightarrow \mathcal{F}_{ae}(\Gamma\mathbb{F}^{\mathbf{a}_1}, \mathbb{C})_\star. \tag{376}$$

Any element of the form $\delta_{f\cdot\varepsilon+\lambda e} = \delta_{f\varepsilon} * \delta_{\lambda e}$ is mapped by pr_*^1 into

$$\mathrm{pr}_*^1(\delta_{f\varepsilon_1+\lambda e}) = \delta_{\mathrm{pr}^1(f\varepsilon_1+\lambda e)} = \delta_{\mathcal{O}}$$

where $\mathcal{O}$ is the zero section of $\Gamma\mathbb{F}^{\mathbf{a}_1}$. The kernel of pr_*^1 is determined by equation (357).

Since the $*$-isomorphism Φ_{ae} maps center to center, we conclude by (357) that

$$\Phi_{ae}(\mathcal{F}_{ae}(C_{X_1}, \mathbb{C})_\star) = \mathcal{F}_{ae}(C_{X_2}, \mathbb{C})_\star$$

where C_{X_1} and C_{X_2} are the centers of the (infinite dimensional) Heisenberg groups $\mathcal{H}_{X_1}^\infty$ and $\mathcal{H}_{X_2}^\infty$, respectively. Therefore, the form of $\ker \mathrm{pr}_*^1$ and $\ker \mathrm{pr}_*^2$ (cf. (358)) yields

$$\Phi_{ae}(\ker \mathrm{pr}_*^1) = \ker \mathrm{pr}_*^2. \tag{377}$$

(Here $\mathrm{pr}^2 : \mathcal{H}_X^\infty \longrightarrow \Gamma\mathbb{F}^{\mathbf{a}_2}$ is defined accordingly as pr^1.) Thus Φ_{ae} factors to a map

$$\mathcal{F}_{ae}(\Gamma\mathbb{F}^{\mathbf{a}_1}, \mathbb{C})_\star \longrightarrow \mathcal{F}_{ae}(\Gamma\mathbb{F}^{\mathbf{a}_2}, \mathbb{C})_\star,$$

again denoted by Φ_{ae}. By proposition 79 we therefore conclude at first that

$$\Phi_{ae} \circ j_1(\Gamma\mathbb{F}^{\mathbf{a}_1}) = j_2(\Gamma\mathbb{F}^{\mathbf{a}_2}) \tag{378}$$

and, in turn, the existence of a group homomorphism

$$\Phi : \mathcal{H}_{X_1}^\infty \longrightarrow \mathcal{H}_{X_2}^\infty,$$

say, such that

$$\Phi_{ae} = \Phi_*.$$

Together with theorem 77 we thus have the following:

PROPOSITION 81. *Under the conditions of proposition 80 any continuous $*$-isomorphism*

$$\Phi_{ae} : \mathcal{F}_{ae}(\mathcal{H}_{X_1}^\infty, \mathbb{C})_\star \longrightarrow \mathcal{F}_{ae}(\mathcal{H}_{X_2}^\infty, \mathbb{C})_\star$$

yields a unique group isomorphism

$$\Phi : \mathcal{H}_{X_1}^\infty \longrightarrow \mathcal{H}_{X_2}^\infty$$

such that $\Phi_{ae} = \Phi_*$ and vice versa. Hence the vector fields X_1 and X_2 are equivalent and have identical field strength functions. This means that $\mathcal{F}_{ae}(\Gamma\mathbb{F}^{\mathbf{a}}, \mathbb{C})_\star$ is characteristic for any singularity free vector field $X = (id, \mathbf{a})$ on M.

The converse obviously is true, as well.

6.5. The C^*-Group Algebra and the Twisted Convolution, the Weyl Algebra and the Poisson Algebra

We emphasize here on the C^*-Heisenberg group algebra and on the C^*-Weyl algebra since the algebras both will be basic ingredients of our field quantization. The rather technical proofs will be omitted here. We follow closely [**14**], [**13**] and [**55**]. For the various general treatments and application of C^*-algebras we refer to [**3**], [**70**] and [**80**].

6.5.1. C^*-Group Algebra of a Group.

Let G be either one of the groups $\Gamma\mathbb{F}^a, \Gamma\mathbf{H}^{\mathbf{a}}$ or $\mathcal{H}_X^\infty$. For the group G the algebra group $\mathcal{F}_{ae}(G, \mathbb{C})_\star$ carries a natural norm (cf. [**13**] and [**55**]), namely:

$$\| f \| := \sup\{\| \Pi(f) \|^0 \,|\Pi \text{ is a } *\text{-representation of } \mathcal{F}_{ae}(G, \mathbb{C})_\star\} \tag{379}$$

for any $f \in \mathcal{F}_{ae}(G, \mathbb{C})_\star$. Here $\| \; \|^0$ the norm of the (Π-dependent) Hilbert space $\mathcal{H}$ on which Π represents $\mathcal{F}_{ae}(G, \mathbb{C})_\star$ in the $*$-algebra $\mathbf{B}(\mathcal{H})$ of all *bounded operators* of $\mathcal{H}$. The *involution* applied to the bounded operator is the *adjoint*. Moreover, the term *$*$-representation* means that it is an algebra representation satisfying $\Pi(*f) = *\Pi(f)$ for all $f \in \mathcal{F}_{ae}(G, \mathbb{C})_\star$.

To indicate that $\| \; \|$ exists on $\mathcal{F}_{ae}(G, \mathbb{C})_\star$ we represent any $f \in \mathcal{F}_{ae}(G, \mathbb{C})_\star$ in the basis $j(G)$ consisting of Dirac maps as in (352). The norm

$$\| f \|^1 := \sum_{s'} |f(s')| \tag{380}$$

obviously exists. It extends to $l^1(G)$, the completion of $\mathcal{F}_{ae}(G, \mathbb{C})_\star$ with respect to $\| \; \|^1$. The expression defined by the right-hand side of (379) satisfies

$$\| f \| \leq \| f \|^1 \qquad \forall f \in \mathcal{F}_{ae}(G, \mathbb{C})_\star. \tag{381}$$

The completion $l^1(G)$ is a Banach $*$-algebra which possesses enough $*$-representations Π into $\mathbf{B}(\mathcal{H})$. "Enough" means that to any $s \in l^1(G)$ there is a representation Π for which $\Pi(s) \neq 0$. Hence we conclude that $\| \ \|$ exists and is a norm (again cf. [**13**] and [**55**]).

The *completion* of $\mathcal{F}_{ae}(G, \mathbb{C})_\star$ by the norm $\| \ \|$ is denoted by C^*G. The involution $*$ on $\mathcal{F}_{ae}(G, \mathbb{C})_\star$ extends to C^*G and hence C^*G is a C^**-algebra*, the C^**-group algebra* or just the C^*-algebra of G.

Any C^*-algebra admits a faithful representation. Hence it is isometrically isomorphic to a self-adjoint algebra of operator on a Hilbert space.

Back to the Heisenberg group of the vector field X on M, we have the following continuous inclusions

$$\mathcal{F}_{ae}(\mathcal{H}_X^\infty, \mathbb{C}) \subset \overline{\mathcal{F}_{ae}(\mathcal{H}_X^\infty, \mathbb{C})}^1 \subset \overline{\mathcal{F}_{ae}(\mathcal{H}_X^\infty, \mathbb{C})} = C^*\mathcal{H}_X^\infty$$

where the index 1 indicates the completion with respect to the norm $\| \ \|^1$. These inclusions hold accordingly if $\mathcal{H}_X^\infty$ is replaced by $\Gamma\mathbf{H}^{\mathbf{a}}$. The C^*-algebra $C^*\mathcal{H}_X^\infty$ is called the C^**-Heisenberg group algebra of* X.

The following theorem developed in [**55**] reveals the importance of the C^*-group algebra:

THEOREM 82. *$C^*\mathcal{H}_X^\infty$ is the unique C^*-algebra (up to $*$-isometry) generated by $\mathcal{F}_{ae}(\mathcal{H}_X^\infty, \mathbb{C})_\star$ such that every $\omega^{\Gamma\mathbb{F}^{\mathbf{a}}}$-projective unitary representation $(\pi, \mathcal{H}_\pi)$ of $\mathcal{H}_X^\infty$ arises from a unitary Hilbert space representation $(\Pi, \mathcal{H}_\Pi)$ with*

$$\pi(\mathbf{s} + \lambda e) = \Pi(\delta_{\mathbf{s}} + \lambda e)$$

for all $\mathbf{s} + \lambda \cdot e \in \mathcal{H}_X^\infty$; i.e. for all $\mathbf{s} \in \Gamma\mathbb{H}^{\mathbf{a}}$ and all $\lambda \in \mathbb{R}$.

6.5.2. C^*-Isomorphisms. Suppose

$$\phi : C^*\mathcal{H}_{X_1}^\infty \longrightarrow C^*\mathcal{H}_{X_2}^\infty$$

is a C^**-isomorphism*. Are the two vector fields $X_1 = (\mathrm{id}, \mathbf{a_1})$ and $X_2 = (\mathrm{id}, \mathbf{a_2})$ equivalent, i.e. are the bundles $\mathbb{F}^{\mathbf{a_1}}$ and $\mathbb{F}^{\mathbf{a_2}}$ isomorphic (cf. definition 15 in section 3.2.3)?

To investigate this question we consider the projection homomorphism

$$\mathrm{pr}_*^r : C^*\mathcal{H}_{X_r}^\infty \longrightarrow C^*\Gamma\mathbb{F}^{\mathbf{a}_r} \qquad r = 1, 2,$$

i.e. the extensions of (376) for $r = 1, 2$ to the completions of domains and ranges with respect to the norm given by (379). These two homomorphisms are continuous by the closed graph theorem (cf. [**82**]). Both are surjections. The center $C_{C^*\mathcal{H}_{X_r}^\infty}$ of $C^*\mathcal{H}_{X_r}^\infty$ and the center $C_{C^*\Gamma\mathbb{F}^{\mathbf{a}_r}}$ of $C^*\Gamma\mathbb{F}^{\mathbf{a}_r}$ are the $\| \ \|$-completions of the respective centers of $\mathcal{F}_{ae}(\mathcal{H}_X^\infty, \mathbb{C})$ and of $\mathcal{F}_{ae}(\Gamma\mathbb{F}^{\mathbf{a}_r}\, \mathbb{C})$ for $r = 1, 2$. Hence for the respective centers

$$\phi(C_{C^*\mathcal{H}_{X_1}^\infty}) = C_{C^*\mathcal{H}_{X_2}^\infty}$$

holds true. Therefore, we may conclude that

$$\phi(\ker^* \mathrm{pr}_*^1) = \ker^* \mathrm{pr}_*^2$$

by (377) and in turn that ϕ factors to

$$\phi : C^*\Gamma\mathbb{F}^{\mathbf{a}} \longrightarrow C^*\Gamma\mathbb{F}^{\mathbf{a_2}},$$

where $\ker^* \mathrm{pr}_*^r \subset C^*\mathcal{H}_{X_r}^\infty$ is the kernel of pr_*^r for $r = 1, 2$.

Now the main goal in the section is to show that

$$\phi = \Phi_*$$

for some $\mathbb{C}$-linear isomorphism

$$\Phi : \Gamma\mathbb{F}^{\mathbf{a}_1} \longrightarrow \Gamma\mathbb{F}^{\mathbf{a}_2}.$$

First we will reconstruct (by means of Pontryagin's duality again) $\Gamma\mathbb{F}^{\mathbf{a}}$ out of $C^*\Gamma\mathbb{F}^{\mathbf{a}}$ for some singularity free vector field $X = (\mathrm{id}\,, \mathbf{a})$ on M. To this end we consider

$$j : \Gamma\mathbb{F}^{\mathbf{a}} \longrightarrow C^*\Gamma\mathbb{F}^{\mathbf{a}}$$

and $\mathrm{Hom}\, C^*\Gamma\mathbb{F}^{\mathbf{a}}$, the collection of all *continuous* $\mathbb{C}$-valued algebra homomorphisms of $C^*\Gamma\mathbb{F}^{\mathbf{a}}$. Clearly, any $h \in \mathrm{Hom}\, C^*\Gamma\mathbb{F}^{\mathbf{a}}$ yields the character $h \circ j \in \chi(\Gamma\mathbb{F}^{\mathbf{a}})$ (cf. section 6.4.3). Since for any two $f_1, f_2 \in \mathcal{F}_{ae}(\Gamma\mathbb{F}^{\mathbf{a}}, \mathbb{C})$ represented as

$$f_r = \sum_i z_r^i \cdot \delta_{s_i+\lambda ie} \qquad r = 1, 2$$

we have

$$|\chi(f_1) - \chi(f_2)| \leq \sum |z_1^i - z_2^i| \cdot |\chi(\delta_{s_i+\lambda ie})| \leq \| f_1 - f_2 \|^1,$$

any character $\chi \in \chi(\Gamma\mathbb{F}^{\mathbf{a}})$ is continuous with respect to $\| \; \|^1$ and hence with respect to $\| \; \|$. The map

$$j^* : \mathrm{Hom}\, C^*\Gamma\mathbb{F}^{\mathbf{a}} \longrightarrow \chi(\Gamma\mathbb{F}^{\mathbf{a}}) \tag{382}$$

is a bijection since any character in $\chi(\Gamma\mathbb{F}^{\mathbf{a}})$ extends uniquely to a continuous homomorphism of $C^*\Gamma\mathbb{F}^{\mathbf{a}}$ since $\mathcal{F}_{ae}(\Gamma\mathbb{F}^{\mathbf{a}}, \mathbb{C})$ is dense in $C^*\Gamma\mathbb{F}^{\mathbf{a}}$. Pulling back the group operations of $\chi(\Gamma\mathbb{F}^{\mathbf{a}})$ to $\mathrm{Hom}\, C^*\Gamma\mathbb{F}^{\mathbf{a}}$ by j^* yields a *commutative group.*

To show that $\mathrm{Hom}\, C^*\Gamma\mathbb{F}^{\mathbf{a}}$ equipped with the topology of *pointwise convergence* is compact we recall that the character group $\chi(\Gamma\mathbb{F}^{\mathbf{a}})$ is endowed with the topology of pointwise convergence. Since the group $\Gamma\mathbb{F}^{\mathbf{a}}$ carries the discrete topology the topological group $\chi(\Gamma\mathbb{F}^{\mathbf{a}})$ is compact (cf. section 6.4.3). Unfortunately, we cannot use Pontryagin's duality directly for $\mathrm{Hom}\, C^*\Gamma\mathbb{F}^{\mathbf{a}}$ to conclude compactness. Thus we will show that j^* is a homeomorphism. To do so, we first observe that any $g \in \mathcal{F}_{ae}(\Gamma\mathbb{F}^{\mathbf{a}}, \mathbb{C})$ causes the continuous map

$$\breve{g} : \mathrm{Hom}\, \mathcal{F}_{ae}(\Gamma\mathbb{F}^{\mathbf{a}}, \mathbb{C})_\star \longrightarrow \mathbb{C}$$

by setting

$$\breve{g}(h) := h(g)$$

for all $h \in \mathrm{Hom}\, \mathcal{F}_{ae}(\Gamma\mathbb{F}^{\mathbf{a}}, \mathbb{C})_\star$. For $g = \sum_r z^r \delta_{s_r}$ in $\mathcal{F}_{ae}(\Gamma\mathbb{F}^{\mathbf{a}}, \mathbb{C})_\star$,

$$\sup_h |\breve{g}(h)| = \sup_h |h(g)| = \sup_h |\sum_r z^r \chi_h(s_r)| \leqq \sum_r |z^r|$$

holds true. Hence any $\breve{g}$ on $\mathrm{Hom}\, \mathcal{F}_{ae}(\Gamma\mathbb{F}^{\mathbf{a}}, \mathbb{C})_\star$ is bounded. We may set

$$\| g \|' := \sup_h |\breve{g}(h)| \qquad \forall g \in \mathcal{F}_{ae}(\Gamma\mathbb{F}^{\mathbf{a}}, \mathbb{C})$$

and observe that

$$\| g \|' \leqq \| g \|^1 .$$

Now let us turn to $\operatorname{Hom} C^*\Gamma\mathbb{F}^{\mathbf{a}}$ and show that it is compact. Any $h \in \operatorname{Hom} C^*\Gamma\mathbb{F}^{\mathbf{a}}$ can be interpreted as a one-dimensional representation on the Hilbert space $\mathbb{C}$. Thus $h_{\mathcal{H}}$ enters into (379). The definition of $\| \ \|$ extends verbatim to $C^*\Gamma\mathbb{F}^{\mathbf{a}}$ since $\mathcal{F}_{ae}(\Gamma\mathbb{F}^{\mathbf{a}}, \mathbb{C})$ is a dense subset of $C^*\Gamma\mathbb{F}^{\mathbf{a}}$. Therefore, any $f \in C^*\Gamma\mathbb{F}^{\mathbf{a}}$ defines a bounded continuous function $\check{f}$ on $\operatorname{Hom} C^*\Gamma\mathbb{F}^{\mathbf{a}}$, saying that

$$C^*\Gamma\mathbb{F}^{\mathbf{a}} \subset C_0(\operatorname{Hom} C^*\Gamma\mathbb{F}^{\mathbf{a}}) \supset C_0(\operatorname{Hom} \mathcal{F}_{ae}(\Gamma\mathbb{F}^{\mathbf{a}}, \mathbb{C}))$$

where C_0 denotes the collection of all *bounded*, $\mathbb{C}$-valued continuous functions on $\operatorname{Hom} C^*\Gamma\mathbb{F}^{\mathbf{a}}$. Hence the restriction map res

$$\operatorname{res} : \operatorname{Hom} C_0(\operatorname{Hom} C^*\Gamma\mathbb{F}^{\mathbf{a}}) \longrightarrow \operatorname{Hom} C^*\Gamma\mathbb{F}^{\mathbf{a}}$$

is surjective. Here Hom denotes the collection of all continuous $\mathbb{C}$-valued homomorphisms. Since the domain of res is *compact* if equipped with the *topology of pointwise convergence* (cf. [**8**]) we conclude that $\operatorname{Hom} C^*\Gamma\mathbb{F}^{\mathbf{a}}$ is *compact*, as well. Thus j^* in (382) has to be an isomorphism. Hence both $\operatorname{Hom} C^*\Gamma\mathbb{F}^{\mathbf{a}_r}$ and $\chi(\Gamma\mathbb{F}^{\mathbf{a}_r})$ are isomorphic compact groups, yielding the following reconstruction lemma (cf. proposition 79):

LEMMA 83. *For any singularity free vector field $X = (id, \mathbf{a})$ on M the maps*

$$\Gamma\mathbb{F}^{\mathbf{a}} \xrightarrow{\ j\ } \chi(\chi(\Gamma\mathbb{F}^{\mathbf{a}})) \xrightarrow{\ j^{**}\ } \chi(\operatorname{Hom} C^*\Gamma\mathbb{F}^{\mathbf{a}})$$

consist of group isomorphisms of discrete groups. Again all character groups carry the compact open topology. Hence $C^*\Gamma\mathbb{F}^{\mathbf{a}}$ determines $\Gamma\mathbb{F}^{\mathbf{a}}$ uniquely.

To answer the question raised at the beginning of this subsection we use Pontryagin's duality (cf. [**8**]) to conclude the existence of Φ such that $\phi = \Phi_*$ (as in the case of proposition 80). Vice versa any Φ yields a C^*-isomorphism ϕ and we may finally state the main theorem in this section characterizing any singularity free vector field X on M in terms of $C^*\mathcal{H}_X^\infty$:

THEOREM 84. *For any two singularity free vector fields $X_r = (id, \mathbf{a}_r)$ on M, $r = 1, 2$, any C^*-isomorphism*

$$\phi : C^*\Gamma\mathcal{H}_{X_1}^\infty \longrightarrow C^*\Gamma\mathcal{H}_{X_2}^\infty$$

determines a $\mathbb{C}$-linear map

$$\Phi : \Gamma\mathbb{F}^{\mathbf{a}_1} \longrightarrow \Gamma\mathbb{F}^{\mathbf{a}_2}$$

such that

$$\phi = \Phi_*$$

and vice versa. If, in addition,

$$\phi(\delta_{f\cdot s}) = \delta_{f\Phi(s)} \qquad \forall\, f \in \mathcal{S}(M, \mathbb{C}) \text{ and } \forall\, s \in \Gamma\mathbb{F}^{\mathbf{a}_1},$$

Φ yields the uniquely determined bundle isomorphism

$$\Phi_{1,2} : \mathbb{F}^{\mathbf{a}_1} \longrightarrow \mathbb{F}^{\mathbf{a}_1}$$

such that

$$\Phi(s) = \Phi_{1,2} \circ s \qquad \forall\, s \in \Gamma\mathbb{F}^{\mathbf{a}_1}$$

and vice versa, saying that X_1 and X_2 are equivalent and have the same field strength function.

Remark. Theorem 84 holds accordingly for $\Gamma\mathbb{F}^{\mathbf{a}}$ as well as for $\Gamma\mathbf{H}^{\mathbf{a}}$ with a difference concerning the vector field $X = (\mathrm{id}\,,\mathbf{a})$; however, The C^*- algebra of $\Gamma\mathbb{F}^{\mathbf{a}}$ determines the equivalence class of X while the C^*-algebra of $\Gamma\mathbf{H}^{\mathbf{a}}$ is characteristic for X since it involves $\omega^{\mathbf{a}}$ (cf. section 3.2.1).

6.5.3. The Twisted Convolution and the Twisted Group Algebra. Given a singularity free vector field $X = (\mathrm{id}\,,\mathbf{a})$ on M the space of sections $\Gamma\mathbf{H}^{\mathbf{a}} = \Gamma(\mathbb{F}^{\mathbf{a}} \oplus \mathbb{R}\cdot\varepsilon)$ is an infinite dimensional group. Let $\delta_{\mathbf{s}_1}, \delta_{\mathbf{s}_2} \in \mathcal{F}_{ae}(\Gamma\mathbf{H}^{\mathbf{a}}, \mathbb{C})$. The *twisted convolution* $\star_\omega$ of $\delta_{\mathbf{s}_1}$ and $\delta_{\mathbf{s}_2}$ in $\mathcal{F}_{ae}(\Gamma\mathbf{H}^{\mathbf{a}}, \mathbb{C})$ is defined to be

$$\delta_{\mathbf{s}_1} \star_\omega \delta_{\mathbf{s}_2} := e^{-\frac{i}{2}\omega(\mathbf{s}_1,\mathbf{s}_2)}\delta_{\mathbf{s}_1} \star \delta_{\mathbf{s}_2} \qquad \forall\, \mathbf{s}_1, \mathbf{s}_2 \in \Gamma\mathbf{H}^{\mathbf{a}}. \tag{383}$$

Here for simplicity we set ω instead of $\omega^{\Gamma\mathbf{H}^{\mathbf{a}}}$ (cf. section 6.2.1). The product $\star_\omega$ extends to all of $\mathcal{F}_{ae}(\Gamma\mathbf{H}^{\mathbf{a}}, \mathbb{C})$ by

$$\begin{aligned}(\sum_m z_m\delta_{\mathbf{s}_m}) \star_\omega (\sum_n z_n\delta_{\mathbf{s}_n}) \quad &:= \quad \sum_{m,n} z_m \cdot z_n \cdot \delta_{\mathbf{s}_m} \star_\omega \delta_{\mathbf{s}_n} \\ &= \quad \sum_{m,n} z_m \cdot z_n \cdot e^{-\frac{i}{2}\omega(\mathbf{s}_m,\mathbf{s}_n)} \cdot \delta_{\mathbf{s}_m} \star \delta_{\mathbf{s}_n}.\end{aligned}$$

$\mathcal{F}_{ae}(\Gamma\mathbf{H}^{\mathbf{a}}, \mathbb{C})$ endowed with this multiplication is called the *discrete Weyl algebra* of the vector field X and will in the sequel be denoted by $W_{ae}\Gamma\mathbf{H}^{\mathbf{a}}$.

Let us examine some special cases of the multiplication in $W_{ae}\Gamma\mathbf{H}^{\mathbf{a}}$. Here we have $\mathbf{s} = s + \lambda\cdot\varepsilon$ with $\mathbf{s} \in \Gamma\mathbf{H}^{a}$, i.e. $s \in \Gamma\mathbb{F}^{\varepsilon}$ and $\lambda \in \mathcal{S}(\mathbb{R}, \mathbb{R})$:

a) $\quad \delta_{s+\lambda\cdot\varepsilon} = \delta_s \star \delta_{\lambda\cdot\varepsilon} = \delta_s \star_\omega \delta_{\lambda\cdot\varepsilon},$

b) $\quad \delta_{s_1+s_2} = \delta_{s_1} \star \delta_{s_2} = e^{\frac{i}{2}\omega(s_1,s_2)} \cdot \delta_{s_1} \star_\omega \delta_{s_2},$

c) $\quad \delta_{s_1+\lambda_1\varepsilon} \star \delta_{s_1+\lambda_2\varepsilon} = e^{\frac{i}{2}\omega(s_1,s_2)\cdot\varepsilon} \cdot \delta_{(\lambda_1+\lambda_2)\cdot\varepsilon} \star_\omega \cdot\delta_{s_1} \star_\omega \delta_{s_2}.$

To demonstrate the relation between $\star_\omega$ and the *Fourier transform* we restrict to the following case:

$$\mathbf{F}(\delta_{\mathbf{s}_1+\mathbf{s}_2}) = \mathbf{F}(\delta_{\mathbf{s}_1} \star \delta_{\mathbf{s}_2}) = \mathbf{F}(\delta_{\mathbf{s}_1}) \cdot \mathbf{F}(\delta_{\mathbf{s}_2}) = e^{2\pi i <\mathbf{s}_1,\ldots>} \cdot e^{2\pi i <\mathbf{s}_2,\ldots>}$$

for all $\delta_{\mathbf{s}_1}, \delta_{\mathbf{s}_2} \in \mathcal{F}_{ae}(\Gamma\mathbf{H}^{\mathbf{a}}, \mathbb{C})_*$. Therefore, by (383)

$$\mathbf{F}(\delta_{\mathbf{s}_1+\mathbf{s}_2}) = e^{2\pi i\omega(\mathbf{s}_1,\mathbf{s}_2)} \cdot \mathbf{F}(\delta_{\mathbf{s}_1} \star_\omega \delta_{\mathbf{s}_2}) \tag{384}$$

holds true.

A slightly different picture of $W_{ae}\Gamma\mathbf{H}^{\mathbf{a}}$ is obtained as follows: For each $\mathbf{s} \in \Gamma\mathbf{H}^{\mathbf{a}}$ let

$$W(\mathbf{s}_1) : \Gamma\mathbf{H}^{\mathbf{a}} \longrightarrow \mathbb{C} \qquad (\Gamma\mathbf{H}^{\mathbf{a}} \quad \text{regarded as a set})$$

be defined by

$$W(\mathbf{s}_1)(\mathbf{s}_2) := \delta_{\mathbf{s}_1,\mathbf{s}_2} \qquad \forall\, \mathbf{s}_1, \mathbf{s}_2 \in \Gamma\mathbf{H}^{\mathbf{a}}. \tag{385}$$

Here $\delta_{\mathbf{s}_1,\mathbf{s}_2}$ is the Kronecker symbol, thus $W(\mathbf{s}_1)$, called a *Weyl element*, extends to a $\mathbb{C}$-linear map of $W_{ae}\Gamma\mathbf{H}^{\mathbf{a}}$ (still denoted by $W(\mathbf{s}_1)$). Hence

$$W(\mathbf{s}_1) \in L(W_{ae}\Gamma\mathbf{H}^{\mathbf{a}}, \mathbb{C}) \tag{386}$$

where $L(W_{ae}\Gamma\mathbf{H}^{\mathbf{a}}, \mathbb{C})$ denotes the *algebraic dual* of $W_{ae}\Gamma\mathbf{H}^{\mathbf{a}}$, i.e. the collection of all $\mathbb{C}$-valued linear maps from $W_{ae}\Gamma\mathbf{H}^{\mathbf{a}}$. The linear $\mathbb{C}$-hull of $\{W(\mathbf{s})|\mathbf{s} \in \Gamma\mathbf{H}^{\mathbf{a}}\}$ as a linear space is obviously isomorphic to $W_{ae}\Gamma\mathbf{H}^{\mathbf{a}}$ and is also denoted by $W_{ae}\Gamma\mathbf{H}^{\mathbf{a}}$. The product in $W_{ae}\Gamma\mathbf{H}^{\mathbf{a}}$ expressed in terms of Weyl elements takes the form

$$W(\mathbf{s}_1) \cdot W(\mathbf{s}_2) = e^{-\frac{1}{2}\omega(\mathbf{s}_1,\mathbf{s}_2)i} \cdot W(\mathbf{s}_1 + \mathbf{s}_2) \qquad \forall\, \mathbf{s}_1, \mathbf{s}_2 \in \Gamma\mathbf{H}^{\mathbf{a}} \tag{387}$$

(cf. (350)) turning $W_{ae}\Gamma\mathbf{H^a}$ to an *(infinite dimensional) $*$-algebra*, the star being the involution defined as in (351).

The definition of the norm $\| \ \|$ on $W_{ae}\Gamma\mathbf{H^a}$ here is according to that in (379), namely

$$\| g \| := \sup\{\| \Pi(g) \|^0 \ | \Pi \ *\text{-representation of } W_{ae}\Gamma\mathbf{H^a}\}.$$

Its completion with respect to the norm $\| \ \|$ (cf. (380)), referred to by $W\Gamma\mathbf{H^a}$, is a *C^*-algebra* with $W(0)$ as identity.

$W\Gamma\mathbf{H^a}$ is called a *twisted C^*-group algebra* with respect to the *multiplier* $\frac{1}{2}\omega^{\Gamma\mathbf{H^a}}$ or simply the *C^*-Weyl algebra* of X.

If $\omega^{\Gamma\mathbf{H^a}} = \omega_X^\infty$ on $\Gamma\mathbf{H^a}$ (cf. section 6.2.1) is deformed into $\frac{\hbar}{2}\omega^{\Gamma\mathbf{H^a}}$ for $\hbar \in \mathbb{R}$, the respective C^*-Weyl algebra is denoted by $W^\hbar\Gamma\mathbf{H^a}$. Equations (383) to (386) hold accordingly.

6.5.4. The Poisson Algebra of a Singularity Free Vector Field. The *classical part* of the quantization of X to be used in chapter 10 is a *commutative* Weyl algebra. It is constructed as follows: The bundle $\mathbf{H^a}$ over M is regarded as a split rank-three vector bundle, i.e. as $\mathbf{H^a} = \mathbb{F}^\mathbf{a} \oplus \mathbb{R} \cdot e$. Let us *neglect* the fibrewisely given presymplectic form $\omega^\mathbf{a}$ on $\mathbf{H^a}$ and replace it by the fibrewisely *trivial* presymplectic structure denoted by 0. The collection $\Gamma\mathbf{H^a}$ of all Schwartz sections is an infinite dimensional $\mathbb{R}$-linear space and, in particular, a *commutative group* under the pointwisely defined operations. As such it determines the group algebra $\mathcal{F}_{ae}(\Gamma\mathbf{H^a}, \mathbb{C})_\star$. Thus the convolution $\star$ on $\mathcal{F}_{ae}(\Gamma\mathbf{H^a}, \mathbb{C})_\star$ twisted by 0 yields the trivial twisted convolution algebra, i.e. the $*$-algebra $\mathcal{F}_{ae}(\Gamma\mathbf{H^a}, \mathbb{C})_\star$ where $\Gamma\mathbf{H^a}$ is regarded as a commutative group (cf. (387)). This shows that this $*$-algebra does not reflect the vector field since ω_X^∞ (cf. section 6.2.1) does not enter into the convolution determined by the additive group $\Gamma\mathbf{H^a}$. However, it reflects ε due to $\Gamma\mathbb{F}^\varepsilon$. The Weyl algebra determined by this convolution algebra is denoted by $W_{ae}^0\Gamma\mathbf{H^a}$. Its elements are called $W^0(\mathbf{s})$ for $\mathbf{s} \in \Gamma\mathbf{H^a}$.

Its completion with respect to the norm $\| \ \|$ (cf. (379)) shall be referred to by $W^0\Gamma\mathbf{H^a}$. This Weyl algebra is a commutative C^*-algebra (cf. [**14**]).

Moreover, it contains a natural *Poisson algebra*: In fact,

$$\mathcal{F}_{ae}(\Gamma\mathbf{H^a}, \mathbb{C}) \subset W^0\Gamma\mathbf{H^a} \tag{388}$$

with the pointwisely defined $\mathbb{C}$-linear structure and endowed with the *Poisson bracket* given by

$$\{W^0(\mathbf{s}_1), W^0(\mathbf{s}_2)\} := \omega^{\Gamma\mathbf{H^a}}(\mathbf{s}_1, \mathbf{s}_2) \cdot W^0(\mathbf{s}_1 + \mathbf{s}_2) \tag{389}$$

for any pair $\mathbf{s}_1, \mathbf{s}_2 \in \Gamma\mathbf{H^a}$ forms a Poisson algebra. The definition of this bracket is purely algebraic; however, it can be calculated by means of the differential calculus presented in [**18**], as done in [**12**]. This bracket extends to $W_{ae}^0\Gamma\mathbf{H^a}$ (cf. [**15**]). The $*$-operation defined by (351) leaves the bracket in (389) invariant. We therefore call this Poisson algebra a *$*$-Poisson algebra $P^\mathbf{a}$*.

The vector field $X = (\mathrm{id}, \mathbf{a})$ enters into (389) again by ω_X^∞. This form is degenerate along $\Gamma\mathcal{L} = \mathcal{S}(\mathbb{R}, \mathbb{C}) \cdot \varepsilon$. The fibrewisely given symplectic structure $\omega^\mathbf{a}$ on the complex line bundle $\mathbb{F}^\mathbf{a}$ can be reconstructed from ω_X^∞. By the remark in section 6.5.2 we have the following lemma reformulating $X = (\mathrm{id}, \mathbf{a})$ in terms of the Poisson algebra:

LEMMA 85. *The Poisson algebra* $P^{\mathbf{a}} \subset W^0\mathbf{H}^{\mathbf{a}}$ *uniquely determines the (singularity free) vector field* $X = (id, \mathbf{a})$.

Remark. The lemma above is still valid if $\Gamma\mathbf{H}^{\mathbf{a}}$ is replaced by the linear space generated by smooth sections with compact support in charts on which $\mathbb{F}^{\mathbf{a}}$ trivializes.

6.5.5. The Heisenberg Group Algebra and Weyl-Algebras. There is a natural map from the C^*-Heisenberg group algebra $C^*\mathcal{H}_X^\infty$ to the C^*-Weyl algebra $W\Gamma\mathbf{H}^{\mathbf{a}}$, namely

$$\begin{aligned} \theta_{W\Gamma\mathbf{H}^{\mathbf{a}}} : C^*\mathcal{H}_X^\infty &\longrightarrow W\Gamma\mathbf{H}^{\mathbf{a}} \\ \mathbf{s} + \lambda e &\longmapsto e^{i\lambda} W(\mathbf{s}). \end{aligned} \tag{390}$$

Obviously, this map is onto. Moreover, it is a C^*-homomorphism. Both C^*-algebras $C^*\mathcal{H}_X^\infty$ and $W\Gamma\mathbf{H}^{\mathbf{a}}$ characterize the vector field X, as it is formulated in theorem 84 and lemma 85.

We have the identity

$$W^0_{ae}\Gamma\mathbf{H}^{\mathbf{a}} \equiv \mathcal{F}_{ae}(\Gamma\mathbf{H}^{\mathbf{a}}, \mathbb{C})_\star.$$

This yields the splitting

$$\mathcal{F}_{ae}(\mathcal{H}^\infty, \mathbb{C})_\star = W^0_{ae}\Gamma\mathbf{H}^{\mathbf{a}} \oplus \mathbb{R} \cdot e$$

of $*$-algebras which is preserved under the completion with respect to the norm $\| \ \|$.

As noticed in section 6.5.3, deforming $\omega^{\Gamma\mathbf{H}^{\mathbf{a}}} = \omega_X^\infty$ on $\Gamma\mathbf{H}^{\mathbf{a}}$ into $\hbar \cdot \omega_X^\infty$ yields a new twisted product $*_{\hbar\cdot\omega}$ on $\mathcal{F}_{ae}(\Gamma\mathbf{H}^{\mathbf{a}}, \mathbb{C})$ which in turn yields the Weyl algebra $W^\hbar\Gamma\mathbf{H}^{\mathbf{a}}$ for any value of the deformation parameter $\hbar \in \mathbb{R}$. Thus we obtain a family of $\hbar$-dependent C^*-Weyl algebras, closely related to $C^*\mathcal{H}_X^\infty$. To see this we need a field of C^*-Weyl algebras, to be introduced next:

A *field of* C^**-Weyl algebras* will consist of a special class $\mathcal{W}$ of maps from $\mathbb{R}$ into the Cartesian product of Weyl algebras, i.e.

$$\mathcal{W} \subset \prod_{\hbar \in \mathbb{R}} W^\hbar \Gamma\mathbf{H}^{\mathbf{a}}$$

carrying the *product norm*. $\mathcal{W}$ can be regarded as a trivial bundle of Weyl algebras over $\mathbb{R}$. Hence the space $\Gamma\mathcal{W}$ of all sections into $\mathcal{W}$ is a $*$-algebra under the pointwisely defined operations. The subcollection $\Gamma_0\mathcal{W} \subset \Gamma\mathcal{W}$ consisting of all *bounded, continuous sections* into $\mathcal{W}$ is a C^*-subalgebra of $\Gamma\mathcal{W}$. The *norm* of $K \in \Gamma_0\mathcal{W}$ is defined by

$$\| K \|_{\sup} := \sup_{\hbar \in \mathbb{R}} \| K(h) \|_\hbar$$

where $\| \ \|_\hbar$ is the norm of $W^\hbar\Gamma\mathbf{H}^{\mathbf{a}}$.

A *continuous field* $\mathcal{K}$ is a sub-C^*-algebra of sections in $\Gamma_0\mathcal{W}$ such that the following holds true:

i): The $\mathbb{R}$-valued map defined on $\mathbb{R}$ by $\hbar' \mapsto \| K(\hbar)' \|_{\hbar}$ is continuous for all $K \in \Gamma_0 \mathcal{W}$.

ii): $\{K(\hbar) | K \subset \mathcal{K}\} = W^{\hbar}\Gamma\mathbf{H^a}$ for any $\hbar \in \mathbb{R}$.

iii): $\mathcal{K}$ is locally complete, i.e. $K \in \Gamma\mathcal{W}$ belongs to $\mathcal{K}$ iff for each $\hbar' \in \mathbb{R}$ and any $\varepsilon > 0$ there is some $H \in \mathcal{K}$ such that

$$\| K(\hbar') - H(\hbar) \|_{\hbar} < \varepsilon$$

holds true in an open neighborhood of $\hbar'$ for all choices of $\hbar \in \mathbb{R}$.

The basic theorem for our setting guarantees existence and uniqueness (cf. [**14**], [**15**] and [**55**]):

THEOREM 86. *There exists a unique continuous field $\mathcal{K}$ for a family of C^*-Weyl algebras $\{\mathcal{W}^{\hbar}\Gamma\mathbf{H^a}\}_{\hbar\in\mathbb{R}}$ formed with the presymplectic structure $\hbar\cdot\omega^{\Gamma\mathbb{E}}$ for any $\hbar \in \mathbb{R}$ such that the map assigning the Weyl element $W^{\hbar}(\mathbf{s})$ to each $\hbar \in \mathbb{R}$ belongs to $\mathcal{K}$ for any $\mathbf{s} \in \Gamma\mathbf{H^a}$.*

In fact, the continuous field mentioned in the above theorem is the sup-norm $\| \; \|_{\text{sup}}$-completion $C^*_{WF}\Gamma\mathbf{H^a}$ of span $\{\hbar \mapsto e^{-\lambda\cdot\hbar\cdot\varepsilon} \cdot W^{\hbar}(\mathbf{s}) | \lambda, \hbar \in \mathbb{R}, \mathbf{s} \in \Gamma\mathbf{H^a}\}$. The main theorem relating $C^*\mathcal{H}^{\infty}_X$ and $C^*_{WF}\Gamma\mathbf{H^a}$ can now be formulated as follows:

THEOREM 87. *There is a unique $*$-isomorphism*

$$\begin{aligned} \theta : C^*\mathcal{H}^{\infty}_X &\longrightarrow C^*_{WF}\Gamma\mathbf{H^a} \\ \mathbf{s} + \lambda\cdot e &\longmapsto \{\hbar \longmapsto e^{-i\lambda\hbar} \cdot W^{\hbar}(\mathbf{s})\} \end{aligned} \tag{391}$$

for any $\mathbf{s} \in \Gamma\mathbf{H^a}$ and any $X \in \mathbb{R}$. Denoting by

$$\begin{aligned} \alpha_{\hbar} : C^*_{WF}\Gamma\mathbf{H^a} &\longrightarrow W^{\hbar}\Gamma\mathbf{H^a} \\ (\hbar' \mapsto e^{-i\lambda\hbar'} \cdot W^{\hbar'}(\mathbf{s})) &\longmapsto e^{-i\lambda\hbar} \cdot W^{\hbar}(\mathbf{s}) \end{aligned}$$

the point evaluation for fixed $\hbar$, then

$$\theta_{\hbar} : C^*\mathcal{H}^{\infty}_X \longrightarrow W^{\hbar}\Gamma\mathbf{H^a}$$

defined by

$$\theta_{\hbar} := \alpha_{\hbar} \circ \theta \qquad \forall\, \hbar \in \mathbb{R} \tag{392}$$

is a surjective homomorphism mapping any $(\mathbf{s} + \lambda\cdot e) \in C^*\mathcal{H}^{\infty}_X$ into $e^{-i\lambda\hbar} \cdot W^{\hbar}(\mathbf{s})$. Moreover, θ is a $*$-isometry.

For a proof we again refer to [**14**], [**15**] and [**55**].

CHAPTER 7

The Schrödinger Representation and the Metaplectic Representation

To make a link between the three-dimensional Heisenberg group, respectively, Heisenberg algebra to signal analysis and quantization we need to look at the Schrödinger representation of the Heisenberg group and the representation of the metaplectic group, i.e. the metaplectic representation. In this chapter we will develop these topics. In doing so we will construct a unitary projective representation of the symplectic group $\mathrm{Sp}(F^a)$ from the Schrödinger representation and pass on to a representation of the twofold covering of $\mathrm{Sp}(F^a)$, the metaplectic representation.

7.1. Definition of the Schrödinger Representation and Phase Space

7.1.1. The Notion of a Representation. In chapter 1 we used the spin representation. Here we will deal with representations in more detail and start with the general notion of a representation: A *representation* of a group G is a homomorphism into the group $GL(V)$ of all isomorphisms of a linear space V. Here, the scalar field of the linear spaces under consideration is mostly $\mathbb{C}$. The dimension of the linear space V is denoted by $\dim_{\mathbb{C}} V$ if it is finite.

Two representations

$$\rho_1 : G \longrightarrow GL(V) \quad \text{and} \quad \rho_2 : G \longrightarrow GL(W)$$

on the $\mathbb{C}$-linear spaces V and W, respectively, are called *equivalent* if there is a linear isomorphism $l : V \longrightarrow W$ such that the following diagram commutes:

$$\begin{array}{ccc} V & \xrightarrow{\rho_1(g)} & V \\ {\scriptstyle l}\downarrow & & \downarrow{\scriptstyle l} \\ W & \xrightarrow{\rho_2(g)} & W \end{array} \tag{393}$$

for all $g \in G$. The isomorphism l is called an *intertwiner.*

A representation $\rho : G \longrightarrow GL(V)$ is called *reducible* to $W \subset V$, say, if $\rho(g)(W) \subset W$ for all $g \in G$. The linear space W is called an *invariant subspace.* If the only invariant subspaces are $\{0\}$ and V, the representation is called *irreducible.*

In the first few sections of this chapter, we are mainly interested in the Schrödinger representation ρ of the three-dimensional Heisenberg group $\mathcal{H}^a$.

7.1.2. The Schrödinger Representation. The Schrödinger representation ρ represents the *reduced Heisenberg group* (cf. section 5.3.2)

$$\mathcal{H}^a_{\text{red}} = U^a(1) \oplus F^a,$$

formed for $a \in E$ on the Hilbert space $L^2(\mathbb{R})$ of *square integrable*, *complex-valued functions* on $\mathbb{R}$, equipped with the usual Hermitian product:

$$(\psi_1 \mid \psi_2) := \int_{\mathbb{R}} \overline{\psi}_1 \cdot \psi_2 \quad d\lambda \qquad \forall\, \psi_1, \psi_2 \in L_2(\mathbb{R}).$$

(For simplicity we write $U^a(1)$ instead of $U^{\frac{a}{|a|}}(1)$, introduced in section 1.2.2.) The complex Hilbert space $L^2(\mathbb{R})$ contains the Schwartz space $\mathcal{S}(\mathbb{R}, \mathbb{C})$ of all smooth rapidly decreasing $\mathbb{C}$-valued functions on $\mathbb{R}$ as a dense subspace (cf. [**38**]).

Let $U(L^2(\mathbb{R}))$ be the *unitary group* of the Hilbert space $L^2(\mathbb{R})$. As we will see the Schrödinger representation ρ maps $\mathcal{H}^a_{\text{red}}$ homomorphically into $U(L^2(\mathbb{R}))$. Thus ρ is called *unitary*.

The basic geometric idea in constructing ρ is to consider the plane F^a as a *phase space*, this is to say as an $\mathbb{R}$-linear space isometric to a Cartesian product $\mathbb{R} \times \mathbb{R}$ with the *position* variable q in the first and the *momentum* variable p in the second factor. This amounts to choosing some unit vector $e_{\bar{q}} \in F^a$ and set $e_{\bar{p}} := e_{\bar{q}} \times \frac{a}{|a|}$ yielding a *coordinate system* with coordinates q and p, respectively. Thus

$$F^a = \mathbb{R} \cdot e_{\bar{q}} + \mathbb{R} \cdot e_{\bar{p}}, \tag{394}$$

and any $h \in F^a$ splits into

$$h = p \cdot e_{\bar{p}} + q \cdot e_{\bar{q}}.$$

The line $\mathbb{R} \cdot e_{\bar{q}}$ is considered as the *configuration space* whose *tangent bundle* is F^a. Thus the scalar product on it is a Riemannian metric turning F^a into a bundle isomorphic to the *cotangent bundle* of $\mathbb{R} \cdot e_{\bar{q}}$, the *canonical symplectic form* of which is ω^a with $a \in S^2$ (cf. [**2**]). Differing from this picture, we identify the velocity with the momentum and regard F^a as a *phase space* equipped with the symplectic structure ω^a if $a \in \dot{E}$ (cf. section 1.3.4).

Moreover, as we observed in section 1.3 that this coordinate system turns F^a into a field isomorphic to $\mathbb{C}$.

Based on this specific *coordinate system* we define the *Schrödinger representation* ρ of the reduced Heisenberg group $\mathcal{H}^a_{\text{red}} = U^a(1) \oplus F^a$ (cf. section 5.3) as follows:

$$\rho(z + q \cdot e_{\bar{q}} + p \cdot e_{\bar{p}})(\psi)(x) := z^{-1} \cdot e^{-|a| \cdot \frac{p \cdot q}{2} \cdot i} \cdot e^{|a| \cdot p \cdot x \cdot i} \cdot \psi(x - q) \tag{395}$$

for all $z \in U^a(1)$, for all $p, q \in \mathbb{R}$, all ψ in the Schwartz space $\mathcal{S}(\mathbb{R}, \mathbb{C})$, and any $x \in \mathbb{R}$. A slightly different way of writing (395) is

$$\rho(z + q \cdot e_{\bar{q}} + p \cdot e_{\bar{p}})(\psi)\left(\xi + \frac{q}{2}\right) = z^{-1} \cdot e^{|a| \cdot p \cdot \xi \cdot i} \cdot \psi\left(\xi - \frac{q}{2}\right) \tag{396}$$

where p, q and $\xi := x - \frac{q}{2}$ vary in $\mathbb{R}$. This form of the Schrödinger representation will provide us with an important tool in signal analysis in chapter 8.

To include arbitrary but specified *frequencies* $\nu \neq 0$ we generalize ρ to ρ_ν defined for any $\xi \in \mathbb{R}$ by

$$\rho_\nu\left(e^{\tau \cdot \frac{a}{|a|}} + q \cdot e_{\bar{q}} + p \cdot e_{\bar{p}}\right)(\psi)\left(\xi + \frac{q}{2}\right) = e^{-\nu \cdot (\tau - |a| \cdot p \cdot \xi) \cdot i}\psi\left(\xi - \frac{q}{2}\right) \tag{397}$$

for all $\tau \in \mathbb{R}$ and all $p, q \in \mathbb{R}$. Writing $h = q \cdot e_{\bar{q}} + p \cdot e_{\bar{p}}$ and using the symplectic form ω^a the above equation now reads

$$\begin{aligned}
&\rho_\nu \left(e^{\tau \cdot \frac{a}{|a|}} + h\right)(\psi)\left(\xi + \frac{q}{2}\right) \\
(398)\qquad &= e^{-\nu \cdot \tau \cdot i} \cdot e^{-\frac{\nu}{2} \cdot \omega^a(p \cdot e_{\bar{p}}, q \cdot e_{\bar{q}}) \cdot \frac{a}{|a|}} \cdot e^{\nu \cdot |a| \cdot p \cdot \xi \cdot i} \cdot \psi\left(\xi - \frac{q}{2}\right).
\end{aligned}$$

We will show next that both ρ and ρ_ν are group homomorphisms. To this end let

$$k_s = e^{\tau_s \cdot \frac{a}{|a|}} + q_s \cdot e_{\bar{q}} + p_s \cdot e_{\bar{p}} \qquad \text{for} \quad s = 1, 2$$

with

$$h_s = q_s \cdot e_{\bar{q}} + p_s \cdot e_{\bar{p}} \quad \text{for} \quad s = 1, 2.$$

On the one hand,

$$\begin{aligned}
&\rho(k_1) \circ \rho(k_2)(\psi)(x) \\
&= e^{-\tau_2 \cdot i} \cdot \rho(k_1)(\rho(h_2)(\psi))(x) \\
&= e^{-(\tau_1 + \tau_2) \cdot i} \cdot e^{-\frac{|a|}{2} \cdot p_1 \cdot q_1 \cdot i} \cdot e^{|a| \cdot p_1 \cdot x \cdot i} \cdot \rho(h_2)(\psi)(x - q_1) \\
&= e^{-(\tau_1 + \tau_2) \cdot i} \cdot e^{-\frac{|a|}{2} \cdot p_1 \cdot q_1 \cdot i} \cdot e^{|a| \cdot p_1 \cdot x \cdot i} \cdot e^{-\frac{|a|}{2} \cdot p_2 \cdot q_2 \cdot i} \cdot e^{|a| \cdot p_2 (x - q_1) \cdot i} \\
&\qquad \cdot \psi(x - (q_1 + q_2)) \\
&= e^{-(\tau_1 + \tau_2) \cdot i} \cdot e^{-\frac{|a|}{2} \cdot (p_1 \cdot q_1 + p_2 \cdot q_2 + 2 \cdot p_2 \cdot q_1) \cdot i} \cdot e^{|a| \cdot (p_1 + p_2) \cdot x \cdot i} \cdot \psi(x - (q_1 + q_2))
\end{aligned}$$

and on the other hand,

$$\begin{aligned}
&\rho(k_1 \cdot k_2)(\psi)(x) \\
&= \rho\left(e^{((\tau_1 + \tau_2) + \frac{\omega^a}{2}(h_1, h_2)) \cdot \frac{a}{|a|}} + h_1 + h_2\right)(\psi)(x) \\
&= e^{-((\tau_1 + \tau_2) - \frac{\omega^a}{2}(h_1, h_2)) \cdot i} \cdot e^{\frac{|a|}{2} \cdot (p_1 + p_2) \cdot (q_1 + q_2) \cdot i} \cdot e^{|a| \cdot (p_1 + p_2) \cdot x \cdot i} \cdot \psi(x - (q_1 + q_2))
\end{aligned}$$

for any $x \in \mathbb{R}$. By (268) and (269) we have the identity

$$\begin{aligned}
&-\frac{1}{2} \cdot \omega^a(h_1, h_2) - \frac{|a|}{2} \cdot (p_1 + p_2) \cdot (q_1 + q_2) \\
&\quad = \frac{|a|}{2} \cdot (p_1 \cdot q_2 - q_1 \cdot p_2 - p_1 \cdot q_1 - p_1 \cdot q_2 - p_2 \cdot q_1 - p_2 \cdot q_2) \\
&\quad = -\frac{|a|}{2} \cdot (2 \cdot (p_1 \cdot q_2 + p_2 \cdot q_1) + p_2 \cdot q_2)
\end{aligned}$$

implying

$$\rho(k_1) \circ \rho(k_2) = \rho(k_1 \cdot k_2) \qquad \forall\, k_1, k_2 \in \mathcal{H}^a.$$

Since in the definition of ρ_ν all the exponents in the phase factors in (395) have to be multiplied by ν it is evident that ρ_ν is a homomorphism, too.

Mostly we will concentrate on $\nu = 1$ and write ρ instead of ρ_1.

On the *Heisenberg group* $\mathcal{H}^a$ the *Schrödinger representation* ρ acts by

$$(399)\quad \rho(\lambda \cdot a + q \cdot e_{\bar{q}} + p \cdot e_{\bar{p}})(\psi)(x) := e^{-\lambda \cdot i} \cdot e^{-|a| \cdot q \cdot p} \cdot e^{|a| \cdot p \cdot x} \cdot \psi(x - q)$$

for all $x, q, p, x \in \mathbb{R}$ and any $\psi \in \mathcal{S}(\mathbb{R}, \mathbb{C})$. The representation ρ_ν on $\mathcal{H}^a$ is introduced in the obvious way.

7.1.3. The Schrödinger Representation for the Rescaled Heisenberg Group. Since $U^a(1)$ operates on any circle centered at 0 in $\mathbb{C}^a$, all these circles inherit a group structure by fixing a point as the unit element. Thus we may replace $U^a(1)$ by $\lambda \cdot U^a(1)$ in the definition of the reduced Heisenberg group (the group operations act in $U^a(1)$). The unity is $\lambda \cdot e$. The resulting Heisenberg group is called the *rescaled* Heisenberg group (cf. [**68**]).

Accordingly, we define the rescaled Schrödinger representation for $\mathcal{H}^a$.

7.1.4. The Commutator. Finally, let us determine the commutator subgroup of $\mathcal{H}^a$. By a commutator of any two elements $h_1, h_2 \in \mathcal{H}^a$ we mean here the product $h_1 \cdot h_2 \cdot h_1^{-1} \cdot h_2^{-1}$. The group generated by all commutators of $\mathcal{H}^a$ is called the *commutator subgroup* and is denoted here by $Com\ \mathcal{H}^a$. The quotient $\mathcal{H}^a / Com\ \mathcal{H}^a$ is an *abelian subgroup*. Clearly, any homomorphism of $\mathcal{H}^a$ into an abelian subgroup factors over $\mathcal{H}^a / Com\ \mathcal{H}^a$.

To identify $Com\ \mathcal{H}^a$ more precisely we observe that a commutator has the form

$$h_1 \cdot h_2 \cdot h_1^{-1} \cdot h_2^{-1} = \omega^a(h_1, h_2) \cdot \frac{a}{|a|} \qquad \forall\, h_1, h_2 \in \mathcal{H}^a.$$

Thus

$$Com\ \mathcal{H}^a = \mathbb{R} \cdot \frac{a}{|a|}$$

and hence

$$\mathcal{H}^a / Com\ \mathcal{H}^a = F^a.$$

In fact, ρ is quadratically integrable on F^a (cf. [**71**]).

As we will see in the next section, $Com\ \mathcal{H}^a$ classifies the equivalence classes of Schrödinger representations of $\mathcal{H}^a$ (cf. section 7.1.1).

7.1.5. The Stone-von Neumann Theorem. It is a remarkable fact that a representation of $\mathcal{H}^a_{\text{red}}$ is either one-dimensional or infinite dimensional. Up to equivalence an irreducible representation is uniquely determined by how it acts on the center of $\mathcal{H}^a_{\text{red}}$. This is verbally formulated the content of the celebrated *Stone-von Neumann theorem*:

THEOREM 88. *Up to the equivalence the only irreducible representations of $\mathcal{H}^a_{\text{red}}$ for each $a \in \dot{E}$ are*

$$\rho_\nu : \mathcal{H}^a_{\text{red}} \longrightarrow U(L^2(\mathbb{R})) \qquad \forall\, \nu \in \dot{\mathbb{R}}$$

or else

$$\rho_{b,h'} : \mathcal{H}^a_{\text{red}} \longrightarrow U^a(1) \qquad \forall\, (b, h') \in \mathbb{R}$$

defined by

$$\rho_{b,h'}(z) := e^{(b \cdot q + h' \cdot p) \cdot \frac{a}{|a|}} \cdot z \qquad \forall\, z \in U^a(1).$$

For a proof we refer to [**37**] or [**71**].

Remark. *Theorem 88 holds accordingly for $\mathcal{H}^a$.*

Let us point out that the equivalence class of ρ does not depend on the coordinate system in F^a. By the Stone-von Neumann theorem any symplectic coordinate transformation of F^a causes an intertwiner for any element in $\mathcal{H}^a_{\text{red}}$ (cf. section 7.1.1). This is the topic of the next section.

7.1.6. The Action of the Symplectic Group. An important application of the Stone-von Neumann theorem is the following: Let $A \in \mathrm{Sp}(F^a)$ be given. A shall be extended to the center by the identity. Hence A determines an automorphism

$$A' : \mathcal{H}^a_{\mathrm{red}} \longrightarrow \mathcal{H}^a_{\mathrm{red}}$$

defined by

$$A'\left(e^{\lambda \cdot \frac{a}{|a|}} + h\right) := e^{\lambda \cdot \frac{a}{|a|}} + A(h) \qquad \forall\, \lambda \in \mathbb{R} \quad \text{and} \quad \forall\, h \in F^a.$$

The image of A' is a reduced Heisenberg group. Clearly, A' changes a given coordinate system on F^a. Given ρ_ν, the representation $\rho_\nu \circ A$ acts on $L_2(\mathbb{R})$ and is an irreducible unitary representation of $\mathcal{H}^a_{\mathrm{red}}$. Due to the Stone-von Neumann theorem the representations ρ_ν and $\rho_\nu \circ A'$ are equivalent, since

$$\rho|_{U^a(1)} = \rho \circ A'|_{U^a(1)}.$$

Thus the transformation A causes an intertwiner (cf. section 7.1.1).

7.1.7. The Contragradient Schrödinger Representation. The *contragradient representation* ρ^* satisfies per definition that

$$(\rho(z+h)\varphi \mid \psi)_{L_2} = (\varphi \mid \rho^*(z+h)\psi)_{L_2}.$$

To determine it we use the definition of $\rho(z+h)$ where $h = z + q \cdot e_{\bar{q}} + p \cdot e_{\bar{p}}$ for two orthonormal vectors $e_{\bar{q}}$ and $e_{\bar{p}}$ in F^a, and use the fact that ρ is unitary.

This is to say that

$$(\rho(z+h)(\varphi) \mid \rho(z+h)(\psi))_{L_2} = (\varphi \mid \psi)_{L_2} \qquad \forall\, \varphi, \psi \in \mathcal{S}(\mathbb{R}, \mathbb{C})$$

and hence

$$(\varphi \mid \rho^*(z+h)(\rho(z+h)(\psi))_{L_2} = (\varphi \mid \psi)_{L_2} \qquad \forall\, \varphi, \psi \in \mathcal{S}(\mathbb{R}, \mathbb{C})$$

showing

$$\rho^*(z+h)(\rho(z+h)(\psi)) = \psi \qquad \forall\, \psi \in \mathcal{S}(\mathbb{R}, \mathbb{C}).$$

This immediately implies

$$\rho^*(z+h) = \rho(z+h)^{-1} = \rho(z^{-1} - h)$$

for all $z \in U^a(1)$ and any $h \in F^a$.

ρ^* applied to $\varphi \in \mathcal{S}(\mathbb{R}, \mathbb{C})$ in terms of the coordinates p and q yields, for any $x \in \mathbb{R}$,

$$\begin{aligned} \rho^*(z+h)(\varphi)(x) &= \rho\left(z^{-1} + (-(q \cdot e_{\bar{q}} + p \cdot e_{\bar{p}}))\right)(\varphi)(x) \\ &= z \cdot e^{-\frac{q \cdot p}{2} \cdot i} \cdot e^{-px} \cdot \varphi(x+q) \end{aligned} \tag{400}$$

for $a \in S^2$.

The restriction of ρ^* to $U^a(1)$ is the central character χ^* given by

$$\chi^*(z) = z^{-1} \qquad \forall z \in U^a(1).$$

Equation (400) suggests introducing the Heisenberg group

$$\mathcal{H}^{-a} := U^{-a}(1) \oplus F^a$$

with ω^{-a} as the symplectic structure on F^a. This plane shall be equipped with the coordinate system with the axis $-\mathbb{R} \cdot e_{\bar{q}}$ and $-\mathbb{R} \cdot e_{\bar{p}}$. Hence ρ^* on $\mathcal{H}^a$ is identical with the Schrödinger representation of type ρ_ν with frequency $\nu = 1$ on $\mathcal{H}^{-a}$ or $\nu = -1$ on $\mathcal{H}^a$.

Clearly, ρ^*_ν is defined accordingly and is a Schrödinger representation, as well, for all $\nu \in \mathbb{R}\backslash\{0\}$.

The representations ρ and ρ^* as introduced are not equivalent: In fact, on $U^a(1)$

$$\rho(z)(\psi) = z^{-1} \cdot \psi \qquad \text{and} \qquad \rho^*(z)(\psi) = z \cdot \psi$$

for all $z \in U^a(1)$ and all $\psi \in L^2(\mathbb{R})$.

7.2. Characteristic Ingredients of the Schrödinger Representation

7.2.1. Goal of this Section. In this section we will exhibit two natural ingredients of the Schrödinger representation ρ_ν of the reduced Heisenberg group $\mathcal{H}^a_{\text{red}}$ with $a \in S^2$.

The first of these ingredients is expressed in the theorem of Stone-von Neumann: The restriction of ρ_ν to the center $U^a(1)$ of the Heisenberg group $\mathcal{H}^a_{\text{red}}$ specifies the equivalence class of ρ_ν. This class is obviously determined by the frequency ν for ρ_ν as in (397).

By definition, the representation ρ_ν depends on *coordinates* in F^a. This feature will enter into the formulation of the second characteristic ingredient which is of quite a different nature.

This ingredient is based on the Kirillov description of the equivalence class of ρ_ν by means of *coadjoint orbits* of $\mathcal{H}^a_{\text{red}}$. As we have shown in proposition 64 in chapter 5, any coadjoint orbit for a fixed frequency ν is of the form $\nu \cdot \frac{a}{|a|} + F^a$. Every element in F^a corresponds to a Schrödinger representation of type ρ_ν and vice versa (cf. [**71**]). In what follows we will describe this correspondence in detail. More precisely, we will show how a fixed vector w in F^a is related to ρ_ν. The pair (ν, w) will be recognized as characteristic for ρ_ν.

7.2.2. The Schrödinger Representation of the Heisenberg Group as a Semidirect Product. The definition of the Schrödinger representation depends on a specified coordinate system. As we will see next this means that ρ_ν depends on the *semidirect product decomposition* of $\mathcal{H}^a_{\text{red}}$ (cf. section 5.4).

As mentioned in section 5.4 the Heisenberg group $\mathcal{H}^a$ is isomorphic to a *semidirect product* $N^a \boxed{\text{s}} K^a$ with

$$N^a = U^a(1) + \mathbb{R} \cdot w \cdot a$$

and

$$K^a = \mathbb{R} \cdot a$$

(cf. section 5.4.1). This product depends on a choice of a non-trivial vector $w \in F'$ to write $F^a = w \cdot \mathbb{C}^a$ (cf. (293)). The isomorphism

$$A : \mathcal{H}^a \longrightarrow N^a \boxed{\text{s}} K^a$$

mentioned is given by equation (310):

$$A(z + \mu \cdot w \cdot a + \lambda \cdot w) := z \cdot e^{\mu \cdot \lambda \cdot a} + h.$$

Given the semidirect product $N^a \boxed{\text{s}} K^a$ and representations S' and S'' of N^a and K^a (on the same representation space), respectively, there is a natural representation T of $N^a \boxed{\text{s}} K^a$ which subduced to N^a and K^a yields S' and S'' again. This representation T shall be called the *semidirect product representation* determined by S' and S''(cf. [**22**]). In our case

$$S'' : K^a \longrightarrow U(L^2(\mathbb{R}, \mathbb{C}))$$

shall be defined for any $\lambda \in \mathbb{R}$, as

$$S''(\lambda, w)(\psi)(t) := \psi(t - \lambda) \qquad \forall\, t \in \mathbb{R},$$

while

$$S' : N^a \longrightarrow U(L^2(\mathbb{R}, \mathbb{C}))$$

is given for any $\mu \in \mathbb{R}$ by

$$S'(z + \mu \cdot w \cdot a)(\psi)(t) = z^{-1} \cdot e^{-\frac{\mu}{2} \cdot t \cdot i}\psi(t) \qquad \forall\, t \in \mathbb{R}.$$

The representation T of the semidirect product is defined as

$$T(z + \mu \cdot w \cdot a + \lambda \cdot w)(\psi)(t) := S'(\lambda \cdot w \cdot (z + \mu \cdot w \cdot a)(-\lambda \cdot w)) \circ S''(\lambda \cdot w)(\psi)(t)$$

for all $t \in \mathbb{R}$ and any choice of $\lambda, \mu \in \mathbb{R}$. Since for any $z \in U^a(1)$ and any $\lambda, \mu \in \mathbb{R}$,

$$\lambda w \cdot (z + \mu \cdot a)(-\lambda \cdot w) = z \cdot e^{\lambda \cdot \mu \cdot a} + \mu \cdot a \cdot w$$

holds true, the representation T takes the form

$$T(z + \mu \cdot w \cdot a + \lambda w)(\psi)(t) = z^{-1} \cdot e^{-\frac{\lambda \cdot \mu}{2} \cdot i} \cdot e^{\mu \cdot t \cdot i} \cdot \psi(t - \lambda) \qquad \forall\, \lambda, \mu, t \in \mathbb{R},$$

which is the Schrödinger representation ρ.

The choice of both groups N^a and K^a depends on the vector w. The two vectors w and $w \cdot a$ uniquely define an orthogonal coordinate system in F with respective coordinates λ and μ. Thus given ν, any vector w in F corresponds to a Schrödinger representation ρ_ν and vice versa. The vector $w = 0$ corresponds to a character times the trivial representation assigning the identity to every element in the Heisenberg group $\mathcal{H}^a_{\text{red}}$. Thus the pair

$$(\nu, w)$$

is characteristic for the representation ρ.

In the following two subsections we will show that the pair (ν, w) is characteristic for ρ in the context of induced representations, as well. This view point sheds new light on the Schrödinger representations of Heisenberg group $\mathcal{H}^a_{\text{red}}$ as a semidirect product.

7.2.3. The Bundle Language of the Induced Representation. In this and the following subsection we will verify that the Schrödinger representation ρ is *induced up* by some representation of a factor in the semidirect product decomposition of $\mathcal{H}^a_{\text{red}}$ (cf. section 5.4). In showing this we will follow the geometric construction of the induced representation.

The general scheme shall be sketched before going into the details concerning ρ_ν; it is based on the construction of vector bundles associated to principal bundles outlined in section 2.3.6: Let a Lie subgroup H of a Lie-group G be given, as well as a finite dimensional vector space V and a representation r of H on V. The smooth principal bundle G over G/H with H as structure group yields the *joint action*

$$\begin{array}{rcl} (G \times V) \times H & \longrightarrow & G \times V \\ ((g, v), n) & \longmapsto & (g \cdot n, n^{-1} \cdot v) \end{array}$$

where $n^{-1} \cdot v := r(n^{-1})(v)$. The quotient of this action is denoted by $G \times_H V$ and is a smooth *homogeneous vector bundle* over G/H. This is to say that each $g \in G$ acts naturally on the vector bundle $G \times_H V$ (associated to the principal bundle G over G/H (cf. [**45**])) as well as on G/H by means of the natural action. These actions commute with the projection from $G \times_H V$ to G/H. To construct the representation space let $\Gamma(G \times_H V)$ be the *space of Schwartz sections* of the *vector bundle* $G \times_H V$ as sketched at the beginning of chapter 6. For any $s \in \Gamma(G \times_H V)$ and any $g \in G$ the natural action on $G \times_H V$ is determined by

$$g \cdot \{g', s(y)\} := \{g' \cdot g, s(y)\} \qquad \forall\, y \in G/H \text{ and } \forall\, g' \in G.$$

Here $\{\ ,\ \}$ denotes the equivalence class. Passing to the quotient we obtain the *equivariant action*

$$(401) \quad s(g \cdot y) = g \cdot s(y) \qquad \forall\, g \in G, \forall\, y \in G/H \text{ and } \forall\, s \in \Gamma(G \times_H V).$$

Thus G operates on $\Gamma(G \times_H V)$ as a group of automorphisms which means that G is represented on $\Gamma(G \times_H V)$. This operation can be uniquely extended to the space $\Gamma_{L_2}(G \times_H V)$ of all L_2-sections of $G \times_H V$ (cf. chapter 6). This representation is called the *induced representation* of r. This scheme shall now be used in the next subsection to exhibit the Schrödinger representation as an induced representation.

7.2.4. Explicit Construction for the Schrödinger Representation. In the following we will exhibit the Schrödinger representation ρ_ν as a representation *induced up* by a character χ of N^a.

At first we recall the natural decomposition of $\mathcal{H}^a_{\text{red}}$ into two *Abelian subgroups* N^a and K^a of $\mathcal{H}^a_{\text{red}}$ as done in section 5.4.2, namely

$$\mathcal{H}^a_{\text{red}} = N^a + K^a.$$

The groups N^a and K^a are, respectively, defined to be $U^a(1) + \mathbb{R} \cdot w \cdot a$ and $\mathbb{R} \cdot w$. Here w is a fixed, non-vanishing vector in F^a, the orthogonal complement of a in the oriented Euclidean space E. Both subgroups of $\mathcal{H}^a_{\text{red}}$ are constructed with the help of a coordinate system in F^a determined by a point w different from the origin. This yields in turn the isomorphism

$$\mathcal{H}^a_{\text{red}} \cong N^a \boxed{\text{s}}\, K^a$$

as expressed in (312).

The Schrödinger representation ρ_ν of $\mathcal{H}^a_{\text{red}}$ or equivalently of $N^a \boxed{\text{s}}\, K^a$ will be obtained as a representation induced by a specified representation χ on N^a. We will describe the induction process on both groups, $\mathcal{H}^a_{\text{red}}$ and $N^a \boxed{\text{s}}\, K^a$.

To get an idea how the Schrödinger representation ρ_ν of $\mathcal{H}^a_{\text{red}}$ can be induced up we consider $\rho_\nu|_{N^a}$ and $\rho_\nu|_{K^a}$. Obviously, for any $\varphi \in \mathcal{S}(\mathbb{R}, \mathbb{C})$,

$$(402) \qquad \rho_\nu(q \cdot w)(\varphi)(x) = \varphi(x - q) \qquad \forall\, x \in \mathbb{R} \text{ and } \forall\, q \in \mathbb{R}$$

and

$$(403) \qquad \rho_\nu(e^{\lambda \cdot i} + p \cdot w \cdot a)(\varphi)(x) = e^{-\nu\lambda i} \cdot e^{\nu \cdot p \cdot x \cdot i} \cdot \varphi(x)$$

for all $\lambda, x \in \mathbb{R}$ and for any $p \in \mathbb{R}$. Here we have identified $\mathbb{R} \cdot w$ and $\mathbb{R} \cdot w \cdot a$ with the q- and the p-axis, respectively (cf. (395)).

We are now prepared to present the induction process. At first the *representation space* shall be constructed. It will be a *space of smooth sections* of a suitable vector bundle over the basis K^a, an associated bundle to the natural *principal bundle* over the basis K^a with $\mathcal{H}^a_{\text{red}}$ as total space and structure group N^a.

The projection

$$\mathcal{H}^a_{\text{red}} \xrightarrow{\text{pr}} \mathcal{H}^a_{\text{red}}/N^a = K^a$$

of this principal bundle is the natural projection; it is smooth. The *structure group* is N^a. The *associated vector bundle* mentioned is $\mathcal{H}^a_{\text{red}} \times_{N^a} F^a$ which we will now describe in more detail.

The principal fibre of the vector bundle $\mathcal{H}^a_{\text{red}} \times_{N^a} F^a$ is $F^a \subset \mathcal{H}^a_{\text{red}}$, regarded as a $\mathbb{C}^a$-linear space; here $\mathbb{C}^a$ is identified with $\mathbb{C}$ (cf. section 1.2.1). On F^a the structure group N^a of $\mathcal{H}^a_{\text{red}}$ operates via a representation χ of N^a. Since N^a is commutative, any representation is a *character* and hence is a map

$$\chi : N^a \longrightarrow U(F^a) \cong U^a(1).$$

The direct sum decomposition of N^a (cf. (303)) requires χ to split into

$$\chi = \chi^a \cdot \chi^{w \cdot a} \tag{404}$$

where the factors on the right-hand side have values in $U^a(1)$ and are of the form

$$\chi^a(e^{t \cdot a}) := e^{-\nu \cdot t \cdot a} \qquad \forall\, t \in \mathbb{R} \tag{405}$$

on the center $U^a(1)$ of $\mathcal{H}^a_{\text{red}}$ and

$$\chi^{a \cdot w}(\mu \cdot w \cdot a) := e^{\nu' \cdot \mu \cdot a} \qquad \forall\, \mu \in \mathbb{R} \tag{406}$$

on the axis $\mathbb{R} \cdot w \cdot a$, respectively, where ν and ν' are real numbers. ν is called the *central frequency*. The meaning of ν' will become apparent during the construction.

By definition the vector bundle $\mathcal{H}^a_{\text{red}} \times_{N^a} F^a$ *associated* to the principal bundle $\mathcal{H}^a_{\text{red}}$ over K^a is the quotient of the *joint action*

$$\Phi : (\mathcal{H}^a_{\text{red}} \times F^a) \times N^a \longrightarrow \mathcal{H}^a_{\text{red}} \times F^a$$

defined in this case by

$$\Phi((z + k, h), n) := ((z + h) \cdot n, \chi(n^{-1})(h)). \tag{407}$$

Here $z + k \in \mathcal{H}^a$ with $z \in U^a(1)$ and $k \in F^a$, moreover, h and n vary in F^a, respectively, in N^a.

Since $\mathcal{H}^a_{\text{red}}$ regarded as a principal bundle over $K^a = \mathcal{H}^a_{\text{red}}/N^a$ is trivial (cf. (306)), the associated bundle $\mathcal{H}^a_{\text{red}} \times_{N^a} F^a$ is trivial over K^a, as well. To verify this we split $z + k \in \mathcal{H}^a_{\text{red}}$ into

$$z + k = n_{z+k} + \lambda_{z+k} \cdot w$$

for some $n_{z+k} \in N^a$ with $n_{z+k} = z + \mu_{z+k} \cdot w \cdot a$ and real numbers μ_{z+k} and λ_{z+k}. Next we take any $n \in N^a$ and observe that

$$(z + k) \cdot n = n_{z+k} \cdot n + \lambda_{z+k} \cdot w.$$

Due to the decomposition

$$\mathcal{H}^a_{\text{red}} \times F^a = K^a \times (N^a \times F^a), \tag{408}$$

the pair $\Big((z+h)\cdot n, \chi(n^{-1})(h)\Big) \in \mathcal{H}^a_{\text{red}} \times F^a$, regarded as an element of the left-hand side $K^a \times (N^a \times F^a)$ of (408), can be rewritten as

$$\Big((z+h)\cdot n, \chi(n^{-1})(h)\Big) = (\lambda_{z+k}\cdot w, (n_{z+k}, \chi(n^{-1})(h))). \tag{409}$$

This formula holds for all $z \in U^a(1)$, any $h \in F^a$ and any $n \in N^a$. Therefore, the equivalence classes of both sides of this equation are

$$\{(z+h)\cdot n, \chi(n^{-1})(h)\} = \{(\lambda_{z+k}\cdot w, (e,h)\}$$

or, in short,

$$\{(z+h)\cdot n, \chi(n^{-1})(h)\} = \{(\lambda_{z+h}\cdot w, h)\}. \tag{410}$$

This establishes the triviality of $\mathcal{H}^a_{\text{red}} \times_{N^a} F^a$ over K^a. Here we made use of

$$\{(z+h)\cdot n, h\} = \{z+h, \chi(n)(h)\}$$

which reads in terms of (409) and (410) as

$$\begin{aligned} \{(z+h)\cdot n, h\} &= \{(\lambda_{z+h}\cdot w, (e, \chi(n)(h))\} \\ &= (\lambda_{z+h}\cdot w, \chi(n)(h)) \end{aligned} \tag{411}$$

for the fixed vector w in F^a, all $n \in N^a$, any $z+h \in \mathcal{H}^a$ and any k in F^a.

So far we have constructed $\mathcal{H}^a_{\text{red}} \times_{N^a} F^a$. The representation space is $\Gamma(\mathcal{H}^a_{\text{red}} \times_{N^a} F^a)$, the space of all Schwartz sections of $\mathcal{H}^a_{\text{red}} \times_{N^a} F^a$.

Now we start to describe the induction process. The induced representation of χ will be caused by a natural action of $\mathcal{H}^a_{\text{red}}$ on the vector bundle $\mathcal{H}^a_{\text{red}} \times_{N^a} F^a$. Therefore, in the next step we will exhibit this associated bundle as a *homogeneous vector bundle* (cf. [**18**]).

The action of $\mathcal{H}^a_{\text{red}}$ on the basis $\mathcal{H}^a_{\text{red}}/N^a = K^a$ of $\mathcal{H}^a$ is the natural action of the group $\mathcal{H}^a_{\text{red}}$ on its quotient $\mathcal{H}^a_{\text{red}}/N^a$, reading in this case

$$(z+k)(\lambda'\cdot w) = (\lambda_{z+k}+\lambda')\cdot w \tag{412}$$

for any $z \in U^a(1)$, any $k \in F^a$ and any two $\lambda_{z+h}, \lambda' \in \mathbb{R}$.

The action of $\mathcal{H}^a_{\text{red}}$ on the total space of $\mathcal{H}^a_{\text{red}} \times_{N^a} F^a$ is more involved. We will work with $N^a \boxed{s} K^a$ at first. This action formulated for the semidirect product (cf. (308)) reads

$$\begin{aligned} (z+h)(z'+h') : &= (n+\lambda w)\boxed{s}(n'+\lambda'\cdot w) \\ &= (\lambda+\lambda')\cdot w + \lambda'\cdot w\cdot n\cdot(-\lambda'\cdot w)\cdot n' \end{aligned}$$

and thus for $n = z + \mu\cdot w\cdot a$,

$$(z+h)(z'+h') = (\lambda+\lambda')\cdot w + n'\cdot(z\cdot e^{\lambda'\cdot\mu\cdot a} + \mu\cdot w\cdot a) \tag{413}$$

where we have used the commutativity of N^a. Setting

$$\hat{n} := z\cdot e^{\lambda'\cdot\mu\cdot a} + \mu\cdot w\cdot a \tag{414}$$

we have

$$(z+h)(z'+h') = (\lambda+\lambda')\cdot w + n'\cdot\hat{n} \tag{415}$$

finally yielding in terms of (410)

$$(z+h)\{z'+h', h\} = \{(\lambda+\lambda')\cdot w, \chi(\hat{n})(h)\}. \tag{416}$$

The induced representation ρ_ν of χ is now constructed as follows: Let us write $\mathbb{R}$ instead of $K^a = \mathbb{R}\cdot w$. Given any Schwartz section $\sigma : \mathbb{R} \longrightarrow \mathbb{R}\times F^a$ which is uniquely represented by a function $\varphi : \mathbb{R} \longrightarrow F^a$ (the principal part of the section σ), we write for any $z + k \in \mathcal{H}^a_{\text{red}}$,

$$\rho_\nu(z+k)(\sigma)(x) := (z_\nu + k)(x, \varphi(x)) \qquad \forall\, x \in \mathbb{R} \tag{417}$$

where we have set $z_\nu := e^{\nu\cdot\lambda a}$ for $z = e^{\lambda\cdot a}$ (cf. (401)). Hence ρ_ν is determined by the action of $\mathcal{H}^a_{\text{red}}$ on the vector bundle $\mathcal{H}^a_{\text{red}} \times_{N^a} F^a$. Using (404), (405) and (416) for $k = q\cdot w + p\cdot w\cdot a$ we obtain alternatively,

$$\rho_\nu(z+k)(\varphi)(x) = \chi(\hat{n})(\varphi(x+q))$$

where we replaced σ by the principal part φ on both sides of the equation. Since $\hat{n}$ in (414) takes the form

$$\hat{n} = z_\nu \cdot e^{p\cdot x} + p\cdot w\cdot a$$

and χ operates on N^a according to (406) by

$$\chi(\hat{n}) = z_\nu^{-1}\cdot e^{\nu\cdot px\cdot a}\cdot e^{\nu'\cdot p\cdot a}, \tag{418}$$

the definition (417) of ρ_ν due to (405) and (406) implies

$$\rho_\nu(z+k)(\varphi)(x) = z_\nu^{-1}\cdot e^{-\nu p\cdot x\cdot a}\cdot e^{\nu' p\cdot a}\cdot\varphi(x+q) \tag{419}$$

for all $z \in U^a(1)$, all $p, q \in \mathbb{R}$ and any $\varphi \in \mathcal{S}(\mathbb{R},\mathbb{C})$. We obtain the form (396) of ρ_ν by the coordinate transformation

$$x' := x + \frac{q}{2}, \qquad \text{i.e.} \quad x = x' - \frac{q}{2}$$

setting, moreover, $\nu' = 0$ in (406) and replacing q by $-q$. It finally reads

$$\rho_\nu(z+k)(\varphi)(x') = z_\nu^{-1}\cdot e^{-\frac{\nu}{2}\cdot p\cdot q\cdot a}\cdot e^{\nu p x'}\cdot\varphi(x - \frac{q}{2}) \qquad \forall\, x' \in \mathbb{R},$$

which is equivalent to (396).

Instead of working with the semidirect product in the construction of ρ_ν we could have worked with $\mathcal{H}^a_{\text{red}}$ itself. Indeed, given any $z + k \in \mathcal{H}^a_{\text{red}}$ we set

$$(z+k)\{z'+h', h\} := \{(z+k)\cdot(z'+k'), h\}$$

for all $z'+k' \in \mathcal{H}^a_{\text{red}}$ and all $h \in F^a$. To convert the product $(z+k)(z'+k')$ in $\mathcal{H}^a_{\text{red}}$ into an expression involving N^a we set

$$z' + k' = n' + \lambda'\cdot w$$

for $n' \in N^a$ and $\lambda' \in \mathbb{R}$. The following is straightforward:

$$(z+k)\cdot(z'+k') \quad = \quad (\lambda+\lambda')\cdot w + z\cdot z'\cdot e^{\frac{1}{2}(\lambda\cdot\mu' - \mu\cdot\lambda')\cdot a} + (\mu+\mu')\cdot w\cdot a$$

and due to (411), this determines the equivalence class

$$\{(z+k)\cdot(z'+k'), h\} = \{(z+k)\cdot(z'+k'), \chi(\hat{n})(h)\}. \tag{420}$$

where

$$\hat{n} := z\cdot e^{-\frac{1}{2}\mu\cdot\lambda'} + (\mu+\mu')\cdot w\cdot a \tag{421}$$

Finally, the last two equations together yield

$$(z+k)\{z'+h', h\} = \{(\lambda+\lambda')\cdot w, \chi(\hat{n}(h))\} \tag{422}$$

which shows that the $\mathcal{H}^a_{\text{red}}$-action on $\mathcal{H}^a_{\text{red}} \times_{N^a} F^a$ commutes with the projection of $\mathcal{H}^a_{\text{red}} \times_{N^a} F^a$ to $\mathcal{H}^a_{\text{red}}/N^a$ being identical to (416). Replacing λ' by x and λ by q turns (422) into

$$(z+k)\{z+h', h\} = \{(x+q)\cdot w, \chi(\hat{n}(h))\}. \tag{423}$$

Now one proceeds as above in the construction of ρ_ν.

By the Stone-von Neumann theorem any irreducible representation of $\mathcal{H}^a_{\text{red}}$ is equivalent to the Schrödinger representation ρ_ν as defined in (395) provided the action on the center is prescribed.

In this context we observe from (404) and (405) that the *characteristic ingredients* of ρ_ν are given by the pair

$$(\nu, w) \tag{424}$$

identified with

$$\nu\cdot a + w \in \nu \cdot a + F^a$$

(cf. (301) and (293)); by a theorem of Dixmier and Kirillov (cf. [**71**]) the points in the coadjoint orbit $a + F^a$ characterize ρ_ν up to equivalence.

To get ρ_ν a point $w \in F^a$ has to be chosen, or equivalently a coordinate system in F^a has to be specified. Obviously, replacing w by another unit vector yields an equivalent representation on $\mathcal{H}^a_{\text{red}}$ due to the Stone-von Neumann theorem.

Finally, we will interpret the parameter $\nu' \neq 0$ in (406). We may assume that ν' is positive; otherwise the orientation has to be changed. Given $w \in F^a\backslash\{0\}$ there is a scalar product $<\,,\,>'$ on F^a such that $|w|' = 1$. Hence any pair

$$(\nu, w)$$

characterizes a Schrödinger representation, equivalent with ρ_ν. Thus ν' in (406) marks a change in the scalar product, hence a scale change in F^a.

7.2.5. Coordinate Systems and Transvections. The Schrödinger representation ρ_ν in section 7.1.2 has been defined with respect to the coordinate system, here determined by the axes $\mathbb{R}\cdot w$ and $\mathbb{R}\cdot w\cdot a$ for a fixed non-vanishing vector w in F^a. This coordinate system is also generated by the transvection $A_{|v|,w}$ as defined in section 5.1.4. In this case it might be written as

$$A_{|v|,w}(h) = h - \omega^a(h,w)\cdot w \qquad \forall\, h \in F^a.$$

Thus $w \in F^a$ is an *eigenvector* of the transvection which in turn determines $w \cdot a$.

On the other hand, any $u \in F^{\mathbf{a}}$ determines a character χ_{F^a} which is of the form

$$\chi_{F^a} = e^{<u,\ldots>\cdot a},$$

and hence corresponds to a linear functional in $(F^a)'$, the real dual space of F^a. Clearly, χ_{F^a} yields a one-dimensional representation of $\mathcal{H}^a_{\text{red}}$, if trivially extended to $U^a(1)$.

This character is in fact a representation of the Abelian group N^a. Since any irreducible representation of N^a is one-dimensional we obtain all irreducible representations of N^a in this way. By setting $\nu = |v|$ and varying $w \in F^a$ we obtain the whole equivalence class of ρ_ν in terms of transvections on F^a, as was seen in sections 7.2.2 and 7.2.4.

7.2.6. Representation Induced by a Central Character. By a central character we mean a homomorphism

$$\chi_\nu : U^a(1) \to U(1)$$

mapping $e^{\theta \cdot a}$ into $e^{\nu \cdot \theta \cdot i}$ for any $\theta \in \mathbb{R}$ and a fixed $\nu \in \mathbb{R}$. Here $a \in S^2$ and $U^a(1) \subset \mathbb{C}^a$ is the unitary group of $\mathbb{C}^a$. Obviously, $U^a(1) \subset \mathcal{H}^a_{\text{red}}$ is a subgroup of $\mathbb{C}^a$. The composition $(i^a)^{-1} \circ \chi_\nu$ shall be called χ^a_ν, its values are in $U(1)$ (cf. section 1.2.2).

The central character χ_ν will be induced up to a representation ρ_{χ_ν} of all of $\mathcal{H}^a_{\text{red}}$. By the Frobenius reciprocity theorem (cf. [**78**]) the representations χ_ν and $\rho_{\chi_\nu}|_{U^a(1)}$ will be equivalent. To prepare the details for this construction we first focus on the geometric grounds.

The representation ρ_{χ_ν} will be constructed by means of the principal bundle

$$\text{pr: } \mathcal{H}^a_{\text{red}} \longrightarrow \mathcal{H}^a_{\text{red}}/U^a(1)$$

where pr denotes the natural projection. The quotient $\mathcal{H}^a_{\text{red}}/U^a(1)$ is naturally isomorphic to $F^a \subset \mathcal{H}^a_{\text{red}}$, the subspace perpendicular to a in E. This is easily seen by

$$(z + h) \cdot U^a(1) = h \cdot U^a(1)$$

since

$$z \cdot h = h \cdot \bar{z}$$

where $z \in U^a(1)$ and $h \in F^a$. Therefore, $(z + h) \cdot U^a(1)$ is uniquely determined by $h \in F$, yielding an isomorphism from $\mathcal{H}^a_{\text{red}}/U^a(1)$ to F^a by which we obtain a natural identification of $\mathcal{H}^a_{\text{red}}/U^a(1)$ with F^a. We hence regard $\mathcal{H}^a_{\text{red}}$ as the total space of the principal bundle over the base F^a, expressed by the diagram

$$\text{pr} : \mathcal{H}^a_{\text{red}} \longrightarrow F^a.$$

Even more, this bundle is trivial since $\mathcal{H}^a_{\text{red}} = U^a(1) \oplus F^a$.

The character

$$\chi_\nu : U^a(1) \longrightarrow U(1)$$

mapping $e^{\lambda \cdot a}$ into $e^{\nu \cdot \lambda \cdot i}$ for any $\lambda \in \mathbb{R}$, a representation of $U^a(1)$ on $\mathbb{C}^a$, yields the bundle

$$\mathcal{H}^a_{\text{red}} \times_{U(1)} \mathbb{C}^a$$

over F^a, associated to the principal bundle $\mathcal{H}^a_{\text{red}}$ over F^a. This bundle is trivial since $\mathcal{H}^a_{\text{red}}$ is a trivial principal bundle over F^a. Of course

$$\mathcal{H}^a_{\text{red}} \times_{U(1)} \mathbb{C}^a = \mathbb{H}.$$

Here $\mathbb{H}$ denotes the skew field of quaternions regarded as a $\mathbb{C}$-linear space (cf. (40)). Hence the space $\Gamma\mathbb{H}$ of all Schwartz sections can be identified with $\mathcal{S}(F^a, \mathbb{C})$, the collection of all $\mathbb{C}$-valued Schwartz functions.

The character χ_ν will be induced up to a unitary representation

$$\rho_{\chi_\nu} : \mathcal{H}^a_{\text{red}} \longrightarrow U(\Gamma_{L_2}\mathbb{H})$$

a character as will be seen in the next section. In doing so, we follow again the recipe outlined in section 7.2.3. To construct the induced representation we need an equivariant action of $\mathcal{H}^a_{\text{red}}$ on the induced vector bundle $\mathbb{H}$. This action will be established next.

To this end we will show that $\mathcal{H}^a_{\text{red}}$ naturally acts the quaternions $\mathbb{H}$. This action will provide us with the representation ρ_{χ_ν} mentioned above.

The Heisenberg group acts on F^a from the right; the action is defined by

$$\begin{array}{rcl} F^a \times \mathcal{H}^a_{\text{red}} & \longrightarrow & F^a \\ (h_1, z_1 + h) & \mapsto & h_1 + h \end{array}$$

for any $z_1 \in U^a(1)$ and any pair $h, h_1 \in F^a$. This is to say that we have an action of $\mathcal{H}^a_{\text{red}}$ on F given by

$$h_1 \Diamond (z_1 + h) := h_1 + h.$$

The central character χ_ν represents the typical fibre $U^a(1)$ (of the total space $\mathcal{H}^a_{\text{red}}$ of the principal bundle $\mathcal{H}^a_{\text{red}}$ over $\mathcal{H}^a_{\text{red}}/U^a(1) = F^a$) on the unitary group $U(1)$ of the $\mathbb{C}$-linear space $\mathbb{C}^a$. We hence have the associated vector bundle $\mathcal{H}^a_{\text{red}} \times_{U(1)} \mathbb{C}^a$ with total space $\mathbb{H}$ and

$$\text{pr} : \mathcal{H}^a_{\text{red}} \times_{U^a(1)} \mathbb{C}^a \longrightarrow F^a$$

the orthogonal projection of $\mathbb{H}$ onto F^a.

From the construction of the bundle $\mathbb{H}$ associated to the principal bundle $\mathcal{H}^a_{\text{red}}$ we observe that the Heisenberg group $\mathcal{H}^a_{\text{red}}$ acts via the character χ_ν on $\mathbb{H}$ as follows (cf. section 5.6.3):

$$\begin{array}{rcl} \mathbb{H} \times \mathcal{H}^a_{\text{red}} & \stackrel{\psi_{\chi_\nu}}{\longrightarrow} & \mathbb{H} \\ ((z_1 + h_1), (z + h)) & \mapsto & (z_1 + h_1) \Diamond (z + h) \end{array}$$

where

$$(z_1 + h_1) \Diamond (z + h) := z_1 \cdot \chi_\nu(z \cdot e^{\frac{1}{2}\omega^a(h_1, h)\cdot a}) + h_1 + h.$$

The *fixed group* of any quaternion is trivial. An orbit consists of the sum (formed in $\mathbb{H}$) of a circle in $\mathbb{C}^a$ centered at 0 and an affine plane in E parallel to F^a.

In fact, the bundle $\mathbb{H}$ over F^a is homogeneous over F^a. This is to say that $\mathcal{H}^a$ acts both on $\mathbb{H}$ and on F^a. In addition, $\text{pr} : \mathbb{H} \longrightarrow F^a$ preserves the action of $\mathcal{H}^a_{\text{red}}$ on $\mathbb{H}$ and on F^a.

With these preparations we are ready to determine the representation space of ρ_{χ_ν}. This space shall consist of a space of sections of the bundle $\mathbb{H}$ over F^a. Since the total space of the vector bundle $\mathbb{H}$ splits into $\mathbb{H} = \mathbb{C}^a \oplus F^a$, each section s is of the form

$$s(k) = f(k) + k \qquad \forall\, k \in F^a \tag{425}$$

for a well determined function $f : F^a \to \mathbb{C}^a$. The section s is a *Schwartz section* provided $f \in \mathcal{S}(F^a, \mathbb{C}^a)$. The collection of all Schwartz sections is denoted by $\Gamma\mathbb{H}$.

The representation ρ_{χ_ν} induced up by χ_ν is defined on each section $s \in \Gamma\mathbb{H}$ by

$$(\rho_{\chi_\nu}(z + h)(s))(k) := s(k + h) \Diamond (z + h)^{-1}$$

for all $z \in U^a(1)$ and any $k \in F^a$ (cf. [**68**]). Representing s as expressed by (425) we may write for each $z \in U^a(1)$ and any $h \in F^a$,

$$(\rho_{\chi_\nu}(z + h)(s))(k) = (f(k) + h) \cdot (z + h)^{-1}$$

and thus

$$(\rho_{\chi_\nu}(z + h)(s))(k) = f(k) \cdot \chi_\nu(z)^{-1} \qquad \forall\, k \in F^a.$$

Clearly, ρ_{χ_ν} extends to $\Gamma_{L_2}\mathbb{H}$ (cf. section 7.2.3) and is a unitary representation.

The representation ρ_{χ_ν} subduced to $U^a(1)$ yields

$$\rho_{\chi_\nu}(z)(s(k)) := \chi^a_\nu(z)\cdot s(k) \qquad \forall\, z\in U^a(1), \forall\, s\in \Gamma\mathbb{H} \text{ and } \forall\, k\in F^a$$

which, however, is equivalent to χ_ν on $U^a(1)$. Moreover, ρ_{χ_ν} is reducible.

7.3. The Infinitesimal Schrödinger Representation and Phase Space

As we learned in section 7.1.2 the Schrödinger representation of the reduced Heisenberg group $\mathcal{H}^a_{\text{red}}$ relies on a phase space structure of F^a. This structure is not fully required for the infinitesimal Schrödinger representation $d\rho_\nu$. We begin with the introduction of $d\rho_\nu$, which is used in quantum mechanics (cf. chapter 9 below).

7.3.1. Infinitesimal Schrödinger Representation. The tangent map $d\rho_\nu$ of ρ_ν in (397) at the unit element $e\in\mathcal{H}^a_{\text{red}}$ is called here the *infinitesimal Schrödinger representation.* Obviously, $T_eG = \mathfrak{h}^a$.

To compute it let us take a smoothly parameterized family

$$e^{\nu\cdot\tau(t)\cdot\frac{a}{|a|}} + q(t)\cdot e_{\bar q} + p(t)\cdot e_{\bar p}$$

of elements in $\mathcal{H}^a_{\text{red}}$. Here t varies in an interval centered at $0\in\mathbb{R}$. Hence the tangent vector of this family at e is

$$\frac{d}{dt}\left(e^{\nu\cdot\tau(t)\cdot\frac{a}{|a|}} + q(t)\cdot e_{\bar q} + p(t)\cdot e_{\bar p}\right)\Big|_{t=0} = \nu\cdot\dot\tau(t)\cdot\frac{a}{|a|}\cdot e^{\nu\cdot\tau(0)\cdot\frac{a}{|a|}} + \dot q(0)\cdot e_{\bar q} + \dot p(0)\cdot e_{\bar p}.$$

Thus, if $\tau(0)=0$, $q(0)=0$, $p(0)=0$ and $\dot\tau(0)=\vartheta$, the right-hand side of the above equation turns into

$$\begin{aligned}&\frac{d}{dt}\left(e^{-\nu\cdot\tau(t)\cdot i}\cdot e^{-\nu\cdot|a|\cdot q(t)\cdot p(t)i}\cdot e^{\nu|a|\cdot p(t)\cdot x\cdot i}\right)(\psi)(x-q(t))\Big|_{t=0}\\ &= -\nu\cdot\vartheta\cdot i\cdot\psi(x) + \nu\cdot|a|\cdot x\cdot\dot p(0)\cdot i\cdot\psi(x) - \psi'(x)\cdot\dot q(0).\end{aligned}$$

In other words, given $\vartheta\cdot\frac{a}{|a|} + h\in\mathfrak{h}^a$, where $h = \dot q(0)\cdot e_{\bar q} + \dot p(0)\cdot e_{\bar p}$,

$$\begin{aligned}&d\rho_\nu\left(\vartheta\cdot\frac{a}{|a|}+h\right)(\psi)(x)\\ &= -\nu\vartheta i\cdot\psi(x) + \nu|a|i\cdot x\cdot\dot p(0)\cdot\psi(x) - \psi'(x)\cdot\dot q(0)\end{aligned}\tag{426}$$

for all $x\in\mathbb{R}$ and any $\psi\in\mathcal{S}(\mathbb{R},\mathbb{C})$. Here we abbreviated $\frac{d}{dx}\psi|_x$ by $\psi'(x)$ for all $x\in\mathbb{R}$.

7.3.2. Representation of the Heisenberg Algebra. Since the brackets in $\mathfrak{h}^a$ are given by

$$\left[\vartheta_1\cdot\frac{a}{|a|}+h_1, \vartheta_2\cdot\frac{a}{|a|}+h_2\right] = \omega^a(h_1,h_2)\cdot\frac{a}{|a|}$$

for any two $\vartheta_1,\vartheta_2\in\mathbb{R}$ and any pair $h_1,h_2\in F^a$, equation (426) yields

$$d\rho_\nu\left(\left[\vartheta_1\cdot\frac{a}{|a|}+h_1, \vartheta_2\cdot\frac{a}{|a|}+h_2\right]\right) = -\nu\cdot\omega^a(h_1,h_2)\cdot i\cdot \mathrm{id}_{\mathcal{S}(\mathbb{R},\mathbb{C})}.\tag{427}$$

The right-hand side of (426) does not rely on a phase space structure of F^a. However, if F^a carries such a structure, (427) implies, in particular,

$$d\rho_\nu([q\cdot e_{\bar q}, p\cdot e_{\bar p}]) = -\omega^a(q\cdot e_{\bar q}, p\cdot e_{\bar p})\cdot i\cdot\mathrm{id}_{L^2(\mathbb{R})}\tag{428}$$

(cf. (269)) showing that

$$d\rho_\nu([q \cdot e_{\bar{q}}, p \cdot e_{\bar{p}}]) = |a| \cdot q \cdot p \cdot i \cdot \mathrm{id}_{L^2(\mathbb{R})} \tag{429}$$

and obviously

$$d\rho_\nu([p \cdot e_{\bar{p}}, q \cdot e_{\bar{q}}]) = -|a| \cdot q \cdot p \cdot i \cdot \mathrm{id}_{L^2(\mathbb{R})}. \tag{430}$$

The last two equations (429) and (430) will play a role in establishing the connection with the *uncertainty relations* in signal analysis in chapter 8.

7.4. Projective Representations of the Symplectic Group Constructed via the Schrödinger Representation

7.4.1. The Projective Representation of $\mathrm{Sp}(F)$**.** The projective representation presented in this section will yield the metaplectic representation in the following, which is used in physics. The notion of a projective representation of the symplectic group $\mathrm{Sp}(F)$ (cf. section 5.1.3 and [**37**], [**31**] and [**32**]) will emanate from a Schrödinger representation of a (reduced) Heisenberg group

$$\mathcal{H}_{\mathrm{red}} = U(1) \oplus F.$$

Here and throughout this chapter $F \subset E$ is a two-dimensional $\mathbb{R}$-linear subspace of the oriented Euclidean space E, perpendicular to a unit vector $a \in S^2$, say. The symplectic structure on F is ω^a, as introduced in (263); for the sake of simplicity we just write ω and identify $U^a(1)$ with $U(1)$ (cf. section 1.2.2). In the linear space F, equipped with the symplectic form ω two orthogonal unit vectors $v, w \in F$, say, shall be specified for which we assume $\omega(v, w) = 1$. Hence F is naturally identified with $\mathbb{R}^2$ with the orthogonal coordinate axis $\mathbb{R} \cdot v$ and $\mathbb{R} \cdot w$ and the symplectic group

$$\mathrm{Sp}(F) = \{A \in \mathrm{GL}(F) \mid \omega(A(h_1)), (A(h_2)) = \omega(h_1, h_2) \quad \forall\, h_1, h_2 \in F\}$$

(cf. section 5.1.1) is naturally identified with $\mathrm{Sp}(2, \mathbb{R})$ if convenient (here we often identify a map A with its matrix $M(A)$).

Each $B \in \mathrm{Sp}(F)$ yields an automorphism of $\mathcal{H}_{\mathrm{red}} = U(1) \oplus F$, (again denoted by B) whose restriction to $U(1)$ is the identity. Given a Schrödinger representation ρ the composition $\rho \circ B$ is an irreducible unitary representation of $\mathcal{H}_{\mathrm{red}}$, too. Due to the Stone-von Neumann theorem and the commuting diagram (393) in section 7.1.1 for any $h \in \mathcal{H}_{\mathrm{red}} = U(1) \oplus F$,

$$\rho(h) \circ B = U_{\mathrm{Sp}}(B) \circ \rho(h) \circ U_{\mathrm{Sp}}(B)^{-1} \tag{431}$$

where $U_{\mathrm{Sp}}(B)$ is an element of $U(\mathcal{H})$, the unitary group of the Hilbert space $\mathcal{H} := L^2(\mathbb{R}, \mathbb{C})$. The unitary operator $U_{\mathrm{Sp}}(B)$ in (431) is again called an intertwiner (cf. section 7.1.1). It is determined up to a *phase factor* in $U(1)$ only, due to Schur's lemma (cf. [**37**], [**58**] and [**78**]). Hence

$$U_{\mathrm{Sp}}(B_1 \circ B_2) = c(B_1, B_2) \cdot U_{\mathrm{Sp}}(B_1) \circ U_{\mathrm{Sp}}(B_2) \tag{432}$$

for any $B_1, B_2 \in \mathrm{Sp}(F)$; here $c(B_1, B_2)$ is in $U(1)$ (cf. [**71**]). The $U(1)$-valued function c on $\mathrm{Sp}(F) \times \mathrm{Sp}(F)$ satisfies a cocycle condition; it is a *two-cocycle* of the group $\mathrm{Sp}(F)$ (cf. [**58**]). This condition emanates from applying U_{Sp} to threefold products $B_1 \circ B_2 \circ B_3$ of symplectic maps B_1, B_2, B_3 by respecting the distributive law of the multiplication (cf. [**58**]). The map U_{Sp} is therefore called a *projective representation* of $\mathrm{Sp}(F)$. Each projective representation of a Lie group, say, can be turned into a linear representation of some other Lie group (cf. [**58**]). In our case,

U_{Sp} can be lifted to a unitary representation of the double cover $\mathrm{Mp}(F)$ of $\mathrm{Sp}(F)$, the metaplectic group of F (cf. [**31**], [**32**] and [**50**]). (The fact that $\mathrm{Mp}(F)$ covers $\mathrm{Sp}(F)$ twice is hidden in the phase factor i^m in (451) and is made explicit in the projection in (469) below.) This representation of $\mathrm{Mp}(F)$ is called the *metaplectic representation* which we will develop and study below.

In section 7.4.3 below we will calculate U_{Sp} (up to a phase factor) for some matrices generating $\mathrm{Sp}(F)$. The group $\mathrm{Mp}(F)$ and the metaplectic representation will be determined explicitly and studied in the next section 7.5.

7.4.2. Convenient Generators of the Symplectic Group. To compute U_{Sp} we will exhibit convenient generators of $\mathrm{Sp}(F)$. The following lemma is basic for our next studies on U_{Sp} and the forthcoming metaplectic representation. For a proof we refer to [**50**]:

LEMMA 89. *The symplectic group* $\mathrm{Sp}(2,\mathbb{R})$ *is generated by*

$$A_1(t) := \begin{pmatrix} t & 0 \\ 0 & t^{-1} \end{pmatrix}, \quad A_2(t) := \begin{pmatrix} 1 & 0 \\ t & 1 \end{pmatrix}, \quad M(J) := \begin{pmatrix} 0 & 1 \\ -1 & 0 \end{pmatrix} \tag{433}$$

for which the matrix products $A_1(t_1) \circ A_1(t_2)$ *and* $A_2(t_1) \circ A_2(t_2)$ *satisfy*

$$A_k(t_1) \circ A_k(t_2) = A_k(t_1 + t_2)$$

for $t, t_1, t_2 \in \mathbb{R}$ *and* $k = 1, 2$.

A convenient symplectic map to work with is

$$A_2' := \begin{pmatrix} 1 & t \\ 0 & 1 \end{pmatrix} = \tilde{A}_2(t)$$

(where $\sim$ denotes the transpose). It is identical to

$$A_2' = J_r \circ A_2'' \circ J_r^{-1} \qquad \text{with } M(A_2'') = \begin{pmatrix} 1 & 0 \\ -t & 1 \end{pmatrix} \tag{434}$$

with J_r as in (266) or, differently formulated,

$$A_2' = J^{-1} \circ A_2'' \circ J \tag{435}$$

(cf. section 5.7).

7.4.3. The Projective Representation U_{Sp}. We will now determine *up to a phase factor* the value of the projective representation U_{Sp} on symplectic maps A_1, A_2 and J all operating on F.

In doing so we follow [**37**] and [**38**]. We will identify a linear map with its matrix here. The symplectic map $A_1(t)$ applied to a pair of coordinates p and q is given by

$$A_1(t)(q,p) = (tq, t^{-1}p)$$

for any pair $q, p \in F$. Setting for any $f \in \mathcal{S}(\mathbb{R},\mathbb{C})$ and any $x \in \mathbb{R}$,

$$\begin{aligned} \rho(A_1(t)(q,p))(f)(x) &= \rho(t\cdot q, t^{-1}p)(f)(x) \\ &= e^{-\frac{q\cdot p}{2}\cdot i} \cdot e^{t^{-1}\cdot x\cdot p} \cdot f(x - t\cdot q) \\ &= e^{-\frac{q\cdot p}{2}\cdot i} \cdot e^{t^{-1}\cdot x\cdot p} \cdot f(t(t^{-1}\cdot x - q)) \end{aligned} \tag{436}$$

holds true. Letting m_t be the multiplication by t on $\mathbb{R}$ we have for all pairs $q,p \in F$,

$$\begin{aligned} \rho(A_1(t)(q,p))(f)(x) &= \rho(q,p)(f \circ m_{t^{-1}})(x) \\ &= U_{\mathrm{Sp}}(A_1(t)) \circ \rho(q,p) \circ U_{\mathrm{Sp}}(A_1^{-1}(t))(f)(t^{-1} \cdot x). \end{aligned} \tag{437}$$

Defining $U_{\mathrm{Sp}}(A_1(t))(f)$ for all $f \in \mathcal{S}(\mathbb{R},\mathbb{C})$ by

$$U_{\mathrm{Sp}}(A_1(t))(f)(x) := t^{-\frac{1}{2}} \cdot f \circ m_t(x) \qquad \forall x \in \mathbb{R}$$

and inserting it into (437) immediately yields (431) for $A_1(t)$. The factor $t^{\frac{1}{2}}$ is inserted to make $U_{\mathrm{Sp}}(A_1(t))$ unitary.

For A_2 we verify for all $f \in \mathcal{S}(\mathbb{R},\mathbb{C})$ and any $x \in \mathbb{R}$ that

$$\begin{aligned} \rho(A_2(q,p))(f)(x) &= \rho(q, t\cdot q + p)(f)(x) \\ &= e^{-\frac{1}{2}\cdot q(t\cdot q+p)\cdot i} \cdot e^{(tq+p)\cdot x\cdot i} \cdot f(x-q) \\ &= e^{t\cdot\frac{x^2}{2}\cdot i} \cdot e^{\frac{-q\cdot p}{2}\cdot i} \cdot e^{p\cdot x\cdot i} \cdot e^{-\frac{t}{2}(x-q)^2\cdot i} \cdot f(x-q). \end{aligned} \tag{438}$$

Setting for any $f \in \mathcal{S}(\mathbb{R},\mathbb{C})$,

$$U_{\mathrm{Sp}}(A_2)(f)(x) = e^{t\cdot\frac{x^2}{2}\cdot i} \cdot f(x) \qquad \forall x \in \mathbb{R} \tag{439}$$

satisfies (431).

Finally, let us compute $U_{\mathrm{Sp}}(J^{-1})$. To this end we evaluate $\rho(q,p) \circ \mathcal{F}(f)$ for any $f \in \mathcal{S}(\mathbb{R},\mathbb{C})$. Here $\mathcal{F}$ denotes the Fourier transform.

The *Fourier transform* and its *cotransform*, i.e. the *inverse Fourier transform*, are defined by

$$\mathcal{F}(f)(\omega) = \int f(x) \cdot e^{-i\omega x} dx \tag{440}$$

with *frequency* ω varying in $\mathbb{R}$ and

$$f(x) = \frac{1}{2\pi} \cdot \int \mathcal{F}(f)(\omega) \cdot e^{i\omega x} dx \tag{441}$$

for a given function $f \in \mathcal{S}(\mathbb{R},\mathbb{C})$ and all x in $\mathbb{R}$.

One easily verifies for all $(q,p) \in F$,

$$\begin{aligned} \rho(q,p) \circ \mathcal{F}(f)(x) &= e^{-\frac{q\cdot p}{2}\cdot i} \cdot e^{p\cdot x\cdot i} \cdot \mathcal{F}(f)(x-q) \\ &= e^{-\frac{q\cdot p}{2}\cdot i} \cdot e^{p\cdot x\cdot i} \cdot \int_{\mathbb{R}} e^{-(x-q)(y+p)\cdot i} f(y+p) dy \\ &= \int_{\mathbb{R}} e^{-\frac{q\cdot p}{2}\cdot i} \cdot e^{p\cdot x\cdot i} \cdot e^{-(x-q)(y+p)\cdot i} \cdot f(y+p) dy \\ &= \int e^{-x\cdot y\cdot i} \cdot e^{\frac{q\cdot p}{2}\cdot i} \cdot e^{q\cdot y} f(y+p) dy \\ &= \int_{\mathbb{R}} e^{-x\cdot y\cdot i} \rho(-p,q) f(y) dy \\ &= \mathcal{F} \circ \rho(-p,q)(f)(x). \end{aligned}$$

Thus

$$\rho(J^{-1}(q,p)) = \mathcal{F}^{-1} \circ \rho(q,p) \circ \mathcal{F} \qquad \forall\, q,p \in F \tag{442}$$

and up to a phase factor

$$U_{\mathrm{Sp}}(J^{-1}) = \mathcal{F}^{-1}$$

as well as

$$U_{\mathrm{Sp}}(J) = \mathcal{F}. \tag{443}$$

Hence $h \in F$ and $J(h) \in F$ are here called *Fourier conjugate* variables.

7.4.4. Infinitesimal Representations. For the remainder of this section we will deal with

$$\dot{U}_{\mathrm{Sp}} := \frac{d}{dt} U_{\mathrm{Sp}}(\exp t \cdot X)\bigg|_{t=0}$$

for any $X \in \mathrm{sp}(2, \mathbb{R})$, the Lie algebra of $\mathrm{Sp}(2, \mathbb{R})$. The following matrices generate $\mathrm{sp}(2, \mathbb{R})$:

$$E_1 := \begin{pmatrix} 0 & 1 \\ 0 & 0 \end{pmatrix}, \; E_2 := \begin{pmatrix} 0 & 0 \\ 1 & 0 \end{pmatrix} \text{ and } E_3 := \begin{pmatrix} 1 & 0 \\ 0 & -1 \end{pmatrix}.$$

Moreover, for any $t \in \mathbb{R}$,

$$\begin{aligned} \exp(tE_1) &:= \begin{pmatrix} 1 & t \\ 0 & 1 \end{pmatrix} = A_2', \\ \exp(tE_2) &:= \begin{pmatrix} 1 & 0 \\ t & 1 \end{pmatrix} = A_2, \\ \exp(tE_3) &:= \begin{pmatrix} e^t & 0 \\ 0 & e^{-t} \end{pmatrix} = A_1. \end{aligned}$$

In addition,

$$\begin{pmatrix} 1 & t \\ 0 & 1 \end{pmatrix} = J^{-1} \circ \begin{pmatrix} 1 & 0 \\ t & 1 \end{pmatrix} \circ J$$

holds true for any $t \in \dot{\mathbb{R}}$.

The following calculations are easily verified:

PROPOSITION 90. *For any $f \in \mathcal{S}(\mathbb{R}, \mathbb{C})$ the matrices E_1, E_2 and E_3 are, respectively, represented as*

$$\begin{aligned} \dot{U}_{\mathrm{Sp}}(E_1)(f)(x) &= \frac{d}{dt} U_{\mathrm{Sp}}(J^{-1} \circ \exp t \cdot E_2 \circ J)|_{t=0}(f)(x) \\ &= \mathcal{F}^{-1}\left(\frac{d}{dt} e^{\frac{t}{2}x^2 \cdot i}|_{t=0} \cdot \mathcal{F}\right)(f) \\ &= \frac{i}{2}\mathcal{F}^{-1}\left(\frac{x^2}{2} \cdot i \cdot \mathcal{F}\right)(f) \\ &= \frac{i}{2}\frac{d^2 f(x)}{dx^2}, \end{aligned}$$

$$\begin{aligned} \dot{U}_{\mathrm{Sp}}(E_2)(f)(x) &= \frac{d}{dt} U_{\mathrm{Sp}}\left(\begin{pmatrix} 1 & 0 \\ t & 1 \end{pmatrix}\right)\bigg|_{t=0}(f)(x) \\ &= \frac{d}{dt}\left(e^{\frac{t}{2} \cdot x^2 \cdot i} \cdot f(x)\right)\bigg|_{t=0} \\ &= \frac{x^2}{2} \cdot i \cdot f(x) \qquad \forall x \in \mathbb{R} \end{aligned}$$

and

$$\begin{aligned}\dot{U}_{\mathrm{Sp}}(E_3)f(x) &= \frac{d}{dt}U_{\mathrm{Sp}}\left(\begin{pmatrix} e^t & 0 \\ 0 & e^{-t}\end{pmatrix}\right)\bigg|_{t=0}(f)(x) \\ &= \frac{d}{dt}\left(e^{-\frac{t}{2}}\cdot f(e^t\cdot x)\right)\bigg|_{t=0} \\ &= -\frac{1}{2}f(x) + x\frac{df(x)}{dx} \qquad \forall x\in\mathbb{R}.\end{aligned}$$

Finally, we state for any $f \in \mathcal{S}(\mathbb{R},\mathbb{C})$ and $x \in \mathbb{R}$,

$$\begin{aligned}\dot{U}_{\mathrm{Sp}}(E_2 - E_1)(f)(x) &= -\left(\frac{i}{2}\frac{d^2f(x)}{dx^2} - \frac{x^2}{2}\cdot i\right)(f)(x) \\ &= -\frac{i}{2}\left(\frac{d^2f(x)}{dx^2} - x^2\right)(f)(x) \qquad \forall x\in\mathbb{R}.\end{aligned}$$

Passing to the complexification $\mathrm{sp}(2,\mathbb{R})\otimes\mathbb{C}$ of $\mathrm{sp}(2,\mathbb{R})$ and extending $\dot{U}$ to it yields

$$\dot{U}_{\mathrm{Sp}}(i(E_2 - E_1))(f)(x) = -\frac{1}{2}\left(\frac{d}{dx^2} - x^2\right)(f)(x)$$

for any $f \in \mathcal{S}(\mathbb{R},\mathbb{C})$ and any $x \in \mathbb{R}$.

7.5. A Realization of the Metaplectic Group and the Metaplectic Representation

7.5.1. The Metaplectic Group. In the previous section we have studied U_{Sp}, a representation of $\mathrm{Sp}(F)$ defined up to a phase factor, called a *projective representation*. To turn this into a linear representation we have to introduce a new group in order to overcome the indeterminacy of the phase factor. This indeterminacy diminishes if we pass on to the *twofold covering* of the symplectic group $\mathrm{Sp}(F)$; it is a Lie group, called the *metaplectic group*.

To this end we observe that the maximal compact subgroup of $\mathrm{Sp}(F)$ is $\mathrm{SO}(F)$, a circle which has $\mathbb{Z}$ as a fundamental group. By a standard theorem in Lie theory, $\mathbb{Z}$ is also the *fundamental group* of $\mathrm{Sp}(F)$, as can be seen from the Cartan decomposition of $\mathrm{Sp}(F)$ (cf. [**32**] and [**33**]). By general principles we therefore have a *twofold covering* of $\mathrm{Sp}(F)$ which is a Lie group again, the *metaplectic group* of F. It is denoted by $\mathrm{Mp}(F)$; the covering projection to $\mathrm{Sp}(F)$ is called $\mathrm{pr}_{\mathrm{Mp}}$. To realize this twofold covering as a group of unitary operators, the introduction of the metaplectic operators is the goal of the following sections.

7.5.2. The Quadratic Fourier Transform. The metaplectic representation of $\mathrm{Mp}(F)$ inherits the right sort of phase factor which was variable in the projective representation of $\mathrm{Sp}(F)$. In this subsection we will approach the notion of these *metaplectic operators* by introducing the *quadratic Fourier transform* (cf. [**31**] and [**32**]) determined by the non-singular operator $A \in \mathrm{End}\ F$ with the two-by-two matrix of the form

(444) $$M(A) = \begin{pmatrix} P & K \\ 0 & Q\end{pmatrix} \text{ with } K \neq 0.$$

(For simplicity we used here the general notation scheme of [**31**].) To this end we consider the non-degenerate quadratic form on $\mathbb{R}^2$ given by

$$W_A(x,y) = \frac{1}{2}(Px^2 - 2Kx\cdot y + Qy^2) \qquad \forall\,(x,y)\in\mathbb{R}^2 \tag{445}$$

for any choice of $P, Q, K \in \mathbb{R}$ with $K \neq 0$ (cf. [**31**], [**32**] and [**50**]).

As we will see below, the *complex square root* $\sqrt{K}$ will be of special interest for the quadratic Fourier transform. In fact,

$$\sqrt{K} = \pm|K|^{\frac{1}{2}} \qquad \text{if } K > 0, \tag{446}$$

$$\sqrt{K} = \pm i|K|^{\frac{1}{2}} \qquad \text{if } K < 0, \tag{447}$$

or, in other words,

$$\sqrt{K} = i^m|K|^{\frac{1}{2}}$$

with

$$m = 0 \text{ or } 2 \qquad \text{if } K > 0$$

and with

$$m = 1 \text{ or } 3 \qquad \text{if } K < 0,$$

for example. Clearly, these two types of choices in (448) are subsumed by the formula:

$$m\cdot\pi = \arg\sqrt{K}. \tag{448}$$

The above choice of m can be rephrased by the simple rule

$$\begin{array}{ll} m \text{ even} & \text{iff } K > 0, \\ m \text{ odd} & \text{iff } K < 0. \end{array} \tag{449}$$

Of course we may take $m \in \mathbb{Z}_4$.

Equation (448) in connection with (449) suggests we consider the tuples

$$(\mathbb{A}, m) \qquad \text{with } \mathbb{A} := (P, K, Q) \quad K \neq 0 \qquad \text{and } m \in \mathbb{Z} \tag{450}$$

for all $P, Q \in \mathbb{R}$ for which $PQ \neq 0$. (Clearly, $\mathbb{A}$ corresponds to a matrix of type (444).) Thus $(\mathbb{A}, m)$ determines $\sqrt{K}$ up to the *sign*.

The *quadratic Fourier transform* of $(\mathbb{A}, m)$ is defined to be

$$U_{(\mathbb{A},m)}(\varphi)(x) := i^m\left(\frac{1}{2\pi i}\right)^{\frac{1}{2}}|K|^{\frac{1}{2}}\cdot\int_{\mathbb{R}} e^{iW_A(x,y)}\varphi(y)dy \tag{451}$$

for all $\varphi \in \mathcal{S}(\mathbb{R},\mathbb{C})$. Clearly,

$$\left(\frac{1}{2\pi i}\right)^{\frac{1}{2}} = e^{-\frac{\pi i}{4}} \tag{452}$$

up to 2π. Given A, for m even or m odd the respective operators differ by the sign.

As a simple example we may consider $P = 0, K = 1, Q = 0$ and $m = 0$, i.e.

$$(\mathbb{A}, m) = (P, K, Q, m) = (0, 1, 0, 0).$$

The quadratic Fourier transform is

$$U_{(\mathbb{A},0)} = i^{-\frac{1}{2}} \cdot \mathbf{F} \tag{453}$$

where $\mathbf{F}$ denotes $(\frac{1}{2\pi})^{\frac{1}{2}}$ times the Fourier transform $\mathcal{F}$ (cf. (480) in section 8.1.2, where we will deal with the Fourier transform in more detail).

Next we restrict this general sort of quadratic transformation to *symplectic matrices* in order to prepare the metaplectic representation used in quantum mechanics (cf. chapter 9).

7.5.3. The Quadratic Fourier Transform of a Symplectic Transformation. Here is the recipe of constructing the quadratic Fourier transform of a symplectic linear map on a oriented Euclidean plane F equipped with an orthogonal coordinate system. Associated with a given *symplectic* transformation $A \in \operatorname{Sp}(F)$ with matrix

$$M(A) := \begin{pmatrix} \alpha & \beta \\ \gamma & \delta \end{pmatrix} \qquad \text{with } \beta \neq 0 \tag{454}$$

(called a free symplectic matrix (cf. [**31**] and [**32**])) we first form according to (444) the matrix

$$M = \begin{pmatrix} P & K \\ 0 & Q \end{pmatrix} \tag{455}$$

where

$$P := \delta \cdot \beta^{-1} \text{ and } K := \beta^{-1} \tag{456}$$

together with

$$Q := \beta^{-1} \cdot \alpha. \tag{457}$$

To the matrix $\begin{pmatrix} P & K \\ 0 & Q \end{pmatrix}$ we associate the quadratic Fourier transform, a unitary operator $U_{(\mathbb{A},m)}$ acting on $\varphi \in \mathcal{S}(\mathbb{R}, \mathbb{C})$ by setting

$$U_{\mathbb{A},m}(\varphi)(x) := i^m e^{-\frac{\pi}{4} i} \cdot |\beta|^{-\frac{1}{2}} \cdot (2\pi)^{-\frac{1}{2}} \cdot \int_{\mathbb{R}} \varphi(y) \cdot e^{iW(x,y)} dy \tag{458}$$

where W is computed by means of (445) expressed in terms of the entries of the symplectic matrix (454) as

$$W(x, y) := \frac{1}{2}(\delta \cdot \beta^{-1} \cdot x^2 + \beta^{-1} \cdot \alpha \cdot y^2 - 2\beta^{-1} x \cdot y) \tag{459}$$

for all pairs of coordinates $(x, y) \in F$. Here m is even if $\beta > 0$ and odd if $\beta < 0$. In either case i^m has two values (cf. [**50**], [**37**], [**38**], [**31**] and [**32**]). Thus $U_{(\mathbb{A},m)}$ depends on γ only by $\det \begin{pmatrix} \alpha & \beta \\ \gamma & \delta \end{pmatrix} = 1$.

$U_{(\mathbb{A},m)}$ is called a *metaplectic* operator associated with A. Note that id on the Hilbert space is not a quadratic Fourier transform since the matrix of the identity on F is not free. (A symplectic matrix is called *free* if it is of the form

$$\begin{pmatrix} A & B \\ C & D \end{pmatrix}$$

with $n \times n$-matrices A, B, C, D and $\det B \neq 0$.)

For $A, B \in \mathrm{Sp}(F)$ and $m, m' \in \mathbb{Z}$ a straightforward calculation shows the following multiplication rule:

$$U_{(\mathbb{A},m)} \circ U_{(\mathbb{B},m')} = U_{(\mathbb{A}\circ\mathbb{B},m+m')}. \tag{460}$$

The *inverse* of $U_{\mathbb{A},m}$ associated with the triplet (P, K, Q) is hence

$$(U_{(\mathbb{A},m)})^{-1} = U_{(\mathbb{A}',m')} \tag{461}$$

where

$$\mathbb{A}' = (-P, -K, -P) \text{ and } m' = 2 - m \pmod 4) \tag{462}$$

as can be calculated directly.

In fact,

$$U_{(\mathbb{A},m)} = -U_{(\mathbb{A}_1,m+2)} \text{ and } U_{(\mathbb{A},m)} = U_{(\mathbb{A},m')} \text{ iff } m = m' \pmod 4.$$

$(U_{\mathbb{A},m})$ is *unitary* on $\mathcal{S}(\mathbb{R},\mathbb{C})$ (and extends unitarily to all of $L^2(\mathbb{R},\mathbb{C})$) as can be seen as follows: Let

$$W(x,y) = \frac{1}{2}(Rx^2 - 2Lx \cdot y + Sy^2) \qquad \forall\, (x,y) \in \mathbb{R}^2$$

be a quadratic polynomial with $K \neq 0$. Associated with this polynomial is the symplectic map A in $\mathrm{Sp}(F)$ with matrix

$$M(A) \begin{pmatrix} L^{-1}S & L^{-1} \\ RL^{-1}S - L & L^{-1}R \end{pmatrix}. \tag{463}$$

Introducing the linear operators

$$U_{(L,m)}(\varphi)(x) := i^m \cdot |L|^{\frac{1}{2}} \cdot \varphi(L \cdot x), \tag{464}$$

$$V_R(\varphi)(x) = e^{-i\cdot\frac{1}{2}(Rx)\cdot x} \cdot \varphi(x) \tag{465}$$

and

$$V_S(\varphi)(x) = e^{-i\cdot\frac{1}{2}(Sy)\cdot y} \cdot \varphi(y) \tag{466}$$

for all $\varphi \in \mathcal{S}(\mathbb{R},\mathbb{C})$, and all $x, y \in \mathbb{R}$, then by using (451), (464) and (466) the quadratic Fourier transform factors into the product

$$U_{(\mathbb{A},m)} = V_{-R} \circ U_{L,m} \circ \mathbf{F} \circ V_{-S} \tag{467}$$

of unitary operators for all quadruples $(\mathbb{A}, m) = (R, L, S, m)$. Equation (467) is easily derived from the identity

$$e^{iW(x,y)} f(y) = V_{-R}(e^{-iK\cdot x\cdot y} \cdot V_{-S} f(y)) \qquad \forall\, x, y \in \mathbb{R} \tag{468}$$

and by insertion of $\mathbf{F}$ (cf. (453)).

Setting $S = 0$ in (463) reveals the operator V_{-R} as a product of quadratic Fourier transforms.

7.5.4. A Realization of the Metaplectic Group. In this subsection we will turn the projective representation of $\mathrm{Sp}(F)$ in section 7.4.1 into a representation of the *metaplectic group* $\mathrm{Mp}(F)$ which, by definition, is the *double covering* of $\mathrm{Sp}(F)$ with *covering projection* $\mathrm{pr}_{\mathrm{Mp}}$, say. This twofold covering exists since the fundamental group of $\mathrm{Sp}(F)$ is isomorphic to $\mathbb{Z}$ (cf. [**32**] or [**33**] and [**50**]). For simplicity we fix a symplectic basis on F.

The collection $\mathcal{A}$ shall consist of all tuples $(\mathbb{A}, m)$ defined by (455) in conjunction with (450), (456) and (457) for all $A \in \mathrm{Sp}(F)$.

Let $\mathcal{S}(F) \subset U(L^2(\mathbb{R}, \mathbb{C}))$ be the group of unitary operators generated by

$$\{\mathbf{F}, U_{(\mathbb{A},m)}, V_R \mid (\mathbb{A}, m) \in \mathcal{A} \text{ and } \forall\, R \in \mathbb{R}\}.$$

$\mathcal{S}(F)$ is called *realization* of the *metaplectic group* $\mathrm{Mp}(F)$; it is a topological subgroup of $U(L^2(\mathbb{R}, \mathbb{C}))$. In fact, $\mathcal{S}(F)$ is generated by the subset

$$\{\mathbf{F}, V_R, U_{(K,m)} \mid R, K \in \mathbb{R} \text{ with } K \neq 0\} \subset U(L^2(\mathbb{R}, \mathbb{C}))$$

(cf. equations (464), (465) and (466)) and is a Lie group. Here $U_{(K,m)}$ is the metaplectic operator associated to the symplectic matrix

$$M = \begin{pmatrix} P & K \\ 0 & Q \end{pmatrix} \qquad \text{for } K \neq 0.$$

(cf. (455)). Equation (461) allows us to define a *natural projection*

$$\mathrm{pr}_{\mathcal{S}} : \mathcal{S}(F) \longrightarrow \mathrm{Sp}(F) \tag{469}$$

by setting in accordance with (467)

$$\mathrm{pr}_{\mathcal{S}}(\mathbf{F}) = J, \quad \mathrm{pr}_{\mathcal{S}}(U_{K,m}) = A_2'(K) \quad \text{and} \quad \mathrm{pr}_{\mathcal{S}}(V_-R) = A \tag{470}$$

for all $R, K \in \mathbb{R}$ and for any $m \in \mathbb{Z}$. Here $A_2'(K)$ is the linear operator A_2' formed from

$$M(A_2'(K)) = \begin{pmatrix} 1 & K \\ 0 & 1 \end{pmatrix}$$

(cf. (434)) and $A \in \mathrm{Sp}(F)$ has the matrix

$$M(A) = \begin{pmatrix} 1 & 0 \\ R & 1 \end{pmatrix}. \tag{471}$$

Due to $U_{K,m} \in S(F)$ this projection is two-to-one and locally a diffeomorphism. Thus $\mathcal{S}(F)$ is a *twofold covering* of $\mathrm{Sp}(F)$ with $\mathrm{pr}_{\mathcal{S}}$ as the covering projection (cf. (469), [**31**] and [**32**]). Therefore, any unitary projective representation of $\mathrm{Sp}(F)$ on $L^2(\mathbb{R}, \mathbb{C})$ *lifts uniquely* to a representation of the *metaplectic group* $\mathrm{Mp}(F)$, the twofold covering of $\mathrm{Sp}(F)$, with projection $\mathrm{pr}_{\mathrm{Mp}}$ (cf. section 7.5.1). The values of these representations by construction are in $\mathcal{S}(F)$. Thus we have the commuting diagram

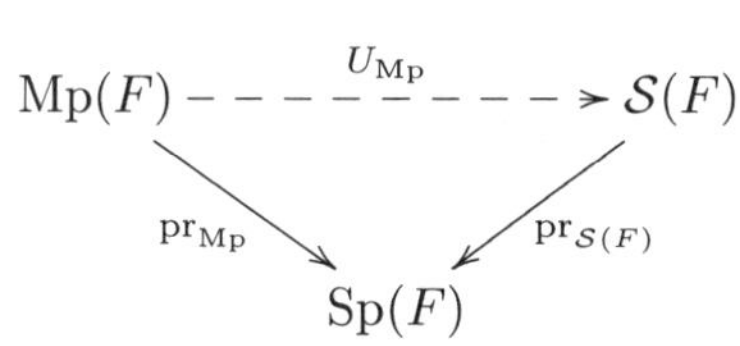

of group homomorphisms. By standard results of covering spaces, the homomorphism U_{Mp} is uniquely determined as can be seen from any text in homotopy theory (cf. [**56**]); it is a smooth *isomorphism* and is called the *metaplectic representation*.

Finally, let us consider the notion of the *Maslov Index* in the setting of the *metaplectic representation* (cf. [**31**] and [**32**]). It relies on the product of two quadratic Fourier transform and the *Gauss formula* (448) concerning the argument of the determinant of a symmetric matrix. At first we point out that any element $S \in \mathrm{Mp}(F)$ is the product of two quadratic Fourier transforms, i.e.

$$S = S_{(\mathbb{A},m)} \circ S_{(\mathbb{A}',m')}.$$

Then the integer

$$m(S) := m + m' - \mathrm{Inert}(P' + Q) \pmod 4$$

does not depend on the factorization on S. Here the *inertia index* $\mathrm{Inert}(R)$ of a *symmetric* and *invertible* matrix R is defined to be

$$\mathrm{Inert}\, R := \text{ number of } \textit{negative} \text{ eigenvalues of } R.$$

The *Gauss formula* for $\arg\det R$ is

$$\arg\det R = \pi \cdot \mathrm{Inert}(R)$$

(cf. [**31**] and [**32**]). In case S is a quadratic Fourier transform given by $(\mathbb{A}'', m'')$, then

$$m'' = m + m' - \mathrm{Inert}(P + Q') \pmod 4.$$

The function

$$\begin{aligned} m : \mathrm{Mp}(F) &\longrightarrow \mathbb{Z}_4 \\ S &\longmapsto m(S) \end{aligned}$$

is the *Maslov index*. We omit the rather technical proof here and refer to [**31**], [**32**] and [**33**].

7.5.5. Examples. *Example 1:* If $A \subset \mathrm{Sp}(F)$ with $M(A) = \left(\begin{smallmatrix} 1 & 0 \\ t & 1 \end{smallmatrix}\right)$ and $m = 0$, then by (451) for any $\varphi \in \mathcal{S}(\mathbb{R}, \mathbb{C})$,

$$V_t(\varphi)(x) := e^{-t \cdot \frac{x^2}{2} \cdot i} \cdot \varphi(x) \qquad \forall x \in \mathbb{R} \tag{472}$$

(cf. lemma 89 and (439)). Clearly,

$$\mathrm{pr}_{\mathcal{S}}(V_t) = A.$$

Example 2: As a second example we consider $J \in \mathrm{End}\ F$ with $M(J) = \left(\begin{smallmatrix} 0 & 1 \\ -1 & 0 \end{smallmatrix}\right)$ and $m = 1$.

$$U_{(J,1)}(\varphi)(x) = i \cdot e^{-\frac{\pi i}{4}} \cdot (2\pi)^{-\frac{1}{2}} \cdot \int_{\mathbb{R}} e^{i \cdot x \cdot y} \varphi(y) dy = \frac{1}{\sqrt{2\pi}} \int_{\mathbb{R}} e^{i \cdot x \cdot y} \varphi(y) dy.$$

Hence

$$U_{(J.1)} = \mathbf{F} \tag{473}$$

(cf. (443)). Clearly,

$$\mathrm{pr}_{\mathcal{S}}(U_{(J,1)}) = J.$$

For $m = 3$ we obtain

$$U_{(J,3)}(\varphi)(x) = i^3 \cdot e^{-\frac{\pi i}{4}} \cdot \int_{\mathbb{R}} e^{i \cdot x \cdot y} \varphi(y) dy = -i \cdot \frac{1}{\sqrt{2\pi}} \int_{\mathbb{R}} e^{i \cdot x \cdot y} \varphi(y) dy.$$

Example 3: For $A_1 \in \mathrm{Sp}(F)$ with matrix

$$M(A_1) = \begin{pmatrix} \alpha & 0 \\ 0 & \frac{1}{\alpha} \end{pmatrix} \qquad \text{where } \alpha \neq 0$$

(which is not free) we write $A_1 = B \circ B'$, where B and B' are free symplectic maps in F with respective matrices $\left(\begin{smallmatrix} \alpha & \eta\cdot\alpha \\ 0 & \frac{1}{\alpha} \end{smallmatrix}\right)$ and $\left(\begin{smallmatrix} 1 & -\eta \\ 0 & 1 \end{smallmatrix}\right)$, to verify for $\eta > 0$,

$$\begin{pmatrix} \alpha & 0 \\ 0 & \frac{1}{\alpha} \end{pmatrix} = \begin{pmatrix} \alpha & \eta\cdot\alpha \\ 0 & \frac{1}{\alpha} \end{pmatrix} \cdot \begin{pmatrix} 1 & -\eta \\ 0 & 1 \end{pmatrix}. \tag{474}$$

Here we use the rule (460) to associate a metaplectic operator $\mathcal{S}$ to A_1 and to handle the indices $m_1, m_2 \in \mathbb{Z}$. We set

$$\mathcal{S}_{(\mathbb{A}_1, m_1+m_2)}(A_1) := U_{(\mathbb{B}, m_1)} \circ U_{(\mathbb{B}', m_2)}.$$

For $\alpha_1 = 1$ and $m = 1$ equation (474) yields the identity. Clearly,

$$\mathrm{pr}_{\mathcal{S}}(\mathcal{S}(\mathbb{A}_1, m_1 + m_2)(A_1)) = A_1.$$

Since the linear maps A_1, B_1 and $J \in \mathrm{Sp}(F)$ with matrices

$$M(A_1) = \begin{pmatrix} \alpha & 0 \\ 0 & \frac{1}{\alpha} \end{pmatrix}, \qquad M(B_1) = \begin{pmatrix} 1 & \beta \\ 0 & 1 \end{pmatrix}, \text{ and } \quad M(J) = \begin{pmatrix} 0 & 1 \\ -1 & 0 \end{pmatrix}$$

generate $\mathrm{Sp}(F)$ (cf. [**31**]), any metaplectic operator associated with any symplectic map A is a product of powers of $\mathcal{S}_{(\mathbb{A}_1, m_1)}, U_{(\mathbb{B}_1, m_2)}$ and $U_{(J, m_3)}$ with the restriction for the product as in (460) reading

$$\mathcal{S}_{(\mathbb{A}_1, m_1)} \circ U_{(\mathbb{B}_1, m_2)} \circ U_{(J, m_3)} \text{ with } m = m_1 + m_2 + m_3 \pmod 4. \tag{475}$$

Comparing U_{Mp} and the projective representation U_{Sp} in this section and in section 7.4.3, one can see that their respective values on a metaplectic operator and its projection to $\mathrm{Sp}(F)$ differ by a multiplicative phase factor.

7.5.6. The Metaplectic Representation in Higher Dimensions. At first we specify an $2n$-dimensional Euclidean space F with a non-degenerate skew-form ω, a symplectic form. The scalar product shall be denoted by $<,>$ as usual. Enlarging F by one dimension $\mathbb{R}$, say, we obtain a Heisenberg group

$$\mathcal{H}_{2n} = F \oplus \mathbb{R}$$

with the operation defined as in (292), generalized in the obvious way. By the theorem of Stone-von Neumann (cf. [**37**]) the only infinite dimensional irreducible, unitary representation of $\mathcal{H}_{2n}$ up to equivalence is the Schrödinger representation

$$\rho : \mathcal{H}_{2n} \longrightarrow U(L^2(F_1, \mathbb{C}))$$

defined by

$$\rho(t + q + p)(\psi)(x) := e^{-t\cdot i} \cdot e^{-\frac{i}{2}\omega(q,p)} \cdot e^{i\cdot <p,x>} \cdot \psi(x - q)$$

provided the restriction to $\mathbb{R}$ is specified. Here F is assumed to split orthogonally into $F = F_1 \oplus F_2$ with $\dim F_1 = \dim F_2 = n$; this is to say that F is turned into a phase space for $t \in \mathbb{R}$, $x, q \in F_1$ and $p \in F_2$.

Since the symplectic group

$$\mathrm{Sp}(F) := \{A \in GL(F) \mid \omega(A(h_1), A(h_2)) = \omega(h_1, h_2) \quad \forall\, h_1, h_2 \in F\}$$

operates on F we construct a projective unitary representation

$$U_{\mathrm{Sp}} : \mathrm{Sp}(F) \longrightarrow U(L^2(F_1, \mathbb{C}))$$

as in section 7.4 on the n-dimensional space F_1.

As in the previous section U_{Sp} will be lifted to the metaplectic representation to be defined below. For this definition we study $\mathrm{Sp}(F)$ a little further.

The topological nature of $\mathrm{Sp}(F)$ is visible from the statement that $\mathrm{Sp}(F)$ is homeomorphic to $U(n) \times \mathbb{R}^{n(n+1)}$ (cf. [**31**]); the result in section 5.5.3 extends to this general situation. The fundamental group hence is

$$\Pi_1(\mathrm{Sp}(F)) = \Pi_1(U(n)) = \mathbb{Z}$$

as in the case of $n = 1$. Therefore, $\mathrm{Sp}(F)$ admits a twofold covering group $\mathrm{Mp}(F)$, the *metaplectic group* of F.

The *metaplectic representation* is constructed by adapting the quadratic Fourier transform as well as the operators V_P and V_Q to our case here (cf. section 7.5.4). To do so we fix a symplectic basis of F. Let $T \in \mathrm{Sp}(F)$ have a free matrix, i.e. a matrix of the form

$$M(T) = \begin{pmatrix} A & B \\ C & D \end{pmatrix} \qquad \text{with } \det B \neq 0$$

and $n \times n$-matrices A, B, C, D (cf. [**31**] and [**32**]). The matrices A, B, C, and D generalize $\alpha, \beta, \gamma, \delta \in \mathbb{R}$, respectively, in (454). Writing the *generating function* W_T in the form

$$W_T(x, y) = \frac{1}{2}(< DB^{-1}x, x > +2 < B^{-1}x, y > + < B^{-1}A\, y, y >) \quad \forall\, x, y \in F_1$$

(cf. (459)), the quadratic Fourier transform in n dimensions reads

$$U_{(\mathrm{T},m)}(\varphi)(x) := i^m e^{-\frac{\pi}{4}i}(2\pi)^{-\frac{1}{2}} \cdot \int_{F_1} \varphi(y) \cdot e^{iW_T(x,y)} dy$$

for all $x \in \mathbb{R}$ and any $\varphi \in \mathcal{S}(F_1, \mathbb{C})$, the Schwartz space of F_1. It is a unitary mapping.

Here

$$m \text{ even if } \det B > 0,$$
$$m \text{ odd if } \det B < 0.$$

Setting

$$P := D \circ B^{-1} \quad \text{and} \quad K := B$$

together with

$$Q := B^{-1} \circ A$$

yields the matrix

$$\begin{pmatrix} P & K \\ 0 & Q \end{pmatrix}.$$

Clearly, P and Q are symmetric. This generalizes (455).

The unitary operators V_P and V_Q are defined by

$$V_P(\varphi)(x) := \exp(-i < P(x), x >) \cdot \varphi(x) \qquad \forall x \in F_1 \tag{476}$$

and

$$V_Q(\varphi)(x) = \exp(-i < Q(x), x >) \cdot \varphi(x) \qquad \forall x \in F_1, \tag{477}$$

respectively, for all $\varphi \in \mathcal{S}(F_1, \mathbb{C})$. Moreover,

$$U_{(\mathbb{K},m)} = i^m |\det K|^{-\frac{1}{2}} \cdot \varphi(K(x))$$

for symmetric $n \times n$-matrices P and Q and an invertible matrix $K = B$ (cf. (464)). The *realization* $\mathcal{S}(F) \subset U(L^2(F_1, \mathbb{C}))$ *of the metaplectic group* $\mathrm{Mp}(F_1)$ by definition is generated by

$$\{\mathbf{F}, V_P, U_{(\mathbb{K},m)} \mid P = \tilde{P}, \det K \neq 0, m \in \mathbb{Z}\}.$$

Again one verifies that $\mathcal{S}(F)$ is a twofold covering of $\mathrm{Sp}(F)$ (cf. [**31**] and [**32**]). Moreover, the metaplectic representation in

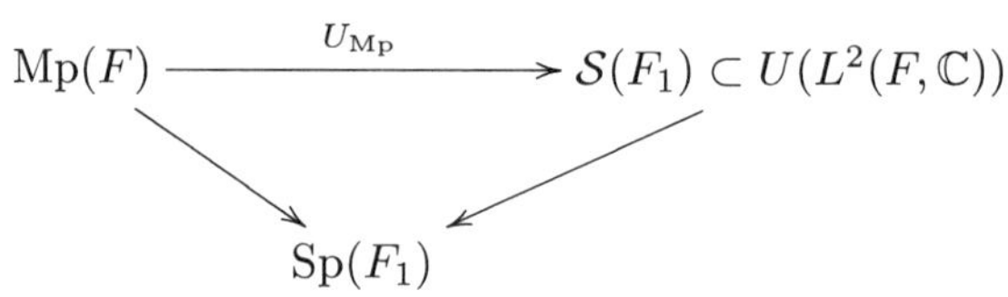

is constructed according to U_{Mp} in section 7.5.4; the maps in this diagram with range $\mathrm{Sp}(F_1)$ are the respective covering projections and $U(L^2(F, \mathbb{C}))$ denotes the group of unitary operators of $L^2(F, \mathbb{C})$. The Maslov index can be extended to these general metaplectic operators, too (cf. [**32**]).

7.5.7. A Special Case. Let $F^a \subset E$ be the orthogonal complement of a in S^2, the two-sphere in E. Since $\mathrm{Sp}(F^a)$ determines its double cover by (470) and ω^a determines the Lorentz metric $g^a_{\mathcal{H}}$ on $\hbar^a$ (cf. (328), section 5.7), the relation between the connected groups $\mathrm{Mp}(F^a), \mathrm{Sp}(F^a)$ and the component $O_0(1,2)$ of id in the Lorentz group of $g^a_{\mathcal{H}}$ is of interest (cf. (329) and (330) in section 5.8.2). The following diagram expresses it:

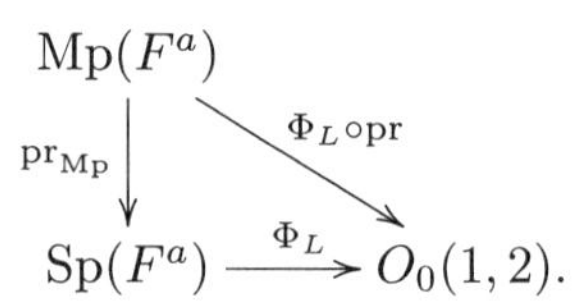

Since pr and the conjugation map Φ_L are both two-to-one maps, the composition $\Phi \circ$ pr is four-to-one and hence $\mathrm{Mp}(F^a)$ is a fourfold covering of $O_0(1,2)$, the component of the identity in the Lorentz group.

$g^a_{\hbar^a}$ is naturally constructed out of the Minkowski metric on $\mathbb{H}$ (cf. (324)). From this point of view it is the symmetry group $O_0(1,2)$ of the Lorentz metric $g^a_{\mathcal{H}^a}$ which causes the metaplectic group.

On the other hand, the vector $a \in S^2$ determines the symplectic form ω^a on F^a. The group $\mathrm{SO}(F^a)$ is a maximal compact subgroup of $\mathrm{Sp}(F^a)$. This group is twofold covered by a subgroup $\widetilde{\mathrm{SO}}(F^a)$ of $\mathrm{Mp}(F^a)$, which is related (within the metaplectic group $\mathrm{Mp}(F^a)$) to $U^a(1)$.

In fact, there is a smooth monomorphism $\tilde{\iota}$ of $U^a(1)$ onto $\widetilde{\mathrm{SO}}(F^a)$ determined by the map τ as can be seen from the diagram:

$$\begin{array}{ccc} U^a(1) & \xrightarrow{\tilde{\iota}} & \mathrm{Mp}(F^a) \\ {\scriptstyle\tau}\downarrow & & \downarrow{\scriptstyle \mathrm{pr}} \\ \mathrm{SO}(F^a) & \xrightarrow{\iota} & \mathrm{Sp}(F^a) \end{array} \tag{478}$$

(cf. section 1.4.1). Here ι is the inclusion map and $\tilde{\iota}$ is the uniquely determined lift of the map $\iota \circ \tau$ mapping unity to unity. The metaplectic representation of $\tilde{\iota}(U^a(1))$ is given by $U_{\mathrm{Mp}}|_{\tilde{\iota}(U^a(1))}$.

Since $U^a(1)$ is commutative, all its irreducible representations are (complex) one-dimensional and thus all are equivalent.

CHAPTER 8

The Heisenberg Group: A Basic Geometric Background of Signal Analysis and Geometric Optics

In the following sections we will emphasize and demonstrate that well-known methods of signal analysis are based on the geometric background provided by the *Heisenberg group* and its Lie algebra, i.e. the *Heisenberg algebra* (cf. chapter 5) together with the *Schrödinger representation* (cf. chapter 7). Hence the aim of this chapter is not to present an introduction to signal analysis. There are excellent texts on this subject such as [**47**]. Here we will present only the very fundamentals of signal analysis which can be brought into this geometric context.

As a very first and simple example in this spirit, we consider in *time-frequency analysis* (or, for that matter, *position–momentum analysis*) the inherent *uncertainty principle* of Heisenberg and relate it to the infinitesimal Schrödinger representation. We will present and study some of the basic notions and methods based on tools like the *cross-ambiguity function*, the *Wigner transformation* and relatives thereof. These functions are essentially transformations of a signal. They may depend on a window, i.e. a reference function. Under a certain, quite general condition, the original signal can be reconstructed from these transformations. Moreover, these tools are interconnected with the Heisenberg group, the Heisenberg algebra and the Schrödinger representation (and prove to be very fundamental, for example, in optics and quantization, as will be seen later on). In fact, a fundamental observation is that the geometric background mentioned above is made visible by means of the cross-ambiguity function formulated in terms of the Schrödinger representation.

As one of the examples of how the cross-ambiguity function is applied in signal theory we will present the use of the *radar ambiguity function*. As a further example of signal processing we will discuss some of the fundamental observations in *geometric optics*. In this context, the initial signal, i.e. here the initial light distribution, is mapped by light rays, which pass an optical system, to an image distribution of light on a target plane. The underlying point mapping (caused by the light rays) from the first plane to the second one is a symplectic transformation. The image of the light distribution (obtained by means of Fresnel theory) is calculated by using the metaplectic representation. It uses the *eikonal*, which complements the method using the Wigner function (cf. [**24**]). In our geometric view point it is the Stone-von Neumann theorem which provides a link to Fresnel theory.

The forward propagation of signals follows the center of the Heisenberg algebra $\mathbb{R} \cdot a$, called the *information channel*, with $a \in S^2$, the unit sphere in the oriented Euclidean space E, say. Information to be sent off is encoded in the plane F^a and is transmitted by means of a signal. We will base quantization in the next chapter on this view point.

8.1. The Notion of a Signal

The purpose of this short section is to provide the very basic notions concerning signals.

8.1.1. The Notion of a Signal. A *signal* (in its abstract formulation) is a $\mathbb{C}$-valued map φ defined on a subset of $\mathbb{R}$. The values in $\mathbb{C}$ are time- or position-dependent pieces of information. If the domain is a discrete set, φ is referred to as a *discrete signal.* Mostly, the signals under consideration in this chapter are in $\mathcal{S}(\mathbb{R},\mathbb{C})$, the *Schwartz space of rapidly decreasing* $\mathbb{C}$*-valued functions* of $\mathbb{R}$ (cf. [**38**]). Here are a few examples of signals:

i) an $\mathbb{R}$-valued function on a finite or infinite discrete domain in $\mathbb{R}$. The values may be interpreted as pulses, costs, etc.,
ii) a time-dependent and position-dependent electrical current,
iii) a position-dependent amplitude of a light or a sound wave,
iv) the evolution of a distribution of light passing through an optical system.

8.1.2. Hermitian Product of Signals. On the Schwartz space $\mathcal{S}(\mathbb{R},\mathbb{C})$ there is a natural *Hermitian product* given by the one on $L^2(\mathbb{R},\mathbb{C})$ denoted by $(\ |\ )_{L^2}$. It is defined by

$$(\varphi \mid \psi)_{L^2} := \int_{\mathbb{R}} \varphi(t)\cdot\overline{\psi}(t)dt \tag{479}$$

for any two $\varphi,\psi \in \mathcal{S}(\mathbb{R},\mathbb{C})$, and is extended to the completion $L^2(\mathbb{R},\mathbb{C})$ of $\mathcal{S}(\mathbb{R},\mathbb{C})$ (cf. (339) in chapter 6 for a real analogue).

Within the realm of Hermitian spaces of complex-valued functions there are a variety of physical interpretations of mathematical expressions, a few of which shall be presented below in order to get some very elementary impressions for the modeling in this field.

A physical interpretation of a signal $\varphi \in \mathcal{S}(\mathbb{R},\mathbb{C})$ is that of an *amplitude* together with a *phase factor*, depending on the time t. The *power* of a signal is proportional to

$$\| \varphi \|_{L_2}^2 := (\varphi \mid \varphi)_{L_2}.$$

Its *total energy* or *total power* is hence proportional to $\int_{\mathbb{R}} |\varphi(t)|^2 dt$. The total energy is finite iff $\varphi \in L^2(\mathbb{R},\mathbb{C})$.

Recall from section 7.4.3 that the *Fourier transform* and its *cotransform*, i.e. the *inverse Fourier transform*, are defined by

$$\mathcal{F}(\varphi)(\omega) = \int \varphi(t)\cdot e^{-i\omega t}dt$$

with a *frequency* ω and

$$\varphi(t) = \frac{1}{2\pi}\cdot\int \mathcal{F}(\varphi)(\omega)\cdot e^{i\omega t}dt \tag{480}$$

for a given signal $\varphi \in \mathcal{S}(\mathbb{R},\mathbb{C})$. Equation (440) shows that φ is an infinite superposition of elementary periodic functions. Here we call t and ω a *Fourier conjugate pair*.

Moreover, $|\varphi(t)|^2$ is called the energy density and $|\mathcal{F}(\varphi)|$ the *energy density spectrum*. These two notions are related by *Plancherel's theorem* stating

$$2\pi \cdot |\varphi|_{L^2}^2 = |\mathcal{F}(\varphi)|_{L^2}^2. \tag{481}$$

The factor 2π depends on the type of Fourier transform used. If the exponent of the phase factor in (440) and (480) is $2\pi i w t$ (up to the sign), the proportionality factor 2π in (481) vanishes (cf. [**38**]).

For the rest of this section let us concentrate on electromagnetic signals φ being amplitudes (for simplicity) depending on time t.

At first, we consider a simple form of *transmission* of signals. It might, for example, be performed by an electrical system. This transmission shall be mathematically modeled by linear operators L on $\mathcal{S}(\mathbb{R}, \mathbb{C})$ being of the form

$$L(\varphi) = \varphi \star \kappa \tag{482}$$

for all $\varphi \in \mathcal{S}(\mathbb{R}, \mathbb{C})$ and some well-determined $\kappa \in \mathcal{S}(\mathbb{R}, \mathbb{C})$. Here $*$ denotes the *convolution* defined by

$$\varphi \star \psi(t) := \int_{-\infty}^{\infty} \varphi(s) \cdot \psi(t-s) ds$$

which has both a smoothing as well as a filtering effect; of course we may also consider (t_0, ∞) as the domain of our signals (cf. (348) for a discrete analogue). Equation (482) can be rewritten by applying the Fourier transform to it:

$$\mathcal{F}(L(\varphi)) = \mathcal{F}^{-1}(\kappa) \cdot \mathcal{F}(\varphi) \qquad \forall \varphi \in \mathcal{S}(\mathbb{R}, \mathbb{C}). \tag{483}$$

The function $\mathcal{F}^{-1}(\kappa)$ is called the *system function*, while κ is referred to as the *impulse response*. If φ is replaced by the Dirac δ-distribution, κ is the output of the input φ, a *pulse*. Equation (483) allows us to reconstruct $\mathcal{F}(\varphi)$ provided $\mathcal{F}^{-1}(\kappa)$ does not vanish (cf. [**38**]).

A signal φ is called *band limited* if it vanishes outside an interval $[-\omega_0, \omega_0] \subset \mathbb{R}$. The real ω_0 is the width of the band. If a given φ is band limited, the following sampling theorem holds true (cf. [**38**]): Suppose $\varphi \in \mathcal{S}(\mathbb{R}, \mathbb{C})$ and $\varphi(\omega) = 0$ for all $|\omega| > \omega_0$. Then φ is completely determined by its values on $t_n := \frac{n \cdot \pi}{\omega_0}$ for $n = 0, \pm 1, \pm 2 \ldots$. In fact,

$$\varphi(t) = \sum_{-\infty}^{\infty} \varphi\left(\frac{n\pi}{\omega_0}\right) \cdot \operatorname{sinc}(\omega_0 \cdot t - n \cdot \pi)$$

where

$$\operatorname{sinc} x := \frac{\sin x}{x} \qquad \forall x \in \dot{\mathbb{R}}; \qquad \operatorname{sinc} 0 = 1.$$

This is Shannon's famous *sampling theorem.*

8.2. Time-Frequency Analysis and the Uncertainty Principle

8.2.1. The Idea of Time-Frequency Analysis. Given a signal $\varphi \in \mathcal{S}(\mathbb{R}, \mathbb{C})$ depending on the time t, say, its Fourier transform $\mathcal{F}(\varphi)$ depends on some other variable, the *frequency* ω, say. The functions φ and $\mathcal{F}(\varphi)$ provide us with two quite different types of information. The signal φ itself gives no information about the frequency, while $\mathcal{F}(\varphi)$ tells nothing about the time dependence of φ.

For instance, $|\varphi(t)|^2$ is up to a proportionality factor the power of φ at t; the map $t \mapsto |\varphi(t)|^2$ is the *power distribution* or *power density*. On the other hand, $|\mathcal{F}(\varphi)(\omega)|^2$ expresses the power in terms of frequency: The map $\omega \mapsto |\mathcal{F}(\varphi)(\omega)|^2$ is the *frequency spectrum*. Power density and frequency spectrum do not agree, they are different maps but related as expressed in Plancherel's theorem (cf. (481)) stated above.

To understand and analyze the signal both sorts of information are needed, the one from φ and the one from $\mathcal{F}(\varphi)$. To do such a *time-frequency analysis* of a signal both variables t and ω have to be considered at once. This is to say we consider the *time-frequency plane* $\mathbb{R}^2$, and construct some tools to extract both sorts of information. As an example quite often presented, we look at the musical score printed on a sheet of paper representing $\mathbb{R}^2$ (cf. introduction). The time coordinate and the frequency coordinate give us different information: The rhythm is horizontally expressed by $\varphi(t)$ while the tone pitch (depending on the frequency) is written vertically and is expressed by $\mathcal{F}(\varphi)$. The music is a composite of both and represents a time-frequency analysis (cf. [**47**]).

The time frequency plane $\mathbb{R}^2$ is also called the *phase space*. In connection with this terminology, the domain of the signal is called the *configuration space*. Of course the phase space in our general setting is an oriented plane F, embedded as a linear subspace in an oriented Euclidean space E, say. F is orthogonally split into

$$F = \mathbb{R} \cdot e_{\bar{q}} + \mathbb{R} \cdot e_{\bar{p}},$$

where we assumed that both $e_{\bar{q}}$ and $e_{\bar{p}}$ are unit vectors. The coordinates are Fourier conjugate pairs: The signal is defined on $\mathbb{R} \cdot e_{\bar{q}}$, while its Fourier transform is given on $\mathbb{R} \cdot e_{\bar{p}}$.

The reason for the chosen orthogonality of q- and p-axis is given by the metaplectic representation (cf. section 7.5) which converts a rotation about $\frac{\pi}{2}$ into the Fourier transform.

Time-frequency analysis will be done on F. Its primary idealistic goal is to resolve a point in the plane arbitrarily well. However, as it is well-known, there is a principal *obstruction* to do so, namely the *Heisenberg uncertainty relations*, which we will look at now.

8.2.2. The Uncertainty Relations. Time-frequency analysis in a plane F^a is principally limited in determining a point in F^a due to the *uncertainty relations*. It makes the concept of *instantaneous time* and *frequency determination* impossible. The uncertainty relations, also called *uncertainty principle*, are based on the non-commutativity of self-adjoint operators. Both are formulated in the following lemma to be found in [**37**] and [**47**]:

LEMMA 91. *Let A_1 and A_2 be (possibly unbound) self-adjoint operators on a complex Hilbert space $\mathcal{H}$. Then*

$$\| (A_1 - \lambda_1 \cdot id)(\varphi) \| \cdot \| (A_2 - \lambda_2 \cdot id)(\varphi) \| \geq \frac{i}{2}([A_1, A_2](\varphi) \mid \varphi) \tag{484}$$

for all $\varphi \in \mathcal{H}$ in the domain of the commutator $[A_1, A_2]$ of A_1 and A_2 and some specified parameters $\lambda_1, \lambda_2 \in \mathbb{R}$. Equality holds true iff for some real λ,

$$(A_1 - \lambda_1 \cdot id)(\varphi) = i \cdot \lambda \cdot (A_2 - \lambda_2 \cdot id)(\varphi) \qquad \forall \varphi \in \mathcal{H} \tag{485}$$

is satisfied. Here $(\ |\ )$ *is the Hermitian product of the Hilbert space* $\mathcal{H}$.

PROOF. We closely follow [**37**] and [**47**]. Let φ vary in the domain of the commutator in $\mathcal{H}$. From

$$
\begin{aligned}
(486)\quad & ([A_1 - \lambda_1 \cdot \mathrm{id}\,, A_2 - \lambda_2 \cdot \mathrm{id}\,](\varphi)|\varphi) \\
= \quad & (((A_1 - \lambda_1 \cdot \mathrm{id}\,) \cdot (A_2 - \lambda_2 \cdot \mathrm{id}\,) - (A_2 - \lambda_2 \cdot \mathrm{id}\,)(A_1 - \lambda_1 \cdot \mathrm{id}\,))(\varphi)|\varphi) \\
= \quad & ((A_2 - \lambda_2 \cdot \mathrm{id}\,)(\varphi)|(A_1 - \lambda_1 \cdot \mathrm{id}\,)(\varphi)) - ((A_1 - \lambda_1 \cdot \mathrm{id}\,)(\varphi)|(A_2 - \lambda_2 \cdot \mathrm{id}\,)(\varphi))
\end{aligned}
$$

we obtain

$$([A_1, A_2](\varphi)|\varphi) \quad = \quad 2 \cdot i \cdot \mathrm{Im}((A_2 - \lambda_2 \cdot \mathrm{id}\,)(\varphi)|(A_1 - \lambda_1 \cdot \mathrm{id}\,)(\varphi)).$$

By the Cauchy-Schwarz inequality applied to the right-hand side,

$$
\begin{aligned}
|([A_1, A_2](\varphi)|\varphi)| \quad &\leq \quad 2((A_2 - \lambda_2 \cdot \mathrm{id}\,)(\varphi)|(A_1 - \lambda_1 \cdot \mathrm{id}\,)(\varphi)) \\
(487) \qquad\qquad &\leq \quad 2 \parallel (A_2 - \lambda_2 \cdot \mathrm{id}\,)(\varphi) \parallel \cdot \parallel (A_1 - \lambda_1 \cdot \mathrm{id}\,)(\varphi) \parallel .
\end{aligned}
$$

Observe that equality holds in both of the inequalities stated above iff for the first inequality

$$\big((A_2 - \lambda_2 \cdot \mathrm{id}\,)(\varphi)|(A_1 - \lambda_1 \cdot \mathrm{id}\,)(\varphi)\big) \in \mathbb{R} \cdot i$$

and for the second inequality

$$(A_1 - \lambda_1 \cdot \mathrm{id}\,)(\varphi) = z \cdot (A_2 - \lambda_2)(\varphi) \text{ with } z \in \mathbb{C},$$

holds true, respectively. Thus both statements hold true simultaneously iff $z \in \mathbb{R} \cdot i$. □

The results in the lemma above may be formulated as follows (cf. [**37**]):

THEOREM 92. *Any* $\varphi \in \mathcal{S}(\mathbb{R}, \mathbb{C})$ *and any* $\lambda_1, \lambda_2 \in \mathbb{R}$ *satisfy*

$$(488) \qquad \frac{1}{2} \parallel \varphi \parallel_{L^2}^2 \leq \parallel (id_{\mathbb{R}} - \lambda_2) \cdot \varphi \parallel_{L^2} \cdot \parallel (id_{\mathbb{R}} - \lambda_1) \cdot \varphi \parallel_{L^2}$$

or, more explicitly,

$$(489) \quad \frac{1}{2} \parallel \varphi \parallel_{L^2}^2 \leq \left(\int (t - \lambda_2)^2 |\varphi(t)|^2 dt \right)^{\frac{1}{2}} \cdot \left(\int (\omega - \lambda_1)^2 \cdot |\mathcal{F}(\varphi)(\omega)|^2 d\omega \right)^{\frac{1}{2}}.$$

Equality holds iff

$$(490) \qquad \varphi(t) = z \cdot e^{i\lambda_2 t i} \cdot e^{-\frac{r}{2} \cdot (t - \lambda_1)^2} \qquad \forall\, t \in \mathbb{R}$$

for some $z \in \mathbb{C}$ *and* $\lambda_1, \lambda_2, r \in \mathbb{R}$ *with* $r > 0$.

8.2.3. Schrödinger Representations and Uncertainty Relations. Next we link the uncertainty relations to the infinitesimal Schrödinger representation $d\rho$ on F^a in the Heisenberg algebra $\mathcal{H}^a = F^a + \mathbb{R} \cdot a$ for $a \in S^2$.

The plane F^a is turned into a phase space by equipping F^a with a coordinate system with axes $\mathbb{R} \cdot e_{\bar{q}_0}$ and $\mathbb{R} \cdot e_{\bar{q}_0} \cdot a$ where $|e_{\bar{q}_0}| = 1$. The coordinates are called ζ and η. Let $h \in F$ be represented as

$$(491) \qquad h = \zeta \cdot e_{\bar{q}_0} + \eta \cdot e_{\bar{q}_0} \cdot a.$$

The points in the plane $F^a \subset \mathcal{H}^a$ are regarded as pieces of information to be modulated on signals by means of $i \cdot d\rho$ with $\mathcal{S}(\mathbb{R}, \mathbb{C})$ (and in turn $L^2(\mathbb{R}, \mathbb{C})$) as representation spaces of ρ.

The degree of resolution of points in $F^a \subset \mathcal{H}^a$ by $d\rho$ is subject to (484). To make this inequality accessible to $d\rho$ we first observe that $i \cdot d\rho$ is self-adjoint since ρ is unitary. We set

$$A_1 := i \cdot d\rho(\zeta \cdot e_{\bar{q}_0}) \tag{492}$$

and

$$A_2 := i \cdot d\rho(\eta \cdot e_{\bar{q}_0} \cdot a) \tag{493}$$

and observe immediately that

$$[A_1, A_2] = -d\rho([\zeta \cdot e_{\bar{q}_0}, \eta \cdot e_{\bar{q}_0} \cdot a])$$

where the bracket on the right-hand side is the one of $\mathcal{H}^a$ (cf. (429)). Since

$$[\zeta \cdot e_{\bar{q}_0}, \eta \cdot e_{\bar{q}_0} \cdot a] = \zeta \cdot \eta \cdot a,$$

we have

$$[A_1, A_2](\varphi) = \zeta \cdot \eta \cdot i \cdot \varphi \qquad \forall \varphi \in \mathcal{S}(\mathbb{R}, \mathbb{C}).$$

These data inserted into (484) and (486) immediately yield the uncertainty relations formulated by means of $d\rho$ (cf. (428), (429) and (430)).

From lemma 91 with $\lambda_1 = \lambda_2 = 0$ and $\alpha \cdot \beta \leq \frac{1}{2}(\alpha^2 + \beta^2)$ we further obtain:

COROLLARY 93. *Any $\varphi \in \mathcal{S}(\mathbb{R}, \mathbb{C})$ satisfies*

$$\| \varphi \|^2 \leq \| d\rho(\zeta \cdot e_{\bar{q}_0})(\varphi) \|^2 + \| d\rho(\eta \cdot e_{\bar{q}_0} \cdot a)(\varphi) \|^2 \tag{494}$$

with equality iff

$$\varphi(t) = z \cdot e^{-\frac{t^2}{2}} \qquad \forall\, t \in \mathbb{R} \text{ and } z \in \mathbb{C}. \tag{495}$$

The functions (490) and (495) are called *coherent states* (cf. [**37**] and [**4**]).

Setting $\lambda_2 = \zeta$ and $\lambda_1 = \eta$ it is clear that the pair (ζ, η) in F^a, i.e. the point $h = \zeta \cdot q_0 + \eta \cdot q_0 \cdot a$ cannot be resolved by means of a signal φ due to (489) and (490), no matter how small the support of φ might be.

The uncertainty relations therefore are inherent of general signal analysis.

8.3. Further Tools of Time-Frequency Analysis

Signals can be analyzed in various ways. One of these methods compares two signals in terms of the Hermitian product on $L^2(\mathbb{R}, \mathbb{C})$. We will essentially present this method to arrive at the Wigner function and its close relative the *cross-ambiguity function*, from which the *auto-ambiguity function* will be defined.

It will be the cross-ambiguity function which will be directly related to the Schrödinger representation.

8.3.1. The Short Time Fourier Transform.

The first tool for the time-frequency analysis we shall introduce is the *short time-frequency Fourier transform* (cf. [**47**]). The idea behind this transformation is to restrict a time-dependent signal $\varphi \in \mathcal{S}(\mathbb{R}, \mathbb{C})$ to a short closed interval I, compare it with a rectangle function ψ supported exactly in that interval and to take the Fourier transform of the *cut off* $\varphi \cdot \bar{\psi}$.

The map ψ is called the *window*. To avoid loss of smoothness, ψ is taken as a smooth function with support in the interval. The interval shall be centered about a fixed time τ, say. Hence

$$\varphi(t) \cdot \overline{\psi}(t-\tau) \tag{496}$$

is the map expressing φ in terms of the window ψ for all $t \in I$. In other words, the signal $\varphi(t)$ is *filtered* by means of $\overline{\psi}(t-\tau)$. The Fourier transform

$$V(\varphi,\psi)(\tau,\omega) := \int_{\mathbb{R}} \varphi(t)\overline{\psi}(t-\tau)e^{-\omega t i}dt \tag{497}$$

depends on both the time τ and a frequency variable ω. Thus it depends on a point (τ,ω) in the phase space $\mathbb{R}^2$. Clearly,

$$V : \mathcal{S}(\mathbb{R},\mathbb{C}) \times \mathbb{R}^2 \longrightarrow \mathbb{C},$$

called the *short time Fourier transform*, operates on φ for any $\psi \in \mathcal{S}(\mathbb{R},\mathbb{C})$. It has a series of nice properties. For example, it fulfills an uncertainty principle and admits an inversion formula (cf. [**47**]). We will show below that there is an orthogonality relation, as well.

It is not hard to verify that (497) can be reformulated into

$$V(\varphi,\psi)(\tau,\omega) = e^{-i\frac{\tau}{2}\omega} \cdot \int_{\mathbb{R}} \varphi\left(t+\frac{\tau}{2}\right)\overline{\psi}\left(t-\frac{\tau}{2}\right)e^{-it\omega}dt \tag{498}$$

by means of a coordinate transformation $t \mapsto t' + \frac{\tau}{2}$. In the following we will omit the domain $\mathbb{R}$ of integration.

We will rewrite (498) in terms of a tensor product as follows: Let

$$\varphi \otimes \psi(\tau,t) := \varphi(\tau) \cdot \psi(t)$$

and consider the coordinate transform defined by

$$\mathcal{T}(\psi)(\tau,t) := \psi(t,t-\tau)$$

for all variables in the domain of ψ. In addition, we need the Fourier transform $\mathcal{F}_2$ in the second variable given by

$$\mathcal{F}_2(\psi)(\tau,w) := \int \psi(\tau,t) \cdot e^{i\omega t}dt.$$

Then (498) is reformulated as

$$V(\varphi,\psi) = \mathcal{F}_2 \circ \mathcal{T}(\varphi \otimes \psi).$$

Therefore, with Plancherel's theorem (cf. (481)) we have

$$\begin{aligned}
\big(V(\varphi_1,\psi_1)|V(\varphi_2,\psi_2)\big)_{L^2} &= \big(\mathcal{F}_2 \circ \mathcal{T}(\varphi_1 \otimes \psi_1)|\mathcal{F}_2 \circ \mathcal{T}(\varphi_2 \otimes \psi_2)\big)_{L^2} \\
&= 2\pi\big(\mathcal{T}(\varphi_1 \otimes \psi_1)|\mathcal{T}(\varphi_2 \otimes \psi_2)\big)_{L^2} \\
&= 2\pi\big(\varphi_1 \otimes \psi_1|\varphi_2 \otimes \psi_2\big)_{L^2}
\end{aligned}$$

and hence

$$(V(\varphi_1|\psi_1)|V(\varphi_2|\psi_2))_{L^2} = 2\pi \cdot (\varphi_1|\varphi_2)_{L^2} \cdot (\psi_1|\psi_2)_{L^2} \tag{499}$$

for any $\varphi_1, \varphi_2, \psi_1, \psi_2 \in \mathcal{S}(\mathbb{R},\mathbb{C})$. This formula obviously yields the mentioned *orthogonality relation* for V.

8.3.2. The Wigner Function. The integral on the right-hand side of (498) already contains time-frequency information in terms of the coordinates τ and ω. Omitting the phase factor in front of the integral yields the so-called *Wigner transform* or *Wigner function*. Specifying the Fourier conjugate pair to be *position* and *momentum* (as often done), i.e. to x and p, say, the *Wigner function*

$$\mathcal{W} : \mathcal{S}(\mathbb{C}, \mathbb{R}^2) \times \mathbb{R}^2 \longrightarrow \mathbb{C}$$

of any two signals $\varphi, \psi \in \mathcal{S}(\mathbb{R}, \mathbb{C})$ is defined by

$$\mathcal{W}(\varphi,\psi)(x,p) := \int e^{-px'i}\varphi\left(x + \frac{x'}{2}\right)\overline{\psi}\left(x - \frac{x'}{2}\right) dx'. \tag{500}$$

Here x and p vary on $\mathbb{R}$, say; in this formula x' is an integration parameter (cf. [**4**]). If we replace $\mathbb{R}^2$ by F, then for any two $\varphi, \psi \in \mathcal{S}(\mathbb{R}, \mathbb{C})$,

$$\mathcal{W}(\varphi, \psi) : F \longrightarrow \mathbb{C}$$

is smooth on the phase space F (equipped with an orthogonal coordinate system). If $\psi = \varphi$, the map $W(\varphi, \varphi)$ is called the *Wigner distribution*. The substitution $x'' := \frac{x'}{2}$ yields

$$\mathcal{W}(\varphi,\psi)(x,p) = 2\int e^{-2p\cdot x''\cdot i}\varphi(x + x'')\cdot\overline{\psi}(x - x'')dx''$$

and hence

$$\mathcal{W}(\varphi,\psi)(x,p) = 2\int e^{-2p\cdot x''\cdot i}\varphi(x'' + x)\cdot\overline{\widetilde{\psi}}(x'' - x)dx'' \tag{501}$$

where

$$\widetilde{\psi}(y) := \psi(-y) \qquad \forall\, y \in \mathbb{R}. \tag{502}$$

A remarkable property of the Wigner distribution $\mathcal{W}(\varphi, \varphi)$ is stated in the following support property of $\mathcal{W}(\varphi, \varphi)$ (cf. [**47**]). The support of φ is denoted by $\operatorname{supp}\varphi$.

LEMMA 94. *Let $\varphi \in \mathcal{S}(\mathbb{R}, \mathbb{C})$. Then if* $\operatorname{supp}\varphi \subset [\alpha_1, \alpha]$, *we have*

$$\mathcal{W}(\varphi,\varphi)(x,p) = 0 \qquad \textit{for } x \notin [\alpha_1, \alpha_2]$$

and accordingly for $\operatorname{supp}\mathcal{F}(\varphi) \subset [\beta_1, \beta_2]$ *we have*

$$\mathcal{W}(\mathcal{F}(\varphi), \mathcal{F}(\varphi))(x,p) = 0 \textit{ if } p \notin [\beta_1, \beta_2].$$

PROOF. Clearly, $\mathcal{W}(\varphi, \varphi)(x,p) \neq 0$ only if $x + \frac{x'}{2} \in [\alpha_1, \alpha_2]$ and $x - \frac{x'}{2} \in [\alpha_1, \alpha_2]$. Therefore, $x = \frac{1}{2}(x + \frac{x'}{2}) + \frac{1}{2}(x - \frac{x'}{2}) \in [\alpha_1, \alpha_2]$, and consequently, $x \notin [\alpha_1, \alpha_2]$ yields $\mathcal{W}(\varphi, \varphi)(x,p) = 0$. □

The second statement is immediate from the following:

LEMMA 95. *Given $\varphi \in \mathcal{S}(\mathbb{R}, \mathbb{C})$,*

$$\mathcal{W}(\varphi,\varphi)(x,p) = \mathcal{W}(\mathcal{F}(\varphi), \mathcal{F}(\varphi))(-p, x)$$

for any $x, p \in \mathbb{R}$.

The proof (cf. [**47**]) is a straightforward calculation and we leave it to the reader.

The Wigner distribution allows us to compute the total energy $\| \varphi \|_{L^2}$ and the energy distribution spectrum $\| \mathcal{F}(\varphi) \|_{L^2}$ of a signal $\varphi \in \mathcal{S}(\mathbb{R}, \mathbb{C})$:

LEMMA 96. *Let* $\varphi \in \mathcal{S}(\mathbb{R}, \mathbb{C})$. *Then*

$$\int \mathcal{W}(\varphi, \varphi)(x, p)dx = \| \mathcal{F}(\varphi) \|_{L^2}^2$$

and

$$\int \mathcal{W}(\mathcal{F}(\varphi), \mathcal{F}(\varphi)d\omega = \| \varphi \|_{L^2}^2 .$$

PROOF. $\varphi \in \mathcal{S}(\mathbb{R}, \mathbb{C})$ implies $\mathcal{F}(\varphi) \in L^2(\mathbb{R}, \mathbb{C})$ as well as that both $\mathcal{W}(\varphi, \varphi)(x, \ldots)$ and $\mathcal{W}(\varphi, \varphi)(\ldots, \omega)$ are in $L^1(\mathbb{R}, \mathbb{C})$. For $y := x + \frac{1}{2}x'$ and $z := x - \frac{1}{2}x'$ we have

$$e^{-px'i} = e^{-2(y-x)\cdot pi} = e^{-2(x-z)\cdot i}$$

and

$$\begin{aligned} \int \mathcal{W}(\varphi, \varphi)(x, \omega)dx &= \int\int e^{-px'i}\varphi\left(x + \frac{x'}{2}\right)\overline{\varphi}\left(x - \frac{x'}{2}\right)dx'dx \\ &= 2\int\int e^{-2ypi}\varphi(y)e^{2z\cdot p\cdot i}\overline{\varphi}(z)dydz \\ &= \big(\mathcal{F}(\varphi)|\mathcal{F}(\varphi)\big)_{L^2} \\ &= \| \mathcal{F}(\varphi) \|_{L^2}^2 . \end{aligned}$$

The second statement is verified by using the Fourier inversion formula (480). □

In ending this subsection we prove the *orthogonality relation* for the Wigner function which is called *Moyal's formula*:

$$\big(\mathcal{W}(\varphi_1, \psi_1) \mid \mathcal{W}(\varphi_2, \psi_2)\big)_{L^2} = 2\pi(\varphi_1 \mid \varphi_2)_{L^2} \cdot (\psi_1 \mid \psi_2)_{L^2}.$$

At first we rewrite the Wigner function $\mathcal{W}$ as

$$\mathcal{W}(\varphi, \psi)(x, \omega) = \int \varphi\left(x + \frac{t}{2}\right)\overline{\psi}\left(x - \frac{t}{2}\right)e^{-i\omega t}dt,$$

use the coordinate transformation $2t = u - x$ and find

$$\begin{aligned} \mathcal{W}(\varphi, \psi)(x, \omega) &= 2\int \varphi(x)\bar{\psi}(-(u - 2x))e^{2i\omega(u-x)}du \\ &= 2e^{-i\omega x}\int \varphi(x)\tilde{\bar{\psi}}(u - 2x)e^{i\omega u}du \end{aligned}$$

for all $x, \omega \in \mathbb{R}^2$ and any $\varphi, \psi \in \mathcal{S}(\mathbb{R}, \mathbb{C})$ yielding

$$\mathcal{W}(\varphi, \psi)(x, \omega) = 2e^{i\omega x} \cdot V(\varphi, \widetilde{\psi})(2x, 2w).$$

Using (499) we hence conclude

$$\big(\mathcal{W}(\varphi_1, \psi_1) \mid \mathcal{W}(\varphi_2, \psi_2)\big)_{L_2} = (V(\varphi_1, \psi_1) \mid V(\varphi_2, \psi_2))_{L_2}$$

and thus by (499) obtain the orthogonality relation

$$\big(\mathcal{W}(\varphi_1, \psi_1) \mid \mathcal{W}(\varphi_2, \psi_2)\big)_{L^2} = 2 \cdot \pi \cdot (\varphi_1 \mid \varphi_2)_{L^2} \cdot (\psi_1 \mid \psi_2)_{L^2} \tag{503}$$

for any $\varphi_1, \varphi_2, \psi_1, \psi_2 \in \mathcal{S}(\mathbb{R}, \mathbb{C})$.

8.3.3. The Cross-Ambiguity Function. Let us turn back to F with orthonormal coordinates ζ and η, regarded as time and frequency or position and momentum variables, etc. The purpose of this subsection is to show that the cross-ambiguity function is up to the factor 2 and a sign change of variables identical to the Wigner function. The Wigner function is widely used in *optics* and *quantum optics* (cf. [**73**], [**23**], [**27**] and [**26**]) while the cross-ambiguity function is also common in *electrical engineering* such as *radar* (cf. [**71**]).

For any two signals $\psi, \varphi \in \mathcal{S}(\mathbb{R}, \mathbb{C})$ the *cross-ambiguity function* H is defined by

$$H(\varphi, \psi)(\zeta, \eta) := \int_{\mathbb{R}} \varphi\left(t - \frac{\zeta}{2}\right) \overline{\psi}\left(t + \frac{\zeta}{2}\right) e^{\eta t i} dt. \tag{504}$$

Here ζ and η are real numbers, called *time* and *frequency shift*, respectively, provided we use the time-frequency plane $\mathbb{R}^2$. If another Fourier dual pair is used, the shifts ζ and η are called accordingly.

Let us point out that H is not sensitive to the multiplication of both functions φ and ψ by the same overall phase factor in $U(1)$.

To relate the Wigner function $\mathcal{W}(\varphi, \psi)$ of signals φ and ψ to the cross-ambiguity function we observe by comparison of (501) and (504) that

$$\mathcal{W}(\varphi, \psi)(x, p) = 2 \cdot H(\varphi, \tilde{\psi})(-2x, -2p) \tag{505}$$

where $\tilde{\psi}$ is defined as in (502) for any $x, p \in \mathbb{R}$. This formula will be fundamental in section 8.8 on geometric optics (cf. [**24**]).

Moreover,

$$\mathcal{W}(e^{i\lambda}\varphi, e^{i\lambda}\psi) = \mathcal{W}(\varphi, \psi) \qquad \forall\, \varphi, \psi \in \mathcal{S}(\mathbb{R}, \mathbb{C}) \text{ and any } \lambda \in \mathbb{R}$$

and

$$H(e^{i\lambda}\varphi, e^{i\lambda}\psi) = H(\varphi, \psi) \qquad \forall\, \varphi, \psi \in \mathcal{S}(\mathbb{R}, \mathbb{C}) \text{ and any } \lambda \in \mathbb{R}.$$

8.3.4. Moyal's Formula and Cross-Ambiguity Function. We will shortly stop to verify *Moyal's formula* for $H(\varphi, \psi)$. By using the orthogonality relations (499) for the Wigner function we immediately deduce Moyal's formula

$$(H(\varphi_1, \psi_1) \mid H(\varphi_2, \psi_2))_{L^2} = 2\pi (\varphi_1 \mid \varphi_2)_{L^2} \cdot (\psi_1 \mid \psi_2)_{L^2} \tag{506}$$

for all $\varphi_1, \varphi_2, \psi_1, \psi_2 \in \mathcal{S}(\mathbb{R}, \mathbb{C})$. Due to

$$(\varphi_1 \mid \varphi_2)_{L^2} \cdot (\psi_1 \mid \psi_2)_{L^2} = (\varphi_1 \otimes \varphi_2 \mid \psi_1 \otimes \psi_2)_{L^2},$$

we rephrase (506) as

$$(H(\varphi_1, \psi_1) \mid (H(\varphi_2, \psi_2))_{L^2} = 2\pi \parallel \varphi_1 \otimes \varphi_2 \parallel_{L_2} \cdot \parallel \psi_1 \otimes \psi_2 \parallel_{L^2} . \tag{507}$$

8.3.5. The Cross-Ambiguity Function is Surjective. The cross-ambiguity function

$$\begin{aligned} H : \mathcal{S}(\mathbb{R}, \mathbb{C}) \times \mathcal{S}(\mathbb{R}, \mathbb{C}) &\longrightarrow \mathcal{S}(\mathbb{R}^2\mathbb{R} \times \mathbb{R}, \mathbb{C}) \\ (\psi, \varphi) &\longmapsto H(\psi, \varphi) \end{aligned} \tag{508}$$

is surjective and sesquilinear (cf. [**71**]).

The related *auto-ambiguity function*

$$\mathcal{A} : \mathcal{S}(\mathbb{R}, \mathbb{C}) \longrightarrow \mathcal{S}(\mathbb{R} \times \mathbb{R}, \mathbb{C})$$

is defined by

$$\mathcal{A}(\varphi)(\zeta, \eta) := H(\varphi, \varphi)(\zeta, \eta)$$

for any $\varphi \in \mathcal{S}(\mathbb{R}, \mathbb{C})$ and any pair $(\zeta, \eta) \in \mathbb{R}^2$.

It is important in signal theory to notice that this restriction of H to the diagonal in $\mathcal{S}(\mathbb{R}, \mathbb{C}) \times \mathcal{S}(\mathbb{R}, \mathbb{C})$ in its domain is not surjective (cf. [**71**]).

The auto-ambiguity function $\mathcal{A}$, for instance, can be used to identify a signal in a noisy background (cf. [**47**]).

8.4. Reconstruction Formulae

8.4.1. Reconstruction Formulae for the Wigner Function. From the Wigner function

$$\mathcal{W}(\varphi, \psi) : \mathbb{R}^2 \longrightarrow \mathbb{C}$$

on the phase space $\mathbb{R}^2$ (with coordinates x and p) the signal φ can be reconstructed. To see this we consider the definition (500) of the Wigner function $\mathcal{W}(\varphi, \psi)$. The Fourier inversion formula yields

$$(509) \qquad \frac{1}{2\pi} \int \mathcal{W}(\varphi, \psi)(x, p) e^{px'i} dp = \varphi\left(x + \frac{x'}{2}\right) \overline{\psi}\left(x - \frac{x'}{2}\right).$$

Setting $x = \frac{x'}{2}$ provides us with the following reconstruction formula

$$(510) \qquad \varphi(x')\overline{\psi}(0) = \frac{1}{2\pi} \cdot \int_{\mathbb{R}} \mathcal{W}(\varphi, \psi)\left(\frac{x'}{2}, p\right) e^{2p \cdot x' i} dp$$

for any $x' \in \mathbb{R}$. In particular, the energy density of ψ at 0 is

$$|\psi(0)|^2 = \frac{1}{2\pi} \cdot \int_{\mathbb{R}} \mathcal{W}(\psi, \psi)(0, p) dp$$

(cf. section 8.2.1) so that

$$(511) \qquad \varphi(x) = \frac{\psi(0)}{|\psi(0)|} \cdot \frac{\int_{\mathbb{R}} \mathcal{W}(\varphi, \psi)(\frac{x}{2}, p) e^{p \cdot x \cdot i} dp}{|\int_{\mathbb{R}} \mathcal{W}(\psi, \psi)(0, p) dp|^{\frac{1}{2}}}$$

provided $\psi(0) \neq 0$; in this case the above expression is valid for all $x \in \mathbb{R}$. Clearly, $\frac{\psi(0)}{|\psi(0)|}$ is an overall *phase factor* depending on $\varphi(0)$. If $\psi(0) = 0$ we determine a phase factor by adding a non-vanishing constant to $\varphi(0)$. Often such an overall phase factor does not matter for the reconstructed signal, it can be omitted (cf. [**23**]). Obviously, the phase factor does not affect the intensity $\| \psi \|_{L_2}$ for any $\psi \in \mathcal{S}(\mathbb{R}, \mathbb{C})$.

8.4.2. Reconstruction formula for the Cross-Ambiguity Function. From this reconstruction formula for the Wigner function we immediately deduce a similar *reconstruction formula* for the ambiguity function H. On the other hand, we may proceed directly from (505) to obtain

$$(512) \qquad \varphi(x) = \frac{2}{\overline{\psi}(0)} \int_{\mathbb{R}} H(\varphi, \widetilde{\psi})(-2x, -2p) e^{2pxi} dp.$$

Hence the signal φ can also be reconstructed from $H(\varphi, \psi)$ for any ψ with $\psi(0) \neq 0$.

8.5. The Geometry Underlying Time-Frequency Analysis

Now we come to the first highlight in this chapter where the general concepts of signal analysis introduced in the previous sections will be linked with *harmonic analysis*. The aim here is to show that time-frequency analysis done by means of the Wigner function is entirely determined by the Heisenberg group and its Schrödinger representation (cf. [**71**]).

8.5.1. Cross Ambiguity Functions and Schrödinger Representation.

In this subsection, it will be shown that the *cross-ambiguity function*, a close relative of the Wigner function as we have seen (cf. (505)), is entirely determined by the Schrödinger representation (cf. [**71**] and [**17**]), a simple but fundamental observation. The tool to be used for this goal is the *coefficient function* of the Schrödinger representation, to be defined below.

Let us sketch this relation for the Heisenberg group $\mathcal{H}^a$ with $a \in S^2$. The phase space F as mentioned in section 8.2.1 is $F^a \subset \mathcal{H}^a$ which is orthogonal to a. The symplectic structure ω^a on F^a is a consequence of the geometric splitting of the oriented Euclidean space E into $E = \mathbb{R} \cdot a \oplus F^a$ (cf. section 1.3.5).

Remark. Notice that the Schrödinger representation modulates both the Heisenberg group $\mathcal{H}^a$ as well as the reduced Heisenberg group $\mathcal{H}^a_{\text{red}}$ onto a space of signals $\mathcal{S}(\mathbb{R}, \mathbb{C})$, say.

Let ρ be the Schrödinger representation with frequency $\nu = 1$, say, of the Heisenberg group $\mathcal{H}^a$ as defined in section 7.1.2. Its *coefficient function* is defined by

$$c_{\rho,\psi,\varphi}(z + h) := (\rho(z + h)(\psi) \mid \varphi)_{L^2} \tag{513}$$

for all $z \in U^a(1)$, all $h \in F^a$ and any fixed pair $\varphi, \psi \in \mathcal{S}(\mathbb{R}, \mathbb{C})$. (The Hermitian product is assumed to be linear in the first slot (cf. (479)).

In order to compute $c_{\rho,\varphi,\psi}(h)$ for any $h \in F^a$ we choose a unit vector $q_0 \in F^a$ and write h in coordinates as

$$h = \zeta q_0 + \eta q_0 \cdot a \tag{514}$$

where ζ and η vary in $\mathbb{R}$. (Here the $\cdot$ denotes the multiplication in $\mathbb{H}$). By using the definition of the Schrödinger representation, the coefficient function is expressed as

$$\begin{aligned} c_{\rho,\varphi,\psi}(h) &= \int_{\mathbb{R}} e^{\eta t' i} e^{-\frac{\eta\zeta i}{2}} \varphi(t' - \zeta)\overline{\psi}(t')dt' \\ &= \int_{\mathbb{R}} e^{\eta(t' - \frac{\zeta}{2})i} \varphi(t' - \zeta)\overline{\psi}(t')dt'. \end{aligned}$$

Changing t' into $t = t' - \frac{\zeta}{2}$ yields

$$c_{\rho,\varphi,\psi}(h) = \int_{\mathbb{R}} e^{\eta t i} \varphi\left(t - \frac{\zeta}{2}\right) \overline{\psi}\left(t + \frac{\zeta}{2}\right) dt \tag{515}$$

or, expressed in terms of the cross-ambiguity function H,

$$c_{\rho,\varphi,\psi}(h) = H(\varphi, \psi)(\zeta, \eta) \qquad \forall\, \varphi, \psi \in \mathcal{S}(\mathbb{R}, \mathbb{C}) \tag{516}$$

for any h represented as in (514), a result which can be found in the fundamental work [**71**]. Clearly, this formula holds for $\mathcal{H}^a_{\text{red}}$, as well. Obviously, (516) has an according formulation in terms of the Wigner function.

So far we know that the Schrödinger representation determines the cross-ambiguity function via the coefficient function. Let us demonstrate the converse: The cross-ambiguity function is given by

$$H(\varphi, \psi)(\zeta, \eta) = \int e^{-\eta t i} \varphi \left(t - \frac{\zeta}{2} \right) \overline{\psi} \left(t + \frac{\zeta}{2} \right) dt$$

for all $\varphi, \psi \in \mathcal{S}(\mathbb{R}, \mathbb{C})$ and any pair $\zeta, \eta \in \mathbb{C}$. The coordinate transformation

$$t' = t + \frac{\zeta}{2}$$

immediately yields

$$H(\varphi, \psi)(\zeta, \eta) = \int e^{\eta t' i} e^{-\frac{\nu \zeta}{2} i} \varphi(t' - \zeta) \overline{\psi}(t') dt'$$

which obviously is identical to

$$H(\varphi, \psi)(\zeta, \eta) = (\rho(h)(\varphi) \mid \psi)_{L^2}. \tag{517}$$

Hence $\varphi|_{F^a}$ can be determined from the right-hand side of (517). Thus the representation ρ (with frequency $\nu = 1$) is not determined by the cross-ambiguity function; it has to be extended to the center of $\mathcal{H}^a$ (cf. section 7.1.5); we do so by setting

$$\rho(z)\varphi := \frac{1}{z} \cdot \varphi \qquad \forall \varphi \in \mathcal{S}(\mathbb{R}, \mathbb{C})$$

for all $z \in U^a(1)$ and any given $\varphi \in \mathcal{S}(\mathbb{R}, \mathbb{C})$.

This shows that whatever is hidden in the cross-ambiguity function and for that matter in the Wigner function is, in fact, encoded in the Heisenberg group $\mathcal{H}^a$ and any of its Schrödinger representations (with frequency one). In other words, the geometry of $\mathcal{H}^a$ or equivalently, the geometry of $\mathcal{H}^a$ transferred to $\mathcal{S}(\mathbb{R}, \mathbb{C})$ via a Schrödinger representation determines the cross-ambiguity and the Wigner function, respectively, and vice versa if the value of the Schrödinger representations on the center is prescribed. ρ provides us therefore with an important tool for time-frequency analysis. Thus we have:

THEOREM 97. *Time-frequency analysis as described here in terms of either the Wigner function or the cross-ambiguity function is determined by the Heisenberg group and any of its Schrödinger representations of frequency one. The converse is true if the value of the Schrödinger representations on the center is specified.*

8.5.2. The Contragradient Schrödinger Representation and the Cross-Ambiguity Function. How does the contragradient Schrödinger representation (cf. section 7.1.6) enter? We answer this question by the observation that

$$H(\varphi, \psi)(-\zeta, -\eta) = (\varphi \mid \rho^*(\psi))_{L^2}$$

for all $\zeta, \eta \in \mathbb{R}$ and for any $\varphi, \psi \in \mathcal{S}(\mathbb{R}, \mathbb{C})$ (cf. section 7.1.6) holds true. Here $\rho^*(\psi)$ is the adjoint of $\rho(\psi)$. Let us denote $H(\varphi, \psi)(-\zeta, -\eta)$ by $H^-(\varphi, \psi)(\zeta, \eta)$ for any choice of $\zeta, \eta \in \mathbb{R}$. Equation (505) therefore immediately yields

$$\mathcal{W}(\varphi, \psi)(\zeta, \eta) = 2 \cdot (\rho(-2\zeta, -2\eta)(\varphi) \mid \psi)_{L^2} \, .$$

With this notation the equation

$$(\rho(\varphi_1) \mid \varphi_2)_{L^2} \cdot (\rho^*(\psi_1) \mid \psi_2)_{L^2} = (\rho \otimes \rho^*(\varphi_1 \otimes \varphi_2) \mid \psi_1 \otimes \psi_2)_{L^2} \tag{518}$$

is reformulated by means of (517) as

$$H(\varphi_1,\psi_1)\cdot H^-(\varphi_2,\psi_2) = (\rho\otimes\rho^*(\varphi_1\otimes\varphi_2) \mid \psi_1\otimes\psi_2)_{L^2}. \tag{519}$$

8.6. The Radar Ambiguity Function

As an application of the preceding constructions we will make use of the ambiguity function in a well studied example, namely in *radar detection* (cf. [**71**] and [**37**]). We will essentially follow [**37**].

Suppose a radar apparatus transmits an electromagnetic signal φ that is reflected by a target and returns to the apparatus. This signal shall be of the form

$$\varphi(t) = \varphi_0(t)e^{i\omega t} \qquad \forall\, t\in\mathbb{R}.$$

We assume that the *frequencies* of φ are essentially concentrated around some large ω_0, and that the *amplitude* $\varphi_0(t)$ varies with t rather slowly in comparison with $e^{i\omega t}$.

The distance from the apparatus to the target shall be r. The *radial* velocity of the target shall be $v := \frac{dr}{dt}$. The domain Δt of the signal is assumed to be large with respect to the period $2\pi/\omega$ but small enough such that v is assumed to be constant in that interval.

The reflected signal arrives back at the apparatus with time delay $\tau := 2r/c$ where c is the speed of light. The frequency dilatation is $1 - 2v/c$ due to the Doppler effect. However, since we assumed that the frequencies of φ are essentially near ω_0, we can take the frequency shift to be $\eta := 2\omega v/c$. The returned signal $\varphi_{\tau,\eta}$ then has the form

$$\varphi_{\tau,\eta}(t) := \varphi_0(t-\tau)e^{-i\eta t}. \tag{520}$$

The parameters τ and η are coordinates in a plane F^a perpendicular to $\mathbb{R}\cdot a$ for $a\in S^2$. The coordinate axes are $\mathbb{R}\cdot e_{\bar q}$ and $\mathbb{R}\cdot e_{\bar q}\times a$. The axis $\mathbb{R}\cdot a$ is the direction the signal φ takes. Hence we have a Heisenberg group $\mathcal{H}^a$; it describes the basic geometry of radar detection, as we will see next.

To obtain some means by which the target can be identified, we first consider two targets with return signals φ_{τ_1,η_1} and φ_{τ_2,η_2}, to which the same sort of signal has been sent. If these two signals are similar, they are hard to distinguish. We thus study the mean squared difference

$$\int |\varphi_{\tau_1,\eta_1} - \varphi_{\tau_2,\eta_2}|^2 dt = 2\int |\varphi|^2 dt - 2\,\mathrm{Re}(\varphi_{\tau_1,\eta_1} \mid \varphi_{\tau_2,\eta_2})_{L_2}. \tag{521}$$

If the left-hand side is large the returned signals are rather characteristic of the targets.

The two targets influence only the second term on the right-hand side. Since the signal φ is given by $\varphi(t) = \varphi_0(t)e^{\pi i\omega t}$, we have

$$(\varphi_{\tau_1,\eta_1} \mid \varphi_{\tau_2,\eta_2})_{L^2} = e^{i\omega(\tau_2-\tau_1)}\int \varphi_0(t-\tau_1)\overline{\varphi}_0(t-\tau_2)e^{i\omega(\eta_2-\eta_1)t}dt \tag{522}$$

with a rapidly oscillating phase factor in front of the integral. If the mean squared difference is large and stable under small perturbations of τ_1 and τ_2, the real number

$|(\varphi_{\tau_1,\eta_1} \mid \varphi_{\tau_2,\eta_2})|$ has to be small. Setting $\tau = \tau_1 - \tau_2$ and $\eta = \eta_1 - \eta_2$, equation (522) becomes

$$(\varphi_{\tau_1,\eta_1} \mid \varphi_{\tau_2,\eta_2})_{L^2} = e^{i\eta\tau} \int \varphi_0(t) \cdot \overline{\varphi}_0(t+\tau) e^{i\eta t} dt. \tag{523}$$

Switching the two targets nothing essential changes (523). Moreover, since only the absolute value of the left-hand side matters, we may replace the phase factor in front of the integral to obtain

$$A(\tau, \eta) := \int \varphi(t) \cdot \overline{\varphi}(t+\tau) e^{-i\eta t} dt.$$

Moreover, replacing τ and η by $-\tau$ and $-\eta$ yields

$$A(-\tau, -\eta) = \int \varphi(t) \cdot \overline{\varphi}(t-\tau) e^{i\eta(t-\tau)} dt$$

and hence we arrive at

$$\int e^{i\eta t'} \varphi_0\left(t' + \frac{\eta}{2}\right) \cdot \overline{\varphi}_0\left(t' - \frac{\eta}{2}\right) dt' = H(\varphi_0, \varphi_0)(\tau, \eta)$$

where we set $t - \frac{\tau}{2} = t'$. Hence $H(\varphi_0, \varphi_0)(\tau, \eta)$ provides a tool to identify the target by comparison of the transmitted signal to the reflected one. Many more details can be found in (cf. [**71**])

Thus the Heisenberg group $\mathcal{H}^a$ together with the Schrödinger representation determine the geometry for the detection of a target by means of radar.

8.7. The Stone-von Neumann Theorem in Time-Frequency Analysis

The cross-ambiguity function H as defined in (504) depends on a coordinate system in F^a. Thus it is natural to ask what effect a coordinate transformation has to this function. In order to leave H invariant, the transformation should preserve the symplectic structure in F^a, because otherwise the Heisenberg group and the Schrödinger representation change. In this section we will thus study the effect of symplectic coordinate transformations to the cross-ambiguity function.

8.7.1. The Effect of a Symplectic Map to the Cross-Ambiguity Function. As we observed in section 8.5, one of the most basic ingredients in signal analysis, namely the cross-ambiguity function or for that matter the Wigner function, depends on the Schrödinger representation ρ of the Heisenberg group $\mathcal{H}^a$. In what follows we base our construction on a fixed $a \in S^2 \subset E$.

Given a symplectic map $A \in \mathrm{Sp}(F^a)$, due to (517) and (431) the ambiguity function H satisfies

$$H(\varphi, \psi)\,(A(x,p)) = H\big(U_{\mathrm{Sp}}(A^{-1})(\varphi) | U_{\mathrm{Sp}}(A^{-1})(\psi)\big)(x,p) \tag{524}$$

for any $x, p \in \mathbb{R}$ (cf. section 7.5). By (431) overall phase factors in front of the unitary projective representation U_{Sp} do not affect H (cf. (513) and (516)). Hence (524) holds true if we replace $A \in \mathrm{Sp}(F^a)$ by $(\tilde{A}) \in \mathrm{Mp}(F^a)$ (cf. section 7.5.1).

To express $U_{\mathrm{Sp}}(A^{-1})$ as an evolution operator we set

$$\varphi := U_{\mathrm{Sp}}(A^{-1})(\varphi_0) \quad \text{and} \quad \psi := U_{\mathrm{Sp}}(A^{-1})(\psi_0);$$

here the pair φ, ψ is determined only up to a common phase factor. This yields for any $A \in \mathrm{Sp}(F^a)$ and initial data φ_0 and ψ_0 the relations

$$H(\varphi_0, \psi_0)\,(A(\xi, \eta)) = H(\varphi, \psi) \tag{525}$$

and

$$\mathcal{W}(\varphi_0, \psi_0)\,(A(x,p)) = \mathcal{W}(\varphi, \psi)(x,p) \qquad \forall\, x, p \in \mathbb{R} \tag{526}$$

between the two pairs φ_0, φ and ψ_0, ψ in terms of the cross-ambiguity function and the Wigner function.

8.7.2. Reconstruction of $A(\varphi_0)$. Combining (511) and (526) under the condition $\varphi_0(0) \neq 0$ yields for $\tilde{A} \in \mathrm{Mp}(F^a)$ which shall project to A and $U_{\mathrm{Mp}}(\tilde{A})(\varphi)$,

$$U_{\mathrm{Mp}}(\tilde{A})(\varphi) = \frac{\varphi(0)}{|\varphi(0)|} \cdot \frac{\int \mathcal{W}_{(\varphi_0,\varphi_0)}\left(A(\frac{x}{2}, p)\right) \cdot e^{p\cdot x \cdot i} dp}{|\int_{\mathbb{R}} \mathcal{W}(\varphi_0, \varphi_0)\,(A(0,p))\,dp|^{\frac{1}{2}}} \tag{527}$$

since

$$\omega^a(A(x \cdot e_{\bar{p}_0}), A(p \cdot e_{\bar{q}_0})) = x \cdot p \cdot \omega^a(e_{\bar{q}_0}, e_{\bar{p}_0} s) = x \cdot p \tag{528}$$

for any pair $x, p \in \mathbb{R}$ of coordinates .

If we reconstruct up to an *overall phase factor* only, the above formula yields

$$U_{\mathrm{Sp}}(A)(\varphi) = \frac{\int_{\mathbb{R}} \mathcal{W}_{(\varphi_0,\varphi_0)} A(\frac{x}{2}, p) \cdot e^{p\cdot x\cdot i} dp}{|\int_{\mathbb{R}} \mathcal{W}(\varphi_0, \varphi_0)\,(A(0,p))\,dp|^{\frac{1}{2}}}. \tag{529}$$

This is usually good enough for many practical purposes (cf. [**23**]–[**28**]).

8.8. Geometric Optics

Here we come to a second highlight in this chapter: In the last section we have observed that time-frequency analysis based on the cross-ambiguity function is determined by a Heisenberg group and its Schrödinger representation (of frequency one). We also studied the effect of symplectic coordinate transformations to the symplectic plane F^a. In this section we will show that these transformations have a rather severe impact on optics; it amounts to applying an optical system to the signal. Vice versa, any optical system is described by applying a symplectic transformation to initial data in F^a (cf. [**41**]). This is the starting point of what is called *Fresnel optics* (cf. [**50**] and [**41**]).

8.8.1. Symplectic Transformations and the Metaplectic Representation. The geometric basis of Fresnel optics as we will present it here consists of two parallel planes F^1 and F^2 in an oriented Euclidean space E together with the axis $\mathbb{R}\cdot a$ perpendicular to both planes for fixed $a \in S^2$. The axis $\mathbb{R}\cdot a$ intersects the two planes in two points being the respective *origins* of F^1 and F^2. Thus, these two planes are linear spaces.

In between these two planes an *optical system* shall be placed. In *linear optics* the image in F^2 of a point in F^1 mapped by a *light ray* passing the given optical system is described by a *linear map* $A' : F^1 \longrightarrow F^2$ (cf. [**41**]). This map is characteristic of the optical system. The concatenation of two optical systems corresponds to the composition of the respective linear maps.

To describe the quality of this map more precisely let $F \subset E$ be a two-dimensional linear subspace parallel to both F^1 and F^2. Hence

$$E = \mathbb{R} \cdot a \oplus F.$$

The vector a defines the symplectic form ω^a. Thus $F = F^a$ and so the geometric setting for geometric optics is the Heisenberg algebra $\mathfrak{h}^a$ together with the Heisenberg group $\mathcal{H}^a$.

The image of a point $q \in F^1$ caused by a light ray (passing the optical system) is thus performed by mapping it to $A(q)$ in F^1 and shifting this image to F^2 by a parallel translation along $\mathbb{R} \cdot a$.

Hence any linear map $A' : F^1 \longrightarrow F^2$ can equivalently be described by some A in the general linear group $GL(F^a)$. Indeed, let $T : F^1 \longrightarrow F^2$ be the shift parallel to $\mathbb{R} \cdot a$ from F^1 to F^2. Obviously, there is some $A \in GL(F^a)$ such that

$$A' = T \circ A \circ T^{-1}.$$

It is a remarkable fact that the map A caused by an optical system *preserves* ω^a. Hence the optical system is characterized by some $A \in \mathrm{Sp}(F^a)$. This is shown by computing the matrix of fundamental optical systems and noticing that they are all *symplectic* (cf. [**27**], [**23**], [**25**], [**41**], [**50**] and [**88**]).

We extend $A \in \mathrm{Sp}(F^a)$ linearly to all of $\mathfrak{h}^a$ by setting

$$A(a) := a,$$

so that $A : \mathfrak{h}^a \rightarrow \mathfrak{h}^a$ is a Lie algebra isomorphism. Since

$$[A(\lambda_1 \cdot a + h_1), A(\lambda_2 \cdot a + h_2)] = \omega^a(A(h_1), A(h_2)) \cdot a = \omega^a(h_1, h_2) \cdot a$$

for any $\lambda_1, \lambda_2 \in \mathbb{R}$ and any $h_1, h_2 \in F^a$, any automorphism of $\mathfrak{h}^a$ is of this form.

The plane F^a shall be equipped with a rectangular coordinate system. The coordinates q and p shall be replaced by the Fourier dual pair position and frequency, respectively. Thus

$$F^a = \mathbb{R} \cdot e_{\bar{q}} + \mathbb{R} \cdot e_{\bar{q}} \cdot a$$

for some unit vector $e_{\bar{q}} \in F^a$.

Any linear map $A \in \mathrm{Sp}(F^a)$ has the matrix

$$M(A) = \begin{pmatrix} \alpha & \beta \\ \gamma & \delta \end{pmatrix} \text{ with } \det A = 1$$

formed with respect to the coordinate system determined by $e_{\bar{q}}$ and $e_{\bar{p}}$. A acts on F^a by

$$\begin{pmatrix} \alpha & \beta \\ \gamma & \delta \end{pmatrix} \begin{pmatrix} q_1 \\ p_1 \end{pmatrix} = \begin{pmatrix} \alpha q_1 & + & \beta p_1 \\ \gamma q_1 & + & \delta p_1 \end{pmatrix} = \begin{pmatrix} q_2 \\ p_2 \end{pmatrix}. \tag{530}$$

Thus given q_1 and q_2, the coordinates p_1 and p_2 are determined by the equations

$$p_1 = \beta^{-1}(q_2 - \alpha \cdot q_1) \tag{531}$$

and

$$p_2 = \gamma q_1 + \delta p_1. \tag{532}$$

This shows that q_1 and the q-coordinate q_2 of the image of $(q_1, p_1) \in F^a$ determine both coordinates p_1 and p_2, respectively. Hence given A, it suffices to specify only

q_1 and its image coordinate q_2 to specify the symplectic map. Thus we are in the realm of *signal analysis* again.

The technique presented here is used, for example, in [**23**] and [**28**] where the light distribution of the optical image is computed by means of the Wigner function, yielding the symplectic matrix on F. This technique uses essentially equation (529).

8.8.2. Some Basic Examples. In this subsection we assume that light propagates along rays in an *empty* (oriented) Euclidean space E.

The initial data of the rays are initial positions x_0 and initial frequencies ν_0 implemented with respect to fixed points $e_{\bar{q}}$ and $e_{\bar{q}} \cdot a$ in a plane F^a, respectively. The data of the image vary in the plane F^a, too. Our first goal is to associate linear maps in $\mathrm{Sp}(F^a)$ with some familiar optical settings. In doing so we follow [**24**] and [**50**].

Let us consider an initial datum

$$h_0 = x_0 e_{\bar{q}} + \nu_0 e_{\bar{q}} \cdot a$$

in position and frequency coordinates evolving into

$$h = x_0 e_{\bar{q}} + \lambda\zeta \cdot \nu_0 e_{\bar{q}} \cdot a.$$

Here $\lambda = \frac{c}{\nu}$ is the wave length with the speed of light c. Moreover, ζ is a real. Comparing h_0 and h there is a *linear* map $A^{\lambda\zeta} \in \mathrm{End}(F^a)$ for which

$$h = A^{\lambda\zeta}(h_0).$$

The matrix of $A^{\lambda\zeta}$ is

$$M(A^{\lambda\zeta}) = \begin{pmatrix} 1 & \lambda\zeta \\ 0 & 1 \end{pmatrix},$$

a *transvection* (cf. (281)). Hence $\det A^{\lambda\zeta} = 1$, which means that $A^{\lambda\zeta} \in \mathrm{Sp}(F)$.

Next let us assume that the light ray passes a thin half-lens at ζ_1 with $\mathbb{R} \cdot a$ as axis of rotation and focal length f. The image h of h_0 shall be in the plane $\zeta_2 a + F^a$. According to [**50**], [**28**] and [**41**], for example, we know that

$$h = A^{\zeta_2} \circ A_{\lambda \cdot f} \circ A^{\zeta_1}(h_0)$$

where $A_{\lambda f} \in \mathrm{Sp}(F)$ has

$$M(A_{\lambda f}) = \begin{pmatrix} 1 & 0 \\ -\frac{1}{\lambda \cdot f} & 1 \end{pmatrix}$$

as its matrix, the *transpose of a transvection.*

Finally, we describe the $2f$*-array* of a thin lens. The left-hand side focus shall be at the origin $0 \in F$ while the right-hand side focus shall be at the position $\lambda f \cdot a$. Here the initial datum h_0 evolves according to

$$h = A_{\lambda f} \circ A_f \circ A_{\lambda f}(h_0)$$

where A_f has

$$M(A_f) = \begin{pmatrix} 0 & \lambda \cdot f \\ -\frac{1}{\lambda \cdot f} & 0 \end{pmatrix}$$

as its matrix (cf. [**50**] and [**24**]). Hence $A_f \in \mathrm{Sp}(F)$ again. Thus A_f is a rotation followed by a dilatation.

Using the notation of section 7.4.2 we observe that

$$\begin{array}{rcll} M(A^{\lambda\zeta}) & = & A_2^{-1} & \text{for } \frac{1}{\lambda f}=t, \\ M(A_f) & = & M(J)\circ A_1 & \text{for } \frac{1}{\lambda f}=-t, \\ M(A_f) & = & J & \text{for } \lambda f = 1. \end{array}$$

As it is easily seen,

$$M\left(A^{\lambda\zeta}\right)^{-1} = M\left(A^{-\lambda\zeta}\right)$$

and

$$M(A^{\lambda\zeta})^{\sim} = J^{-1}\circ M(A^{\lambda\zeta})\circ J = \begin{pmatrix} 1 & 0 \\ \lambda\zeta & 1 \end{pmatrix}$$

Here $\sim$ denotes the transpose. Clearly, $A^{\lambda\zeta}$, A_f and J generate $\mathrm{Sp}(F^a)$. Therefore, we conclude:

THEOREM 98. *Any element in* $\mathrm{Sp}(F^a)$ *corresponds to an optical system.*

8.8.3. Mapping Light Distributions. In the previous section we studied linear mappings in $\mathrm{Sp}(F)$ caused by the light rays passing from F^1 to F^2. The optical system corresponding to the symplectic transformation maps a given light distribution in F^1 to a light distribution in F^2, i.e. associated with any of these symplectic transformations is a mapping of a given light distribution. In geometric optics the image distribution is computed by means of the Huygens' principle. This method, called the Fresnel-Huygens-theory or just *Fresnel optics*, yields the so-called *Fresnel integrals* as is shown in [**41**]. These integrals are directly related to the metaplectic representation which we described in sections 7.5.3 and 7.5.4 (cf. [**50**] and [**32**]).

In this subsection we present the *metaplectic representation* to describe the evolution of a given light distribution in F^1 i.e. the image *light distribution* in F^2.

Given the optical system determining $A \in \mathrm{Sp}(F)$ with free matrix (cf. section 7.5.3)

$$A = \begin{pmatrix} \alpha & \beta \\ \gamma & \delta \end{pmatrix} \qquad \text{with } \beta \neq 0,$$

the *optical path* between $x \in F^1$ and its image point $y \in F^2$, expressed in the q-coordinates q_1 and its image coordinate q_2, is given by

$$W(q_1, q_2) = \delta\cdot\beta^{-1}(\alpha q_1^2 + \delta q_2^2 + 2q_1\cdot q_2) + d \tag{533}$$

(cf. [**41**] or [**50**] and section 7.5.3). Here d is the *optical distance* between the two planes F^1 and F^2 on the axis $\mathbb{R}\cdot a$. The function W is called the *eikonal*. It crucially depends on A, i.e. on its matrix in the coordinate system introduced above.

Given the light distribution φ on F^1, the one on F^2 caused by the optical system will be called ψ. It is expressed by Fresnel's formula (or Fresnel's integral) as

$$\psi(q_2) = e^{-\frac{\pi i}{4}}(\beta\lambda)^{-\frac{1}{2}}\int \varphi(q_1)e^{\frac{2\pi i}{\lambda}W(q_1,q_2)}dq \tag{534}$$

for any $q_2 \in F^2$ (cf. [**41**]), a quadratic Fourier transform (cf. (458)). Here we omitted d since it would have caused only a fixed overall phase factor. We will do so in the sequel, as well. For simplicity we set $\lambda = 2\pi$, as done in (458), which amounts to choosing a unit of length.

Clearly, φ depends on the coordinates (q_1, p_1) and ψ on the image coordinates (q_2, p_2) where the p-coordinates p_1 and p_2 are calculated by the formulae (531) and (532). This, however, shows that in the case of $\beta \neq 0$,

$$\psi = U_{\mathrm{Mp}}(A)(\varphi) \tag{535}$$

where U_{Mp} is the *metaplectic representation* (cf. section 7.5.3, equation (461) and section 7.5.4).

A transformation $A \in \mathrm{Sp}(F)$ with matrix

$$M(A) = \begin{pmatrix} \alpha & 0 \\ \gamma & \frac{1}{\alpha} \end{pmatrix}$$

describes the transfer of points and hence of the light distribution along $\mathbb{R} \cdot a$. The optical path in this case is given by the eikonal

$$W(q_1) = d - \frac{\alpha}{2\gamma} \cdot q_1^2 \tag{536}$$

(cf. [**41**]). Hence we again obtain

$$\psi = U_{\mathrm{Mp}}(A)(\varphi)$$

for a light distribution φ on F^1. This formula can alternatively be calculated by means of (474).

Remark. *Hence we can view the form of W in the integrand of (451) of the quadratic Fourier transform associated with a symplectic matrix (454) in the light of Fresnel optics; it is a consequence of Huygens' principle which determines (533).*

*If only the reconstruction up to an overall phase factor is needed, then the light distribution φ can be reconstructed by either (511) or (512) (cf. [**24**]) or by means of (527) or (529).*

The optical path the light takes produces a point-dependent phase which is encoded in the eikonal. A *photographic image* of the image distribution would suppress this phase. However, up to an overall phase factor, it can be reconstructed by means of *holography*, as we will see in the next section.

8.9. Holography

One of the most important concepts of modern optics is *holography*. It allows us to reconstruct the *phase* of an *intensity measurement* of a wave. As an example of such a measurement we may take a photograph. The photograph is the distribution of the square magnitude of the amplitude. The phase is completely lost. How it can be reconstructed is sketched in the sequel. The reconstruction can be performed by using the metaplectic representation, as it is beautifully done in [**50**]. Here we will follow the traditional path as it is more intuitive in our setting.

8.9.1. A Simple Example and the Definition of a Hologram. We begin with a simple setup to produce a hologram as presented in [**52**]. Let us split the Euclidean space E into $E = \mathbb{R} \cdot a \oplus F$ with some $a \subset S^2$. Two *plane waves* Φ_R and Φ_O with the same amplitude φ_O are supposed to travel through E. These two waves are called *reference wave* and *object wave*, respectively. The reference wave Φ_R shall travel along $\mathbb{R} \cdot a$ with F parallel to the planes of equal phase. The wave vector k_O of the plane wave Φ_O shall be *inclined* to $\mathbb{R} \cdot a$. It shall intersect $\mathbb{R} \cdot a$ in some non-vanishing angle δ. Both Φ_R and Φ_O are assumed to have the same phase α at the origin $O \in F$.

The two vectors a and k_O span a plane perpendicular to F which intersects F in an axis passing through $O \in F$, the x-axis. Now suppose we had two rays of Φ_O, one passing through O, the other one through x_1, say. The plane of equal phase of Φ_O containing O and the one containing x_1 have the same phase if the distance between the two planes is a multiple n of the wave length λ. Hence

$$\lambda = x_1 \cdot \sin\delta$$

if the x_1 is such that $n = 1$. For a general x the phase difference between the origin and x is hence given by

$$\frac{\alpha(x)}{2\pi} = \frac{x}{x_1}$$

and thus

$$\alpha(x) = \frac{2\pi}{\lambda} \cdot x \cdot \sin\delta$$

or, if we replace $\frac{2\pi}{\lambda}$ by $|k_O|$,

$$\alpha(x) = |k_O| \cdot x \cdot \sin\delta.$$

Clearly, $\alpha(x)$ is the relative phase between φ_R and φ_O.

The wave Φ_R has a constant phase all along F. Hence we may write the two plane waves at x and time t as

$$\Phi_R(x,t) = \varphi_R \cdot e^{i(\omega t - \alpha(x))},$$

being the *reference wave*, and

$$\Phi_O(x,t) = \varphi_O \cdot e^{i(\omega t - \alpha - \alpha(x))},$$

the wave from the object, the *object wave*. Its phase is written relative to the phase of the reference wave.

The superposition of both waves is

$$(\Phi_R + \Phi_O)(t,x) = \varphi_O \cdot e^{i\omega t} \cdot e^{-i\alpha}(1 + e^{-i\alpha(x)}).$$

To write $1 + e^{-i\alpha(x)}$ in polar coordinates, as well, we observe that

$$|1 + e^{i\alpha(x)}|^2 = 2(1 + \cos\alpha(x)).$$

Since

$$1 + \cos\alpha(x) = 2\cos^2\frac{\alpha(x)}{2},$$

the absolute value of $1 + e^{i\alpha(x)}$ is

$$|1 + e^{i\alpha(x)}| = 2\cos\frac{\alpha(x)}{2}$$

and hence

$$1 + e^{i\alpha(x)} = 2\cos\frac{\alpha(x)}{2} \cdot e^{i\gamma(x)}.$$

Comparing real parts yields

$$\cos\frac{\alpha(x)}{2} = \cos\gamma(x)$$

and hence

$$1 + e^{i\alpha(x)} = 2\cos\frac{\alpha(x)}{2} \cdot e^{\frac{i\alpha(x)}{2}}.$$

Therefore,

$$(\Phi_R + \Phi_O)(x,t) = 2 \cdot A \cdot \cos \frac{\alpha(x)}{2} \cdot e^{i(\omega t - \alpha - \frac{\alpha(x)}{2})}$$

holds true. Here A is a real number.

The hologram we will define below is based on the *intensity* of the light distribution on F caused by the superposition of reference and object wave. The intensity I of a light wave is defined to be half of the square of the magnitude of the amplitude (cf. [**21**]). The square of the magnitude of the amplitude is

$$|\phi_{R(x)} + \phi_O(x)|^2 = 4A^2 \cos^2 \frac{\alpha(x)}{2}$$

and the intensity $I(x)$ of $\varphi_R + \varphi_O$ at x is hence

$$I(x) = 2 \cdot A^2 \cos^2 \frac{\alpha(x)}{2}$$

or, formulated in terms of $\alpha(x)$,

$$I(x) = A^2 + A^2 \cdot \cos \alpha(x).$$

The *intensity distribution* of $\varphi_R + \varphi_O$ thus defines essentially a cosine-grid (cf. section 5.1.4) on F namely $A^2 \cdot \cos \alpha$ which is independent of the y-coordinate. This distribution is called the *hologram*. It might be realized on a (transparent) thin film.

As mentioned above we just constructed a rather simple sort of hologram. Now let us pass to a general case. A much more interesting setting is determined, for example, by a photograph. For the sake of simplicity, we assume that the photograph exhausts all of F. The reference and object wave have, in general, much more complicated *phase distributions*. The reference wave shall have the form

$$\Phi_R(x,y) = \varphi_R(x,y) \cdot e^{i(\nu t + \alpha_R(x,y))}$$

for all x, y of the hologram in F. The object wave is written as

$$\Phi_O(x,y,t) = \varphi_O(x,y) \cdot e^{i(\nu t + \alpha_O(x,y))} \tag{537}$$

holding true at any x, y.

The hologram on F is the *intensity distribution* of the superposition $\Phi_R + \Phi_O$.

8.9.2. Reconstruction. The goal now is to *reconstruct* Φ_O from the reference wave Φ_R and the hologram.

For this purpose the amplitude φ_R of the reference wave is assumed to be *constant*. For the reconstruction of the object wave the reference wave Φ_R is supposed to be *inclined* by the angle δ to the plane containing the hologram. Thus the phase of the reference wave is

$$\alpha_R(x,y) = \alpha_0(x,y) + t\frac{2\pi}{\lambda} \cdot x \cdot \sin \delta, \tag{538}$$

where $\alpha_0(x,y)$ is a constant, i.e. $\alpha_R(x,y)$ is independent of y. Hence the reference wave Φ_R is of the form

$$\Phi_R(x,y,t) = \varphi_R(x,y) \cdot e^{(\nu \cdot t + \alpha_R(x,y)) \cdot i} \qquad \forall\, x, y \in F \text{ and } \forall\, t \in \mathbb{R}$$

with $\alpha_R(x,y)$ as in (538). The object wave Φ_O in (537) stays unchanged.

The *reconstruction wave* Φ_{RE} is the superposition of Φ_R and Φ_O and hence is of the form

$$\Phi_{RE}(x,y,t) = \Phi_R(x,y) + \Phi_O(x,y,t) \qquad \forall\, x,y \in F \text{ and } t \in \mathbb{R}.$$

The reconstruction of the amplitude φ_O is based on the observation that the *transmission function* of the translucent film is assumed to be *proportional* to the intensity $I(x,y)$ for all $x,y \in F$. This function models the "transparency" of the film. Thus the amplitude of the wave passed through the film is

$$\phi_{RE}(x,y) = \frac{1}{2} \cdot |\varphi_O \cdot e^{\alpha_O(x,y)\cdot i} + \varphi_R \cdot e^{\alpha_R(x,y)\cdot i}|^2 \cdot \varphi_R \tag{539}$$

up to the proportionality factor emanating from the portion of light transmitted through the film. The intensity (computed by the Hermitian product) is proportional to

$$\begin{aligned} &|\Phi_R(x,y,t) + \Phi_O(x,y,t)|^2 \\ &\quad = |(\varphi_O(x,y)\cdot e^{i(\nu t+\alpha_O(x,y))} + \varphi_R(x,y)\cdot e^{i(\nu t+\alpha_R(x,y))}|^2 \\ &\quad = |\varphi_O(x,y)|^2 + |\varphi_R(x,y)|^2 + \varphi_O(x,y)\cdot\varphi_R(x,y)\cdot e^{i(\alpha_O(x,y)-\alpha_R(x,y))} \\ &\qquad + \varphi_R(x,y)\cdot\varphi_O(x,y)\cdot e^{-i(\alpha_O(x,y)-\alpha_R(x,y))}, \end{aligned}$$

i.e. it is proportional to

$$\begin{aligned} &|\varphi_R(x,y) + \varphi_O(x,y)|^2 \\ &\quad = (\varphi_R^2 + \varphi_O^2)(x,y) + \varphi_R\cdot\varphi_O(x,y)\cdot(\cos(\alpha_R(x,y) - \alpha_O(x,y))). \end{aligned}$$

Denoting the *spatial part* of the wave Φ_{RE} by $\Phi^t(x,y) := \Phi_{RE}(x,y,t)$ for fixed t, it is hence

$$\begin{aligned} \Phi^t(x,y) &= \varphi_R^2\cdot\varphi_R(x,y) + \varphi_O^2\cdot\varphi_R(x,y) \\ &\quad +\varphi_R\cdot\varphi_O(x,y)\cdot e^{i(\alpha_R(x,y)-\alpha_O(x,y))}\cdot\varphi_R(x,y) \\ &\quad +\varphi_R\cdot\varphi_O(x,y)\cdot e^{i(\alpha_O(x,y)-\alpha_R(x,y))}\cdot\varphi_R(x,y) \end{aligned}$$

or, finally,

$$\begin{aligned} \Phi_{RE}(x,y,t) &= 2\varphi_R^3\cdot e^{i(\nu t+\alpha_R(x,y))} + \varphi_O^2(x,y)\cdot\varphi_R\cdot e^{i(\nu t+\alpha_R(x,y))} \\ &\quad +\varphi_R\cdot\varphi_O(x,y)\cdot e^{i(\nu t+2\alpha_R(x,y)-\alpha_O(x,y))} \\ &\quad +\varphi_R^2\cdot\varphi_O(x,y)\cdot e^{i(\nu t+\alpha_O(x,y))} \end{aligned}$$

for all $(x,y) \in \mathbb{R}^2$. The first term on the right-hand side of this equation is pointwisely proportional to the reconstruction wave. The last term contains, up to a constant multiple $|\varphi_R|^2$, the wave scattered by the object, i.e. the object wave. The second term is

$$\varphi_R \cdot e^{2i\alpha_R} \cdot \varphi_O,$$

i.e. φ_O rotated by $e^{2i\alpha_R(x,y)}$. Thus this rotated wave and the one of the last term are separated, yielding two images of the object, the *primary* and the *secondary* one. The primary one is called the *virtual* image while the secondary one is called the *real image*, i.e. the image produced by means of holography. In other words, we have reconstructed the *phase* from an *intensity distribution*.

We have given here a very short introduction to the quite illustrative traditionally formulated holography. There are many excellent texts in modern optics such as [**21**] and [**52**] which describe this very fundamental field (in theory and application) in much more detail.

It is rather interesting to know that within the scope of these notes holography can be formulated in the framework of Fresnel optics. As mentioned above, it can also be described by means of the metaplectic representation (cf. [**50**]).

CHAPTER 9

Quantization of Quadratic Polynomials

The mathematical apparatus developed in the previous chapters allows us not only to model signal transmission but is also a basic device to describe quantization. In this chapter we hence describe some fundamentals of quantization and pay some attention to the connection with optics.

9.1. Elementary Observations on Information and its Transmission

In this section we will introduce the notion of information and make precise what an information preserving transmission means. This is particularly important since the preservation of information causes quantum mechanics as we will see below. However, we will not go on from here to treat quantum information where we would rather refer to [**65**].

9.1.1. Information. Suppose we had a photograph in a plane F, say, and want to transmit it from one place to another. Naturally the photograph will be transmitted in a three-dimensional $\mathbb{R}$-linear space E. For the sake of simplicity, the transmission shall follow an axis transverse to the plane F. This axis shall be the *channel of information*, i.e. the transmission of information will follow along this channel, as we will make more precise later.

Describing this process requires analyzing the *information* that the photograph presents, the *geometry* underlying the transmission, the *encoding* in *signals* and their *detection* as well as the *preservation of information* during the transmission.

In this section we will mainly concentrate on the analysis of the notion of information, on the geometry of the transmission and, most important, on the *preservation* of the information during the transmission. The latter will directly yield the quantization of inhomogeneous quadratic polynomials on F, as we will see in the sequel.

9.1.2. Information Density and Entropy. At first let us concentrate on the notion of *information.* The information the photograph contains is encoded in *gray scales*. The distribution of these gray scales is modeled by a *positive density function*, i.e. a non-vanishing positive Schwartz function

$$f : F \longrightarrow \mathbb{R}$$

expressing the gray shade at each point $q \in F$. Its value is 1 if the point is a black spot and is close to 0 if the spot is almost white.

A *bit* of information is a point $q \in F$ together with a gray value $f(q)$. Thus f is here called an *information density*.

We define the *total amount* j *of information* as

$$j := \int_F f \cdot \omega \tag{540}$$

where ω is a (non-vanishing) fixed constant volume form on F. It is a *symplectic form*.

Obviously,

$$\frac{f}{j} : F \to \mathbb{R} \tag{541}$$

is a *probability density*. Associated with it is one of the most basic ingredients of information theory, namely Shannon's *entropy*

$$\zeta := -\log \frac{f}{j} \tag{542}$$

(cf. [**30**]). Roughly speaking, the entropy provides a measure of uncertainty. In our case it tells us how reliable the transmitted information will be (cf. [**30**]).

An *information density* might be defined by some $\psi \in \mathcal{S}(F, \mathbb{C})$ split into

$$f = \psi \cdot \overline{\psi}.$$

In this case $\frac{\psi}{\sqrt{j}}$ is called a *probability amplitude density*.

9.1.3. The Heisenberg Algebra. Next let us turn to the *geometry* underlying the *information transmission*. To this end E will be equipped with a scalar product constructed from the symplectic form ω on $F \subset E$. This will exhibit ω to be of the form ω^a (cf. section 1.3.5) and turn E into a *Heisenberg algebra* with $\omega^a \cdot \frac{a}{|a|}$ as commutator. It will be this algebra which allows us to study the preservation of information during a transmission. In the sequel a will be a unit vector.

The scalar product $< \, , \, >$ on E mentioned above will be constructed on

$$E = \mathbb{R} \cdot a \oplus F$$

for some a transversal to F. To do so, a symplectic basis $\{e_1, e_2\}$ on F is chosen, satisfying

$$\omega(e_1, e_2) = 1.$$

This basis yields the scalar product $< \, , \, >$ on E defined by

$$< e_1, e_1 > = < e_2, e_2 > = < a, a >= 1$$

and

$$< e_1, e_2 > = < e_1, a > = < e_2, a > = 0.$$

ω together with the oriented basis $\{e_1, a, e_2\}$ determines a unique volume form μ_E on E given by

$$\mu_E(e_1, a, e_2) = \omega(e_1, e_2) = 1.$$

Thus

$$F = F^a \qquad \text{and} \qquad \omega = \omega^a$$

(cf. (19) in section 1.2.1 and section 1.3.5) and E is oriented and Euclidean.

Hence

$$\mathfrak{h}^a := \mathbb{R} \cdot a \oplus F$$

is a Heisenberg algebra equipped with the symplectic structure ω on F^a. The axis $\mathbb{R} \cdot a$ is our *information channel* along which information shall be transmitted.

Clearly, in more general physical settings, the transmission of information follows much more general curves than straight lines. However, the setting chosen here is general enough to provide us with some of the most elementary observations on information transmission.

Resuming the above so far, the setting consists of a plane, an information density, the total amount of information (which determines a constant volume form on F, a symplectic form) and the information channel. All together naturally yield a Heisenberg algebra $\mathfrak{h}^a$. We will treat information transmission and its preservation in $\mathfrak{h}^a$.

9.2. Preservation of Information

9.2.1. Information and Transmission. As we have said so far, the plane F^a containing the *encoded information* in terms of a photograph, say, shall be embedded in the Heisenberg algebra

$$\mathfrak{h}^a = \mathbb{R} \cdot a \oplus F^a$$

and the distribution of the information is described by the *information density*

$$f : F^a \to \mathbb{R},$$

a smooth positive function. The transmission of the information encoded in F^a shall follow along $\mathbb{R} \cdot a$.

In *mathematical* terms, transmission means that the points in F^a shall be smoothly mapped into the planes $t \cdot a + F^a$ for any $t \in \mathbb{R}$. Without loss of generality, this map can be composed by a smooth map

$$\Phi(t) : F^a \to F^a$$

smoothly depending on t, followed by a *translation*

$$\Phi_t : F^a \longrightarrow t \cdot a + F^a$$

for any $t \in \mathbb{R}$. In contrast to the translation Φ_t, which is certainly information preserving, the map $\Phi(t)$ rearranges the information in F^a. If we wish the information to be preserved, $\Phi(t)$ has to be at least a *diffeomorphism* for all $t \in \mathbb{R}$. For technical reasons we require that $\Phi(t)$ extends to a smooth diffeomorphism to the one-point compactification, being diffeomorphic to S^2. This requirement could be weakened, however, the assumption made suits our purpose.

In order to investigate further requirements for $\Phi(t)$ which guarantee information preservation we study

$$f(t) := f \circ \Phi(t) \tag{543}$$

for any $t \in \mathbb{R}$, mapping F^a to $\mathbb{R}$. The total information $j(t)$ in $t \cdot a + F$ is defined by

$$j(t) := \int_{F^a} f(t) \cdot \omega(t)$$

(cf. (540)). Here $\omega(t)$ is the volume form on F^a caused by the diffeomorphism $\Phi(t)$ on F^a, this is to say that

$$\omega(t) = \Phi(t)^* \omega \qquad \forall\, t \in \mathbb{R}. \tag{544}$$

Expressed in more detail, (544) reads

$$\omega(t)(v,w) = \omega(D\Phi(t)(v), D\Phi(t)(w))$$

or, rewritten in terms of the determinant,

$$\omega(t)(v,w) = \det D\Phi(t) \cdot \omega(v,w) \tag{545}$$

for any $v, w \in F$ and any $t \in \mathbb{R}$. Here D is the (total) differentiation with respect to the variables in F^a. The *initial* condition for f shall be

$$f(0) = f \quad \text{and} \quad \Phi(0) = \mathrm{id}\,. \tag{546}$$

Thus the total information can be reformulated as

$$j(t) = \int_{F^a} f(t) \cdot (\det D\Phi(t)) \cdot \omega \qquad \forall\, t \in \mathbb{R}. \tag{547}$$

Obviously, $j(t)$ is preserved iff

$$\frac{dj(t)}{dt} = 0 \qquad \forall\, t \in \mathbb{R}.$$

9.2.2. The Continuity Equation. Unfortunately, the *mean gray value* expressed by the total information is not very informative. More informative is the stronger requirement that the *volume form* $f(t)\cdot\omega(t)$ is preserved for all t (this form expresses the pointwise information contributing to the total information). This is to say that for any $h \in F^a$,

$$\frac{d}{dt} f(t) \cdot \omega + f(t) \cdot \frac{d}{dt}(\det D\Phi(t))(h) \cdot \omega = 0 \tag{548}$$

has to be satisfied for any $t \in \mathbb{R}$ (here we used the assumption for $\Phi(t)$ made above). This equation is called the *continuity equation* for f (cf. [**11**]). Clearly, (548) requires $j(t)$ to be constant in t.

Since both the information density $f(t)$ and the total information $j(t)$ are positive for any $t \in \mathbb{R}$, the *entropy*

$$\zeta(t) := -\log \frac{f(t)}{j(t)} \tag{549}$$

is well defined. Thus equation (548) becomes

$$\frac{d\zeta(t)}{dt} = \operatorname{tr} D\Phi(t)^{-1} \circ \frac{d}{dt} D\Phi(t)(h) \qquad \forall\, t \in \mathbb{R}$$

for any $h \in F^a$. Here we call the transmission *information preserving* iff f satisfies (548) and the *entropy* ζ in (549) is preserved. Since

$$\frac{df(t)}{dt} = 0 \Leftrightarrow \frac{d\zeta(t)}{dt} = 0 \qquad \forall\, t \in \mathbb{R},$$

the information is preserved iff for any $h \in F^a$,

$$\frac{d}{dt} \det D\Phi(t)(h) = 0 \qquad \forall\, t \in \mathbb{R},$$

which, due to (546), requires

$$\det D\Phi(t)(h) = 1 \qquad \forall\, t \in \mathbb{R}.$$

This equation, however, means that for any $h \in F^a$,

$$D\Phi(t)(h) \in \mathrm{Sp}(F^a) \qquad \forall\, t \in \mathbb{R}.$$

We therefore may state:

THEOREM 99. *The information density f — supposed to satisfy (548) — is preserved iff for any $h \in F^a$,*

$$D\Phi(t)(h) \in \mathrm{Sp}(F^a) \qquad \forall\, t \in \mathbb{R}. \tag{550}$$

Let us point out that $\mathrm{Sp}(F^a)$ is independent of the particular (constant) symplectic form since any such form is a real multiple of ω.

Clearly, the translation Φ_t preserves the symplectic form. Therefore, theorem 99 implies:

COROLLARY 100. *If f satisfies (548), the transmission is information preserving iff for any $h \in F^a$ the linear map*

$$D(\Phi_t \circ \Phi(t))(h) : F^a \longrightarrow t \cdot a + F^a$$

is symplectic *for any $t \in \mathbb{R}$.*

This content of this corollary resembles the situation in geometric optics as described in section 8.8. There the light rays passing F^a and an optical system behind it yield a symplectic map from F^a to an image plane $t \cdot a + F^a$, say. The rearrangement of the information there is caused by the optical system; this rearrangement is *linear* and symplectic; it does not depend on points in F^a. Here general information preserving transmission causes the same geometric setting as in geometric optics but on an *infinitesimal level*, i.e. with *infinitesimal transformations* involved.

9.2.3. The Schrödinger Representation and Information.

The goal of this section is to reformulate entropy preservation in terms of the Schrödinger representation. To this end we need a Heisenberg group. The plane F^a together with $a \in S^2$ (cf. section 1.2.3) defines a Heisenberg group $\mathcal{H}^a = \mathbb{R} \cdot a \oplus F^a$ where F^a carries ω^a as a symplectic form. F^a shall be equipped with an *orthonormal coordinate system* with unit vectors $e_{\bar{q}}$ and $e_{\bar{p}}$.

The Schrödinger representation ρ (with frequency $\nu = 1$) modulates the group $\mathcal{H}^a$ onto each signal $\varphi \in \mathcal{S}(\mathbb{R}, \mathbb{C})$ (cf. section 7.1). This is to say that, in particular, it modulates the points of F^a onto each signal φ. Referring to a point $h = q \cdot e_{\bar{q}} + p \cdot e_{\bar{p}}$ in F^a as the locus of a bit of information (encoded in the probability density (541)), we can say that ρ_ν modulates this locus onto any signal $\varphi \in \mathcal{S}(\mathbb{R}, \mathbb{C})$ by

$$\rho_\nu(q \cdot e_{\bar{q}} + p \cdot e_{\bar{p}})(\varphi)(x) = e^{-\nu \cdot \frac{p \cdot q}{2} \cdot i} \cdot e^{\nu \cdot p \cdot \zeta \cdot i} \cdot \varphi(x - q)$$

for any $\zeta \in \mathbb{R}$.

To study the effect of a volume preserving rearrangement of information on a Schrödinger representation of the Heisenberg algebra $\mathcal{H}^a$, we extend any $A \in \mathrm{Sp}(F^a)$ to all of $\mathcal{H}^a$ by setting it equal to the identity on the center $\mathbb{R} \cdot a$. This extension is called A again. Hence

$$A : \mathcal{H}^a \longrightarrow \mathcal{H}^a$$

is a group automorphism of $\mathcal{H}^a$.

As an example we may consider $A = D\Phi(t)(h) \in \mathrm{Sp}(F^a)$ for $h \in F^a$, because by theorem 99 the infinitesimal transmission of information is a symplectic map on F^a, provided the volume form $f(t) \cdot w(t)$ is preserved for any $t \in \mathbb{R}$, regardless of the specific nature of the density f (cf. section 9.2).

If the information is modulated on $\mathcal{S}(\mathbb{R},\mathbb{C})$ by means of the Schrödinger representation, the preservation of information is expressed in terms of the projective representation U_{Sp} which is based on the Stone-von Neumann theorem (cf. theorem 88 in section 7.1.5) and equation (431) together with (432) in section 7.4.1. The cocycle in (432) does not affect (431). We hence may reformulate (431) in terms of the *metaplectic representation* U_{Mp} as

$$\rho \circ \mathrm{pr}_{\mathrm{Mp}}(A) = U_{\mathrm{Mp}}(A) \circ \rho \circ U_{\mathrm{Mp}}(A^{-1}) \tag{551}$$

for any $A \in \mathrm{Mp}(F^a)$. Here

$$U_{\mathrm{Mp}} : \mathrm{Mp}(F) \longrightarrow U(L^2(\mathbb{R},\mathbb{C}))$$

is the metaplectic representation. Both (431) and (551) express the fact that the preservation of information during the transmission is reparaphrased by maintaining the *equivalence class* of the Schrödinger representation during the transmission.

In short, by corollary 100:

PROPOSITION 101. *Suppose f satisfies (548). Then preserving the entropy of the information means preserving the equivalence class of the Schrödinger representation.*

Let us point out that at this stage only $F^a \subset \mathfrak{h}^a$ is used to define the metaplectic representation involved, i.e. it is only the symplectic Euclidean space F^a (as phase space) and not the Heisenberg algebra $\mathfrak{h}^a$ (which carries a characteristic Minkowski metric (cf. section 5.7.2)). This Lie algebra will enter later when the quantization of inhomogeneous quadratic polynomials will be treated.

9.2.4. Summary on Information Transmission and Preservation. Information shall be encoded in a plane F^a, an $\mathbb{R}$-linear subspace of a three-dimensional $\mathbb{R}$-linear space E. The distribution of information on F^a is supposed to be described by a smooth real-valued positive density function f.

The notion of total information j on F^a requires an $\mathbb{R}$-valued non-degenerate bilinear skew-symmetric map ω, a symplectic form, assumed to be constant. The rescaled information density $\frac{f}{j}$ is a probability density with entropy density $-\log \frac{f}{j}$.

The information on F^a is transmitted along the line $\mathbb{R} \cdot a$ transversal to F^a with $a \in E$. It might be detected on the affine plane $t \cdot a + F^a$ for any $t \in \mathbb{R}$, which is a coadjoint orbit of $\mathcal{H}^a$.

The symplectic form ω yields a natural scalar product on F^a which extends to E in a unique fashion such that a is a unit vector. ω is identical with ω^a. Hence the information transmission along $\mathbb{R} \cdot a$ yields the Heisenberg algebra

$$\mathfrak{h}^a := \mathbb{R} \cdot a + F^a$$

which provides the basic geometry of information encoding and transmission.

The information transmission from F^a to $t \cdot a + F^a$, say, is modeled by a smooth diffeomorphism $\Phi(t)$ of F^a and a t-dependent density function $f(t) \in C^\infty(F^a, \mathbb{R})$.

The requirement that the total amount of information is preserved on any region as small as it might be yields the continuity equation for the information density and the entropy density, as well.

By definition information is preserved during the transmission if the density $f(t) = f$ for all $t \in \mathbb{R}$. In fact, the information density is preserved iff the entropy $-\log \frac{f(t)}{j(t)}$ is constant in t which requires the derivative $D\Phi(t)$ of $\Phi(t)$ on F^a to satisfy $D\Phi(t) \in \mathrm{Sp}(F^a)$ for any t and vice versa. In other words, the information is preserved iff $\Phi(t)$ is a *symplectomorphism* for all $t \in \mathbb{R}$. Therefore, the preservation of information reduces the transmission to a study of *symplectic geometry*. The information density is thus no longer in the scope. This resembles parts of geometric optics.

From now on we assume that information is preserved while being transmitted. This is the starting point for *classical quantum mechanics* which we will develop in the following subsections.

9.3. The Poisson Algebra of all Homogeneous Quadratic Polynomials in Two Variables

We are now ready to prepare the quantization of homogeneous quadratic polynomials of a plane. For quantization of this kind of polynomials is of special interest, as we will point out below. As a general reference for quantization of quadratic polynomials we quote [**50**]. This preparation is subdivided into several steps.

As the first step we will represent next each element of the collection $\mathcal{Q}$ of all *homogeneous quadratic polynomials* (with real coefficients) of the plane F^a as a symplectic automorphism of F^a. Here $a \in S^2$. A polynomial in $\mathcal{Q}$ will be mapped into its Hamiltonian vector field, a trace-free map on F^a. In addition, we will observe that both $\mathcal{Q}$ and $\mathrm{sp}(F^a)$ carry natural Heisenberg algebra structures both isomorphic to $\mathfrak{h}^a$ (cf. sections 5.3.3 and 5.5.1). As a general reference we again refer to [**50**].

9.3.1. Hamiltonian Vector Fields of Homogeneous Quadratic Polynomials. Our first goal is to establish a natural isomorphism between the Poisson algebra $\mathcal{Q}$ of all *homogeneous quadratic polynomials* on F^a and $\mathrm{sp}(F^a)$, the Lie algebra of $\mathrm{Sp}(F^a)$.

The *coordinate system* in F^a determined by $e_{\bar{q}}$ and $e_{\bar{p}}$ (cf. section 9.2.3) allows us to introduce the concept of a *polynomial* on F^a in *two variables* q and p. This collection $\mathcal{Q}$ obviously forms an $\mathbb{R}$-vector space.

Given a polynomial pol on F^a let $f_{\mathrm{pol}} : F^a \longrightarrow \mathbb{R}$ be its polynomial function. In the sequel we will identify the collection

$$P_{\mathcal{Q}} := \{f_{\mathrm{pol}} |\, \mathrm{pol} \in \mathcal{Q}\}$$

with $\mathcal{Q}$ by identifying any pol with f_{pol}, if no confusion arises.

At first we will verify that $\mathcal{Q}$ forms a Poisson sub-algebra of $C^\infty(F^a, \mathbb{C})$.

We define a *Poisson bracket* on $C^\infty(F^a, \mathbb{R})$, the $\mathbb{R}$-algebra of all smooth $\mathbb{R}$-valued functions of F^a as follows: Given $f \in C^\infty(F^a, \mathbb{C})$, its *Hamiltonian vector field* X_f is defined by

$$\omega^a(X_f, X) = df(X) \tag{552}$$

for any smooth vector field X on F^a (cf. [**60**]).

A more informative version of the above equation (552) is obtained if the vector field X_f is replaced by their respective principal part $\mathbf{a}_f$. (We will do so throughout this chapter.) Thus (552) turns into

$$\omega^a(\mathbf{a}_f(h), k) = df(h)(k) \qquad \forall\, h, k \in F^a. \tag{553}$$

Here $\mathbf{a}_f(h)$ denotes the principal part of the Hamiltonian vector fields X_f evaluated at $h \in F^a$; moreover, $df(h)(k)$ is the derivative of f at h evaluated at $k \in F^a$.

Equation (553) can be rewritten in terms of the scalar product and in turn yields a link to the gradient of $f \in C^\infty(F^a, \mathbb{R})$. Indeed,

$$\omega^a(\mathbf{a}_f(h), k) =< \mathbf{a}_f(h) \times a, k >= df(h)(k) \qquad \forall\, h, k \in F^a$$

which is reformulated as

$$< \mathbf{a}_f(h) \times a, k >= df(h)(k) =< \text{grad } f(h), k > \qquad \forall\, h, k \in F^a.$$

Since both $\mathbf{a}_f(h)$ and grad $f(h)$ are in F^a, the gradient of f at h is rewritten as

$$\text{grad } f(h) = \mathbf{a}_f(h) \times a = -a \times \mathbf{a}_f(h) \qquad \forall\, h \in F^a.$$

Therefore, the principal part $\mathbf{a}_f$ of the Hamiltonian vector field is

$$\mathbf{a}_f(h) = a \times \text{grad } f(h) \qquad \forall\, h \in F^a$$

or, what is the same,

$$\mathbf{a}_f = a \cdot \text{grad } f \tag{554}$$

where $\cdot$ denotes the multiplication in $\mathbb{H}$. Since

$$\text{grad } f(h) = \frac{\partial f(h)}{\partial q} \cdot e_{\bar{q}} + \frac{\partial f(h)}{\partial p} \cdot e_{\bar{p}} \qquad \forall\, h \in F^a,$$

the principal part of $\mathbf{a}_f(h)$ has the form

$$\mathbf{a}_f(h) = \frac{\partial f(h)}{\partial q} \cdot a \cdot e_{\bar{q}} + \frac{\partial f(h)}{\partial p} \cdot a \cdot e_{\bar{p}},$$

and finally,

$$\mathbf{a}_f(h) = \frac{\partial f(h)}{\partial p} \cdot e_{\bar{q}} - \frac{\partial f(h)}{\partial q} \cdot e_{\bar{p}} \qquad \forall\, h \in F^a \tag{555}$$

due to

$$e_{\bar{p}} = e_{\bar{q}} \cdot a.$$

This formula allows us to compute the principal part of the Hamiltonian vector field X_f of f directly out of f.

9.3.2. Examples. Two elementary examples shall illustrate equation (555):

i) Any constant vector field X on F^a is Hamiltonian. Indeed, let $h_0 \in F^a$ be the value of the principal part of X. The differential of the linear map

$$f_{h_0} : F^a \longrightarrow \mathbb{R}$$

defined by

$$f_{h_0}(h) :=< h, h_0 > \qquad \forall\, h \in F^a$$

is

$$df_{h_0}(h)(k) =< k, h_0 > \qquad \forall\, h, k \in F^a.$$

Thus

$$\text{grad } f_{h_0} = h_0,$$

and therefore,

(556) $$\mathbf{a}_{f_{h_0}} = a \cdot h_0 \qquad \forall\, h_0 \in F^a$$

according to (554). Denoting by f_q and f_p the (linear) *coordinate functions* on F^a determined by $e_{\bar{q}}$ and $e_{\bar{p}}$, respectively, (556) obviously implies

(557) $$\mathbf{a}_{f_q} = a \cdot e_{\bar{q}} = -e_{\bar{p}}$$

and

(558) $$\mathbf{a}_{f_p} = a \cdot e_{\bar{p}} = e_{\bar{q}}.$$

Clearly, we could have obtained (556), (557) and (558) directly out of (555), as well.

ii) Of special interest with respect to our goal of quantizing quadratic polynomials is the linear dependence of $\mathbf{a}_f(h)$ on the position $h \in F^a$ for any $f \in \mathcal{Q}$. By (553) this is to say that for any $f \in \mathcal{Q}$,

(559) $$\mathbf{a}_f = \ell_f$$

for some linear map $\ell_f \in \operatorname{End} F^a$, say. In particular, we find by (555) that

$$\mathbf{a}_f(e_{\bar{q}}) = \ell_f(e_{\bar{q}})e_{\bar{q}} - \ell_f(e_{\bar{q}})e_{\bar{p}}$$

and

$$\mathbf{a}_f(e_{\bar{p}}) = \ell_f(e_{\bar{p}})e_{\bar{q}} - \ell_f(e_{\bar{p}})e_{\bar{p}}$$

and thus the matrix $M(\mathbf{a}_f)$ is

(560) $$M(\mathbf{a}_f) = \begin{pmatrix} \ell_f(e_{\bar{q}}) & \ell_f(e_{\bar{q}}) \\ -\ell_f(e_{\bar{p}}) & -\ell_f(e_{\bar{p}}) \end{pmatrix}.$$

9.3.3. Poisson Bracket. We will show next that $C^\infty(F^a, \mathbb{R})$ is a Poisson algebra, which means that the $\mathbb{R}$-algebra $C^\infty(F^a, \mathbb{R})$ (under pointwisely defined operations) has a Lie algebra structure. To this end let $f, g \in C^\infty(F^a, \mathbb{R})$ with a_f and a_g as the principal parts of their Hamiltonian vector fields. Obviously,

$$\omega^a(\mathbf{a}_f(h), \mathbf{a}_g(h)) = df(h)(\mathbf{a}_g(h))$$

(cf. section 1.3.5). Thus

$$df(\mathbf{a}_g) =< \mathbf{a}_f(h) \times a, \mathbf{a}_g(h) >= - < a, \mathbf{a}_f \times \mathbf{a}_g > .$$

To evaluate the cross product we use equation (555) and obtain

$$\begin{aligned} \mathbf{a}_f \times \mathbf{a}_g &= \left(\frac{\partial f}{\partial p} \cdot e_{\bar{q}} - \frac{\partial f}{\partial q} \cdot e_{\bar{p}}\right) \times \left(\frac{\partial g}{\partial p} \cdot e_{\bar{q}} - \frac{\partial g}{\partial q} \cdot e_{\bar{p}}\right) \\ &= -\frac{\partial f}{\partial p}\frac{\partial g}{\partial q} \cdot e_{\bar{q}} \times e_{\bar{p}} - \frac{\partial f}{\partial q} \cdot \frac{\partial g}{\partial p} \cdot e_{\bar{p}} \times e_{\bar{q}} \end{aligned}$$

and, therefore,

(561) $$\mathbf{a}_f \times \mathbf{a}_g = -\left(\frac{\partial f}{\partial q} \cdot \frac{\partial g}{\partial p} - \frac{\partial f}{\partial p} \cdot \frac{\partial g}{\partial q}\right) \cdot a$$

since

$$e_{\bar{q}} \times e_{\bar{p}} = e_{\bar{q}} \cdot e_{\bar{q}} \cdot a = -a$$

and

$$e_{\bar{p}} \times e_{\bar{q}} = e_{\bar{q}} \cdot a \cdot e_{\bar{q}} = a.$$

Hence

$$df(a_g) = \frac{\partial f}{\partial q} \cdot \frac{\partial g}{\partial p} - \frac{\partial f}{\partial p} \cdot \frac{\partial g}{\partial q}$$

holds true.

Defining the *Poisson bracket* of f and g in $C^\infty(F^a, \mathbb{R})$ by

$$\{f, g\} := \frac{\partial f}{\partial q} \cdot \frac{\partial g}{\partial p} - \frac{\partial f}{\partial p} \cdot \frac{\partial g}{\partial q}, \tag{562}$$

which hence can be expressed as

$$\{f, g\} \cdot a = -\mathbf{a}_f \times \mathbf{a}_g, \tag{563}$$

this yields

$$\{f, g\} \cdot e = a \cdot (\mathbf{a}_f \times \mathbf{a}_g) \qquad \forall\, f, g \in C^\infty(F^a, \mathbb{R}).$$

Clearly,

$$\{f, g\} \cdot a = \omega^a(\mathbf{a}_g, \mathbf{a}_f) \cdot a \tag{564}$$

or

$$\{f, g\} = \omega^a(\mathbf{a}_g, \mathbf{a}_f)$$

holds true for all $f, g \in C^\infty(F^a, \mathbb{R})$. Hence $\{f, g\}$ is bilinear and skew-symmetric in f and g. These brackets satisfy a Jacobi identity, i.e.

$$\{\{f_1, f_2\}, f_3\} + \{\{f_2, f_3\}, f_1\} + \{\{f_3, f_1\} f_2\} = 0.$$

Therefore, $C^\infty(F^a, \mathbb{R})$ is a Lie algebra under the Poisson bracket (562) (cf. [**83**]).

Equation (563) shows that the Poisson bracket on $C^\infty(F^{\mathbf{a}}, \mathbb{R})$ evaluated at points can naturally be related to elements in the center $\mathbb{R} \cdot a$ of the Heisenberg algebra $\mathfrak{h}^a$.

Since $\{f, g\} \in C^\infty(F^a, \mathbb{R})$, we may investigate $\mathbf{a}_{\{f,g\}}$. In fact,

$$< a \cdot d\{f, g\}(h), a >= d\{f, g\}(h)$$

obviously holds true and thus by (563), (554) and the symmetry of the Hessian

$$\begin{aligned} d\{f, g\}(h) &= < d\mathbf{a}_f(h) \cdot \mathbf{a}_g, \mathbf{a} > + < \mathbf{a}_f, d\mathbf{a}_g(h) \cdot a > \\ &= < a \cdot d(\text{grad } f)(h) \cdot \mathbf{a}_g, a > + < a \cdot \mathbf{a}_f \cdot d(\text{grad } g)(h), a > \\ &= - < d(\text{grad } f)(h), \mathbf{a}_g > + < d(\text{grad } g)(h), \mathbf{a}_f > \\ &= - < d(\text{grad } f)(\mathbf{a}_g), h > + < d(\text{grad } g)(\mathbf{a}_f), h > \\ &= - < d\mathbf{a}_f(\mathbf{a}_g), a \cdot h > + < d\mathbf{a}_g(\mathbf{a}_f), a \cdot h > \\ &= < [\mathbf{a}_f, \mathbf{a}_g], a \cdot h > \\ &= -\omega^a([\mathbf{a}_f, \mathbf{a}_g], h). \end{aligned}$$

Here $[\mathbf{a}_f, \mathbf{a}_g]$ denotes the principal part of the *Lie bracket* $[X_f, X_g]$ defined by

$$[\mathbf{a}_f, \mathbf{a}_g] := d\mathbf{a}_g(\mathbf{a}_f) - d\mathbf{a}_f(\mathbf{a}_g).$$

Thus we may state

$$d\{f, g\}(h) = -\omega^a([\mathbf{a}_f, \mathbf{a}_g], h) \qquad \forall\, h \in F^a$$

showing

$$\mathbf{a}_{\{f,g\}} = -[\mathbf{a}_f, \mathbf{a}_g] \qquad \forall\, f, g \in C^\infty(F^a, \mathbb{R}). \tag{565}$$

Denoting by $\operatorname{Ham} F^a$ the collection of all Hamiltonian vector fields on F^a we obtain the linear map

$$\operatorname{ham} : C^\infty(F^a, \mathbb{R}) \to \operatorname{Ham} F^a$$

determined by

$$\operatorname{ham} f := X_f \qquad \forall\, f \in C^\infty(F^a, \mathbb{R})$$

which satisfies (565). The formalism we have used here to introduce the Poisson bracket and its Hamiltonian vector fields is adapted to our setting for Heisenberg algebras. For a general formalism we refer to [**2**] or [**50**].

As a simple example for a Poisson bracket let us consider maps f and g for which df and dg are constant one-forms. This is equivalent to saying that the gradients $\operatorname{grad} f$ and $\operatorname{grad} g$ are both constant. The functions f and g are thus affine maps of the form

$$f(h) = \zeta_f \cdot < h, b_f > + \lambda_f$$

and

$$g(h) = \zeta_g \cdot < h, b_g > + \lambda_g$$

where b_f and b_g are fixed unit vectors in F^a and $\zeta_f, \lambda_f, \zeta_g$ as well as λ_g are constants in $\mathbb{R}$. The respective gradients of f and g are, therefore,

$$\operatorname{grad} f = \zeta_f \cdot b_f$$

and

$$\operatorname{grad} g = \zeta_g \cdot b_g$$

implying by (563)

$$\{f, g\} \cdot a = -\zeta_f \cdot \zeta_g \cdot b_f \times b_g = \zeta_f \cdot \zeta_g \cdot a.$$

The Poisson bracket for these affine functions f and g is hence

$$\{f, g\} = \zeta_f \cdot \zeta_g.$$

In particular, for the coordinate functions f_q and f_p on F^a we find

$$\{f_q, f_p\} = 1.$$

9.3.4. Generators of $\mathcal{Q}$. The $\mathbb{R}$-linear space $\mathcal{Q}$ (cf. section 9.3.1) is a Poisson subalgebra of $C^\infty(F^a, \mathbb{C})$ as we will see below. To this end we point out that $\mathcal{Q}$ is generated by $\frac{q^2}{2}, \frac{p^2}{2}$ and $q \cdot p$.

By (555) the Hamiltonian vector fields of $f_{\frac{q^2}{2}}, f_{\frac{p^2}{2}}$ and $f_{q\cdot p}$ are the following for all $h \in F^a$ of the form $h = q \cdot e_{\bar{q}} + p \cdot e_{\bar{p}}$:

$$\begin{aligned} \mathbf{a}_{f_{\frac{q^2}{2}}}(h) &= -q \cdot e_{\bar{p}} \\ \mathbf{a}_{f_{\frac{p^2}{2}}}(h) &= p \cdot e_{\bar{q}} \\ \mathbf{a}_{f_{q\cdot p}}(h) &= q \cdot e_{\bar{q}} - p \cdot e_{\bar{p}}. \end{aligned} \tag{566}$$

9.3.5. Poisson Bracket in $\mathcal{Q}$. The Poisson brackets $\{f_{\frac{q^2}{2}}, f_{\frac{p^2}{2}}\}, \{f_{q\cdot p}, f_{\frac{q^2}{2}}\}$ and $\{f_{q\cdot p}, f_{\frac{p^2}{2}}\}$ are easily calculated as follows: Due to (563),

$$\{f_{\frac{q^2}{2}}, f_{\frac{p^2}{2}}\} \cdot a = -q \cdot p \cdot e_{\bar{q}} \times e_{\bar{p}} = q \cdot p \cdot a,$$

which means

(567) $$\{f_{\frac{q^2}{2}}, f_{\frac{p^2}{2}}\} = f_{q\cdot p}.$$

Moreover,

$$\{f_{q\cdot p}, f_{\frac{p^2}{2}}\} \cdot a = -(q \cdot e_{\bar{q}} - p \cdot e_{\bar{p}}) \times (q \cdot e_{\bar{p}}) = q^2 \cdot a$$

which is rewritten as

(568) $$\{f_{q\cdot p}, f_{\frac{q^2}{2}}\} = f_{q^2}.$$

Accordingly, we find

(569) $$\{f_{q\cdot p}, f_{\frac{p^2}{2}}\} = f_{p^2}.$$

Hence we have shown:

PROPOSITION 102. *$\mathcal{Q}$ closes under the Poisson brackets.*

All of $\mathbf{a}_{f_{q^2}}, \mathbf{a}_{f_{p^2}}$ and $\mathbf{a}_{f_{q\cdot p}}$ depend linearly on $h \in F^a$ and hence can be *identified* with linear endomorphisms ℓ_{q^2}, ℓ_{p^2} and $\ell_{q\cdot p}$, say, of F^a (cf. 559). Their respective matrices $M(\mathbf{a}_{f_{q^2}}), M(\mathbf{a}_{f_{p^2}})$ and $M(\mathbf{a}_{f_{q\cdot p}})$ formed with respect to $e_{\bar{q}}$ and $e_{\bar{p}}$ are

(570) $$\begin{aligned} M(\mathbf{a}_{f_{\frac{q^2}{2}}}) &= \begin{pmatrix} 0 & 0 \\ -1 & 0 \end{pmatrix}, \\ M(\mathbf{a}_{f_{\frac{p^2}{2}}}) &= \begin{pmatrix} 0 & 1 \\ 0 & 0 \end{pmatrix}, \\ M(\mathbf{a}_{f_{q\cdot p}}) &= \begin{pmatrix} 1 & 0 \\ 0 & -1 \end{pmatrix}. \end{aligned}$$

These matrices are all trace free which means they belong to the Lie algebra $\mathrm{sl}(F^a)$ of $\mathrm{SL}(F^a)$ (cf. subsection 5.1.3), the special linear group of F^a. In fact, they generate $\mathrm{sl}(F^a)$. Due to (280) we, moreover, have

(571) $$\mathrm{sp}(F^a) = \mathrm{sl}(F^a).$$

According to (555) and (560) for any $f \in \mathcal{Q}$ the matrix $M(\ell_f)$ of ℓ_f is

(572) $$M(\ell_f) = \begin{pmatrix} \frac{\partial f}{\partial p}(e_{\bar{q}}) & \frac{\partial f}{\partial p}(e_{\bar{p}}) \\ -\frac{\partial f}{\partial q}(e_{\bar{q}}) & -\frac{\partial f}{\partial q}(e_{\bar{p}}) \end{pmatrix}.$$

Since the endomorphisms $\ell_{\frac{q^2}{2}}, \ell_{\frac{p^2}{2}}$ and ℓ_{pq} of F^a generate $\mathrm{sp}(F^a)$ we have a surjective linear map

(573) $$\begin{aligned} \mathrm{ham} : \mathcal{Q} &\longrightarrow \mathrm{sp}(F^a) \\ \mathrm{ham}(f) &:= \ell_f \quad \forall\, f \in \mathcal{Q}. \end{aligned}$$

ham is a *linear isomorphism* since both the domain and the range are three-dimensional. Moreover, the following holds true:

PROPOSITION 103. ham *is a Lie algebra isomorphism satisfying*

$$\text{ham}\{f,g\} = [\ell_f, \ell_g]_{\text{sl}(F^a)} \qquad \forall\, f,g \in \mathcal{Q}. \tag{574}$$

Here

$$[\ell_f, \ell_g]_{\text{sl}(F^a)} := \ell_f \circ \ell_g - \ell_g \circ \ell_f \qquad \forall\, f,g \in Q$$

is the communicator in $\text{sl}(F^a)$.

PROOF. The validity of (574) is verified by a straightforward calculation: To evaluate the commutator $[\ell_f, \ell_g]$ for any couple $f,g \in \mathcal{Q}$ we first observe that

$$\mathbf{a}_g(h) = d\mathbf{a}_g(h) = \ell_g(h) \qquad \forall\, h \in F^a$$

since $g \in \mathcal{Q}$. Thus

$$d\mathbf{a}_g(d\mathbf{a}_f(h)) = \ell_g(\ell_f(h)) \qquad \forall\, h \in F^a.$$

Therefore, the Lie bracket $[\mathbf{a}_f, \mathbf{a}_g]$ of $\mathbf{a}_f$ and $\mathbf{a}_g$ evaluated at h is

$$[\mathbf{a}_f, \mathbf{a}_g](h) = (\ell_g \circ \ell_f - \ell_f \circ \ell_g)(h) \qquad \forall\, h \in F^a$$

saying that

$$-[\mathbf{a}_f, \mathbf{a}_g] = [\ell_f, \ell_g]_{\text{sl}(F^a)} \qquad \forall\, f,g \in \mathcal{Q} \tag{575}$$

or, differently formulated,

$$\mathbf{a}_{\{f,g\}} = -[\mathbf{a}_f, \mathbf{a}_g] = [\ell_f, \ell_g]_{\text{sl}(F^{\mathbf{a}})} \qquad \forall\, f,g \in \mathcal{Q}. \tag{576}$$

Hence for any $f,g \in \mathcal{Q}$,

$$\text{ham}\{f,g\} = \ell_{\{f,g\}} = [\ell_f, \ell_g]_{\text{sl}(F^a)} \tag{577}$$

showing that ham is a Lie algebra isomorphism. □

To evaluate the commutator $[\ ,\ ]_{\text{sp}(F^a)}$ in more detail it is enough to do so on the generators. Let us use the matrix representation of the maps involved

$$\begin{aligned} M\left([\ell_{\frac{q^2}{2}}, \ell_{\frac{p^2}{2}}]_{\text{sl}(F^a)}\right) &= \begin{pmatrix} 0 & 0 \\ -1 & 0 \end{pmatrix}\begin{pmatrix} 0 & 1 \\ 0 & 0 \end{pmatrix} - \begin{pmatrix} 0 & 1 \\ 0 & 0 \end{pmatrix}\begin{pmatrix} 0 & 0 \\ -1 & 0 \end{pmatrix} \\ &= \begin{pmatrix} 0 & 0 \\ 0 & -1 \end{pmatrix} - \begin{pmatrix} 1 & 0 \\ 0 & 0 \end{pmatrix} \\ &= \begin{pmatrix} 1 & 0 \\ 0 & -1 \end{pmatrix} \end{aligned}$$

showing

$$[\ell_{\frac{q^2}{2}}, \ell_{\frac{p^2}{2}}]_{\text{sl}(F^a)} = \ell_{q \cdot p}. \tag{578}$$

Since

$$\begin{aligned} M([\ell_{\frac{q^2}{2}}, \ell_{q \cdot p}]_{\text{sl}(F^a)}) &= \begin{pmatrix} 0 & 0 \\ -1 & 0 \end{pmatrix}\begin{pmatrix} 1 & 0 \\ 0 & -1 \end{pmatrix} - \begin{pmatrix} 1 & 0 \\ 0 & -1 \end{pmatrix}\begin{pmatrix} 0 & 0 \\ -1 & 0 \end{pmatrix} \\ &= \begin{pmatrix} 0 & 0 \\ -1 & 0 \end{pmatrix} - \begin{pmatrix} 0 & 0 \\ 1 & 0 \end{pmatrix} \\ &= \begin{pmatrix} 0 & 0 \\ -2 & 0 \end{pmatrix} \end{aligned}$$

we find

$$[\ell_{\frac{q^2}{2}}, \ell_{q \cdot p}]_{\text{sl}(F^a)} = \ell_{q^2} \tag{579}$$

and accordingly

$$[\ell_{\frac{p^2}{2}}, \ell_{q\cdot p}]_{\mathrm{sl}(F^a)} = \ell_{p^2}. \tag{580}$$

9.3.6. The Linear Spaces $\mathrm{sp}(F^a)$ and $\mathcal{Q}$ as Heisenberg algebras. At first we will show that $\mathcal{Q}$ carries a natural Heisenberg algebra structure. Let $\mathcal{F} \subset \mathcal{Q}$ be spanned by f_{q^2} and f_{p^2}. To generalize (567) to all $f, g \in \mathcal{F}$ we set

$$\begin{aligned} f &= \alpha_f \cdot f_{\frac{q^2}{2}} + \beta_f \cdot f_{\frac{p^2}{2}}, \\ g &= \alpha_g \cdot f_{\frac{q^2}{2}} + \beta_g \cdot f_{\frac{p^2}{2}} \end{aligned}$$

with real coefficients $\alpha_f, \alpha_g, \beta_f, \beta_g$. The Poisson bracket $\{f, g\}$ is hence of the form

$$\{f, g\} = (\alpha_f \beta_g - \alpha_g \beta_f) \cdot f_{q\cdot p} \tag{581}$$

due to (567). Equation (581) suggests to form the matrix $M_{\{f,g\}}$; it is

$$M_{\{f,g\}} := \begin{pmatrix} \alpha_f & \alpha_g \\ \beta_f & \beta_g \end{pmatrix} \tag{582}$$

by which equation (581) turns into

$$\{f, g\} = \det M_{\{f,g\}} \cdot f_{q\cdot p}. \tag{583}$$

$f_{q\cdot p}$ is transversal to span $\{f_{\frac{q^2}{2}}, f_{\frac{p^2}{2}}\}$. Turning back to (563) we notice that

$$\{f, g\} \cdot a = -\mathbf{a}_f \times \mathbf{a}_g = \det M_{\{f,g\}} \cdot f_{q\cdot p} \cdot a \tag{584}$$

for any $f, g \in$ span $\{f_{\frac{q^2}{2}}, f_{\frac{p^2}{2}}\}$. Due to (558) and (564)

$$\omega^a(\mathbf{a}_g, \mathbf{a}_f) \cdot a = \det M_{\{f,g\}} \cdot f_{q\cdot p} \cdot a \tag{585}$$

holds true. Hence $\det M_{\{,\}}$ is a symplectic structure on the linear space $\mathcal{F} \subset \mathcal{Q}$. The three-dimensional linear space $\mathcal{Q}$ is turned into a Heisenberg algebra in the obvious way; it is called $H\mathcal{Q}$. Therefore, we state:

PROPOSITION 104. *$H\mathcal{Q} = \mathcal{F} + \mathbb{R} \cdot f_{q\cdot p}$ is a Heisenberg algebra with brackets*

$$[f, g]_{H\mathcal{Q}} := \det M_{\{f,g\}} \cdot f_{q\cdot p} = \{f, g\} \quad \forall f, g \in \mathcal{F}$$

together with

$$[f, f_{q\cdot p}]_{H\mathcal{Q}} := 0 \quad \forall f, g \in \mathcal{Q}.$$

Moreover, by (563) and (564)

$$\omega^a(\mathbf{a}_f, \mathbf{a}_g) = \{f, g\} \cdot f_{q\cdot p} \quad \forall f, g \in \mathcal{F} \tag{586}$$

where ω^a is the symplectic structure on F^a.

Now we proceed to show that $H\mathcal{Q}$ is isomorphic to $\mathrm{sp}(F^a)$ endowed with a natural Heisenberg algebra structure. Here natural means that this sort of Heisenberg algebra structure emanates from the Lie product of $\mathrm{sp}(F^a)$ as the one presented in section 5.5.1. This structure is not unique at all, it depends on the splitting of the three-dimensional $\mathbb{R}$-linear space $\mathrm{sp}(F^a)$. As in $\mathrm{sp}(F^a)$ we first state a natural scalar product

$$< \ell_f, \ell_g > := \frac{1}{2} \operatorname{tr} \ell_f \circ \tilde{\ell}_g \qquad \forall f, g \in \mathcal{Q} \tag{587}$$

on $\mathrm{sp}(F^a)$. Here $\tilde{\ell}_g$ is the adjoint of ℓ_g. As one easily verifies the following equations of norms are satisfied:

$$|\ell_{q\cdot p}| = 1 \quad \text{and} \quad |\ell_{\frac{q^2}{2}}| = |\ell_{\frac{p^2}{2}}| = \frac{1}{2} \tag{588}$$

as well as

$$< \ell_{q\cdot p}, \ell_{\frac{q^2}{2}} >=< \ell_{q\cdot p}, \ell_{\frac{p^2}{2}} >= 0.$$

Setting

$$\mathcal{F}^{\ell_{q\cdot p}} := \mathrm{span}\,\{\ell_{\frac{q^2}{2}}, \ell_{\frac{p^2}{2}}\},$$

the linear space $\mathrm{sp}(F^a)$ splits orthogonally into

$$\mathrm{sp}(F^a) = \mathbb{R}\cdot \ell_{q\cdot p} \oplus \mathcal{F}^{\ell_{q\cdot p}}. \tag{589}$$

The constant two-form

$$\omega^{\ell_{q\cdot p}}(\ell_f, \ell_g) := -< [\ell_f, \ell_g]_{\mathrm{sp}(F^a)}, \ell_{q\cdot p} > \qquad \forall\, \ell_f, \ell_g \in \mathcal{F}^{\ell_{q\cdot p}} \tag{590}$$

is obviously a symplectic structure on $\mathcal{F}^{\ell_{q\cdot p}}$ for which

$$\omega^{\ell_{q\cdot p}}(\ell_{q^2}, \ell_{p^2}) = 1 \tag{591}$$

due to equation (567).

Setting

$$[\ell_f, \ell_g]_{H\,\mathrm{sp}(F^a)} :=< [\ell_f, \ell_g]_{\mathrm{sp}(F^a)}, \ell_{q\cdot p} > \qquad \forall\, f, g \in \mathcal{Q}, \tag{592}$$

$[\ell_f, \ell_g]_{H\,\mathrm{sp}(F^a)}$ is a Lie bracket on the linear space $\mathrm{sp}(F^a)$ and yields a natural Heisenberg algebra structure on it. This Heisenberg algebra shall be called $H\,\mathrm{sp}(F^a)$ in the sequel.

This structure on $\mathrm{sp}(F^a)$ can be pulled back to $\mathcal{Q}$ by ham (cf. (573)): For any $f, g \in \mathcal{Q}$ we set

$$< f, g >:= \frac{1}{2}\,\mathrm{tr}\,\ell_f \circ \tilde{\ell}_g$$

and obtain

$$\mathcal{F} = \mathrm{span}\,\{f_{q^2}, f_{p^2}\} = \mathrm{ham}^{-1}\left(\mathcal{F}^{\ell_{q\cdot p}}\right)$$

as well as the orthogonal splitting

$$\mathcal{Q} = \mathbb{R}\cdot f_{q\cdot p} \oplus \mathcal{F}^{f_{p\cdot q}}$$

together with the symplectic structure on $\mathcal{F}^{f_{p\cdot q}}$,

$$\omega^{f_{qp}}(f, g) := \omega^{\ell_{q\cdot p}}(\mathrm{ham}\, f, \mathrm{ham}\, g) \qquad \forall\, f, g \in \mathcal{F}^{f_{q\cdot p}}.$$

Due to equations (578), (579), (580) which describe the Lie product on $\mathrm{sp}(F^a)$ as well as equations (567), (568) and (569) which characterize the Poisson brackets on $\mathcal{Q}$, we finally see from proposition 104:

PROPOSITION 105. ham : $H\mathcal{Q} \longrightarrow H\,\mathrm{sp}(F^a)$ *is a Heisenberg algebra isomorphism since*

$$[f, g]_{H\mathcal{Q}} :=< [\ell_f, \ell_g]_{\mathrm{Hsp}(F^a)}, \ell_{q\cdot p} > \qquad \forall\, f, g \in \mathcal{Q}$$

holds true. Moreover,

$$< \{\ell_f, \ell_g\}, f_{q\cdot p} >= \omega^{f_{q\cdot p}}(f, g) = \{f, g\} \qquad \forall\, f, g \in \mathcal{F}^{f_{q\cdot p}}$$

is valid.

The role of $f_{q,p}$ or $\ell_{q\cdot p}$ in HQ and in $H\,\mathrm{sp}(F^a)$, respectively, is analogous to that of a in $\mathfrak{h}^a$. This suggests introducing

$$\mathcal{R} : H\,\mathrm{sp}(F^a) \longrightarrow \mathfrak{h}^a$$

given on the generators $\ell_{q\cdot p}, \ell_{\frac{q^2}{2}}$ and $\ell_{\frac{p^2}{2}}$ by

$$\begin{aligned} \mathcal{R}(\ell_{q\cdot p}) &:= a, \\ \mathcal{R}\left(\frac{q^2}{2}\right) &:= e_{\bar{q}}, \\ \mathcal{R}\left(\frac{p^2}{2}\right) &:= e_{\bar{p}}. \end{aligned}$$

Since

$$\omega^{\ell_{qp}}(\ell_{\frac{q^2}{2}}, \ell_{\frac{p^2}{2}}) = 1 = \omega^a(e_q, e_p)$$

we conclude by the definition of $\mathfrak{h}^a$ in section 5.2 the following:

PROPOSITION 106. *The linear map $\mathcal{R}$ is also a Heisenberg algebra homomorphism and so*

$$\mathcal{R} \circ \mathrm{ham} : \mathcal{Q} \longrightarrow \mathfrak{h}^a$$

is an isomorphism of Heisenberg algebras. Hence all Heisenberg algebras $H\mathcal{Q}, H\,\mathrm{sp}(F^a)$ and $\mathfrak{h}^a$ are isomorphic.

9.4. The Quantization of Inhomogeneous Quadratic Polynomials

Poisson algebras play a fundamental role in Hamiltonian mechanics (cf. [**2**] and [**60**]). In the setting of *quantum mechanics*, Poisson algebras constitute classical observables and are represented as algebras of *Hermitian operators* on suitable Hilbert spaces. These algebras of operators form the algebras of *quantum observables.* The aim in this section is to carry out this representation for the collection of all *inhomogeneous quadratic polynomials* on F^a. For simplicity we again identify a polynomial with its function. The vector field of a polynomial function is again identified with its principal part. For simplicity we set $\hbar = 1$.

This formalism does not extend beyond the collection of inhomogeneous quadratic polynomials (cf. [**50**]).

For a treatment from the physical point of view we refer to [**19**], [**20**] [**29**], [**64**] and [**87**] to name but a few.

9.4.1. Quantization of Homogeneous Quadratic Polynomials.

A *quantization* in this chapter means an *irreducible representation* on a Hilbert space in terms of *Hermitian operators* of the Poisson algebra $\mathcal{Q}_{\mathrm{inh}}$ of all *inhomogeneous quadratic polynomials* in two variables (cf. [**50**]). At first we restrict to $\mathcal{Q} \subset \mathcal{Q}_{\mathrm{inh}}$, the Poisson algebra of all homogeneous quadratic polynomials on F^a. The *quantization map* Q is defined to be the composition of

$$\mathcal{Q} \overset{\mathrm{ham}}{\longrightarrow} \mathrm{sp}(F^a) \overset{-i\cdot dU_{\mathrm{Mp}}}{\longrightarrow} \mathrm{End}(L^2(\mathbb{R}, \mathbb{C})), \tag{593}$$

i.e.

$$Q := -i \cdot dU_{\mathrm{Mp}} \circ \mathrm{ham}\,. \tag{594}$$

The value of the representation $-i \cdot dU_{\mathrm{Mp}} \circ \mathrm{ham}$ on any polynomial $f \in \mathcal{Q}$ is its *quantization*, a *Hermitian operator* on the Hilbert space $L^2(\mathbb{R}, \mathbb{C})$. The map dU_{Mp}

is the differential of the (unitary) *metaplectic representation* U_{Mp} (cf. section 7.5.4) restricted to a neighborhood of id $\in \mathrm{Mp}(F^a)$, *a leaf* over id $\in \mathrm{Sp}(F^a)$. Both maps in the composition $dU_{\mathrm{Mp}} \circ \mathrm{ham}$ are injective Lie algebra homomorphisms (cf. proposition 103). Hence Q is injective, as well.

The values of dU_{Mp} are *skew-symmetric operators* acting on the Hilbert space $L^2(\mathbb{R},\mathbb{C})$, hence $\mp i \cdot dU_{\mathrm{Mp}}$ maps into the collection of Hermitian operators. The choice of the factor $-i$ in Q (cf. (594)) was made to guarantee

$$Q(\{f,g\}) = [Q(f), Q(g)]. \tag{595}$$

To evaluate the quantization map Q we use the methods collected in sections 9.3.2 and 9.3.5 to determine ham. The values of the representation dU_{Mp} of the Lie algebra $\mathrm{mp}(F^a)$ of the metaplectic group $\mathrm{Mp}(F^a)$ are obtained by means of pieces of a one-parameter groups near id $\in \mathrm{Sp}(F^a)$, lifted to a leaf of id $\in \mathrm{Mp}(F^a)$.

More precisely, the procedure to compute Q is as follows: Q is determined on the generators $\frac{q^2}{2}$, $\frac{q^2}{2}$ and $q \cdot p$ of $\mathcal{Q}$ (cf. section 9.3.4) by finding their Hamiltonian vector fields $a_{\frac{q^2}{2}}$, $a_{\frac{p^2}{2}}$, respectively, $a_{q\cdot p}$ and identifying them as linear maps in $\mathrm{sl}(F^a)$ (cf. (573)); the values of Q on these generators are the Hermitian operators $-i \cdot dU_{\mathrm{Mp}}(\ell_{\frac{q^2}{2}})$, $-i \cdot dU_{\mathrm{Mp}}(\ell_{\frac{p^2}{2}})$ and $-i \cdot dU_{\mathrm{Mp}}(\ell_{q\cdot p})$ on the Hilbert space $L^2(\mathbb{R},\mathbb{C})$ (cf. section 7.5.4). Let us determine them.

At first we compute $dU_{\mathrm{Mp}}(\ell_{\frac{1}{2}q^2})$. To this end we consider the linear map $a_{\frac{q^2}{2}}$ in the Lie algebra

$$\mathrm{sl}(F^a) = \mathrm{sp}(F^a) = \mathrm{mp}(F^a)$$

as the generator of a one-parameter subgroup $\sigma_{\frac{q^2}{2}}$ in $\mathrm{Sp}(F^a)$. The skew-adjoint operator $dU_{\mathrm{Mp}}(a_{\frac{q^2}{2}})$ multiplied with $-i$ is then the *quantization* of $\frac{q^2}{2}$, a Hermitian operator on the Hilbert space $L^2(\mathbb{R},\mathbb{C})$ as mentioned above. Using (570) in section 9.3.5 the operator associated with $a_{\frac{q^2}{2}}$ is

$$-i \cdot dU_{\mathrm{Mp}}(\ell_{\frac{q^2}{2}}) = -i \cdot dU_{\mathrm{Mp}}(\dot\sigma_{\frac{q^2}{2}}(0)).$$

In terms of matrices, $\ell_{\frac{q^2}{2}}$ with matrix

$$M(\ell_{\frac{q^2}{2}}) = \begin{pmatrix} 0 & 0 \\ -1 & 0 \end{pmatrix}$$

(cf. (570)) generates a piece of the one-parameter group with values

$$M(\sigma_{\frac{q^2}{2}}(t)) = \begin{pmatrix} 1 & 0 \\ -t & 1 \end{pmatrix}$$

near id $\in \mathrm{Sp}(F^a)$ defined for all t near zero in $\mathbb{R}$. Clearly, $\sigma_{\frac{q^2}{2}}(t)$ coincides with $A_2(t)$ in (433) in section 7.4.2. The one-parameter group of multiplication operators in $\mathrm{Mp}(F)$ passing through the identity operator and projecting to $\sigma_{\frac{q^2}{2}}$ has values

$$V_{-t}(f)(x) = e^{i\cdot\frac{1}{2}tx^2} \cdot f(x)$$

for each t, for all $f \in \mathcal{S}(\mathbb{R},\mathbb{C})$ and for all $x \in \mathbb{R}$ (cf. section 7.5.5). Hence the operator $-i \cdot dU_{\mathrm{Mp}}(\ell_{\frac{q^2}{2}})$ acts on $\mathcal{S}(\mathbb{R},\mathbb{C})$ by

$$-i \cdot dU_{\mathrm{Mp}}(\ell_{\frac{q^2}{2}})(f)(x) = \frac{x^2}{2} \cdot f(x) \qquad \forall\, f \in \mathcal{S}(\mathbb{R},\mathbb{C}). \tag{596}$$

This operator is the *quantization* of the polynomial $\frac{q^2}{2}$, where the variable of f in $\mathbb{R}$ is denoted by x.

Turning to the quantization of $\frac{1}{2}p^2$, the traceless map $\ell_{\frac{1}{2}p^2}$ quantized (i.e. the representation $-i \cdot dU_{\mathrm{Mp}}$ applied to $\ell_{\frac{1}{2}p^2}$), acts on $\mathcal{S}(\mathbb{R}, \mathbb{C})$ by

$$-i \cdot dU_{\mathrm{Mp}}(\ell_{\frac{p^2}{2}})(f) = -\frac{1}{2} \cdot \frac{\partial^2 f}{\partial x^2} \qquad \forall\, f \in \mathcal{S}(\mathbb{R}, \mathbb{C}). \tag{597}$$

To establish this we first consider the one-parameter group $\sigma_{\frac{p^2}{2}}$ in $\mathrm{Sp}(F^a)$ generated by $\ell_{\frac{p^2}{2}}$ and compute its quadratic Fourier transform $U_{\mathrm{Mp}}(\sigma_{\frac{q^2}{2}})$, a *unitary operator* (cf. (451)). In matrix form $\frac{p^2}{2}$ is

$$M(\sigma_{\frac{p^2}{2}}(t)) = \begin{pmatrix} 1 & t \\ 0 & 1 \end{pmatrix} \qquad \forall\, t \in \mathbb{R}.$$

By (434) we have

$$\begin{aligned} M(\sigma_{\frac{p^2}{2}}(t)) &= J^{-1} \circ \begin{pmatrix} 1 & 0 \\ -t & 1 \end{pmatrix} \circ J \\ &= \begin{pmatrix} 0 & -1 \\ 1 & 0 \end{pmatrix} \circ \begin{pmatrix} 1 & 0 \\ -t & 1 \end{pmatrix} \circ \begin{pmatrix} 0 & 1 \\ -1 & 0 \end{pmatrix} \\ &= J_r^{-1} \circ M(\sigma_{\frac{q^2}{2}}) \circ J_r, \end{aligned}$$

and hence by section (7.5.2) the quadratic Fourier transform (with $m = 1$) is

$$U_{\mathrm{Mp}}((\sigma_{\frac{p^2}{2}}(t))(f)(x) \;=\; \mathbf{F} \circ U_{\mathrm{Mp}} \circ \sigma_{\frac{q^2}{2}}(t) \circ \mathbf{F}^{-1}(f)(x)$$

yielding (597).

Finally, let us determine the quantization of $f_{pq} \in \mathcal{Q}$. The one-parameter group it generates in $\mathrm{Sp}(F^a)$ is described in matrix form by

$$M(\sigma_{p \cdot q}(t)) = \begin{pmatrix} e^t & 0 \\ 0 & e^{-t} \end{pmatrix} \qquad \forall\, t \in \mathbb{R}$$

computed directly from (570). At first we observe that

$$U_{\mathrm{Mp}}(\sigma_{p \cdot q}(t))(f)(x) = e^{-\frac{t}{2}} \cdot f(e^t \cdot x) \qquad \forall\, t \in \mathbb{R}, \forall\, x \in \mathbb{R} \text{ and } \forall\, f \subset \mathcal{S}(\mathbb{R}, \mathbb{C}).$$

Thus the quantization of f_{pq} is the operator computed by using equation (474) in example (iii) in section 7.5.5 to obtain

$$-i \cdot dU_{\mathrm{Mp}}(\ell_{pq}) = i \left(x \cdot \frac{\partial}{\partial x} + \frac{1}{2} \right).$$

The *quantization procedure* as developed by now is thus given on the generators of the Poisson algebra $\mathcal{Q}$ by the following table of correspondences

function	Hermitian operator
q^2	m_{x^2}
pq	$i(x\frac{\partial}{\partial x} + \frac{1}{2})$
p^2	$-\frac{\partial^2}{\partial x^2}$

where m_{x^2} is the operator determined by multiplication with x^2, or more generally expressed as

$$m_\lambda(f)(x) = \lambda \cdot f(x) \qquad \forall x \in \mathbb{R} \text{ and } \forall f \subset \mathcal{S}(\mathbb{R}, \mathbb{C})$$

for any $\lambda \in \mathbb{C}$.

9.4.2. General Quantization Procedure. With the above correspondence we have the following general quantization procedure for a homogeneous quadratic polynomial on the plane F^a (with $a \in S^2$) in the Euclidean space E: Given any $A \in \mathrm{sp}(F^a)$ with matrix

$$M(A) = \begin{pmatrix} \lambda & \alpha \\ \beta & -\lambda \end{pmatrix}, \tag{598}$$

say, the map

$$\begin{aligned} p_A : F^a &\longrightarrow \mathbb{R}, \\ p_A(v) &:= \omega(A(v), v) \qquad \forall v \in F^a \end{aligned}$$

is a homogeneous quadratic polynomial (again we identified a polynomial pol with its function f_{pol} (cf. section 9.3.1)). Indeed, if $v = q \cdot e_q + p \cdot e_{\bar{p}}$, then

$$A(v) = (\lambda q + \alpha \cdot p)e_{\bar{q}} + (\beta \cdot q - \lambda \cdot p)e_{\bar{p}}.$$

To compute $\omega^a(A(v), v)$ (cf. section 1.3.5) we need $A(v) \cdot a$ which is

$$\begin{aligned} A(v) \cdot a &= -(\beta \cdot q - \lambda \cdot p) \cdot e_{\bar{q}} + (\lambda \cdot q + \alpha \cdot p) \cdot e_{\bar{p}} \\ &= (-\beta \cdot q + \lambda \cdot p) \cdot e_{\bar{q}} + (\lambda \cdot q + \alpha \cdot p) \cdot e_{\bar{p}}. \end{aligned}$$

Setting $p_A(q,p) := p_A(v)$ yields

$$p_A(q,p) = \alpha \cdot p^2 + 2\lambda \cdot p \cdot q - \beta \cdot q^2. \tag{599}$$

As is easily seen, the traceless linear map $\ell_{\frac{p_A}{2}}$ corresponding to the Hamiltonian vector field of $\frac{p_A}{2}$ has $M(A)$ as its matrix. This is to say we have the reverse map of ham in (573) formulated as:

PROPOSITION 107. *For any $A \in \mathrm{sp}(F^a)$*

$$M\left(\mathrm{ham}\left(\frac{p_A}{2}\right)\right) = M(\ell_{\frac{p_A}{2}}) = M(A)$$

holds true.

For this one-to-one correspondence between elements of $\mathrm{sp}(F^a)$ and homogeneous quadratic polynomials we also refer to (459).

Therefore, the quantization $-i \cdot dU_{\mathrm{Mp}}(A)$ of any $A \in \mathrm{sp}(F^a)$ with matrix (598) is

$$\begin{aligned} &-i \cdot dU_{\mathrm{Mp}}(A)(\varphi)(x) \\ &= -\alpha \cdot \frac{\partial^2 \varphi}{\partial x^2}(x) + 2i \cdot \lambda x \cdot \frac{\partial \varphi}{\partial x}(x) + i \cdot \lambda \cdot \varphi(x) - \beta \cdot x^2 \varphi(x) \end{aligned} \tag{600}$$

for any $\varphi \in \mathcal{S}(\mathbb{R}, \mathbb{C})$ and any $x \in \mathbb{R}$. Here we identified $\mathrm{sp}(F^a)$ with $\mathrm{mp}(F^a)$, the Lie algebra of the metaplectic group $\mathrm{Mp}(F^a)$.

9.4.3. Quantization of Inhomogeneous Quadratic Polynomials. To obtain a quantization scheme for all *inhomogeneous quadratic polynomials* we again follow [**50**]. In doing so we form the *semidirect product* $\mathrm{Mp}(F^a) \boxed{s} \mathcal{H}^a$ of the metaplectic group $\mathrm{Mp}(F^a)$ and the Heisenberg group $\mathcal{H}^a$ (cf. section 5.4.2). In more detail, we observe that $\mathrm{Mp}(F^a)$ operates on $\mathcal{H}^a$ via $\mathrm{Sp}(F^a)$ (leaving the center $U^a(1)$ fixed). The group operation on this semidirect product is thus by definition

$$(A_1, z_1 + h_1) \cdot (A_2, z_2 + h_2) := (A_1 \circ A_2, (z_1 + h_1) \cdot (A_1(z_2 + h_2)))$$

or, more explicitly,

$$(A_1, z_1 + h_1) \cdot (A_2, z_2 + h_2) = (A_1 \circ A_2, z_1 \cdot z_2 \cdot e^{\frac{1}{2}\omega^a(h_1, A_1(h_2))\cdot a} + h_1 + A_1(h_2))$$

for all $A_1, A_2 \in \mathrm{Mp}(F^a)$, any $z_1, z_2 \in U^a(1)$ and any $h_1, h_2 \in F^a$.

The Lie algebra of $\mathrm{Mp}(F^a) \boxed{s} \mathcal{H}^a$ is identical to the Lie algebra of the direct product and hence is

$$\mathrm{mp}(F^a) \times \mathcal{h}^a = \mathrm{sp}(F^a) \times \mathcal{h}^a.$$

Here again we identified the Lie algebras of $\mathrm{Mp}(F^a)$ and $\mathrm{Sp}(F^a)$. Therefore, the Lie bracket in $\mathrm{sp}(F^a) \times \mathcal{h}^a$ is

$$[(A_1, t_1 \cdot a + h_1), (A_2, t_2 \cdot a + h_2)] = (A_1 \circ A_2 - A_2 \circ A_1, e^{\omega^a(h_1, h_2)\cdot a} \cdot a)$$

or, formulated by means of the commutator of A_1 and A_1,

$$[(A_1, t_1 \cdot a + h_1), (A_2, t_2 \cdot a + h_2)] = ([A_1, A_2], e^{\omega^a(h_1, h_2)\cdot a} \cdot a)$$

where A_1, A_2 vary in the Lie algebra $\mathrm{mp}(F^a) = \mathrm{sl}(F^a)$.

Again let $\mathcal{Q}_{\mathrm{inh}}$ be the Poisson algebra of all *inhomogeneous quadratic polynomials*. Clearly, $\mathcal{Q} \subset \mathcal{Q}_{\mathrm{inh}}$ is a Poisson subalgebra of $\mathcal{Q}_{\mathrm{inh}}$. To extend ham to all of $\mathcal{Q}_{\mathrm{inh}}$ we need to define this extension on f_1, f_p and f_q,. Here f_1 is the constant function with value 1 while f_q and f_p denote the functions of the polynomials q and p, respectively, the coordinate functions. This is to say we have to compute the Hamiltonian vector fields of f_1,f_p and f_q. We thus consider

$$\omega^a(a_{f_p}(h), k) = df_q(h)(k) = 1 \qquad \forall\, h, k \in F^a$$

(cf. (554)). Similarly,

$$\omega^a(a_{f_q}(h), k) = df_p(h)(k) = -1 \qquad \forall\, h, k \in F^a$$

and hence obtain for the principal part of the vector fields

$$a_{f_p}(h) = e_{\overline{q}} \qquad \forall\, h \in F^a$$

as well as

$$a_{f_q}(h) = -e_{\overline{p}} \qquad \forall\, h \in F^a.$$

(cf. (558) and (557)). From

$$\omega^a(a_{f_1}(h), h) = 0 \qquad \forall\, h \in F^a$$

and

(601) $$\omega^a(k_1, a_{f_1}(h)) = < k_1 \times a, a_{f_1}(h) > = - < k_1, a_{f_1}(h) \times a >$$

one finally obtains

$$a_{f_1}(h) = -a$$

which solves equation (601) above. So if ham is extended by

$$\begin{aligned} \mathrm{ham}(1) &:= -a, \\ \mathrm{ham}(f_p) &:= e_q, \\ \mathrm{ham}(f_q) &:= -e_p, \end{aligned}$$

one verifies that

$$\mathrm{ham} : \mathcal{Q}_{\mathrm{inh}} \longrightarrow \mathrm{mp}(F^a) \times \mathfrak{h}^a$$

is a Lie algebra isomorphism.

In constructing the quantization map the next step is to take representations of the constituents of the Lie algebra of the direct product $\mathrm{mp}(F^a)$ and $\mathfrak{h}^a$. These are the infinitesimal metaplectic representation dU_{Mp} and the infinitesimal Schrödinger representation $d\rho$, respectively. Thus

$$(602) \qquad -i \cdot dU \otimes -i \cdot d\rho : \mathrm{mp}(F^a) \times \mathfrak{h}^a \longrightarrow \mathrm{End} \bigotimes^2 L^2(\mathbb{R}, \mathbb{C}),$$

composed with $\mathrm{ham} \otimes \mathrm{id}_{\mathcal{H}^a}$ yields the *quantization map*

$$Q : \mathcal{Q}_{\mathrm{inh}} \longrightarrow \mathrm{End} \bigotimes^2 L^2(\mathbb{R}, \mathbb{C}),$$

i.e.

$$(603) \qquad Q := (-i \cdot dU_{\mathrm{Mp}} \otimes d\rho) \circ (\mathrm{ham} \otimes \mathrm{ham}),$$

a Lie algebra homomorphism.

Since as in (426) we can use the maps m_i and m_p denoting the multiplication by i and p, respectively, this yields

$$\begin{aligned} d\rho(a) &= m_{-i}, \\ d\rho(e_{\bar{p}}) &= m_p, \\ d\rho(e_{\bar{q}}) &= -\frac{\partial}{\partial x}, \end{aligned}$$

and we have the following quantization scheme:

(604)

polynomial	operator
1	m_1
q	m_x
p	$-i\frac{\partial}{\partial x}$
q^2	m_{x^2}
$p \cdot q$	$i(x\frac{\partial}{\partial x} + \frac{1}{2})$
p^2	$-\frac{\partial^2}{\partial x^2}$.

Equations (566) and (603) show that $-i \cdot dU_{\mathrm{Mp}}$ is a constituent of the quantization map. As we know, the metaplectic representation U_{Mp} describes the effect of the optical systems on the light distribution (cf. (535) in section 8.8.3). The representation ρ determines the ambiguity function (cf. (517)) and the Wigner function (cf. (505) in section 8.8.1). Up to the factor $-i$ in this sense $-i \cdot dU_{\mathrm{Mp}}$ refers to *infinitesimal geometric optics*. This sort of quantization extends to more than two variables p and q by using the metaplectic representation of the higher dimensional metaplectic group as described in section 7.5.6 as well as the Schrödinger representation of the higher dimensional Heisenberg group $\mathcal{H}_{2n}$ with F^a replaced by a $2n$-dimensional space F. The Lie algebra of the semidirect product of $\mathrm{Mp}(F)$ and $\mathcal{H}_{2n}$ is isomorphic to the space of inhomogeneous quadratic polynomials of F, given the structure of a phase space $F = F_1 \oplus F_2$ as done in section 7.5.6 (cf. [**50**]). The quantization map Q hence generalizes in the obvious way with quantum observables in $\operatorname{End} L^2(F_1, \mathbb{C})$.

9.4.4. Minkowski Space and Quantization. The purpose of this subsection is to extract a natural Minkowski space from the Lie algebra $Q(\mathcal{Q})$ of all quantized *homogeneous quadratic polynomials*. To this end we choose some symplectic plane F^a (with $a \in S^2$) in the Euclidean space E, use proposition 66 and the Lie algebra isomorphism

$$\mathrm{sp}(F^a) \overset{dU_{\mathrm{Mp}}}{\longrightarrow} \mathrm{End}(L^2(\mathbb{R}, \mathbb{C})).$$

As seen from the general quantization procedure in section 9.4.2, the quantization map $Q = -i \cdot dU_{\mathrm{Mp}} \circ \mathrm{ham}$ in (594) is a Lie algebra monomorphism (cf. (595)) and hence dU_{Mp} is one, too. We extend this monomorphism to a linear injective map (still called dU_{Mp})

$$dU_{\mathrm{Mp}} : \mathbb{R} \cdot \mathrm{id}_F \oplus \mathrm{sp}(F) \longrightarrow \mathrm{End}(L^2(\mathbb{R}, \mathbb{C}))$$

to all of $\operatorname{End} F = \mathbb{R} \cdot \mathrm{id}_F \oplus \mathrm{sp}(F)$ by setting

$$dU_{\mathrm{Mp}}(\mathrm{id}_F) = \mathrm{id}_{\,\mathrm{End}(L^2(\mathbb{R},\mathbb{C})}.$$

Pushing forward the skew field structure of $\operatorname{End} F$ mentioned in proposition 66 yields a structure of a skew field on the image $\operatorname{im} dU_{\mathrm{Mp}}$ of dU_{Mp}. From proposition 106 we know that $\mathcal{Q}$ has a natural Heisenberg algebra structure, as well. The plane $\mathcal{F} \subset \mathcal{Q}$ in section 9.3.6 can be chosen such that $\mathrm{ham}(\mathcal{F})$ coincides with the plane Σ in $\mathrm{sp}(F)$, as introduced in subsection 5.5.1. The skew field structure on $\operatorname{im} dU_{\mathrm{Mp}}$ is transferred to $\mathbb{R} \cdot \mathrm{id}_{\,\mathrm{End}\, L^2(\mathbb{R},\mathbb{C})} \oplus Q(\mathcal{Q})$ by converting each skew-adjoint operator into a self-adjoint one by multiplication with $-i$. Hence

$$\mathbb{R} \cdot \mathrm{id}_{\,\mathrm{End}(L^2(\mathbb{R},\mathbb{C})} \oplus Q(\mathcal{Q})$$

(generated by the quantized polynomials m_{x^2},$-\frac{\partial^2}{\partial x^2}$ and $i(x\frac{\partial}{\partial x} + \frac{1}{2})$ together with $\mathrm{id}_{\,\mathrm{End}(L^2(\mathbb{R},\mathbb{C}))}$) has a skew field structure isomorphic to the quaternions. The resulting scalar product on it shall be called the natural scalar product. Thus the $\mathbb{R}$-linear space $\mathbb{R} \cdot \mathrm{id}_{\,\mathrm{End}(L^2(\mathbb{R},\mathbb{C}))} \oplus\ Q(\mathcal{Q})$ carries a natural Minkowski metric as seen from subsection 5.7.2. Proposition 66 shows that this Minkowski metric is determined by the commutator in $Q(\mathcal{Q})$. Let us formulate these results in the following theorem:

THEOREM 108. *The $\mathbb{R}$-linear space $\mathbb{R} \cdot id_{\mathrm{End}(L^2(\mathbb{R},\mathbb{C}))} \oplus\ Q(\mathcal{Q})$ equipped with the natural scalar product for which the identity $id_{\mathrm{End}(L^2(\mathbb{R},\mathbb{C}))}$ is a unit vector carries the unique structure of a skew field determined by the commutator of this space. Hence it is isomorphic to the quaternions and admits a natural Minkowski metric.*

9.5. The Schrödinger Equation

In deriving the Schrödinger equation we closely follow [**50**]. We depart from the Heisenberg group $\mathcal{H}^a = \mathbb{R} \cdot a \oplus F^a$ and the metaplectic group $\mathrm{Mp}(F^a)$. This later group is represented on $\mathcal{S}(\mathbb{R}, \mathbb{C})$ (of which the L_2-completion is $L^2(\mathbb{R}, \mathbb{C})$) to which the metaplectic representation extends) by the *unitary* representation U_{Mp} derived from U_{Sp} as done in section 7.5. The elements in $L^2(\mathbb{R}, \mathbb{C})$ are called states. This is to say that

$$U_{\mathrm{Mp}}(A)(\varphi) \in \mathcal{S}(\mathbb{R}, \mathbb{C}) \qquad \forall\, \varphi \in \mathcal{S}(\mathbb{R}, \mathbb{C}) \text{ and } \forall\, A \in \mathrm{Mp}(F^a).$$

Let $A(t) \in \mathrm{Mp}(F^a)$ be a one-parameter group for which $U_{\mathrm{Mp}}(t) := U_{\mathrm{Mp}}(A(t))$ is a one-parameter group in the unitary group $U(L^2(\mathbb{R}, \mathbb{C}))$ of the Hilbert space $L^2(\mathbb{R}, \mathbb{C})$, continuous in the strong operator topology (cf. [**86**]), then the derivative $\frac{d}{dt} U_{\mathrm{Mp}}(A(t))(\varphi)|_{t=0}$ exists and $U_{\mathrm{Mp}}(t)$ (where t runs through $\mathbb{R}$) is by Stone's theorem a one-parameter group

$$U_{\mathrm{Mp}}(t) = e^{t \cdot \dot{U}_{\mathrm{Mp}}(0)} \qquad \forall\, t \in \mathbb{R}$$

generated by the *skew Hermitian* operator $\dot{U}_{\mathrm{Mp}}(0)$ being of the form

$$\dot{U}_{\mathrm{Mp}}(0) = i \cdot H \tag{605}$$

for some *Hermitian operator* H on the Hilbert space on $L^2(\mathbb{R}, \mathbb{C})$ (containing $\mathcal{S}(\mathbb{R}, \mathbb{C})$ as a dense subspace) (cf. [**86**]). Since dU_{Mp} in (594) is injective, there is a unique $\ell \in \mathrm{sp}(F^a) = \mathrm{mp}(F^a)$ such that

$$dU_{\mathrm{Mp}}(\ell) = i \cdot H.$$

The quantum observable $-\hbar \cdot H$ corresponds to the energy, thus minus sign in $-\hbar \cdot H$ is directly related to the quantization of p^2 (cf. (604)), which is proportional to the energy (in mechanics) of a particle freely moving on the q-axis. Hence the evolution equation

$$\dot{U}_{\mathrm{Mp}}(t) = i \cdot H \circ U_{\mathrm{Mp}}(t) \tag{606}$$

holds true for any $t \in \mathbb{R}$. This equation is valid on all of $L^2(\mathbb{R}, \mathbb{C})$.

Let us connect this equation more closely with the quantization procedure presented in section 9.4. To this end F^a has to be endowed with a coordinate system given by the orthonormal unit vectors $e_{\bar{q}}$ and $e_{\bar{p}}$, respectively. This provides us with the $\mathbb{R}$-linear space $\mathcal{Q}$ of all *quadratic homogeneous polynomials*, which, equipped with the Poisson bracket, is a *Poisson algebra* (cf. section 9.3). As we learned $\mathrm{ham}\, f_{\mathrm{pol}} \in \mathrm{sp}(F^a) = \mathrm{mp}(F^a)$ for any $f_{\mathrm{pol}} \in \mathcal{Q}$. Given some $f_{\mathrm{pol}} \in \mathcal{Q}$ (depending on the physical problem) the Hermitian operator H in

$$H := -i \cdot dU_{\mathrm{Mp}}(\mathrm{ham}\, f_{\mathrm{pol}})$$

is called the *Hamiltonian*. The associated evolution equation of the one-parameter group on the Hilbert space with parameters in $\mathbb{R}$, therefore, is

$$\dot{U}_{\mathrm{Mp}}(t) = i \cdot H \cdot e^{i \cdot H} \qquad \forall\, t \in \mathbb{R}. \tag{607}$$

The time-dependence of the one-parameter group turns any state $\varphi \in \mathcal{S}(\mathbb{R}, \mathbb{C})$ into a time-dependent one, namely

$$\varphi(t, x) := U_{\mathrm{Mp}}(-t)(\varphi)(x) \qquad \forall\, t \in \mathbb{R}$$

with the unitary operator $U_{\mathrm{Mp}}(t)$ as the evolution operator. This is called a *wave function* or a *state* depending on t. The probability density associated with it is by definition

$$|\varphi(t,x)|^2,$$

provided $\int |\varphi(t,x)|^2 dx = 1$.

Equation (607) applied to this state yields

$$-i \cdot \frac{\partial \varphi}{\partial t}(t,x) + H(\varphi)(t,x) = 0 \qquad \forall t \in \mathbb{R} \text{ and } \forall x \in \mathbb{R}, \tag{608}$$

the *Schrödinger equation* for this sort of wave functions. It describes the evolution of the state $\varphi(0,x)$, say.

As an example let us choose the quadratic polynomial $\mathrm{pol}(q,p) := \frac{p^2}{2} + q^2$ for which

$$Q\left(\frac{p^2}{2} + q^2\right) = -i \cdot dU_{\mathrm{Mp}}(\mathrm{ham}(f_{\frac{p^2}{2}+q^2})) = -\frac{1}{2}\frac{\partial^2}{\partial x^2} + m_{x^2}$$

holds true according to the quantization scheme (cf. (604)). Here x^2 acts as a potential.

$$H := -i \cdot dU_{\mathrm{Mp}}(\mathrm{ham}\, f_{\frac{p^2}{2}+q^2}) = -\frac{1}{2}\frac{\partial^2}{\partial x^2} + m_{x^2}$$

is the Hamilton operator of the *harmonic oscillator*.

The wave function thus is

$$\psi(t,x) \quad = \quad e^{-i \cdot t(\frac{1}{2}\frac{\partial^2}{\partial x^2} - x^2)}(\varphi)(x) \qquad \forall t \in \mathbb{R} \text{ and } \forall x \in \mathbb{R}.$$

9.6. State Spaces and Observables, Elements of Stochastic Interpretation

So far we focussed on the mathematical view point of quantum mechanics. In this section we sketch in rough terms part of the stochastic interpretation needed for a physical interpretation. In doing so we will follow [**29**].

9.6.1. States and the Result of a Measurement. As we observed in section 9.5 every quadratic polynomial on the plane F can be represented on a separable, complex Hilbert space $\mathcal{H}$ equipped with a Hermitian product $< \mid >$ which is antilinear in the first variable and linear in the second (here we use the notation $< \mid >$ common in physics to adapt ket and bra, introduced below). As mentioned above, the elements of $\mathcal{H}$ are called *states*. The aim of this section is to sketch how the stochastic formalism emanates from self-adjoint operators on the Hilbert space. For a detailed treatment of this field we refer to [**29**] and [**64**].

The abstract setting of a physical system S is a Hilbert space $\mathcal{H}$, the collection of states of the system S. At a fixed time t_0 a physical system is specified by a state $\psi(t_0) \in \mathcal{H}$ (cf. [**29**] and [**64**]), in physics literature is often denoted by $\mid \psi(t_0) >$, the *ket* notation of the state $\psi(t_0)$; its dual $< \psi(t_0) \mid$ is called *bra*. The evolution $\psi(t)$ of the state is governed by the Schrödinger equation. To see how physical qualities enter into the stochastic description, for example, into the evolution of a state, is our next task.

A *measurable* physical quantity is described by a *self-adjoint operator* A. The only possible *result of a measurement* of a quantity is one of the *eigenvalues* of A (cf. [**29**]).

A self-adjoint operator A on $\mathcal{H}$ gives rise to a probability distribution: In case the operator A has a finite discrete spectrum, the *probability* $\mathcal{P}(\lambda_n)$ of measuring an eigenvalue λ_n in the *normalized state* $\psi \in \mathcal{H}$ is

$$\mathcal{P}(\lambda_n) := \sum_{j=1}^{m} | < e_n^j | \psi > |^2$$

where m is the *degree of degeneracy* of λ_n and $\{\varphi_n^1, \dots, \varphi_n^m\}$ is an orthonormal basis of the corresponding *eigenspace* $\mathcal{H}_n$. Obviously,

$$\mathcal{P}(\lambda_n) \leq 1.$$

Clearly, $\psi = \sum_n \sum_m z_m^n \cdot \varphi_n^m$ and therefore $\mathcal{P}(\lambda_n)$ can be rephrased by

(609) $$\mathcal{P}(\lambda_n) \ = < \psi | P_n(\psi) > = \ \operatorname{tr} P_n$$

where $P_n : \mathcal{H} \to \mathcal{H}_n$ is the orthogonal projection.

Obviously, for any overall phase factor, i.e. for any $z \in U(1)$,

$$\mathcal{P}(\lambda_n) = < z \cdot \psi | P_n(z \cdot \psi) > = < \psi | P_n(\psi) >$$

holds true. Thus ψ and $z \cdot \psi$ are called *equivalent*.

A is called an *observable* iff the collection of its eigenvectors constitutes an *orthonormal basis* of the state space.

Of a special nature are the *one-dimensional subspaces* of the Hilbert space $\mathcal{H}$ spanned by ψ, say, usually a normed state. The state ψ here is called a *pure state*. If

$$P_\psi : \mathcal{H} \longrightarrow \mathbb{C} \cdot \psi$$

is the orthonormal projection and ψ (being normalized) is in the domain of an observable A, then the map

$$P_\psi \circ A \circ P_\psi : \mathcal{H} \to \mathbb{C} \cdot \psi$$

is Hermitian with eigenvector ψ and eigenvalue λ, say, and

$$< \psi, A \circ P_\psi(\psi) > = \operatorname{tr} A \circ P_\psi = \lambda.$$

Clearly, $\mathcal{P}(\lambda) = 1$.

In case the spectrum of A is not discrete, the measurement of α is a continuum. Let the spectrum be non-degenerate. Given a system of eigenvectors ψ_λ with eigenvalue λ forming a continuous basis, a state $\psi \in \mathcal{H}$ can be decomposed into

(610) $$\psi = \int c(\lambda) \cdot \psi_\lambda d\lambda$$

(cf. [**29**]). The probability density $dP(\lambda)$ is thus

$$dP(\lambda) := \rho(\lambda) \cdot d\lambda$$

with

(611) $$\rho(\lambda) := |c(\lambda)|^2 = < \psi_\lambda | \psi >^2 .$$

Finally, if the measurement of the physical quantity in the state ψ gives λ, the state ψ' immediately after the measurement is

$$\psi' = \frac{P(\psi)}{\sqrt{< \psi, P(\psi) >}}. \tag{612}$$

Here P is the orthogonal projection into the eigenspace of λ. Obviously, this formula is invariant under the multiplication by $z \in \mathbb{C} = \mathbb{C}\backslash\{0\}$ (cf. [**29**]).

The time evolution $\psi(t)$ of a state is governed by the Schrödinger equation

$$i \cdot \hbar \frac{\partial}{\partial t}\psi(t) = H(t)(\psi(t)) \tag{613}$$

where $H(t)$ is the observable associated with the total energy of the system (cf. (608)). This is equivalent to the statement that

$$\psi(t_2) = U(t_1, t_2)\psi(t_1)$$

for an unitary operator on $U(t_1, t_2)$ on $\mathcal{H}$.

9.6.2. Solutions to the Schrödinger Equation. The solution of the Schrödinger equation for a conservative system (which means that the energy does not depend on time) is rather simple. Let the Hamiltonian H have a discrete spectrum. Its eigenvalues shall here be called E_1, E_2, etc. where E stands for energy. Since it is an observable, H has an eigenbasis $\varphi_1, \varphi_2, \ldots$ possibly consisting of infinitely many states

$$i\hbar\frac{d}{dt} < \varphi_n|\psi(t) >=< \varphi_n|H(\psi(t)) > .$$

Therefore,

$$i\hbar\frac{d}{dt} < \varphi_n|\psi(t) > = < H(\varphi_n)|\psi(t) > = \ E_n < \varphi_n|\psi(t) >$$

where E_n is the eigenvalue of H on φ_n. Expanding

$$\varphi(t) = \sum_n \gamma_n(t) \cdot \varphi_n$$

the above differential equation reads

$$i\hbar\frac{d}{dt}\gamma_n(t) = E_n\gamma_n(t)$$

and thus

$$\gamma_n(t) = \gamma_n(0)e^{-\frac{i}{\hbar}(E_n(t-t_0))} \qquad \forall t \in \mathbb{R}.$$

Therefore, $\varphi(t)$ becomes

$$\varphi(t) = \sum_n \gamma_n(0) \cdot e^{-\frac{i}{\hbar}(E_n(t-t_0))} \cdot \varphi_n \qquad \forall t \in \mathbb{R}$$

which is the wanted solution of the Schrödinger equation.

9.6.3. Expectation Value. The *expectation value* $< A >_\psi$ of an observable A in the normalized state ψ is defined as

$$< A >_\psi := < \psi | A(\psi) > .$$

If P denotes the orthogonal projection of $\mathcal{H}$ onto $\mathbb{R} \cdot \psi$, then obviously

$$< A >_\psi = \operatorname{tr} A \circ P \tag{614}$$

holds true. If ψ is not normalized,

$$< A >_\psi = \frac{1}{< \psi | \psi >} \cdot < \psi | A(\psi) > .$$

The *variance* $(< A- < A >>)^2$ thus is

$$(< A- < A >>)^2 = < \psi | A^2(\psi) > - < \psi | A(\psi) >^2$$

since

$$\begin{aligned} < (A- < A >)^2 > &= < A^2 - 2 < A > A + < A >^2 > \\ &= < A^2 > -2 < A >^2 + < A >^2 \\ &= < A^2 > - < A >^2 . \end{aligned}$$

The evolution of the expectation mean value for any t is

$$\frac{d}{dt} < A > = \frac{1}{i\hbar} < [A, H(t)] > + < \frac{\partial A}{\partial t} > \tag{615}$$

(cf. [**29**]). The variance gives rise to the uncertainty relations (cf. [**29**] and [**32**]).

9.6.4. Density matrix. Let us assume that the state ψ of a system with $|\psi| \neq 0$ is perfectly known. Such a state ψ is called a *pure state* (cf. section 9.6.1). For an observable A we have

$$< A >_\psi = \operatorname{tr} A \circ P_\psi \tag{616}$$

where P_ψ is the projection onto $\mathbb{C} \cdot \frac{\psi}{|\psi|}$. The evolution of ψ is given by the Schrödinger equation. Assuming that $\mathcal{H}$ has a basis $\{\varphi_n | n = 1, \ldots\}$ the state ψ is represented by

$$\psi = \sum \gamma_n \cdot \varphi_n \tag{617}$$

and hence

$$\operatorname{tr} A \circ P_\psi = \sum_{m,n} \overline{\gamma}_m \cdot \gamma_n < \varphi_m | A(\varphi_n) > . \tag{618}$$

Assuming that ψ is normalized, the complex number $\overline{\gamma}_m \cdot \gamma_n$ can be thought of as being the matrix elements of an operator ϱ, called *density operator*, i.e.

$$< \varphi_m | \varrho(t)(\varphi_n) > = \overline{\gamma}_n \cdot \gamma_n .$$

This matrix $\overline{\gamma_m} \cdot \gamma_n$ is called the *density matrix*

$$|\psi(t)|^2 = \sum \overline{\gamma}_n \gamma_n = \operatorname{tr} \varrho = 1.$$

Moreover, by (614) we obviously have

$$< A >_{\psi(t)} = \operatorname{tr} \varrho(t) \circ A. \tag{619}$$

This formula allows us to compute A in terms of ϱ.

As is easily shown (cf. [**29**]), the evolution of ϱ is governed by the equation

$$\frac{d}{dt}\varrho(t) = [H(t), \varrho(t)] \qquad \forall t. \tag{620}$$

Finally, the probability $\mathcal{P}(\lambda)$ of measuring an eigenvalue λ is

$$\mathcal{P}(\lambda) := < \psi(t)|P_n(\psi(t)) >,$$

and due to (609),

$$\mathcal{P}(\lambda) = \operatorname{tr} P_n \circ \varrho(t). \tag{621}$$

9.6.5. Statistical Mixtures. Next let us assume that the system S is composed of the systems $S_1, \ldots, S_k$ and let ψ_s be a state of the system S_s occurring with probability p_s for $s = 1, \ldots, k$. Then $\sum_s p_s = 1$. This is called a *statistical mixture*. Given an eigenvalue λ of an observable A, the probability of measuring λ in the state $\sum \psi_s$, called *total state*, is

$$\mathcal{P}(\lambda) = \sum_k p_s \mathcal{P}_s(\lambda) \tag{622}$$

where

$$\mathcal{P}_s(\lambda) = < \psi_s|P_s(\psi_s) >$$

is the probability of measuring the eigenvalue λ in the state ψ_s. Hence

$$\mathcal{P}_s(\lambda) = \operatorname{tr} \varrho \circ P_s$$

with the density operator

$$\varrho := \sum_{s=1}^{k} p_s \varrho_s$$

where ϱ_s is the density operator of the state ψ_s.

9.6.6. Superposition. In contrast to the statistical mixture let us study the *superposition* of states. For simplicity we only take two states ψ_1 and ψ_2, say. We consider the normalized state

$$\psi := \lambda_1 \cdot \psi_1 + \lambda_2 \cdot \psi_2$$

with

$$|\lambda_1|^2 + |\lambda_2|^2 = 1.$$

If ψ_1 and ψ_2 are two eigenstates of an observable A corresponding to two eigenvalues β_1 and β_2, say, the probability of finding β_1 is $|\lambda_1|^2$ and that of finding β_2 is $|\lambda_2|^2$. As shown in [**29**] ψ is not a statistical mixture: The probability of measuring an eigenvalue λ of the observable A in the state φ is

$$\mathcal{P}(\lambda) = | < \varphi|\psi > |^2$$

as mentioned earlier. $< \varphi|\psi >$ is called the *probability amplitude*. Hence

$$\mathcal{P}(\lambda) = (\zeta_1 < \varphi|\psi_1 > + \zeta_2 < \varphi|\psi_2 >)^2$$

yielding

$$\begin{aligned} \mathcal{P}(\lambda) &= |\zeta_1|^2 \cdot | < \varphi|\psi_1 > |^2 + |\zeta_2|^2 \cdot < \varphi|\psi_2 >^2 \\ &\quad + \zeta_1 \cdot \bar{\zeta}_2 < \varphi|\psi_1 > \overline{< \varphi|\psi_2 >} + \zeta_2\bar{\zeta}_1 < \varphi|\psi_2 > \cdot \overline{< \varphi|\psi_1 >}. \end{aligned}$$

Since

$$< \varphi|\psi_s > = \mathcal{P}_s(\lambda) \qquad \text{for } s = 1, 2,$$

we find

$$\mathcal{P}(\lambda) = |\zeta_1|^2 \cdot \mathcal{P}_1(\lambda) + |\zeta_2|^2 \cdot \mathcal{P}_2(\lambda) + 2Re\zeta_1 \cdot \bar{\zeta}_2 < \varphi|\psi_1 > \cdot \overline{< \varphi|\psi_2 >}.$$

The mixed term is responsible for *interferences*. This term is missing if the superposition is treated by mixed statistics introduced in section 9.6.5 as (622) shows.

As an example let ψ_1, ψ_2 be two normalized states superposed to

$$\psi = \frac{1}{\sqrt{2}}(\psi_1 + \psi_2).$$

Obviously,

$$\mathcal{P}(\lambda) = \frac{1}{2} \cdot (\mathcal{P}_1(\lambda) + \mathcal{P}_2(\lambda)) + Re < \varphi|\psi_1 > \cdot \overline{< \varphi|\psi_2 >}. \tag{623}$$

We will come back to this example later.

9.6.7. Tensor Products, Product States and Entanglement. If a physical system S consists of two *components* S_1 and S_2, say, with respective Hilbert spaces $\mathcal{H}_1$ and $\mathcal{H}_2$, the Hilbert space $\mathcal{H}$ of the *total system* S (constituting of both S_1 and S_2) is

$$\mathcal{H} = \mathcal{H}_1 \otimes \mathcal{H}_2$$

where $\otimes$ denotes the tensor product of the Hilbert spaces (cf. [**29**]). If $< \mid >_1$ and $< \mid >_2$ denote the respective Hermitian products of $\mathcal{H}_1$ and $\mathcal{H}_2$, the Hermitian product $< \mid >$ of $\mathcal{H}$ is given by

$$< \varphi_1 \otimes \varphi_2|\psi_1 \otimes \psi_2 > := < \varphi_1|\psi_1 >_1 \cdot < \varphi_2|\psi_2 >_2 \tag{624}$$

for all $\varphi_1, \varphi_2 \in \mathcal{H}_1$ and for all $\psi_1, \psi_2 \in \mathcal{H}_2$. The Hilbert space $\mathcal{H}$ is the completion of the algebraic tensor product of $\mathcal{H}_1$ and $\mathcal{H}_2$ with respect to $< \mid >$. If $\{v_i(1)|i \in I_1\}$ and $\{w_j(2)|j \in I_2\}$, each forms a basis in $\mathcal{H}_1$ and $\mathcal{H}_2$, respectively,

$$\{v_i(1) \otimes w_j(2)|i \in I_1, j \in I_2\} \subset \mathcal{H}_1 \otimes \mathcal{H}_2$$

is a basis of $\mathcal{H}_1 \otimes \mathcal{H}_2$. Here I_1 and I_2 are index sets.

A *product state* $\varphi \otimes \psi$ in $\mathcal{H} = \mathcal{H}_1 \otimes \mathcal{H}_2$ represents a *juxtaposition* of the two systems. In such a state the systems are uncorrelated. This is not the case for a general element in $\mathcal{H} = \mathcal{H}_1 \otimes \mathcal{H}_2$, i.e. if ψ is not a product state, the two sub-systems are *correlated*.

For any Hermitian operator A_1 on $\mathcal{H}_1$, its extension $A_1 \otimes \mathrm{id}_{\mathcal{H}_2}$ to $\mathcal{H}$ is Hermitian again. As is easily seen, the only eigenvectors of $A_1 \otimes \mathrm{id}_{\mathcal{H}_2}$ are of the form $\varphi_1 \otimes \psi_2$ for any eigenvector φ_1 of A_1 and any $\psi_2 \in \mathcal{H}_2$. The spectrum of $A_1 \otimes \mathrm{id}_{\mathcal{H}_2}$ is thus identical to that of A_1 on $\mathcal{H}_1$. If A_1 is an observable so is $A_1 \otimes \mathrm{id}_{\mathcal{H}_2}$. These remarks hold true accordingly for A_2.

Clearly, for two $\mathbb{C}$-valued, $\mathbb{C}$-linear maps ℓ_1 and ℓ_2 of $\mathcal{H}_1$ and $\mathcal{H}_2$, respectively,

$$\ell_1 \otimes \ell_2(\varphi \otimes \psi) = \ell_1(\varphi) \cdot \ell_2(\psi)$$

for any $\varphi \in \mathcal{H}_1$ and any $\psi \in \mathcal{H}_2$.

As an example let us consider the Euclidean space E with a coordinate system with axes $\mathbb{R} \cdot e_s$ for $s = 1, 2, 3$ and coordinates x_1, x_2 and x_3. Given $x \in E$ a state $\varphi \in L^2(E, \mathbb{C})$ depends on all three coordinates. However,

$$L^2(E, \mathbb{C}) = \bigotimes_{s=1}^{3} L^2(\mathbb{R} \cdot e_s, \mathbb{C}) \tag{625}$$

holds true. A state $\varphi \in L^2(E, \mathbb{C})$ is called *decomposable* if

$$\varphi = \varphi_1 \otimes \varphi_2 \otimes \varphi_3 \qquad \text{for } \varphi_s \in L^2(\mathbb{R}e_s, \mathbb{C}) \text{ with } s = 1, 2, 3.$$

Clearly,

$$\varphi(x_1, x_2, x_3) = \varphi_1(x_1) \cdot \varphi_2(x_2) \cdot \varphi_3(x_3) \qquad \forall x_s \in \mathbb{R} \; s = 1, 2, 3.$$

In general, a state φ on E is not decomposable.

Given an observable A on $\mathcal{H}_1 \otimes \mathcal{H}_2$ and an eigenvalue λ, the probability of measuring λ in the normalized state ψ is

$$\mathcal{P}(\lambda) =< \psi | P(\psi) > \tag{626}$$

where P is the projection onto the eigenspace of λ. Let us consider the case $A = A_1 \otimes A_2$ with product state $\varphi = \varphi_1 \otimes \varphi_2$ where both φ_1 and φ_2 are eigenstates of A_1 and A_2 with respective eigenvalue λ_1 and λ_2. Then

$$A(\varphi_1 \otimes \varphi_2) = A_1(\varphi_1) \otimes A_2(\varphi_2) = \lambda_1 \cdot \varphi_1 \otimes \lambda_2 \cdot (\varphi_2). \tag{627}$$

Therefore, if $\psi = \psi_1 \otimes \psi_2$ with normalized $\psi_s \in \mathcal{H}_s$ for $s = 1, 2$, the probability $\mathcal{P}(\lambda)$ of measuring λ is the *product*

$$\mathcal{P}(\lambda) \equiv \mathcal{P}(\lambda_1, \lambda_2) =< \psi_1 | P(\psi_1) >< \psi_2 | P(\psi_2) >$$

of the probabilities of measuring λ_1 and λ_2, respectively, and the expectation value is

$$< A >_{\psi_1 \otimes \psi_2} = (\mathrm{tr}\, A_1 \circ P_1) \cdot \mathrm{tr}(A_2 \circ P_2) \tag{628}$$

(cf. (609)). Here P_1 and P_2 are the orthogonal projection onto the $\mathbb{R} \cdot \psi_1$ and $\mathbb{R} \cdot \psi_2$, respectively.

9.6.8. Entanglement. Again let a physical system S consist of two subsystems S_1 and S_2 with respective state spaces $\mathcal{H}_1$ and $\mathcal{H}_2$. Next we assume that a measurement on S is made by a measurement on S_1 only, disregarding S_2. An observable on $\mathcal{H} = \mathcal{H}_1 \otimes \mathcal{H}_2$ thus is of the form $A_1 \otimes \mathrm{id}_{\mathcal{H}_2}$.

The projection P onto the eigenspace of an eigenvalue λ of $A_1 \otimes \mathrm{id}_{\mathcal{H}_2}$ is $P_1 \otimes \mathrm{id}_{\mathcal{H}_2}$, where P_1 is the projection on to the eigenspace in $\mathcal{H}_1$ of λ_1. Hence the probability $\mathcal{P}(\lambda)$ of finding λ in the state ψ of the composed system is

$$\mathcal{P}(\lambda) =< \psi | P_1 \otimes \mathrm{id}_{\mathcal{H}_2}(\psi) > \tag{629}$$

(cf. (609)). The expectation value of finding the system in the global state ψ is

$$< A_1 \otimes \mathrm{id}_{\mathcal{H}_2} >_\psi = \mathrm{tr}(A_1 \otimes \mathrm{id}_{\mathcal{H}_2}) \circ P_\psi \tag{630}$$

where P_ψ is the projection onto $\mathbb{C} \cdot \psi$. In case

$$\psi = \psi_1 \otimes \psi_2 \qquad \text{with } \psi_s \in \mathcal{H} \text{ for } s = 1, 2,$$

we have the expectation value

$$\begin{aligned} < A_1 \otimes \mathrm{id}_{\mathcal{H}_2} >_{\psi_1 \otimes \psi_2} &= (\operatorname{tr} A \circ P_{\psi_1}) \cdot \operatorname{tr} P_{\psi_2} \\ &= \operatorname{tr} A \circ P_{\psi_1}. \end{aligned} \tag{631}$$

The state ψ immediately after the measurement is

$$\psi' = \frac{P_1 \otimes \mathrm{id}_{\mathcal{H}_2}(\psi)}{\sqrt{< \psi | P_1 \otimes \mathrm{id}_{\mathcal{H}_2}(\psi_2) >}}. \tag{632}$$

Let us consider two simple examples. We assume that the complex Hilbert spaces of the two systems S_1 and S_2 are both *one-dimensional* and are denoted by $\mathcal{H}_1$ and $\mathcal{H}_2$ again. If S_1 and S_2 do not interact, the Hilbert space of the composite system is

$$\mathcal{H} = \mathcal{H}_1 \otimes \mathcal{H}_2.$$

This $\mathbb{C}$-linear space is one-dimensional. Given any

$$\psi := \sum_{i=1}^{m} \varphi_i \otimes \psi_i$$

it can be represented as

$$\psi = z \cdot \varphi_0 \otimes \psi_0 \qquad z \in \mathbb{C}$$

where $\{\varphi_0\}$ and $\{\psi_0\}$ form a basis in $\mathcal{H}_1$ and $\mathcal{H}_2$, respectively. Thus any state in $\mathcal{H}$ is a product state.

For two systems S_1 and S_2, whose respective Hilbert spaces $\mathcal{H}_1$ and $\mathcal{H}_2$ are *two-dimensional*, the complex Hilbert space

$$\mathcal{H} = \mathcal{H}_1 \otimes \mathcal{H}_2$$

of the composite system is *four-dimensional.* This is to say that $\{\varphi_1, \varphi_2\}$ forms a basis in $\mathcal{H}_1$ and so does $\{\psi_1, \psi_2\}$ in $\mathcal{H}_2$. The collection $\varphi_1 \otimes \psi_1, \varphi_1 \otimes \psi_2, \varphi_2 \otimes \psi_1$ and $\varphi_2 \otimes \psi_2$ is a basis of $\mathcal{H}$. Thus if $\psi \in \mathcal{H}$ is a state represented by

$$\sum_{\substack{i=1,2 \\ j=1,2}} z_{i,j} \psi_i \otimes \psi_j \qquad \text{with } z_{i,j} \in \mathbb{C} \text{ for } i,j = 1,2,$$

in general, it is not of the form $\varphi \otimes \psi$ with $\varphi \in \mathcal{H}_1$ and $\psi \in \mathcal{H}_2$. To see this we consider $\varphi_1 \otimes \psi_2 + \varphi_2 \otimes \psi_1$ assuming that

$$\varphi_1 \otimes \psi_2 + \varphi_2 \otimes \psi_1 = \varphi \otimes \psi.$$

Representing φ and ψ by

$$\varphi = \sum_{i=1,2} \alpha_i \varphi_i \qquad \text{and} \qquad \psi = \sum_{j=1,2} \beta_j \psi_j$$

we conclude from

$$\sum \alpha_i \cdot \beta_j \varphi_i \otimes \psi_j - \varphi_1 \otimes \psi_2 - \varphi_2 \otimes \psi_1 = 0$$

that

$$\alpha_1 \cdot \beta_2 = 1 \quad \text{and} \quad \alpha_2 \cdot \beta_1 = 1$$

as well as

$$\alpha_1 \cdot \beta_1 = \alpha_2 \cdot \beta_2 = 0,$$

which is a contradictory system of equations.

However, if we had $\varphi_1 \otimes \psi_1 + \varphi_1 \otimes \psi_2$, we would get

$$\varphi_1 \otimes \psi_1 + \varphi_1 \otimes \psi_2 = \varphi_1 \otimes (\psi_1 + \psi_2),$$

a product state.

Clearly, the states $\varphi_1 \otimes \psi_2$ and $\varphi_2 \otimes \psi_1$ in

$$\varphi_1 \otimes \psi_2 + \varphi_2 \otimes \psi_1$$

are *entangled* while $\varphi_1 \otimes \psi_1$ and $\varphi_1 \otimes \psi_2$ in

$$\varphi_1 \otimes \psi_1 + \varphi_1 \otimes \psi_2$$

are not.

Suppose $\{\varphi_1, \varphi_2\}$ and $\{\psi_1, \psi_2\}$ each form a basis of $\mathcal{H}_1$ and $\mathcal{H}_2$, respectively, the following states called *Bell states* form an orthonormal basis in $\mathcal{H}_1 \otimes \mathcal{H}_2$:

$$\begin{aligned}
&\frac{1}{\sqrt{2}}(\varphi_1 \otimes \psi_1 + \varphi_2 \otimes \psi_2),\\
&\frac{1}{\sqrt{2}}(\varphi_1 \otimes \psi_1 - \varphi_2 \otimes \psi_2),\\
&\frac{1}{\sqrt{2}}(\varphi_2 \otimes \psi_1 + \varphi_1 \otimes \psi_2),\\
&\frac{1}{\sqrt{2}}(\varphi_1 \otimes \psi_2 - \varphi_2 \otimes \psi_1).
\end{aligned} \tag{633}$$

All these states are not representable as a tensor product and therefore are *entangled*.

In quantum mechanics there is a much bigger variety of entangled states than the types we presented here.

We close this section by bringing attention to a generalized quantum theory in [**6**]. This generalized quantum theory is no longer of a geometric nature. However, for instance, the concept of entanglement still makes sense.

CHAPTER 10

Field Theoretic Weyl Quantization of a Vector Field in 3-Space

In this chapter we will apply the field theoretic *Weyl quantization* to a singularity free vector field X in 3-space. This sort of quantization associates with X representations of a characteristic C^*-Weyl algebra as well as field operators naturally emanating from the vector field (cf. [**16**] and [**12**]). In particular, we will study the influence of the topology of the domain of X on its Weyl quantization.

For a comprehensive study of the general techniques in Weyl quantization we refer to [**13**], [**14**], [**15**], and [**55**].

10.1. The Mathematical Setting

At first we will specify the kind of vector fields under consideration: Suppose M' is a submanifold (with or without boundary) of the oriented Euclidean space E and X' is a smooth vector field on it. Deleting all the singularities of X' yields a vector field X on a space M, say. As in chapter 4, we suppose that M is again a *submanifold* or a *submanifold with boundary* in E. In either case we will refer to M as a manifold in E. Then X is *singularity free* on M and is of the form $X = (\mathrm{id}\,, \mathbf{a})$. Here $\mathbf{a}$ is a smooth, E-valued map on M, as before called the principal part of X. As in chapter 6 we furthermore assume

$$\sup_{x\in M} \parallel \mathbf{a}(x) \parallel < \infty.$$

The pointwisely formed orthogonal complements of $\mathbf{a}$ in E yield a rank-two vector bundle $\mathbb{F}^{\mathbf{a}} \subset M \times E$ (cf. theorem 12). In addition, $\mathbf{a}$ defines a natural (fibrewisely given) *symplectic form* $\omega^{\mathbf{a}}$ on $\mathbb{F}^{\mathbf{a}}$ via the fibrewisely formed cross-product (cf. sections 3.2.1 and 3.2.2). Hence $\mathbb{F}^{\mathbf{a}}$ is fibrewisely oriented. The pair $(\mathbb{F}^{\mathbf{a}}, \omega^{\mathbf{a}})$ determines the vector field uniquely. In general, $\mathbb{F}^{\mathbf{a}}$ is not trivial, as can be seen from the Coulomb field on $\dot{E} := E \setminus \{0\}$ (cf. section 3.7.1).

The vector field itself determines a real trivial line bundle $\mathcal{L}^{\mathbf{a}}$ on M for which $\mathbb{F}^{\mathbf{a}} \oplus \mathcal{L}^{\mathbf{a}}$ spans $M \times E$ (cf. section 3.1.4). Moreover, complementing E to the field of quaternions $\mathbb{H}$, the unit vector $\varepsilon(x) := \frac{\mathbf{a}(x)}{|\mathbf{a}(x)|}$ is an *imaginary unit* in $\mathbb{H}$ for any $x \in M$. Hence ε determines the bundle $\mathcal{L}^{\mathbf{a}}$ and a bundle $\mathbb{C}^{\varepsilon}$ of complex subfields of $\mathbb{H}$ on M (cf. section 3.1.5). This bundle $\mathbb{C}^{\varepsilon}$ operates fibrewisely on $\mathbb{F}^{\mathbf{a}}$ and therefore turns $\mathbb{F}^{\mathbf{a}} \equiv \mathbb{F}^{\varepsilon}$ into a *complex line bundle* on M (cf. sections 3.1.3 and 3.1.4).

The vector field X determines a *Heisenberg group bundle*

$$\mathbf{H}^{\mathbf{a}} = \mathbb{F}^{\mathbf{a}} \oplus \mathcal{L}^{\mathbf{a}}$$

on M whose fibres are Heisenberg groups (cf. section 6.1.2); the fibrewisely given centers are the fibres of $\mathcal{L}^{\mathbf{a}}$. In fact, both $\mathbb{F}^{\mathbf{a}}$ and $\mathbf{H}^{\mathbf{a}}$ characterize X (cf. section 6.1.2). Clearly, the space $\Gamma\mathcal{L}^{a}$ of all smooth Schwartz sections (cf. chapter 6) contains X as a section if $\mathbf{a}$ is a Schwartz map.

The collection $\Gamma\mathbf{H}^{\mathbf{a}}$ inherits a presymplectic structure ω_X^{∞} from $\omega^{\mathbf{a}}$ by which

$$\mathcal{H}_X^{\infty} := \Gamma\mathbf{H}^{\mathbf{a}} + \mathbb{R}\cdot e$$

is turned into an infinite dimensional *Heisenberg group* (cf. sections 6.2.2 and 6.2.3). Here again $e \in \mathbb{H}$ is the unit element. The infinite dimensional Heisenberg group $\mathcal{H}_X^{\infty}$ is the algebraic object emanating from the geometric object $\mathbf{H}^{\mathbf{a}}$ and characterizes X (cf. proposition 81 in section 6.4.5).

The collection $\mathcal{F}_{ae}(\mathcal{H}_X^{\infty},\mathbb{C})$ of all $\mathbb{C}$-valued functions on (the discrete topological space) $\mathcal{H}_X^{\infty}$ vanishing almost everywhere is a convolution algebra $\mathcal{F}_{ae}(\mathcal{H}_X^{\infty},\mathbb{C})_{\star}$ with an involution (cf. section 6.4.1). The collection of all $*$-representations of this algebra into the $*$-algebra of all bounded operators of a given (separable) complex Hilbert space defines a natural norm on $\mathcal{F}_{ae}(\mathcal{H}_X^{\infty},\mathbb{C})_{\star}$. Its completion with respect to this norm is a C^*-algebra $C^*\mathcal{H}_X^{\infty}$, called the *$C^*$-Heisenberg group algebra* of X (cf. section 6.4.1).

This Heisenberg group $\mathcal{H}_X^{\infty}$, determining X, is a subgroup of the group algebra $\mathcal{F}_{ae}(\mathcal{H}_X^{\infty},\mathbb{C})_{\star}$ and of the C^*-algebra generated by it (cf. section 6.4.1). The center of $\mathcal{F}_{ae}(\mathcal{H}_X^{\infty},\mathbb{C})_{\star}$ is the group algebra of $\Gamma\mathcal{L}\oplus\mathbb{R}\cdot e$. This remark transfers to the C^*-algebra level which means it holds true for the completions, as well.

As in the case of $\mathcal{H}_X^{\infty}$, the commutative group $\Gamma\mathbf{H}^{\mathbf{a}}$ determines a C^*-group algebra whose convolution deformed by the presymplectic form $\hbar\cdot\omega_X^{\infty}$ (on $\Gamma\mathbf{H}^{\mathbf{a}}$) yields the Weyl C^*-algebra $W^{\hbar}\Gamma\mathbf{H}^{\mathbf{a}}$ (cf. sections 6.5.3 and 6.5.4).

The trivial skew form on $\Gamma\mathbf{H}^{\mathbf{a}}$ causes the C^*-group algebra $C^*\Gamma\mathbf{H}^{\mathbf{a}}$ which contains $\mathcal{F}_{ae}(\Gamma\mathbf{H}^{\mathbf{a}},\mathbb{C})_{\star}$ (cf. section 6.5.4). This subalgebra endowed with a natural Poisson bracket yields the canonical *$*$-Poisson algebra* $P^{\mathbf{a}}$ of X (cf. section 6.5.4).

Since $\mathbb{F}^{\mathbf{a}} = \varepsilon^*(TS^2)$, the first Chern class $\check{c}_1(\mathbb{F}^{\mathbf{a}})$ of $\mathbb{F}^{\mathbf{a}}$ is the pull-back of the Chern class of TS^2 by ε (cf. sections 4.3.4 and 4.3.5). Here $S^2 \subset E$ is the two-sphere. $\check{c}_1(\mathbb{F}^{\mathbf{a}})$ is determined by the curvature two-form $\Omega^{\mathbf{a}}$ of $\mathbb{F}^{\mathbf{a}}$ which is formed with respect to a natural connection form (cf. corollary 25 in section 3.7.1). Thus the relation between the topology of M and the vector field is encoded in $\check{c}_1(\mathbb{F}^{\mathbf{a}})$, or, equivalently, in the *homotopy class* of ε (cf. corollaries 37 and 38 as well as propositions 42 and 45 in sections 4.2.2, 4.3.4 and 4.3.7, respectively). The two-chart trivialization of S^2 yields a two-chart trivialization of $\mathbb{F}^{\mathbf{a}}$ and hence, in particular, a *cocycle* φ_{NS}, say, determining $\check{c}_1(\mathbb{F}^{\mathbf{a}})$ (cf. section 4.3.7). The influence of the topology of M on the Weyl quantization of X will emanate from this cocycle, as we will see later.

10.2. The Idea of Weyl Quantization of X

10.2.1. The Basic Idea of Weyl Quantization. Let us sketch the basic idea of a Weyl quantization of the vector field X. As we have noticed in chapter 9, the quantization map Q in (594) assigns self-adjoint operators on a Hilbert space to quadratic polynomials and hence the *commutator* to the *Poisson brackets* of two quadratic polynomials. This picture shall be generalized in the forthcoming field quantization. The goal, however, is to go further and construct field operators associated with the vector field X which satisfy specific commutation relations, called *canonical commutation relations* (CCR). These relations are hence of basic interest and will be treated in more detail below.

The fact that the CCR are realized by means of the images of Poisson brackets under a quantization map and the subsequent construction of field operators we take as our intuitive picture and as a starting point for the Weyl quantization.

To be more precise on the CCR, let A and B be two *symmetric operators* associated to X. Formally, the *canonical commutation relations* of these operators read

$$[A,B](\psi) = i\lambda\psi \qquad \forall\, \psi \in \mathrm{dom}([A,B]) \tag{634}$$

for some $\lambda \in \mathbb{R}$. Here dom of an operator denotes the *domain of the definition* of the operator. This equation shall be realized here as the *image* of a Poisson bracket of elements in a *Poisson algebra* associated with the vector field X. We will do so by using the *Weyl form* of (634) which formally reads

$$\exp i\lambda' A \cdot \exp i\lambda B = \exp(-i\lambda\lambda'\hbar) \exp i\lambda B \cdot \exp i\lambda' A \tag{635}$$

(cf. [**55**] and [**1**]). Relations of this type are present in *Weyl algebras* associated with X (cf. section 6.5.3). Weyl algebras will, therefore, be the basic ingredients in our quantization formalism.

10.2.2. An Example. At first let us study at a simple example the role of a Weyl algebra in the construction of a field operator in order to understand its relation to quantization (cf. [**55**]).

We consider a plane $F \subset E$ with an orthonormal coordinate system determined by the unit vectors $e_{\bar{q}}$ and $e_{\bar{p}}$ (cf. (394) in section 7.1.2). Any $v \in F$ is represented in the basis $\{e_{\bar{q}}, e_{\bar{p}}\}$ as

$$v = q \cdot e_{\bar{q}} + p \cdot e_{\bar{p}}$$

and is thus identified with a tuple (q,p) of *coordinates*. The symplectic form

$$\omega : F \times F \longrightarrow \mathbb{R}$$

assigns with any pair of tuples $(q_1,p_1),(q_2,p_2)$ of coordinates in F the value

$$\omega((q_1,p_1),(q_2,p_2)) = q_1 \cdot p_2 - q_2 \cdot p_1 \tag{636}$$

(cf. section 1.3.5). Thus F is turned into a *phase space*.

Each vector $v \in E$ yields the δ-function

$$\delta_v : F \longrightarrow \mathbb{C}$$

on F (cf. section 6.4.1). The *Weyl algebra structure* on $\mathcal{F}_{ae}(F,\mathbb{C})$, as introduced in section 6.5.5, yields

$$W(v) \cdot W(w) = e^{-\frac{1}{2}\omega(v,w)} \cdot W(v+w) \tag{637}$$

where $W(v)$ and $W(w)$ re-denote the δ-functions δ_v and δ_w, respectively. The collection $\{W(v) | v \in \mathcal{F}_{ae}(F,\mathbb{C})\}$ is called a *family* of *Weyl elements*. In particular, the following relations hold true:

$$\begin{aligned} W(e_{\bar{q}}) \cdot W(e_{\bar{q}}) &= W(2 \cdot e_{\bar{q}}), \\ W(e_{\bar{p}}) \cdot W(e_{\bar{p}}) &= W(2 \cdot e_{\bar{p}}), \\ W(e_{\bar{q}}) \cdot W(e_{\bar{p}}) &= e^{-\frac{1}{2}\omega(v,w)\cdot i} \cdot W(e_{\bar{q}} + e_{\bar{p}}). \end{aligned} \tag{638}$$

The linear space $\mathcal{F}_{ae}(F,\mathbb{C})$ with this Weyl algebra structure is denoted by $W_{ae}(F)$. It is called the *Weyl algebra* of the *phase space* F.

$W_{ae}(F)$ is, in fact, also a *Poisson algebra* $P_{ae}(F)$, say, with the *brackets* defined by

$$\{W(v), W(w)\} := [W(v), W(w)] = e^{\omega(v,w)\cdot i} \cdot W(v+w) \tag{639}$$

for all $v, w \in F$.

A *realization* of this family of Weyl elements in terms of *unitary operators* on Hilbert space $\mathcal{H}$ is achieved by the map

$$W_R : F \longrightarrow U(\mathcal{H})$$

where $U(\mathcal{H})$ denotes the *unitary group* consisting of all *unitary operators* of $\mathcal{H}$. The map W_R shall satisfy the following conditions:

$$\begin{aligned} W_R(v) \circ W_R(w) &= e^{-\frac{1}{2}\omega(v,w)\cdot i} \cdot W_R(v+w) \qquad \forall\, v, w \in F, \\ W_R(v)^* &= -W_R(v) \qquad \forall\, v \in F \end{aligned} \tag{640}$$

where for the sake of simplicity the operator $W_R(W(v))$ is abbreviated by $W_R(v)$ for all $v \in F$.

Thus W_R extends to a *representation* of the Weyl algebra $W_{ae}(F)$ obviously causing a representation of the Poisson algebra $P_{ae}(F)$, since

$$W_R(\{W(v), W(w)\}) = [W_R(v), W_R(w)] \qquad \forall\, v, w \in F. \tag{641}$$

Let us assume that

$$W_R(\ldots \cdot v) : \mathbb{R} \longrightarrow \mathbf{B}(\mathcal{H})$$

defined by $t \mapsto W_R(t \cdot v)$ is a *strongly continuous* one-parameter group on $\mathbb{R}$ for any $v \in F$; recall that $\mathbf{B}(\mathcal{H})$ is the collection of all *bounded* linear operators on the complex Hilbert space $\mathcal{H}$ (cf. section 6.5.1). "Strongly continuous" here means that

$$W_R(\ldots \cdot v)(\psi) : \mathbb{R} \longrightarrow \mathbf{B}(\mathcal{H})$$

is continuous for any $\psi \in \mathcal{H}$ (cf. [**86**]). Then by Stone's theorem

$$W_R(v) = e^{i\Phi_R(v)} \qquad \forall\, v \in F \tag{642}$$

holds true for a well defined *self-adjoint* operator $\Phi_R(v)$ (cf. [**86**]). From (640) we immediately deduce

$$\Phi_R(v) + \Phi_R(w) = \Phi_R(v+w)$$

and

$$\Phi_R(\lambda v) = \lambda \Phi_R(v) \qquad \forall\, \lambda \in \mathbb{R} \quad \text{and} \quad \forall\, v \in F,$$

i.e. the linearity of Φ_R on F. The operator Φ_R is called the *field operator* on F.

The *commutation relations* of Φ_R on a dense domain in $\mathcal{H}$ are expressed by

$$[\Phi_R(v), \Phi_R(w)] = i \cdot \omega(v,w) \cdot \mathrm{id}_{\mathcal{H}}. \tag{643}$$

In particular, one verifies the *canonical commutation relations*, namely

$$[\Phi_R(e_{\bar{q}}), \Phi_R(e_{\bar{q}})] = [\Phi_R(e_{\bar{p}}), \Phi_R(e_{\bar{p}})] = 0 \tag{644}$$

and

$$[\Phi_R(e_{\bar{q}}), \Phi_R(e_{\bar{p}})] = i \cdot \omega(e_{\bar{q}}, e_{\bar{p}}) \cdot \mathrm{id}_{\mathcal{H}} = i \cdot \mathrm{id}_{\mathcal{H}}. \tag{645}$$

Hence we have established a field operator to the phase space F in terms of a *Weyl algebra*. In fact, all we really needed was the symplectic phase space (F, ω). The basis $\{e_{\bar{q}}, e_{\bar{p}}\}$ could have been replaced by any symplectic coordinate system. It will be this algebraic notion of a Weyl algebra (a Weyl $*$-algebra) which will lead us to the mentioned type of quantization of a singularity free vector field X on a manifold with or without boundary embedded into E, a Euclidean 3-space.

Up to now, we have only constructed a *classical field operator*. To obtain *quantum* field operators additional representations have to be chosen. We omit this quantization step in this example since much more formalism is needed, as we will see in the next section in a more extended setting.

10.3. Weyl Quantization of Singularity Free Vector Fields in 3-Space

To establish the *field theoretic* Weyl quantization of a singularity free vector field in this section, we will use the general quantization scheme associated with an *infinite dimensional test function space* endowed with a presymplectic structure (cf. section 6.1.1), as presented in [**13**] and [**55**]. We will apply it to the special situation of a test function space emanating from the singularity free vector field X on a manifold M (with or without boundary). This space is $\Gamma\mathbf{H^a}$, the space of all Schwartz sections of the Heisenberg group bundle $\mathbf{H^a}$ of X; it is endowed with its presymplectic structure $\omega^{\Gamma\mathbf{h^a}}$, simply denoted by ω_X^∞ (cf. sections 6.1.1, 6.1.2 and 6.2.2). Clearly, the bundle $\mathbf{H^a}$ on M exists for singularity free vector fields only.

10.3.1. Outline of the Construction. In view of the example above (cf. section 10.2.2), in the *Weyl quantization* of a singularity free vector field X on M the phase space is $\Gamma\mathbf{H^a}$ together with the presymplectic structure ω_X^∞ (cf. sections 6.2.1 and 6.2.2). The linear space F in the example mentioned represents *finitely* many degrees of freedom while the section spaces $\Gamma\mathbb{F}^\mathbf{a}$ and $\Gamma\mathbf{H^a}$ expresses *infinitely* many degrees of freedom. The bundles $\mathbb{F}^\mathbf{a}$ and $\mathbf{H^a}$ both model *internal degrees freedom* of X.

Before we outline the quantization procedure in more detail, let us focus on the role of the vector field in the infinite dimensional phase space $\Gamma\mathbf{H^a}$ and in the process of quantization.

In the quantization of X to be presented below, the singularity free vector field keeps its classical nature throughout the quantization procedure as we will see next. This feature is manifested in the infinite dimensional Heisenberg group $\mathcal{H}_X^\infty$ (cf. section 6.2.2) and its Lie algebra $\mathcal{h}_X^\infty$, respectively, both constructed out of Heisenberg group bundle $\mathbf{H^a}$ on M. The bundles $\mathbb{F}^\mathbf{a}$ and $\mathbf{H^a}$ are reflected in $\mathcal{H}_X^\infty$ through the splitting

$$\mathcal{H}_X^\infty = \Gamma\mathbb{F}^\mathbf{a} \oplus (\Gamma\mathcal{L} \oplus \mathbb{R}\cdot e);$$

the multiplication of sections $s_1, s_2 \in \Gamma\mathbb{F}^\mathbf{a}$ is non-commutative due to ω_X^∞.

The vector field enters through the presymplectic structure ω_X^∞ (cf. (335) in section 6.2.1) of $\mathcal{H}_X^\infty$ on the one hand, and on the other hand (up to equivalence), through the *center*

$$C = \Gamma\mathcal{L} \oplus \mathbb{R}\cdot e$$

of $\mathcal{H}_X^\infty$ (cf. section 6.4.1). If the principal part $\mathbf{a}$ of X is in $\Gamma\mathcal{L}$ (cf. theorem 12), obviously X is encoded in C (cf. (138)). Clearly, the principal part of X may not be a Schwartz map and hence may not be in the center; however, up to equivalence it is one. The center C expresses classical ingredients which will *not* be quantized. Thus X keeps its classical nature.

The center of $\mathcal{F}_{ae}(\mathcal{H}_X^\infty, \mathbb{C})_\star$ encodes the vector field algebraically as a classical object and is characterized as the maximal subspace on which the *bicharacter* $e^{i\omega_X^\infty}$ is constant and equal to 1. The subalgebra $\mathcal{F}_{ae}(\Gamma\mathbb{F}^{\mathbf{a}}, \mathbb{C})_*$ is the non-commutative part of $\mathcal{F}_{ae}(\mathcal{H}_X^\infty, \mathbb{C})_\star$ on which the quantization focuses. This part obviously is determined by $\Gamma\mathbb{F}^{\mathbf{a}}$. These statements hold accordingly for the respective completions, the C^*-algebras $C^*\mathcal{H}_X^\infty$ and $C^*\Gamma\mathbb{F}^{\mathbf{a}}$.

Next in these introductory remarks we will sketch the description of the notion of a Weyl quantization. In the context of Weyl quantization for a given $\hbar \in \mathbb{R}$ we pass from the infinite dimensional Heisenberg group C^*-algebra $C^*\mathcal{H}_X^\infty$ to the Weyl algebra $W^\hbar\Gamma\mathbf{H}^{\mathbf{a}}$ by means of a map (cf. theorem 87 on page 160). This algebra is called the *algebra of observables*. It is not commutative (cf. section 6.5.5). Here $\hbar$ is just a real parameter. If it is fixed to the correct value, it will become *Planck's constant.*

The "limit"

$$W^0\Gamma\mathbf{H}^{\mathbf{a}}$$

as $\hbar$ tends to zero no longer involves the bicharacter $e^{i\hbar\cdot\omega_X^\infty}$; it provides the underlying space $W^0_{ae}\Gamma\mathbf{H}^{\mathbf{a}}$ of the *Poisson algebra* $P^{\mathbf{a}}$ of X (cf. section 6.5.4), the collection of *classical observables*; it reflects *classical field theoretic aspects.*

In a *prequantization*, consisting of two steps, we map the canonical Poisson algebra $P^{\mathbf{a}}$ of X, a $*$-algebra, into $C^*\mathcal{H}_X^\infty$; it transfers classical observables in the $*$-Poisson algebra $P^{\mathbf{a}}$ into elements of $C^*\mathcal{H}_X^\infty$. A surjection maps this C^*-algebra to $W^\hbar\Gamma\mathbf{H}^{\mathbf{a}}$. The composition $Q_\hbar$ of these two assignments is called a *quantization map.* An additional $*$-representation Π of $W^\hbar\Gamma\mathbf{H}^{\mathbf{a}}$ yields *physical quantum observables.* This representation involves Planck's constant and provides a physical $*$-algebra inside the collection of all bounded operators on the (infinite dimensional) Hilbert space. A field operator is constructed from Π; it satisfies the CCR. Due to the lack of a Stone-von Neumann theorem (cf. section 7.1.5) for this kind of representation, the quantization of X is not determined (up to equivalence) by its specified value on the center (cf. chapter 9). This feature of field quantization is well known from quantum optics (cf. [**55**] and [**73**]).

Therefore, this quantization procedure, formulated in short terms, is presented in three steps: First we map the Poisson algebra into $C^*\mathcal{H}_X^\infty$, then pass on to $W^\hbar\Gamma\mathbf{H}^{\mathbf{a}}$ and finally represent this C^*-algebra over a Hilbert space in order to construct field operators satisfying the CCR. The representation involves Planck's constant $\hbar$. The limit $\hbar \to 0$ makes sense. This exhibits the Weyl quantization as a *deformation quantization* (cf. [**13**]).

A word on the influence of the topology to the Weyl quantization. The two-chart trivialization of $\mathbb{F}^{\mathbf{a}}$ (cf. section 4.3.7), causing a trivialization of $\mathbf{H}^{\mathbf{a}}$ over the same charts, yields the local C^*-Heisenberg group algebras of the vector field restricted to these charts. These local C^*-algebras are not sensitive to the topology of M since the respective Heisenberg algebra bundles $\mathbf{H}^{\mathbf{a}}{}_{U_N^\varepsilon}$ and $\mathbf{H}^{\mathbf{a}}{}_{U_S^\varepsilon}$ are trivial, where U_N^ε and U_S^ε are the charts of an atlas of M on which $\mathbf{H}^{\mathbf{a}}$ trivializes. However, the local C^*-Heisenberg group algebras determine $C^*\mathcal{H}_X^\infty$. Representations of each local C^*-algebra determine a representation of $C^*\mathcal{H}_X^\infty$ iff they are linked on $U_N \cap U_S$ by a C^*-isomorphism caused by the two-chart cocycle $\varphi_{\nu\sigma}$ (cf. section 4.3.7). This fact, however, is strongly influenced by the topology of M and hence encodes the influence of the topology on the quantization of X.

Now we go into more detail.

10.3.2. The Classical Part, the Poisson Algebra. The *classical part* of the quantization scheme below is the *commutative* Weyl algebra $W^0\Gamma\mathbf{H^a}$. We repeat in short the notion of the Poisson algebra $P^{\mathbf{a}}$ (cf. section 6.5.4). It is constructed as follows: The Heisenberg group bundle $\mathbf{H^a}$ over M is regarded as a split rank-three vector bundle, i.e. if the fibrewisely given presymplectic form $\omega^{\mathbf{a}}$ (cf. section 3.2.1) on the bundle $\mathbf{H^a}$ is *neglected* or differently formulated, it is replaced by the *trivial presymplectic structure* 0. Thus $\Gamma\mathbf{H^a}$, the collection of all Schwartz sections, is an infinite dimensional $\mathbb{R}$-linear space and, in particular, a *commutative group* under the addition. As such it determines the group algebra $\mathcal{F}_{ae}(\Gamma\mathbf{H^a},\mathbb{C})_\star$. The twisted convolution on $\mathcal{F}_{ae}(\Gamma\mathbf{H^a},\mathbb{C})$ for $\hbar = 0$ is the convolution $\star$. Hence $\mathcal{F}_{ae}(\Gamma\mathbf{H^a},\mathbb{C})_\star$ coincides with $W^0_{ae}(\Gamma\mathbf{H^a})$ (cf. section 6.5.3). Its Weyl elements (the δ-functions) are denoted by $W^0(s)$ for $\mathbf{s} \in \Gamma\mathbf{H^a}$. Clearly, each $\mathbf{s} \in \Gamma\mathbf{H^a}$ splits into

$$\mathbf{s} = s + \lambda \cdot e \qquad \lambda \in \mathbb{R} \text{ and } s \in \Gamma\mathbb{F}^{\mathbf{a}}.$$

The completion of $W^0_{ae}(\Gamma\mathbf{H^a})$ with respect to the norm $\| \; \|$ shall be referred to as $W^0\Gamma\mathbf{H^a}$, (cf. section 6.5.4). The product in this algebra is determined by

$$W^0(\mathbf{s}_1) \cdot W^0(\mathbf{s}_2) = W^0(\mathbf{s}_1 + \mathbf{s}_2) \qquad \forall\, \mathbf{s}_1, \mathbf{s}_2 \in \Gamma\mathbf{H^a}. \tag{646}$$

This C^*-Weyl algebra is *commutative* which expresses the *classical nature* of this algebra. So far the vector field does not show up in $W^0\Gamma\mathbf{H^a}$.

However, $W^0_{ae}\Gamma\mathbf{H^a} \subset W^0\Gamma\mathbf{H^a}$ carries a natural *Poisson algebra structure* defined by

$$\{W^0(\mathbf{s}_1), W^0(\mathbf{s}_2)\} := e^{\omega_X^\infty(\mathbf{s}_1,\mathbf{s}_2)i} \cdot W^0(\mathbf{s}_1 + \mathbf{s}_2) \qquad \forall\, \mathbf{s}_1, \mathbf{s}_2 \in \Gamma\mathbf{H^a}. \tag{647}$$

$W^0_{ae}\Gamma\mathbf{H^a}$ endowed with the brackets in (647) is called the *Poisson algebra* of X and is denoted by $P^{\mathbf{a}}$. Its elements are called *classical observables.*

The vector field $X = (\mathrm{id}\,, \mathbf{a})$ enters into the Poisson algebra $P^{\mathbf{a}}$ by $\omega_X^\infty \equiv \omega^{\Gamma\mathbf{h^a}}$ in (647). This skew bilinear map is degenerate along $\mathcal{L}$ and only along $\mathcal{L}$ (since X is singularity free on M). The Poisson algebra determines X as stated in lemma 85, while $W^0\Gamma\mathbf{H^a}$ determines X up to equivalence only.

10.3.3. The Quantization Map. The basic reference of this section is [**13**], we follow it rather closely. First we will introduce the notion of a *prequantization* of the singularity free vector field X. It is a map, called the *quantization map*, defined on the Poisson algebra $P^{\mathbf{a}}$ with values $W^\hbar\Gamma\mathbf{H^a}$, the C^*-Weyl algebra of X for any $\hbar \in \mathbb{R}$ (cf. section 6.5.5). It will involve as an intermediate step a map from the Poisson algebra $P^{\mathbf{a}}$ into either $C^*\mathcal{H}_X^\infty$ or (what is the same up to a unique $*$-isometry) into $C^*_{WF}(\Gamma\mathbf{H^a})$ (cf. theorem 87 in section 6.5.5).

The map

$$\begin{aligned} Q : \mathcal{F}_{ae}(\Gamma\mathbf{H^a},\mathbb{C}) = P^{\mathbf{a}} &\longrightarrow C^*_{WF}(\Gamma\mathbf{H^a}) \\ W^0(\mathbf{s}) &\longmapsto (\hbar \longmapsto W^\hbar(\mathbf{s})) \end{aligned} \tag{648}$$

is $*$ preserving. Clearly, the maps $W^0(\mathbf{s}) = W^\hbar(\mathbf{s})$ coincide with $\delta_{\mathbf{s}}$ for any $\mathbf{s} \in \Gamma\mathbf{H^a}$.

From Q we form the *quantization map*

$$Q_\hbar := \alpha_\hbar \circ Q \qquad \text{with} \quad \hbar \in \mathbb{R} \text{ fixed},$$

where $\alpha_\hbar$ is the point evaluation (cf. section 6.5.5). Thus $Q_\hbar$ extends to $P^{\mathbf{a}}$ as

$$Q_\hbar : P^{\mathbf{a}} \longrightarrow W^\hbar \Gamma \mathbf{H}^{\mathbf{a}}.$$

$Q_\hbar$ expresses a *prequantization*. It is *continuous* and $*$-preserving. Therefore, it is called a *continuous quantization map*. The terms *continuous* refers to the continuity expressed in theorem 86 yielding the continuity of the point evaluation, described in this theorem, too. Since $\hbar$ may vary in $\mathbb{R}$, we call this a *strict deformation quantization* of X because of the following properties of $Q_\hbar$ (cf. [**14**] and [**15**]):

a) *Dirac's condition:*

$$\lim_{\hbar \to 0} (\| \, [Q_\hbar(W^0(\mathbf{s}_1)), Q_\hbar(W^0(\mathbf{s}_2))]_\hbar - Q_\hbar(\{W^0(\mathbf{s}_1), W^0(\mathbf{s}_2)\}) \, \|_\hbar) = 0$$

for all $\mathbf{s}_1, \mathbf{s}_2 \in \Gamma\mathbf{H}$, here $[\ ,\ ]_\hbar = i \cdot \hbar^{-1} \cdot [\ ,\ ]$.

b) *von Neumann's condition:*

$$\lim_{\hbar \to 0} \| \, Q_\hbar(W^0(\mathbf{s}_1)) \cdot Q_\hbar(W^0(\mathbf{s}_2)) - Q_\hbar(W^0(\mathbf{s}_1) \cdot W^0(\mathbf{s}_2)) \, \|_\hbar = 0$$

for all $\mathbf{s}_1, \mathbf{s}_2 \in \Gamma\mathbf{H}$.

c) *Rieffel's condition:*

$$\| \, Q_\hbar(W^0(\mathbf{s})) \, \|_\hbar$$

varies continuously with $\hbar$ for all $\mathbf{s} \in \Gamma\mathbf{H}$.

In addition, we verify that the map $Q_\hbar$ is injective and its image is closed with respect to the product. Hence $Q_\hbar$ is a *strict deformation quantization* in the sense of Rieffel (cf. [**55**]).

Of course we may use $C^*\mathcal{H}_X^\infty$ instead of $C^*_{WF}(\Gamma\mathbf{H}^{\mathbf{a}})$; in this case we take the $*$-isomorphism θ in (391) in theorem 87 and form

$$\theta^{-1} \circ Q : P^{\mathbf{a}} \longrightarrow C^*\mathcal{H}_X^\infty.$$

The map

$$\alpha_\hbar \circ \theta^{-1} \circ Q : P^{\mathbf{a}} \longrightarrow W^\hbar \Gamma \mathbf{H}^{\mathbf{a}}$$

is called a prequantization map, too. This quantization map is a *strict deformation quantization*, as well. Moreover, $\alpha_\hbar \circ \theta^{-1} \circ Q$ is a *surjection*.

10.3.4. The Physical $*$-Algebra. The *physics* enters when we combine $\theta_\hbar$ with a further *representation* of the $*$-algebra $W^\hbar \Gamma \mathbf{H}^{\mathbf{a}}$: Let

$$\Pi : W^\hbar \Gamma \mathbf{H}^{\mathbf{a}} \longrightarrow \mathbf{B}(\mathcal{H})$$

be a $*$-*representation*. Setting

$$\Pi_\hbar := \Pi \circ \theta_\hbar,$$

(cf. theorem 87 on page 160 for $\theta_\hbar$) the $*$-representation

$$\Pi_\hbar : C^*\mathcal{H}_X^\infty \longrightarrow \mathbf{B}(\mathcal{H}) \tag{649}$$

is called a *physical representation* and

$$\Pi_\hbar(C^*\mathcal{H}_X^\infty) \subset \mathbf{B}(\mathcal{H})$$

is called a *physical $*$-algebra* or the *algebra of physical quantum observables*.

The application of this Weyl quantization, for example, in *quantum optics* is rather involved and is beyond the scope of these notes.

So far we have no *field operator* yet. Such an operator will be defined next.

10.3.5. Canonical Field Quantization. At first we deal with the classical part. The test function space $\Gamma\mathbf{H}^{\mathbf{a}}$ yields the *classical Weyl system* (385) of δ-functions regarded as linear maps of $\Gamma\mathbf{H}^{\mathbf{a}}$. The Weyl element $W^0(\mathbf{s})$ (being a δ-function on $\Gamma\mathbf{H}^{\mathbf{a}}$) in the Poisson algebra $P^{\mathbf{a}}$ (cf. section 6.5.4) is represented as

$$W^0(\mathbf{s}) = e^{-i\Phi^0(\mathbf{s})}$$

where $\mathbf{s}$ varies in $\Gamma\mathbf{H}^{\mathbf{a}}$ and

$$\Phi^0(\mathbf{s}) : \Gamma\mathbf{H}^{\mathbf{a}} \longrightarrow \mathbb{R} \qquad \forall\, \mathbf{s} \in \Gamma\mathbf{H}^{\mathbf{a}}$$

is a linear map, as can be easily seen from (646); it is called the *classical field observable* or *classical Weyl functional*. Thus the *classical field system* may be written as

$$\Phi^0 : \Gamma\mathbf{H}^{\mathbf{a}} \longrightarrow (\Gamma\mathbf{H}^{\mathbf{a}})', \qquad \mathbf{s} \longmapsto \Phi^0(\mathbf{s}).$$

The range of Φ^0 is the algebraic dual of the domain.

From the Poisson brackets (647) of the Weyl system in $P^{\mathbf{a}}$ we immediately deduce

(650) $$\{\Phi^0(\mathbf{s}_1), \Phi^0(\mathbf{s}_2)\} = \omega_X^\infty(\mathbf{s}_1, \mathbf{s}_2) \cdot i \cdot \mathbb{1}^{\mathbf{a}}$$

where $\mathbb{1}^{\mathbf{a}}$ is the linear map into $(\Gamma\mathbf{H}^{\mathbf{a}})'$ defined by the constant map on $\Gamma\mathbf{H}^{\mathbf{a}}$ with value 1.

Now we turn to the *quantization* and let Π be a *non-degenerate* $*$-representation of the C^*-Weyl algebra $W^\hbar\Gamma\mathbf{H}^{\mathbf{a}}$ acting on the Hilbert space $\mathcal{H}_\Pi$ (being unitary on $j(\Gamma\mathbf{H}^{\mathbf{a}})$ (cf. (349))). For

(651) $$W_\Pi^\hbar(\mathbf{s}) := \Pi(W^\hbar(\mathbf{s})) \qquad \forall\, \mathbf{s} \in \Gamma\mathbf{H}^{\mathbf{a}}$$

the equation

(652) $$W_\Pi^\hbar(t \cdot \mathbf{s}) \cdot W_\Pi^\hbar(r \cdot \mathbf{s}) = W_\Pi^\hbar((t + r) \cdot \mathbf{s})$$

holds true for all $t, r \in \mathbb{R}$ since $\omega_X^\infty(\mathbf{s}, \mathbf{s}) = 0$; hence $W_\Pi^\hbar(\mathbf{s})$ determines a unitary *one-parameter group*. Let us assume that the representation Π is *regular*. This means that the one-parameter group determined by $W_\Pi^\hbar(s)$ is *strongly continuous* (cf. [**80**] and [**86**]) for each $\mathbf{s} \in \Gamma\mathbf{H}^{\mathbf{a}}$. Consequently, by Stone's theorem there is a generator $\Phi_\Pi(\mathbf{s})$ of the one-parameter group, the *self-adjoint field operator* obtained by differentiation at $t = 0$:

(653) $$\Phi_\Pi(\mathbf{s}) := -i\,\frac{d}{dt} W_\Pi^\hbar(t \cdot \mathbf{s})\bigg|_{t=0} \qquad \forall\, \mathbf{s} \in \Gamma\mathbf{H}^{\mathbf{a}}.$$

For these field operators we find for any two fixed $\mathbf{s}_1, \mathbf{s}_2 \in \Gamma\mathbf{H}^{\mathbf{a}}$ that the CCR

(654) $$[\Phi_\Pi(\mathbf{s}_1), \Phi_\Pi(\mathbf{s}_2)](\psi) = -i\hbar\, \omega_X^\infty(\mathbf{s}_1, \mathbf{s}_2) \cdot \psi$$

are valid for any $\psi \in \operatorname{dom} \Phi_\Pi(\mathbf{s}_1) \cap \operatorname{dom} \Phi_\Pi(\mathbf{s}_2)$. We hence have a *(strong) realization* of the CCR. While the definition of the Weyl systems seems unusual in the first place, with (652) and (654) we now see the reason for doing so.

10.3.6. The Heisenberg Group Quantization. Again we are given a *regular* $*$-representation (being unitary on $j(\Gamma\mathbf{H^a})$)

$$\Pi : W^\hbar\Gamma\mathbf{H^a} \to \mathbf{B}(\mathcal{H})$$

with some non-vanishing $\hbar \in \mathbb{R}$, and form $\Pi_\hbar$ as in (649) in section 10.3.4. These data yield the one-parameter group $V_\Pi^\hbar(\lambda, \mathbf{s}) \in \mathbf{B}(\mathcal{H})$ defined by

$$V_\Pi^\hbar(t\cdot\lambda, t\cdot\mathbf{s}) := e^{-i\cdot t\cdot\lambda\cdot\hbar}\cdot\Pi(W^\hbar(t\cdot\mathbf{s})) \qquad \forall t \in \mathbb{R}$$

for fixed $\lambda \in \mathbb{R}$ and $\mathbf{s} \in \Gamma\mathbf{H^a}$. This one-parameter group is based on the one-parameter group $V(t\cdot(\lambda, \mathbf{s})) \in C^*\mathcal{H}_X^\infty$ given by

$$V(t\cdot(\lambda,\mathbf{s})) := t\cdot s + t\cdot\lambda\cdot e \quad \in C^*\mathcal{H}_X^\infty$$

where t varies in $\mathbb{R}$. In fact,

$$\theta(t\cdot s + t\cdot\lambda\cdot e) = e^{-i\cdot t\cdot\lambda\cdot\hbar}\cdot W^\hbar(t\cdot\mathbf{s}) \qquad \forall t\in\mathbb{R}$$

with θ as in theorem 87. Then the generator is

$$\frac{d}{dt}V_\Pi^\hbar(t\cdot\lambda, t\cdot\mathbf{s})\mid_{t=0} = -\lambda\cdot\hbar\cdot i\cdot \mathrm{id}_{\mathcal{H}} + i\cdot\Phi_\Pi^\hbar(\mathbf{s})$$

for a well defined field operator $\Phi_\Pi^\hbar(\mathbf{s})$. The self-adjoint operator

$$\frac{1}{i}\cdot R_\Phi^\hbar(\lambda,\mathbf{s}) := -\lambda\hbar\cdot\mathrm{id}_{\mathcal{H}} + \Phi_\Pi^\hbar(\mathbf{s})$$

generating $V_\Pi^\hbar$ is in general *unbounded* if $\Phi_\Pi^\hbar(\mathbf{s})$ does not vanish. The field operator $\Phi_\Pi^\hbar(\mathbf{s})$ is of course identical to the one given by (653).

The commutator thus is

$$[R_\Pi^\hbar(\lambda_1,\mathbf{s}_1), R_\Pi^\hbar(\lambda_2,\mathbf{s}_2)] = -i\cdot\hbar\cdot\omega_X^\infty(\mathbf{s}_1,\mathbf{s}_2)\cdot\mathrm{id}_{\mathcal{H}}$$

for any choice of $\lambda_1+\mathbf{s}_1$ and $\lambda_2+\mathbf{s}_2$ in $\mathcal{H}_X^\infty$; a realization of the CCR in terms of $R_\Pi^\hbar$ (cf. [**12**], [**13**] and [**15**]).

10.3.7. Direct Field Quantization. So far we have treated *bounded* physical observables in $\mathbf{B}(\mathcal{H})$. For possibly *unbounded* operators we proceed according to [**13**] as follows: First we select some $\hbar \neq 0$ and a *physically* suitable *regular* representation $\prod_\hbar$ of the C^*-algebra $W^\hbar(\Gamma H^a)$. The *unbounded self-adjoint* field operators are determined by

$$\Phi_{\Pi_\hbar}(\mathbf{s}) := -i\frac{d}{dt}\Pi_\hbar(W^\hbar(\mathbf{s}))\bigg|_{t=0} \qquad \forall\mathbf{s}\in\Gamma H^a.$$

The differentiation takes place in the Hilbert space on which the representation acts. These operators are defined only *densely*, in general. For certain representations there exist *common dense cores*.

For a family $\Pi := \{\Pi_\hbar\}_{\hbar\in\mathbb{R}\setminus\{0\}}$ of *regular* and *faithful* representations $\Pi_\hbar$ of $W^\hbar(\Gamma\mathbf{H^a})$, the Π-dependent quantization extends linearly and continuously as

$$\Phi_{\Pi_\hbar}(W^0(\mathbf{s})) := \Phi_\hbar(W^0(\mathbf{s})) \qquad \forall\mathbf{s}\in\Gamma\mathbf{H^a}.$$

For the *classical case*, i.e. for $\hbar = 0$, and a given locally convex topology τ on $\Gamma\mathbf{H^a}$ yielding the locally convex vector space $(\Gamma\mathbf{H^a})_\tau$, the *field observable*

$$\Phi^0(\mathbf{s}) : (\Gamma\mathbf{H^a})'_\tau \longrightarrow \mathbb{R},$$

defined by $\Phi^0(\mathbf{s})(F) := F(\mathbf{s})$ for any $\mathbf{s} \in \Gamma\mathbf{H}^{\mathbf{a}}$, is obtained in the above scheme by

$$\Phi^0(\mathbf{s})(F) := -i\frac{d}{dt}W^0(t\cdot\mathbf{s})(F)\Big|_{t=0} = F(\mathbf{s})$$

for any given $\mathbf{s} \in \Gamma\mathbf{H}^{\mathbf{a}}$ and for all $F \in (\Gamma\mathbf{H}^{\mathbf{a}})'_\tau$. Here $(\Gamma\mathbf{H}^{\mathbf{a}})'_\tau$ denotes the *topological dual* of the locally convex space $(\Gamma\mathbf{H}^{\mathbf{a}})_\tau$.

In a first step a *direct field quantization* has to replace the classical field expression $\Phi^0(\mathbf{s})$ by the quantum one $\Phi^\hbar_\Pi(\mathbf{s})$ for every $\mathbf{s} \in \Gamma\mathbf{H}^{\mathbf{a}}$. The second step has to specify the quantized expressions for all of the classical field monomials

$$\Phi^0(\mathbf{s}_1)\cdot\ldots\cdot\Phi^0(\mathbf{s}_n) = (-i)^n\,\frac{\partial^n}{\partial t_1\cdots\partial t_n}W^0\left(\sum_{k=1}^n t_k\mathbf{s}_k\right)\Big|_{t_1=\ldots=t_n=0}.$$

The *Weyl quantization* now suggests extending the quantization map $Q^\Pi_\hbar$ from the represented bounded observables $\Pi_\hbar(W^\hbar\Gamma\mathbf{H}^{\mathbf{a}})$ to the unbounded field polynomials by the (complex) linear extension of

$$\begin{aligned} Q^\Pi_\hbar(\Phi^0(\mathbf{s}_1)\cdot\ldots\cdot\Phi^0(\mathbf{s}_n)) &:= (-i)^n\,\frac{\partial^n}{\partial t_1\cdots\partial t_n}Q^\Pi_\hbar\left(W^0(\sum_{k=1}^n t_k\mathbf{s}_k)\right)\Big|_{t_1=\ldots=t_n=0} \\ &:= (-i)^n\,\frac{\partial^n}{\partial t_1\cdots\partial t_n}\Pi_\hbar\left(W^\hbar(\sum_{k=1}^n t_k\mathbf{s}_k)\right)\Big|_{t_1=\ldots=t_n=0}, \end{aligned} \tag{655}$$

where $n \in \mathbb{N}$ and $\mathbf{s}_k \in \Gamma\mathbf{H}^{\mathbf{a}}$ are arbitrary and t_k is a real parameter for any $k = 1,\ldots$. In particular, for $n = 1$ and $n = 2$ we have

$$Q^\Pi_\hbar(\Phi^0(\mathbf{s})) = \Phi^\hbar_\Pi(\mathbf{s})$$

and

$$Q^\Pi_\hbar(\Phi^0(\mathbf{s})\Phi^0(\mathbf{s}')) = \tfrac{1}{2}\left(\Phi^\hbar_\Pi(\mathbf{s})\Phi^\hbar_\Pi(\mathbf{s}') + \Phi^\hbar_\Pi(\mathbf{s}')\Phi^\hbar_\Pi(\mathbf{s})\right).$$

The classical field observables $\Phi^0(\mathbf{s})$ fulfill the Poisson bracket relations

$$\{\Phi^0(\mathbf{s}), \Phi^0(\mathbf{s}')\} = -\omega^\infty_X(\mathbf{s},\mathbf{s}')\cdot \mathrm{id}_{\mathcal{H}} \quad \forall \mathbf{s},\mathbf{s}' \in \Gamma\mathbf{H}^{\mathbf{a}}$$

with the unit function $\mathbb{1}^0$ being identical to one on $(\Gamma\mathbf{H}^{\mathbf{a}})_\tau$, whereas for the quantum field operators $\Phi^\hbar_\Pi(\mathbf{s})$ one has the $\hbar$–scaled commutators

$$[\Phi^\hbar_\Pi(\mathbf{s}), \Phi^\hbar_\Pi(\mathbf{s}')]_\hbar = -\hbar\cdot\omega(\mathbf{s},\mathbf{s}')\cdot \mathrm{id}_{\mathcal{H}} \qquad \forall \mathbf{s},\mathbf{s}' \in \Gamma\mathbf{H}^{\mathbf{a}}.$$

Obviously, these two formulae are formally in accordance with the Dirac condition (cf. section 10.3.3).

In equation (655) the derivatives taken are those of the quantized, i.e. represented Weyl operators. The quantization prescription $Q^\Pi_\hbar$ for the field polynomials is still called *Weyl quantization* or *symmetric quantization.* With the convention (655) we have completed the mathematical description of the formal Weyl quantization procedure in physical quantum field theory (cf. [**13**]) here applied to the test function space $(\Gamma\mathbf{H}^{\mathbf{a}}, \omega^{\Gamma\mathbb{E}})$ arising from the principal part $\mathbf{a} : M \to E$ of our vector field X.

In the literature on *quantum field theory* and *quantum optics* one finds further quantization procedures for fields which fulfill the first quantization step in an unchanged manner: the classical field observable $\Phi^0(f)$ is replaced by the quantum one $\Phi^\hbar_\Pi(f)$ for every test function f. But in the second quantization step for the higher field monomials there is a difference to the Weyl quantization described by equation (655). For example, in [**55**] it is demonstrated how various direct field quantizations arise from continuous deformation quantizations, based on the continuous field of C^*-Weyl algebras from theorem 86 and 87; it is equivalent to the Weyl quantization in the sense of [**59**].

10.4. The Relation to the GNS Representation

In this section we will relate the representations determining field operators in (653) to representations obtained in the setting given by Gelfand, Naimark, and Segal (GNS-setting) and hence relate the Weyl quantization of X based on geometric grounds to a stochastic setting. Here [**55**] is our essential source.

10.4.1. Review of State Spaces. In order to introduce the notion of a *state* and the *state space* in the context of *quantum field theory* we need to enlarge our algebraic concept. This requires us to generalize the twisted group algebra, as introduced in section 6.5.3. To this end we need the following preparations:

Throughout this subsection G denotes a group again and

$$\Sigma : G \times G \longrightarrow \dot{\mathbb{C}}$$

an *antisymmetric bicharacter*. This is to say $\Sigma(g, \ldots)$ and $\Sigma(\ldots, g)$ are characters related by

$$\Sigma(g_1, g_2) = (\Sigma(g_2, g_1))^{-1}$$

(expressing the antisymmetry of Σ) for all choices of $g, g_1, g_2 \in G$. Consequently, $\Sigma(g, g) = 1$ for any $g \in G$. As an example of a bicharacter Σ we consider the additive group $\Gamma\mathbf{H}^{\mathbf{a}}$ and set

$$\Sigma(\mathbf{s}_1, \mathbf{s}_2) := e^{-\frac{i}{2}\hbar\omega_X^\infty(\mathbf{s}_1,\mathbf{s}_2)} \qquad \forall\, \mathbf{s}_1, \mathbf{s}_2 \in \Gamma\mathbf{H}^{\mathbf{a}}$$

(cf. section 10.3.1).

We generalize the notion of a twisted group algebra based on the twisted convolution (cf. (383) in section 6.5.3) and introduce the *twisted group algebra*

$$\mathcal{F}_{ae}(G, \mathbb{C})_{\star_\Sigma}$$

formed with respect to Σ, called a *multiplier*. Here we have replaced $e^{-\frac{i}{2}\hbar\omega_X^\infty(\mathbf{s}_1,\mathbf{s}_2)}$ in the twisted convolution by $\Sigma(\mathbf{s}_1, \mathbf{s}_2)$ for all $\mathbf{s}_1, \mathbf{s}_2 \in \Gamma\mathbf{H}^{\mathbf{a}}$. Letting

$$\delta_g^* := \delta_{g^{-1}} \qquad \forall\, g \in G,$$

this $*$-operation is the same as the one in (351) and extends to all of $\mathcal{F}_{ae}(G, \mathbb{C})_{*_\Sigma}$, too. As its *completion* with respect to the norm $\|\ \|$ introduced in section 6.5.2 we obtain the *C^*-algebra*

$$C^*(G, \Sigma)_{\star_\Sigma}$$

(cf. [**13**]). The above defined $*$-operation

$$f \longmapsto f^* \qquad \forall\, f \in \mathcal{F}_{ae}(G, \mathbb{C})_{\star_\Sigma}$$

extends to all of $C^*(G, \Sigma)$.

In order to define a *state* we consider a Σ-*positive definite* function

$$C : G \longrightarrow \mathbb{C}$$

which by definition satisfies

$$\sum_{j,k=1}^{n} \bar{z}_j \cdot z_k - \overline{\Sigma(g_j, g_k)} \cdot C(g_j^{-1} \cdot g_k) \geq 0 \tag{656}$$

for any $n \in \mathbb{N}$, all $z_j, z_k \in \mathbb{C}$ and all $g_j, g_k \in G$. If, in addition, $C(e) = 1$ (where $e \in G$ is the unit element) the function C is called *normalized.*

$C(G)_\Sigma$ shall be the collection of all *normalized*, Σ*-positive definite* $\mathbb{C}$-valued functions. It is a *convex set* as is easily verified.

Each function $C \in C(G)_\Sigma$ yields a *state* ℓ_C in the *algebraic dual* $\mathcal{F}_{ae}(G,\mathbb{C})'$ of $\mathcal{F}_{ae}(G,\mathbb{C})$ defined by

$$\ell_C\left(\sum_k z_k \cdot \delta_{g_k}\right) := \sum_k z_k C(g_k) \tag{657}$$

for any $f \in \mathcal{F}_{ae}(G,\mathbb{C})_{\star\Sigma}$ which is decomposed into $f = \sum_k z_k \delta_{g_k}$.

The *state space*, i.e. the collection of all states is denoted by $\mathcal{S}$. Vice versa, if ℓ in the algebraic dual space $\mathcal{F}_{ae}(G,\mathbb{C})'$ satisfies the two conditions

i): $\ell(f^* \cdot f) \geq 0 \qquad \forall\, f \in \mathcal{F}_{ae}(G,\mathbb{C})_{\star\Sigma}$,
ii): $\ell(e) = 1$,

then ℓ is of the form ℓ_C for some $C \in C(G)_\Sigma$. Indeed, it is straightforward to verify that the function

$$C_\ell : G \longrightarrow \mathbb{C}$$

determined by the equation

$$\sum_k z_k C_\ell(g_k) := \ell\left(\sum z_k \cdot \delta_{g_k}\right) \qquad \forall\, g_k \in G \tag{658}$$

is Σ-positive definite. The assignment

$$\begin{array}{rcl} C(G)_\Sigma & \longrightarrow & \mathcal{S} \\ C & \longmapsto & \ell_C \end{array} \tag{659}$$

is a *bijection* (cf. [**14**] and [**13**]).

The *physical meaning* of a state ℓ_C is that of an *expectation value* (cf. [**55**]).

By the above defined correspondence (659) the Σ-positive definite function

$$\delta_e \in \mathcal{F}_{ae}(G,\mathbb{C})$$

defines a linear map on $\mathcal{F}_{ae}(G,\mathbb{C})$, the so-called *tracial state* ℓ_{tr}. Here e is the unit element of G. This state defines the square of the norm $\|\ \|_2$ by means of

$$\| f \|_2^2 := \ell_{\text{tr}}(f^* \cdot f) \qquad \forall\, f \in C(G,\Sigma) \tag{660}$$

for which

$$\| f \|_2^2 = \sum_{g\in G} |f(g)|^2 = \sum |z_k|^2 \qquad \forall\, f \in \mathcal{F}_{ae}(G,\mathbb{C}) \tag{661}$$

holds true. Here f is again decomposed into$f = \sum_k z_k \delta_{g_k}$.

The norm $\|\ \|_2$ satisfies

$$\| f \|_2 \leq \| f \| \leq \| f \|^1 \qquad \forall\, f \in \mathcal{F}_{ae}(G,\mathbb{C}) \tag{662}$$

where $\|\ \|$ and $\|\ \|^1$ are defined as in (379) and (380), respectively.

The completions of the linear space $\mathcal{F}_{ae}(G,\mathbb{C})_{\star\Sigma}$ with respect to the norms $\| \|_2$, $\| \|$ and $\| \|^1$ are related as follows:

$$l^1(G) \subseteq C^*G \subseteq l^2(G)$$

which are proper for non-finite G. Here $l^1(G)$ and $l^2(G)$ are the spaces of summable and square summable sequences of G, respectively.

10.4.2. Weyl Systems and Projective Representations. In this subsection we will outline the program relating the algebraic setting of quantization as described in section 10.3.5 with a *stochastic concept* of quantization formulated by means of *GNS-representations.* To this end we focus here on the notion of a *Weyl system* $(W_R, \mathcal{H}_R)$ on $\Gamma\mathbf{H^a}$. Here a Weyl system consists of a collection of *projective unitary representations* W_R acting on a Hilbert space $\mathcal{H}_R$. The space of Schwartz sections $\Gamma\mathbf{H^a}$ equipped with ω_X^∞ determines a C^*-Weyl algebra $W\Gamma\mathbf{H^a}$ of $\Gamma\mathbf{H^a}$ as introduced in section 6.5.5. The Weyl system is obtained by representing $W\Gamma\mathbf{H^a}$. Below we will relate Weyl systems with GNS-representations.

At first we will show that Weyl systems determine *projective group representations* (cf. [**14**]) of $\Gamma\mathbf{H^a}$. Suppose we are given a map

$$W_R : \Gamma\mathbf{H^a} \longrightarrow U(\mathcal{H}_R), \qquad s \mapsto W_R(s),$$

into the unitary group $U(\mathcal{H}_R)$ of the non-trivial Hilbert space $\mathcal{H}_R$. Here R stands for realization. The collection $\{W_R(\mathbf{s}) | \mathbf{s} \in \mathbf{\Gamma H^a}\}$ or equivalently the tuple $(W_R, \mathcal{H}_R)$ is called a (quantum) *Weyl system* (or a *realization* of the *Weyl relations* on $\Gamma\mathbf{H^a}$ (cf. (387))) if W_R satisfies the Weyl relations

$$W_R(\mathbf{s}_1) \circ W_R(\mathbf{s}_2) = e^{-\frac{1}{2}\omega_X^\infty(\mathbf{s}_1,\mathbf{s}_2)\cdot i} \cdot W_R(\mathbf{s}_1 + \mathbf{s}_2) \tag{663}$$

and

$$W_R(\mathbf{s})^* = W_R(-\mathbf{s}) \tag{664}$$

for all $\mathbf{s}, \mathbf{s}_1$ and $\mathbf{s}_2 \in \Gamma\mathbf{H^a}$. The unitary operator $W_R(\mathbf{s})$ is called a *Weyl operator.* Due to the phase factor in (663), the Weyl system W_R is obviously a *projective representation* of the additive group $\Gamma\mathbf{H^a}$.

The Weyl system is called *regular* if the assignment

$$t \longmapsto W_R(t\cdot \mathbf{s}) \qquad \forall t \in \mathbb{R}$$

is *strongly* continuous for each $\mathbf{s} \in \Gamma\mathbf{H^a}$. (In the general theory of Weyl quantization $\Gamma\mathbf{H^a}$ is replaced by any presymplectic space (cf. [**55**]).)

The map W_R extends to all of the Weyl algebra $W\Gamma\mathbf{H^a}$ (here $W_R(\mathbf{s})$ abbreviates $W_R(W(\mathbf{s})$ for all $\mathbf{s} \in \mathbf{\Gamma H^a}$ as seen from (385)) and hence determines a unitary representation of $W\Gamma\mathbf{H^a}$ (cf. section 6.5.3).

Realizations of Weyl relations can be obtained by means of representations of the Weyl algebra $W\Gamma\mathbf{H^a}$ as we will see next.

To establish the link mentioned in the title of this section, we consider a non-degenerated $*$-representation $(\Pi, \mathcal{H}_\Pi)$ of the C^*-algebra $W\Gamma\mathbf{H^a}$ and set

$$W_{R\Pi}(\mathbf{s}) := \Pi(W(\mathbf{s})) \qquad \forall \mathbf{s} \in \Gamma\mathbf{H^a} \tag{665}$$

where $W(\mathbf{s}) \in W\Gamma\mathbf{H^a}$ denotes the Weyl element of $\mathbf{s}$ (cf. section 6.5.3), i.e. the function $\delta_{\mathbf{s}}$ in $\mathcal{F}(\Gamma\mathbf{H^a}, \mathbb{C})\star_{\omega_X^\infty}$. Obviously,

$$\{W_{R\Pi}(\mathbf{s}) \mid \mathbf{s} \in \Gamma\mathbf{H^a}\}$$

forms a Weyl system. This is to say that the $*$-representation Π determines a Weyl system. Of course, given the Weyl system $W_{R\Pi}$ or vice versa equation (665) yields the representation

$$\Pi : W\Gamma\mathbf{H^a} \longrightarrow \mathbf{B}(\mathcal{H}_R) \tag{666}$$

if $W_{R\Pi}$ in (665) is redenoted by W_R and extended to all of $W\Gamma\mathbf{H^a}$. Hence Π is *uniquely determined* by $(W_R, \mathcal{H}_R)$. Therefore, we have the following:

PROPOSITION 109. *Any non-degenerated $*$-representation Π of $W\Gamma\mathbf{H^a}$ yields a Weyl system which vice versa determines the representation Π.*

10.4.3. The Connection Between Weyl Algebras and Positive Definite Functions. The goal of this subsection is to relate the concept of a state (cf. section 10.4.1) with the one of a Weyl system. To this end we will construct from $C \in C(\Gamma\mathbf{H^a})_\Sigma$ a kernel κ and a *unitary* Weyl system W_C together with a normalized vector $\kappa(0)$. In this construction the antisymmetric bicharacter Σ will be chosen such that it is directly related to the presymplectic structure ω_X^∞ on $\Gamma\mathbf{H^a}$ (cf. section 10.3.1). This system W_C determines a representation $(\Pi_C, \mathcal{H}_C)$, as constructed above, unitary on $\Gamma\mathbf{H^a}$. Hence, any state $\ell \in \mathcal{F}(\Gamma\mathbf{H^a}, \mathbb{C})'$ will determine some Σ-positive function via C_ℓ (cf. (658)) and in turn a $*$-representation $(\Pi_{C_\ell}, \mathcal{H}_{C_\ell})$, say, unitary on $\Gamma\mathbf{H^a}$. In fact, Π_{C_ℓ} is a *GNS-representation* (cf. [**14**] and [**55**]). To outline the construction of Π_{C_ℓ} is the goal of this subsection.

First, we consider the presymplectic space $\Gamma\mathbf{H^a}$ endowed with ω_X^∞. (The group structure on $\Gamma\mathbf{H^a}$ is the underlying additive group.) The multiplier Σ used here is

$$\Sigma = e^{-\frac{\hbar}{2}\cdot i\omega_X^\infty};$$

this is to say, the *bicharacter* $e^{-\frac{\hbar}{2}\cdot i\omega_X^\infty}$ applied to any pair $\mathbf{s}_1, \mathbf{s}_2 \in \Gamma\mathbf{H^a}$ takes the value $e^{-\frac{\hbar}{2}\cdot i\omega_X^\infty(\mathbf{s}_1,\mathbf{s}_2)}$. Hence

$$C(\Gamma\mathbf{H^a})_\Sigma = C(\Gamma\mathbf{H^a})_{e^{-\frac{\hbar}{2}\cdot i\omega_X^\infty}}.$$

Let us re-denote $C(\Gamma\mathbf{H^a})_{e^{-\frac{\hbar}{2}\cdot i\omega_X^\infty}}$ by $C(\Gamma\mathbf{H^a})$, for simplicity. The group $\Gamma\mathbf{H^a}$ is equipped with the involution $\mathbf{s}^* := -\mathbf{s}$ for any $\mathbf{s} \in \Gamma\mathbf{H^a}$. Associated with $C \in C(\Gamma\mathbf{H^a})$ is the *kernel*

$$K_C : \Gamma\mathbf{H^a} \times \Gamma\mathbf{H^a} \longrightarrow \mathbb{C}$$

defined by

$$K_C(\mathbf{s}_1, \mathbf{s}_2) := e^{-\frac{i}{2}\hbar\cdot i\omega_X^\infty(\mathbf{s}_1,\mathbf{s}_2)} \cdot C(\mathbf{s}_2 - \mathbf{s}_1).$$

A kernel of this form is named *positive definite.* Hence this kernel can be represented by

$$K_C(\mathbf{s}_1, \mathbf{s}_2) = (\kappa_C(\mathbf{s}_1) \mid \kappa_C(\mathbf{s}_2))$$

for some mapping

$$\kappa_C : \Gamma\mathbf{H^a} \longrightarrow \mathcal{H},$$

a so-called *Kolmogorov decomposition* (cf. [**14**] and [**36**]). (Here $(\ \mid\ )$ denotes the *right-linear Hermitian product* on $\mathcal{H}$.) In fact, this kernel is *minimal*, which means that $\kappa_C(\Gamma\mathbf{H^a})$ is dense in $\mathcal{H}$ and hence κ_C is called the *minimal Kolmogorov decomposition* of C. For a minimal Kolmogorov decomposition the following holds true: There is a unique Σ*-representation*

$$W_C : \Gamma\mathbf{H^a} \longrightarrow U(\mathcal{H}),$$

of the group $\Gamma\mathbf{H^a}$, i.e. a *projective representation* for which

$$(667)\quad W_C(\mathbf{s}_1)(\kappa_C(\mathbf{s}_2)) := e^{\frac{i}{2}\hbar\cdot\omega_X^\infty(\mathbf{s}_1,\mathbf{s}_2)}\cdot\kappa_C(\mathbf{s}_1+\mathbf{s}_2)\qquad\forall\,\mathbf{s}_1,\mathbf{s}_2\in\Gamma\mathbf{H^a}.$$

It is the only *unitary* Σ-representation satisfying (667) (cf. theorem 2.1 in [**14**]).

This projective representation together with $\kappa_C(0)$ determines a Weyl system $\{W_C,\mathcal{H}_C\}$ of $\Gamma\mathbf{H^a}$ satisfying the following conditions:

i): $\{W_C(\mathbf{s})(\kappa_C(0))\mid\mathbf{s}\in\Gamma\mathbf{H^a}\}\subset\mathcal{H}$ is *total*,
ii): $C(\mathbf{s}) = (\kappa_C(0)\mid W_C(\mathbf{s})(\kappa_C(0))\qquad\forall\,\mathbf{s}\in\Gamma\mathbf{H^a}$

(cf. theorem 3.1 in [**14**]). Here $\mathcal{H}$ is the Hilbert space on which the Weyl system acts.

Vice versa let $\{W_R,\mathcal{H}_R\}$ be a Weyl system on the Hilbert space $\mathcal{H}_R$ and let ϱ be a *density operator* on $\mathcal{H}_R$, i.e. an operator of *trace class* with $\operatorname{tr}\varrho = 1$. The function

$$C_\varrho(\mathbf{s}) := tr(\varrho\circ W_R(\mathbf{s}))\qquad\forall\,\mathbf{s}\in\Gamma\mathbf{H^a}$$

is *positive definite*. Indeed, any

$$A := \sum_j z_j\cdot W_R(\mathbf{s}_j)\qquad\text{with } \mathbf{s}_j\in\Gamma\mathbf{H^a}$$

yields

$$0\le\operatorname{tr}(\varrho\circ A^*\circ A) = \sum_{i,j=1} e^{-\frac{\hbar}{2}i\cdot\omega^\infty(\mathbf{s}_1,\mathbf{s}_2)}\cdot\operatorname{tr}\varrho\circ W_R(\mathbf{s}_i-\mathbf{s}_j)$$

and, in turn,

$$\sum_{i,j=1}\bar{z}_i z_j e^{\frac{\hbar}{2}\cdot i\cdot\omega(\mathbf{s}_1,\mathbf{s}_2)}\cdot C_\varrho(\mathbf{s}_i,\mathbf{s}_j)\ge 0.$$

This expresses the positive definiteness of C_ϱ. Moreover, $C_\varrho(0) = 1$, i.e. C_ϱ is normalized.

Thus we have related the stochastic concepts of a state and density operator with the notion of a Weyl system and vice versa.

10.4.4. The Gelfand-Naimark-Segal-Representation of a Weyl Algebra. The GNS-*construction* associates with a state ℓ of a C^*-algebra $\mathcal{A}$ a triple $(\Pi_\ell,\mathcal{H}_\ell,w_\ell)$. Here $\ell\in\mathcal{A}'$ where $\mathcal{A}'$ is the algebraic dual of $\mathcal{A}$, and Π_ℓ is a C^*-algebra representation of $\mathcal{A}$ which acts on a Hilbert space $\mathcal{H}_{\Pi_\ell}$. Moreover, w_ℓ is a normalized *cyclic vector* in $\mathcal{H}_{\Pi_\ell}$. Cyclic means that $\Pi_\ell(\mathcal{A})(w_\ell)\subset\mathcal{H}_{\Pi_\ell}$ is dense. Such a representation always exists and two different GNS-representations of a state are unitarily equivalent (cf. [**36**] and [**55**]).

In the context of the Weyl algebra $W^\hbar\Gamma\mathbf{H^a}$ with $\hbar>0$ (cf. section 6.5.3) for the GNS-construction we consider a C^*-algebra representation Π of $W^\hbar(\Gamma\mathbf{H^a})$ acting on a Hilbert space $\mathcal{H}_\Pi$ together with a density operator ϱ (being a one-dimensional projection). The *functional* ℓ_ϱ on $W^\hbar\Gamma\mathbf{H^a}$ is defined by

$$\ell_\varrho(A) := \operatorname{tr}\varrho\circ\Pi(A)$$

for any $A \in W^\hbar\Gamma\mathbf{H^a}$. The linear map ℓ_ϱ is a state, as is easily verified. A state of this type is called a Π-*normal state.* The collection

$$\mathscr{F}_\Pi := \{\ell_\varrho \in \mathcal{S}(\Gamma\mathbf{H^a}) | \varrho \text{ density operator on } \mathcal{H}_\Pi\},$$

is a convex set, called the *folium* of Π.

For ϱ and a unit vector $\psi \in \operatorname{im}\varrho$ we find

$$\ell_\varrho(A) := (\psi \mid \Pi(A)(\psi)) \qquad \forall\, A \in W^\hbar\Gamma\mathbf{H^a}.$$

$(\,|\,)$ again denotes the right-linear Hermitian product on $\mathcal{H}_\Pi$.

The relation between regular C^*-algebra representations of $W^\hbar\Gamma\mathbf{H^a}$ and GNS-representations is as follows (cf. [**55**]): Let Π be a *regular* representation of $W^\hbar\Gamma\mathbf{H^a}$ meaning that Π is strongly continuous. The C^*-algebra representation Π is *quasi-equivalent* to a representation constituting of GNS-representations Π_ℓ for Π-normal states $\ell \in \mathcal{F}_\Pi$. (Two C^*-algebra representations Π_1 and Π_2 are said to be quasi-equivalent iff $\mathcal{F}_{\Pi_1} = \mathcal{F}_{\Pi_2}$(cf. [**55**]).)

In particular, the representations (666) and W_Π in (651) are made up of GNS representations (cf. [**55**]).

10.5. The Influence of the Topology on the Weyl Quantization

The topology of M influences the type of Heisenberg algebra bundle $\mathbf{H^a}$ on M through the Chern class of $\mathbb{F}^\mathbf{a}$ as mentioned in proposition 41 in chapter 4. Hence the topology affects the Weyl quantization of X as constructed in section 10.3 via the module $\Gamma\mathbf{H^a}$ of Schwartz sections (cf. section 6.2) from which the Weyl algebra $W\Gamma\mathbf{H^a}$ and the Heisenberg group C^*-algebra (cf. section 6.5) are constructed. For simplicity, M is assumed to be an oriented, *compact* ∂*-manifold.* Hence $\Gamma\mathbf{H^a}$ consists of all smooth sections of $\mathbf{H^a}$. In the present chapter we will exhibit the topological influence to this quantization in more detail.

10.5.1. Local C^*-Group-Algebra. Our first goal is to introduce the *local* C^*-Heisenberg group algebra. To this end we first fix the typical fibre of the Heisenberg group bundle $\mathbf{H^a}$ and let F be a two-dimensional linear subspace of the three-dimensional Euclidean space E. Moreover, F carries a symplectic form ω^a for some fixed $a \in S^2$ perpendicular to F. This form is extended trivially to $\mathbb{R}\cdot a$, turning ω^a to a presymplectic form for which $\mathbb{R}\cdot a$ and F are orthogonal. The Heisenberg group structure on $F \oplus \mathbb{R}\cdot a$ is defined accordingly as in section 5.3. Finally, F is endowed with a fixed complex linear structure (cf. section 1.3.2).

To form a local bundle chart of the Heisenberg group bundle $\mathbf{H^a}$ with $F \oplus \mathbb{R}\cdot a$ as a typical fibre we let $U \subset M$ be an open set on which $\mathbf{H^a}$ trivializes as a Heisenberg group bundle, i.e. there is a bundle isomorphism

$$\begin{array}{ccc} \mathbf{H^a}|_{U_\iota} & \xrightarrow{\Phi_U} & U_\iota \times (F \oplus \mathbb{R}\cdot a) \\ {\scriptstyle \mathrm{pr}}\downarrow & & \downarrow{\scriptstyle \mathrm{pr}_1} \\ U_\iota & \longrightarrow & U_\iota \end{array} \tag{668}$$

which is fibrewisely a Heisenberg group isomorphism. Here pr_1 projects to the first component U_ι.

We fix an atlas $\mathcal{A}$ on M consisting of finitely many charts on which $\mathbf{H}^{\mathbf{a}}$ trivializes. In the sequel we write $\mathbf{H}_\iota^{\mathbf{a}}$ instead of $\mathbf{H}^{\mathbf{a}}|_{U_\iota}$ for ι in the index set I of the collection of charts in $\mathcal{A}$. Accordingly, we write Φ_ι instead of Φ_{U_ι}. By $\Gamma_c\mathbf{H}_\iota^a$ we denote the $C_c^\infty(U_\iota,\mathbb{R})$-module of all smooth sections into $\mathbf{H}_\iota^a$ with compact support in U_ι. Thus

$$\Phi_\iota^* : C_c^\infty(U_\iota, \mathcal{H}^a) \longrightarrow \Gamma_c\mathbf{H}_\iota^{\mathbf{a}}, \qquad (s,\lambda) \longmapsto (s,\lambda)\circ\Phi_\iota \tag{669}$$

is a linear isomorphism; here $C_c^\infty(U_\iota, \mathcal{H}^a)$ denotes the collection of all $\mathcal{H}^a$-valued smooth functions with compact support in U_ι.

Obviously, any section $\mathbf{s} = (s,\lambda) \in \Gamma_c\mathbf{H}_\iota^{\mathbf{a}}$ can be viewed as a section in $\Gamma\mathbf{H}^{\mathbf{a}}$. This is to say that

$$\Gamma_c\mathbf{H}_\iota^{\mathbf{a}} \subset \Gamma\mathbf{H}^{\mathbf{a}} \tag{670}$$

for any open set $U_\iota \subset M$ on which $\mathbf{H}^{\mathbf{a}}$ trivializes (in fact, this holds for any open set accordingly).

The linear space

$$\mathcal{H}_{X,U_\iota}^\infty := \Gamma_c\mathbf{H}_\iota^{\mathbf{a}} \oplus \mathbb{R}\cdot e \tag{671}$$

is turned into a Heisenberg group by equipping $\Gamma_c\mathbf{H}_\iota^{\mathbf{a}}$ in (670) with a presymplectic form ω_X^∞. This form ω_X^∞ is given by

$$\begin{aligned}
\omega_X^\infty((\mathbf{s}_1,\lambda_1),(\mathbf{s}_2,\lambda_2)) : &= \int_{U_\iota} \omega^{\mathbf{a}}((\mathbf{s}_1,\lambda_1\cdot e),(\mathbf{s}_2,\lambda_2\cdot e))\mu_M \\
&= \int_M \omega^{\mathbf{a}}((\mathbf{s}_1,\lambda_1\cdot e),(\mathbf{s}_2,\lambda_2\cdot e))\mu_M \\
&= \int \omega^{\mathbf{a}}(\mathbf{s}_1,\mathbf{s}_2)\mu_M.
\end{aligned}$$

Hence $\mathcal{H}_{X,U_\iota}^\infty$ is a Heisenberg subgroup of $\mathcal{H}_X^\infty$. Therefore, we have the commuting diagram

$$\begin{array}{ccc}
\mathcal{F}_{ae}(\mathcal{H}_{X,U_\iota}^\infty,\mathbb{C})_\star & \subset & \mathcal{F}_{ae}(\mathcal{H}_X^\infty,\mathbb{C})_\star \\
\delta\uparrow & & \uparrow\delta \\
\mathcal{H}_{X,U_\iota}^\infty & \subset & \mathcal{H}_X^\infty
\end{array}$$

of group algebras. This diagram is also valid if the convolution $\star$ is replaced by the twisted convolution $\star_{\omega_X^\infty}$ (cf. section 6.5.3).

The completion of $\mathcal{F}_{ae}(\mathcal{H}_{X,U_\iota}^\infty,\mathbb{C})_\star$ with respect to the norm $\| \ \|$ (cf. section 6.5.1) is a C^*-algebra, called the local C^*-algebra of X and denoted by $C^*\mathcal{H}_{X,U_\iota}^\infty$. Since any C^*-algebra representation of $C^*\mathcal{H}_X^\infty$ can be subduced to $C^*\mathcal{H}_{X,U_\iota}^\infty$, the *inclusion*

$$C^*\mathcal{H}_{X,U\iota}^\infty \hookrightarrow C^*\mathcal{H}_X^\infty$$

is continuous as can be seen directly from the construction of a C^*-algebra in section 6.5.1.

We will show next that $C^*\mathcal{H}_X^\infty$ is the inductive limit of $C^*\mathcal{H}_{X,U\iota}^\infty$. Given an open finite covering $\{U_\iota\}_{\iota\in I}$ of M on which $\mathbf{H^a}$ trivializes we find a smooth *partition of unity* $\{\varphi_\iota\}_{\iota\in I}$ of which φ_ι has compact support for each $\iota \in I$ (cf. [**85**]). Clearly,

$$\bigcup_{\iota\in I} \mathbf{H}^{\mathbf{a}}_{U_\iota} \subseteq \mathbf{H^a},$$

and since any $\mathbf{s} \in \Gamma\mathbf{H}^a$ is of the form $\mathbf{s} = \sum_\iota \varphi_\iota \cdot \mathbf{s}_\iota$, a finite sum, the decomposition

$$\bigoplus_\iota \Gamma_c \mathbf{H}^{\mathbf{a}}_\iota = \Gamma\mathbf{H^a}$$

holds true, as well. Therefore,

$$\bigoplus_{\iota\in I} \mathcal{F}_{ae}(\mathcal{H}^\infty_{X,U_\iota}, \mathbb{C})_\star = \mathcal{F}(\mathcal{H}^\infty_X, \mathbb{C})_\star$$

which shows that

$$\lim_{\overrightarrow{\iota\in I}} C^*\mathcal{H}^\infty_{X,U\iota} = C^*\mathcal{H}^\infty_X \tag{672}$$

as $*$-algebras, where $\lim_{\overrightarrow{\iota\in I}}$ denotes the *inductive limit*. Due to section 4.3.7 it suffices to take $I = \{1,2\}$.

Since $\mathbf{H}^{\mathbf{a}}_\iota$ is trivial for any ι, both local objects $\mathbf{H}^{\mathbf{a}}_\iota$ and $C^*\mathcal{H}^\infty_{X,U_\iota}$ are not affected by the topology of M. However, since both global objects $\mathbf{H^a}$ and $C^*\mathcal{H}^\infty_X$ are affected by the Chern class of $\mathbb{F}^{\mathbf{a}}$, the topological influence of M can be expressed by the reconstruction of $C^*\mathcal{H}^\infty_X$ from the local C^*-algebra of the vector field. In addition, any C^*-representation of $C^*\mathcal{H}^\infty_X$ is determined by *local representations* emanating from the covering. Here a representation is called local if it is a representation of $C^*\mathcal{H}^\infty_{X,U_\iota}$ for some $\iota \in I$. Let us resolve this in more detail in the next section.

10.5.2. Topological Influences on the Weyl Quantization of a Vector Field. In this section $X = (\mathrm{id}\,,\mathbf{a})$ is again a *singularity free* vector field on M. As far as the topology of M is concerned, the Heisenberg group bundle $\mathbf{H^a}$ enters as the vector bundle $\mathbf{H}^\varepsilon = \mathbb{F}^\varepsilon \oplus \mathcal{L}$, the Whitney sum of the *complex line bundle* $\mathbb{F}^\varepsilon$ and the *trivial real line* bundle $\mathcal{L}$ (cf. (139)). The non-vanishing field strength $|\mathbf{a}|$ has no influence on the topology. Each fibre $\mathcal{L}_x$ is of the form

$$\mathcal{L}_x = \mathbb{R}\cdot\varepsilon(x) \qquad \forall\, x \in M$$

with $\varepsilon(x) = \frac{\mathbf{a(x)}}{|\mathbf{a(x)}|}$ again.

The group $U^a(1)$ operates on $F \oplus \mathbb{R}\cdot a$ by right multiplication and on $\mathbb{R}\cdot a$ by the identity where $a \in S^2 \subset E$ is a unit vector perpendicular to F. Thus the characteristic bundle $\mathbb{P}^\varepsilon$ of $(\mathrm{id}\,,\varepsilon)$ (cf. section 3.7.1) is contained in $\mathbf{H}^\varepsilon$. Therefore, for the topological discussions we replaced $\mathbf{H^a}$ by $\mathbf{H}^\varepsilon$ as a bundle associated with $\mathbb{P}^\varepsilon$ reading

$$\mathbf{H}^\varepsilon = \mathbb{P}^\varepsilon \times_{U^a(1)} F \oplus \mathbb{R}\cdot a$$

(cf. chapter 4.3). Here $U^a(1)$ is pointwisely canonically isomorphic to $U^{\varepsilon(x)}(1)$ for each $x \in M$ and hence operates on $\mathbb{P}^\varepsilon_x$ via $U^{\varepsilon(x)}(1)$.

These remarks suggest the following approach for the formulation of the topological influence of M on the Weyl quantization of $X = (\mathrm{id}\,,\mathbf{a})$: As far as the Heisenberg group bundle $\mathbf{H}^{\mathbf{a}}$ and its section spaces $\Gamma\mathbf{H}^{\mathbf{a}}$ are concerned, the topological influence of M will be expressed through $\mathbf{H}^{\varepsilon}$ and in turn through $\Gamma\mathbf{H}^{\varepsilon}$. The dependence of $\mathbf{H}^{\varepsilon}$ on the topology of M relies on the Chern–de Rham Class of $\mathbb{F}^{\varepsilon}$ (cf. section 4.3.7) which is expressed in the mapping degree $\deg \Xi$ of Ξ (cf. theorem 48); hence we could concentrate on the vector field $(\mathrm{id}\,,\varepsilon)$. However, for the *Weyl quantization* of the vector field $X = (\mathrm{id}\,,\mathbf{a})$ we use (infinite dimensional) $*$-Poisson algebra $P^{\mathbf{a}}$, the C^*-Heisenberg group algebra $C^*\mathcal{H}_X^\infty$ and the Weyl algebra $W^\hbar\Gamma\mathbf{H}^{\mathbf{a}}$, all depending on the presymplectic form ω_X^∞ on $\Gamma\mathbf{H}^{\varepsilon}$ (cf. 6.2.1). This form, however, is determined by the field strength of the vector field $X = (\mathrm{id}\,,\mathbf{a})$. Hence we will use the principal part $\mathbf{a}$.

The topology of M is hidden in the $*$-algebras $P^{\mathbf{a}}$, $C^*\mathcal{H}_X^\infty$ and $W^\hbar\Gamma\mathbf{H}^{\mathbf{a}}$, all characteristic for X (cf. section 6.5.5). To exhibit its influence on the *Weyl quantization* we will focus on $C^*\mathcal{H}_X^\infty$ and its representations only; the other C^*-algebras and representations involved are treated in the same way. Here are the details:

As shown in section 4.3.7 there is a two-chart trivialization of $\mathbb{F}^{\mathbf{a}}$ and hence of $\mathbf{H}^{\mathbf{a}}$. The charts shall be denoted by U_N^ε and U_S^ε, being pull-backs by ε of the respective charts U_N and U_S. The open sets $U_N^\varepsilon, U_S^\varepsilon$ and $U_N^\varepsilon \cap U_S^\varepsilon$ form a directed system and we may consider the trivialization

$$\Phi_N : \mathbf{H^a}_{U_N^\varepsilon} \longrightarrow U_{N^\varepsilon} \times F^a \oplus \mathbb{R}\cdot a$$

for a specified $a \in S^2$. Accordingly, we have Φ_S and $\Phi_{N\cap S}$. Moreover, $\Phi_{S^\varepsilon}\circ(\Phi_{N^\varepsilon})^{-1}$ is rephrased in terms of the transition function

$$\begin{array}{rcl} \varphi_{NS} : U_{N^\varepsilon} \times F^a \oplus \mathbb{R}\cdot a & \longrightarrow & U_{S^\varepsilon} \times F^{-a} \oplus \mathbb{R}\cdot(-a) \\ (x,h,\lambda\cdot a) & \longmapsto & (x, h\cdot e^{-2g(x)\cdot a} - \lambda\cdot a) \end{array} \tag{673}$$

mentioned in (212) for $\mathbb{F}^{\mathbf{a}}$. Now the antipodal points $a, -a \in S^2$ are identified with the points N and S, respectively. Passing on to the spaces of sections the bundle isomorphism φ_{NS}^ε acts as

$$\varphi_{NS}(\mathbf{s}+\nu\cdot e)(x) := s(x)\cdot e^{-2g(x)\cdot a} + \nu(x)\cdot e + \lambda(x)\cdot a$$

for any $\mathbf{s} = s + \lambda\cdot a \in \Gamma(U_{N\cap S}^\varepsilon \times (F^a + \mathbb{R}\cdot a))$, any $\lambda, \nu \in C^\infty(U_{N\cap S}^\varepsilon)$, and any $x \in U_{N\cap S}^\varepsilon$. The latter space shall be denoted by $\Gamma_{N\cap S}^a$ and accordingly we interpret $\Gamma_{N\cap S}^{-a}$ for $-a \in S^2$.

In view of (671), the atlas consisting of the three charts $U_N^\varepsilon, U_S^\varepsilon, U_N^\varepsilon \cap U_S^\varepsilon$ (for simplicity, denoted by V_N, V_S, and $V_N \cap V_S$) yields the three infinite dimensional Heisenberg groups

$$\begin{array}{rcl} \mathcal{H}_{X,V_N}^\infty & = & \Gamma_c\mathcal{H}_{V_N}^{\mathbf{a}} \oplus \mathbb{R}\cdot e, \\ \mathcal{H}_{X,V_S}^\infty & = & \Gamma_c\mathcal{H}_{V_S}^{\mathbf{a}} \oplus \mathbb{R}\cdot e, \\ \mathcal{H}_{X,V_N\cap V_S}^\infty & = & \Gamma_c\mathcal{H}_{V_N\cap V_S}^{\mathbf{a}} \oplus \mathbb{R}\cdot e. \end{array}$$

The transition map φ_{NS} determines via diagram (668) a Heisenberg group isomorphism denoted by

$$\begin{array}{rcl} \varphi_{NS}^{\mathcal{H}} : \mathcal{H}^\infty{}_{X,V_N\cap V_S} & \longrightarrow & \mathcal{H}_{X,V_N\cap V_S}^\infty \\ s & \longmapsto & \varphi_{NS}(s) \end{array}$$

(cf. section 6.3). In turn φ_{NS} extends to an isomorphism between the local group algebras

$$\varphi_{NS} : \mathcal{F}_{ae}(\mathcal{H}^{\infty}_{X,V_N \cap V_S}, \mathbb{C})_{\star} \longrightarrow \mathcal{F}_{ae}(\mathcal{H}^{\infty}_{X,V_N \cap V_S}, \mathbb{C})_{\star} \tag{674}$$

and yields a $\star$-isomorphism between the respective C^*-algebras (cf. section 6.4.1). Moreover,

$$\varphi_{NS}(\delta_s) := \delta_{\varphi^{\varepsilon}_{NS}(s)} \qquad \forall\, s \in \mathcal{H}^{\infty}_a|_{V^{\varepsilon}_N}.$$

The influence of the topology to the Weyl quantization comes into effect by (674) as follows: The topological aspects of the bundle $\mathbb{F}^{\mathbf{a}}$ and in consequence of $\mathbf{H}^{\mathbf{a}}$ and the completion $C^*\mathcal{H}^{\infty}_X$ of $\mathcal{F}_{ae}(\mathcal{H}^{\infty}_X, \mathbb{C})_{\star}$ are encoded in the Čech cocycle φ_{NS}. This cocycle defines a Čech cohomology class, characteristic for the complex line bundle $\mathbb{F}^{\mathbf{a}}$. The image under the connecting homomorphism (cf. equations (203) and (208)) of the Čech class yields the integral de Rham class of the curvature $\Omega^{\mathbf{a}}$. A fundamental aspect of quantization is manifested in the representations of $C^*\mathcal{H}^{\infty}_X$ (among other C^*-algebras). The Čech cocycle φ_{NS} provides a link between the representations of the global object $C^*\mathcal{H}^{\infty}_X$ and the representations of the local C^*-algebras presented in the previous section. This link reflects the influence of topology to quantization.

In view of (672), the following is now immediate:

THEOREM 110. *Given the local $*$-representations*

$$\Pi_N : C^*\mathcal{H}^{\infty}{}_{X,V_N} \longrightarrow \mathbf{B}(\mathcal{H}) \qquad \text{and} \qquad \Pi_S : C^*\mathcal{H}^{\infty}_{X,V_S} \longrightarrow \mathbf{B}(\mathcal{H}),$$

they define a global representation

$$\Pi : C^*\mathcal{H}^{\infty}_X \longrightarrow \mathbf{B}(\mathcal{H})$$

subducing to Π_N and Π_S iff

$$\Pi_S \circ \varphi_{NS} = \Pi_N \qquad \text{on} \qquad C^*\mathcal{H}^{\infty}_{X,V_N \cap V_S}.$$

Here $\mathbf{B}(\mathcal{H})$ denotes the C^*-algebra of all bounded operators on the Hilbert space $\mathcal{H}$. The two-chart cocycle $\varphi_{NS} : V_N \cap V_S \longrightarrow U^a(1)$ satisfies

$$[\Omega^{\mathbf{a}}] = \iota^* \circ \delta([\varphi_{NS}])$$

where $[\Omega^{\mathbf{a}}] = [\Omega^{\varepsilon}]$ is de Rham class of the curvature $\Omega^{\mathbf{a}}$ of $\mathbb{F}^{\mathbf{a}}$. Moreover, for the compact δ-manifold M,

$$[\Omega^{\varepsilon}] = 2 \cdot \deg \Xi$$

with Ξ as in theorem 48.

By theorem 87 the C^*-algebra $C^*\mathcal{H}^{\infty}_X$ describes the C^*-algebra of Weyl fields out of which the C^*-Weyl algebra $W^{\hbar}\Gamma\mathbf{H}^{\mathbf{a}}$ emanates. This C^*-Weyl algebra is the key to Weyl quantization. An analogue to theorem 110 holds true verbatim for the Weyl algebra $W^{\hbar}(\Gamma\mathbf{H}^{\mathbf{a}})$, as is easily seen.

The quantization procedure for C^*-Weyl algebras presented in this chapter can be generalized to any complex line bundle on M provided an additional real line bundle is specified in order to construct a Heisenberg group bundle.

APPENDIX A

Thermodynamics, Geometry and the Heisenberg Group by Serge Preston

In this appendix we discuss geometrical properties of the thermodynamical phase space ([**A12, A17**]) of a homogeneous thermodynamical system [**A5**].

We study the indefinite metric G in the contact phase space (P, θ) introduced by R. Mrugala. We calculate the curvature tensor, Killing vector fields, and the second fundamental form of Legendre submanifolds of P-constitutive surfaces of different homogeneous thermodynamical systems. We establish an isomorphism of the space (P, θ, G) with the Heisenberg Lie group $\mathcal{H}_n$ endowed with the right-invariant contact structure and the right-invariant indefinite metric. The lift $\tilde{G}$ of the metric G to the symplectization $\tilde{P}$ of the contact space (P, θ), its curvature properties, and its Killing vector fields are studied. Finally, we introduce the *hyperbolic projectivization* of the space $(\tilde{P}, \tilde{\theta}, \tilde{G})$ that can be considered as the natural **compactification** of the thermodynamical phase space (P, θ, G).

A.1. Introduction

Geometrical methods in the study of homogeneous thermodynamical systems were pioneered by J. Gibbs ([**A8**]) and C. Caratheodory ([**A6**]). They were further developed in the works of R. Hermann ([**A12**]), R. Mrugala, P. Salamon and their collaborators, in the dissertations of H. Heemeyer ([**A11**]) and L. Benayoun ([**A3**]). Thermodynamical metrics (TD-metrics) in the form of the Hessian of a thermodynamical potential were explicitly introduced by F. Weinhold ([**A27**]) and, from a different point of view, by G. Ruppeiner ([**A23**]).

More indepth studies by P. Salamon and his collaborators, and by P. Mrugala and H. Janyszek (see review papers [**A7, A18, A24**] and the references therein) clarified principal properties of thermodynamical metrics, relations between different TD-metrics, and their relations to the contact structure of an equilibrium thermodynamical phase space.

G. Ruppeiner (see [**A23**] and the review [**A24**]) has developed a covariant thermodynamical fluctuation theory based on the Riemannian metric η_S defined by the second momenta of entropy with respect to the fluctuations. He also related the curvature of this metric to the correlational volume near the critical points of thermodynamical systems.

In his work [**A17**] (see also the review paper [**A18**]) R. Mrugala introduced the pseudo-Riemannian (indefinite) metric G of signature $(n+1, n)$ in the thermodynamical contact space (P, θ). The restriction of the metric G to the constitutive surfaces defined by different thermodynamical potentials coincides with the corresponding thermodynamical metric (the Weinhold metric for the internal energy and Ruppeiner metric for the entropy).

Studying geometrical properties of the metric G and its relation to the contact structure θ of the thermodynamical phase space (shortly TPS) (P, θ) (cf. [**A22**]) is enlightened enormously by the fact that the TPS P is diffeomorphic, together with its contact structure θ and the compatible indefinite metric G to the Heisenberg group $\mathcal{H}_n$ with the right invariant contact structure $\theta_{\mathcal{H}}$ and the right-invariant indefinite metric $G_{\mathcal{H}}$. The goal of this present appendix is to present the relation of the space (P, θ, G) to the Heisenberg group space $(\mathcal{H}_n, \theta_{\mathcal{H}}, G_{\mathcal{H}})$ and to exploit these relations for the study of geometry of thermodynamical phase spaces.

A short exposition of most results of this appendix is given in [**A29**].

A.2. The Contact Structure of Homogeneous Thermodynamics

A phase space of homogeneous thermodynamics (thermodynamical phase space) is the $(2n+1)$-dimensional vector space $P = \mathbb{R}^{2n+1}$ endowed with the standard contact structure (cf. [**A2, A12**])

$$\theta = dx^0 + \sum_{i=1}^{n} p_l dx^l. \tag{675}$$

The horizontal distribution D_m of this structure is generated by two families of vector fields: P_i, X_i as

$$D_m =< P_i = \partial_{p_l},\ X_i = \partial_{x^i} - p_i \partial_{x^0} > .$$

The two-form

$$\omega = d\theta = \sum_{l=1}^{n} dp_l \wedge dx^l$$

is a non-degenerate, symplectic form on the distribution D.

The Reeb vector field, uniquely defined as the generator ξ of $\ker(d\theta)$ satisfying $\theta(\xi) = 1$, is simply

$$\xi = \partial_{x^0}.$$

A.3. Gibbs Space. Legendre Surfaces of Equilibrium

Constitutive surfaces of concrete thermodynamical systems are determined by their "constitutive equations", which in their fundamental form determine the value of a **thermodynamical potential** $x^0 = E(x^i)$ as the function of n extensive variables x^i. Dual intensive variables are then determined as the partial derivatives of the thermodynamical potential by the extensive variables $p_i = \frac{\partial E}{\partial x^i}$.

Thus a constitutive surface represents the **Legendre submanifold** (i.e. the maximal integral submanifold) Σ_E of the contact form θ projecting diffeomorphically to the space X of variables x^i. The space Y of variables $x^0, x^i,\ i = 1, \dots, n$ is sometimes named the **Gibbs space** of the thermodynamical potential $E(x^i)$. The thermodynamical phase space (P, θ) (or, more precisely, its open subset) appears as the **first jet space** $J^1(Y \to X)$ of the (trivial) line bundle $\pi : Y \to X$. Projection of Σ_E to the Gibbs space Y is the **graph** Γ_E of the constitutive law $E = E(x^i)$.

Another choice of the thermodynamical potential together with the n-tuple of extensive variables leads to another representation of an open subset of TPS P as the 1-jet bundle of the corresponding Gibbs space. It is known that any Legendre submanifold of a contact manifold (X, θ) can locally be presented in this form for some choice of the set of extensive variables and of the thermodynamical potential as the function of these variables.

We will be using a local description of Legendre submanifolds which takes a slightly different form. Let P^{2n+1} be a contact manifold. The following result (locally) characterizes all Legendre submanifolds (V.I. Arnold, [**A1, A2**]).

Choose (local) Darboux coordinates (x^0, x^i, p_j) in which $\theta = dx^0 + p_k dx^k$. Let I, J be a partition of the set of indices $1, \dots, n$, and consider any function $\phi(p_i, x^j)$, $i \in I, j \in J$. Then the following equations define a Legendre submanifold Σ_ϕ:

$$(676)\qquad \begin{cases} x^0 & = & \phi - \sum_{i \in I} p_i \frac{\partial \phi}{\partial p_i}, \\ p_j & = & -\frac{\partial \phi}{\partial x^j},\ j \in J, \\ x^i & = & \frac{\partial \phi}{\partial p_i},\ i \in I. \end{cases}$$

Moreover, every Legendre submanifold is locally given by some choice of a splitting I, J and of a function $\phi(p_i, x^j)$ with $i \in I$ and $j \in J$.

In physics, the most commonly used thermodynamical potentials ϕ are: the internal energy U, the entropy S, the free energy of Helmholtz F, the enthalpy and the free Gibbs energy.

On the intersection of the domains of these representations, corresponding points are related by a **Legendre transformation** (cf. [**A1**]).

EXAMPLE 1. As an example of such a thermodynamical system, consider the van der Waals gas — a system with two thermodynamical degrees of freedom. The space P is 5-dimensional (for 1 mole of gas) with the canonical variables $(s, (-\frac{1}{T}, u), (-\frac{p}{T}, v))$ (internal energy per 1 mole, temperature, entropy per 1 mole, pressure, volume of 1 mole of gas), the contact form is

$$\theta = ds - \frac{1}{T} du - \frac{p}{T} dv,$$

and the fundamental constitutive law ([**A5**]) is taken to be

$$s(u, v) = s_0 + R \cdot \ln[(v - b)(u + \frac{a}{v})^{c_V}].$$

Here R is the Rydberg constant, c_V is the heat capacity at constant volume, a, b are parameters of the gas reflecting the interaction between molecules and the part of volume occupied by molecules, respectively.

A.4. Thermodynamical Metrics of Weinhold and Ruppeiner

A thermodynamical metric η_U (the Weinhold metric) in the space X of extensive variables corresponding to the choice of an internal energy U as the thermodynamical potential E was explicitly introduced by F. Weinhold (cf. [**A27**]) as the Hessian of the internal energy $U(x^i)$,

$$(677)\qquad \eta_{U\ ij} = \frac{\partial^2 U}{\partial x^i \partial x^j}.$$

G. Ruppeiner's metric η_S corresponding to the choice of the entropy S as the thermodynamical potential E was defined by the same formula and intensively studied by Ruppeiner ([**A23**]) in the framework of the fluctuational theory of thermodynamical systems.

Interest in these metrics is partly due to the fact that the definiteness of η_U (positive or negative) at a point $x \in X$ of the constitutive surface delivers the local criteria of stability of the equilibria given by the corresponding point of the surface Γ_E or Σ_E (cf. [**A5**]).

An interesting meaning important in applications was assigned to the length of curves ("processes") in the space X ([**A7**]). Also, it was suggested that the curvature of these metrics is related to the interactions in the (microscopical) system and that singularities of the scalar curvature of these metrics were related to the properties of the system near the surfaces separating different phases and the triple point of the thermodynamical system (cf. [**A23, A24**]).

A.5. Indefinite Thermodynamical Metric G of R. Mrugala

In the paper [**A17**], R. Mrugala defined a pseudo-Riemannian (indefinite) metric G of signature $(n+1, n)$ in the thermodynamical contact space (P, θ). It is given by the formula

$$G = 2dp_k \odot dx^k + \theta \otimes \theta, \tag{678}$$

where $dp_k \odot dx^k = \frac{1}{2}(dp_k \otimes dx^k + dx^k \otimes dp_k)$ is the symmetrical product of 1-forms.

Its physical motivation is twofold. First, it appeared naturally in the framework of statistical mechanics. Second, its reduction to the Legendre submanifolds Σ_ϕ corresponding to the choice of entropy or internal energy as the TD potential coincides with the previously studied Ruppeiner and Weinhold metrics η_ϕ. More specifically, one has

THEOREM 1 (R. Mrugala, [**A17**]). *Let $i_\phi : \Sigma_\phi \to P$ be a constitutive (Legendre) submanifold of (P, θ) corresponding to the TD potential $\phi(x^i)$. Then*

$$i_\phi^*(G) = \eta_\phi.$$

Returning to the space P we see that in the coordinates x^0, p_s, x^i, the matrix of the metric G is

$$G = (G_{ij}) = \begin{pmatrix} 1 & 0 & p_j \\ 0 & 0 & I_n \\ p_i & I_n & p_i p_j \end{pmatrix}. \tag{679}$$

It is easy to see that $\det(G) = (-1)^n$ and, therefore, the metric G is non-degenerate.

It is convenient to introduce the following non-holonomic frame in the tangent bundle $T(P)$ (cf. [**A13**])

$$\xi = \partial_{x^0}, \ P_l = \partial_{p_l}, \ X_i = \partial_{x^i} - p_i \partial_{x^0} \tag{680}$$

whose only non-zero commutator relation is

$$[P_i, X_j] = -\delta_{ij}\xi.$$

In this frame the metric G takes the simple form

$$G = (G_{ij}) = \begin{pmatrix} 1 & 0 & 0 \\ 0 & 0 & I_n \\ 0 & I_n & 0 \end{pmatrix}. \tag{681}$$

The dual coframe of the frame (680) is given by (θ, dp_l, dx^i).

The metric G is compatible with the contact structure θ in the following natural sense (see [**A22**] or [**A21**] for more details):

THEOREM 2. (1) *The indefinite metric G is compatible with the contact structure θ in the following sense. For a natural* ***almost contact structure*** *(ϕ, θ, ξ) on (P, θ) (see [**A17**]) where*

$$\phi = \begin{pmatrix} 0 & p_1, \dots, p_n & 0 \\ 0 & 0_n & I_n \\ 0 & -I_n & 0_n \end{pmatrix}$$

one has

$$G(\phi X, \phi Y) = -G(X, Y) - \theta(X)\theta(Y), \ \ X, Y \in T(P).$$

(2) *(P, θ, G) is the* ***indefinite Sasaki manifold****: The Nijenhuis Tensor of the almost complex structure*

$$J(X, f\partial_t) = (\phi(X) - f\xi, \eta(X)\partial_t)$$

on the symplectization $\tilde{P}$ of (P, θ) is zero.

A.6. Levi-Civitá Connection of the Metric G

In this section, we present the Levi-Civitá connection $\Gamma^{\alpha}_{\beta\gamma}$ of the metric G and calculate the covariant derivatives of vector fields of the frame (680) with respect to the vector fields of the same frame.

We have (for the proof, which is a standard exercise in calculations, see [**A22**]):

PROPOSITION 1. *The non-zero Christoffel coefficients $\Gamma^{\alpha}_{\beta\gamma}$ of metric G are the following:*

$$\Gamma^0_{0x^i} = \frac{1}{2}p_i, \ \Gamma^0_{x^i p_j} = \frac{1}{2}\delta_{ij}, \ \Gamma^0_{x^i x^j} = p_i p_j, \ \Gamma^{p_i}_{0p_j} = \frac{1}{2}\delta^i_j,$$
$$\Gamma^{x^i}_{0x^j} = -\frac{1}{2}\delta^i_j, \ \Gamma^{p_i}_{x^j p_k} = \frac{1}{2}\delta^i_k p_j, \ \Gamma^{x^i}_{x^j x^k} = -\frac{1}{2}\left(\delta^i_j p_k + \delta^i_k p_j\right). \tag{682}$$

The covariant derivative of the connection Γ takes a particularly simple form when expressed in the non-holonomic frame (680):

PROPOSITION 2. *In the canonical frame (ξ, P_i, X_i) the covariant derivatives of metric G are*

$$\begin{aligned} &\nabla_\xi \xi = 0, \nabla_{P_i} P_j = 0, \nabla_{X_i} X_j = 0, \\ &\nabla_\xi P_i = \nabla_{P_i}\xi = \frac{1}{2}P_i, \\ &\nabla_\xi X_j = \nabla_{X_j}\xi = -\frac{1}{2}X_j, \\ -&\nabla_{P_i} X_j = \frac{1}{2}\delta_{ij}\xi = \nabla_{X_i} P_j. \end{aligned} \tag{683}$$

PROOF. Let G be any pseudo-Riemannian metric. Then, for any vector fields X, Y, Z, the Levi-Civitá connection of the metric G satisfies:

$$YG(X,Z) = G(\nabla_Y X, Z) + G(X, \nabla_Y Z),$$
$$ZG(X,Y) = G(\nabla_Z X, Y) + G(X, \nabla_Z Y),$$
$$XG(Y,Z) = G(\nabla_X Y, Z) + G(Y, \nabla_X Z).$$

If we add the first two equations and subtract the third, the result is:

$$2G(X, \nabla_Y Z) = YG(X,Z) + ZG(X,Y) - XG(Y,Z) + G(Y, [X,Z]) + G(Z, [X,Y]) + G(X, [Y,Z]). \tag{684}$$

Here we have used the fact that the Levi-Civitá connection is symmetric; i.e. $\nabla_X Y - \nabla_Y X = [X,Y]$ for all X, Y.

In taking X, Y, Z from the vectors of the frame $\{\xi, P_i, X_j\}$, the scalar products $G(X,Y)$, $G(Y,Z), G(Z,X)$ will all be constant; therefore, the first three terms on the right-hand side of the formula above will vanish, leaving

$$2G(X, \nabla_Y Z) = G(Y, [X,Z]) + G(Z, [X,Y]) + G(X, [Y,Z]). \tag{685}$$

Among the basic vectors, the only pair of vectors with non-zero Lie bracket is $[P_i, X_j] = -\delta_{ij}\xi$. It follows that if we substitute the basis vectors into the equation above, the right-hand side is equal to zero unless two of the vectors are P_i and X_j, respectively. Since their bracket is proportional to ξ, which is orthogonal to the contact distribution, the third vector must be ξ. In particular, we immediately obtain the following relations:

$$\nabla_\xi \xi = 0, \qquad \nabla_{P_i} P_j = 0, \qquad \nabla_{X_i} X_j = 0.$$

Additionally, we see that the only non-zero component of $\nabla_{P_i} X_j$ is the ξ-component, which is found by

$$2G(\xi, \nabla_{P_i} X_j) = G(\xi, \delta_{ij}\xi) = -\delta_{ij}.$$

Therefore, $\nabla_{P_i} X_j = -\frac{1}{2}\delta_{ij}\xi$. Note that interchanging P_i with X_j changes the sign in the right-hand side: $\nabla_{X_j} P_i = \frac{1}{2}\delta_{ij}\xi$.

The next equation to consider is

$$2G(P_i, \nabla_\xi X_j) = G(\xi, [P_i, X_j]) = -\delta_{ij}.$$

It follows that $\nabla_\xi X_j = -\frac{1}{2} X_j$.

Interchanging the roles of P_i, X_j yields $\nabla_\xi P_i = \frac{1}{2} P_i$. By the symmetry of the connection $\nabla_{X_j}\xi = \nabla_\xi X_j$ and $\nabla_{P_i}\xi = \nabla_\xi P_i$ since $[\xi, P_i] = 0$. This ends the proof. □

A.7. Curvature Properties of G

Curvature properties of metric G are collected in the following statement:

THEOREM 3. (1) *The Ricci Tensor of the metric G in the frame $\xi, \partial_{p_j}, \partial_{x^i}$ is*

$$Ric(G) = (R_{ij}) = \begin{pmatrix} -\frac{n}{2} & 0 & -\frac{n}{2}p_j \\ 0 & 0 & \frac{1}{2}\delta_{ij} \\ -\frac{n}{2}p_i & \frac{1}{2}\delta_{ij} & -\frac{n}{2}p_i p_j \end{pmatrix}.$$

(2) *The scalar curvature of the metric G is*

$$R(G) = \frac{n}{2}.$$

(3) *The sectional curvatures $K(X,Y)$, $X,Y = \partial_{x^0}, \partial_{x^i}, \partial_{p_j}$ for non-degenerate planes (X,Y) are*

$$K(X,Y) = 0, \text{ except } K(\partial_{x^i}, \partial_{p_i}) = \frac{3}{4}.$$

Next we describe the Lie algebra of Killing vector fields of the metric G.

PROPOSITION 3. **Killing vector fields: $\mathfrak{iso}_G$** *The Lie algebra $\mathfrak{iso}_G$ of the isometry group $Iso(G)$ of the metric G is the Lie algebra $\mathfrak{gl}(n,\mathbb{R}) \times \mathfrak{h}_n$ — the semidirect product of linear Lie algebra $\mathfrak{gl}(n,R)$ with generators $\{Q^k_l = p_l\partial_{p_k} - x^k\partial_{x^l}\}$ and of the Heisenberg Lie algebra $\mathfrak{h}_n$ with generators $\{\xi = \partial_{x^0}, A_i = \partial_{p_i} + x^i\partial_{x^0}, B_j = -\partial_{x^j}\}$ and commutative relations*

$$[A^i, B_j] = \delta_{ij}\xi, \quad [Q^k_l, A_i] = -\delta_{il}A_k, \quad [Q^k_l, B_j] = \delta_{kj}B_l,$$
$$[Q^k_l, Q^r_s] = \delta_{ks}Q^r_l - \delta_{rl}Q^k_s.$$

All these vector fields are θ-contact with the contact Hamiltonians

$$H_\xi = 1, H_{A_j} = p_j, H_{B_i} = x^i, H_{Q^k_l} = x^k p_l.$$

For the proof of these results and for some additional results, see [**A22**].

A.8. The Heisenberg Group as the Thermodynamical Phase Space

The commutation relations for the vector fields of the frame (680) show that the Heisenberg group $\mathcal{H}_n$ acts at least locally on the space P. Actually, much more is true. In this section we establish an isomorphism of the TPS (P,θ,G) (with its contact structure and indefinite metric G) with the Heisenberg group $\mathcal{H}_n$ endowed with the right invariant contact structure $\theta_{\mathcal{H}}$ and the right invariant indefinite metric $G_{\mathcal{H}}$.

Recall ([**A19**]) that the Heisenberg group $\mathcal{H}_n$ is the nilpotent Lie group of $n \times n$ real matrices

$$g = g(\bar{a},\bar{b},c) = \begin{pmatrix} 1 & \bar{a} & c \\ 0 & I_n & \bar{b} \\ 0 & 0 & 1 \end{pmatrix} \tag{686}$$

with the product

$$gg_1 = \begin{pmatrix} 1 & \bar{a} & c \\ 0 & I_n & \bar{b} \\ 0 & 0 & 1 \end{pmatrix} \cdot \begin{pmatrix} 1 & \bar{a}_1 & c_1 \\ 0 & I_n & \bar{b}_1 \\ 0 & 0 & 1 \end{pmatrix} = \begin{pmatrix} 1 & \bar{a}+\bar{a}_1 & c+c_1+<\bar{a},\bar{b}_1> \\ 0 & I_n & \bar{b}+\bar{b}_1 \\ 0 & 0 & 1 \end{pmatrix}, \tag{687}$$

where $<\bar{a},\bar{b}_1>$ is the Euclidean scalar product of two vectors in $\mathbb{R}^n$ (see, for instance, [**A19**]).

The Lie group $\mathcal{H}_n$ is the central extension of the abelian group $\mathbb{R}^{2n}_{\bar{a},\bar{b}} = \mathbb{R}^n_{\bar{a}} \oplus \mathbb{R}^n_{\bar{b}}$ with the local parameters $\bar{a}$ and $\bar{b}$, respectively, by the one-dimensional abelian group $\mathbb{R}_c$ with the local parameter z:

$$1 \to \mathbb{R}_c \to \mathcal{H}_n \to \mathbb{R}^{2n}_{\bar{a},\bar{b}} \to 1.$$

The Lie algebra $\mathfrak{h}_n$ of the Heisenberg group is formed by the matrices

$$X(\bar{a}, \bar{b}, z) = \begin{pmatrix} 0 & \bar{a} & z \\ 0 & 0 & \bar{b} \\ 0 & 0 & 0 \end{pmatrix} \tag{688}$$

with the conventional matrix bracket as the Lie algebra operation. The Lie algebra $\mathfrak{h}_n$ is mapped diffeomorphically onto $\mathcal{H}_n$ by the exponential mapping

$$\exp \begin{pmatrix} 0 & \bar{a} & z \\ 0 & 0 & \bar{b} \\ 0 & 0 & 0 \end{pmatrix} = \begin{pmatrix} 1 & \bar{a} & c = z + \frac{1}{2} < \bar{a}, \bar{b} > \\ 0 & I_n & \bar{b} \\ 0 & 0 & 1 \end{pmatrix}. \tag{689}$$

The Lie algebra $\mathfrak{h}_n$ is, in turn, the central extension

$$0 \to \mathbb{R}_z \to \mathfrak{h}_n \to \mathbb{R}^{2n}_{\bar{a},\bar{b}} \to 0$$

defined by the 2-cocycle

$$\omega_+ : \mathbb{R}^{2n} \bigwedge \mathbb{R}^{2n} \to \mathbb{R}$$

given by

$$\omega_+(X(\bar{a}, \bar{b}, z), X(\bar{a}', \bar{b}', z')) = \frac{1}{2}(< \bar{a}, \bar{b}' > - < \bar{a}', \bar{b} >),$$

the canonical symplectic form in $\mathbb{R}^{2n}$ (cf. [**A19**]).

We construct the diffeomorphic mapping

$$\chi : H_n \longrightarrow P$$

by defining

$$\chi : \; g = \begin{pmatrix} 1 & \bar{x} & x^0 \\ 0 & I_n & \bar{p} \\ 0 & 0 & 1 \end{pmatrix} \mapsto m = \begin{pmatrix} -x^0 \\ \bar{p} \\ \bar{x} \end{pmatrix}. \tag{690}$$

The action of the group $\mathcal{H}_n$ on itself by the left translation: $L_g : g_1 \to gg_1$ defines the corresponding left action of $\mathcal{H}_n$ on the space P as $T_g : m \to \chi(L_g \chi^{-1}(m))$, or

$$T_g \begin{pmatrix} x^0 \\ \bar{p} \\ \bar{x} \end{pmatrix} := \chi\left(\begin{pmatrix} 1 & \bar{a} & z \\ 0 & I_n & \bar{b} \\ 0 & 0 & 1 \end{pmatrix} \cdot \begin{pmatrix} 1 & \bar{x} & -x^0 \\ 0 & I_n & \bar{p} \\ 0 & 0 & 1 \end{pmatrix} \right) = \begin{pmatrix} x^0 - c - \bar{a} \cdot \bar{p} \\ \bar{p} + \bar{b} \\ \bar{x} + \bar{a} \end{pmatrix}. \tag{691}$$

In these terms, let us find the generators of the left action of the basic one-parameter subgroups of the group $\mathcal{H}_n$ corresponding to the elements $A_i = X(e_i, 0, 0)$, $B_j = X(0, e_i, 0), Z = X(0, 0, 1)$ of the Lie algebra $\mathfrak{h}_n$ on the vectors of P. For any element $X \in \mathfrak{h}_n$ let ξ_X and η_X denote the right and left invariant vector fields on $\mathcal{H}_n$ generated by the left and right translations by $\exp(tX)$, respectively.

In coordinates $(\bar{a}, \bar{b}, c)$ we have

$$\xi_Z = \partial_c, \qquad \xi_{A_i} = \partial_{a_i} + b_i \partial_c, \qquad \xi_{B_j} = \partial_{b_j}.$$

Applying the diffeomorphism χ to these vector fields we get the correspondences

$$\begin{pmatrix} 1 & 0 & t \\ 0 & I_n & 0 \\ 0 & 0 & 1 \end{pmatrix} \cdot \chi^{-1}(m) = \begin{pmatrix} 1 & \bar{x} & -x^0+t \\ 0 & I_n & \bar{p} \\ 0 & 0 & 1 \end{pmatrix} \to \chi_{*m}(\xi_Z) = -\partial_{x^0}, \tag{692}$$

$$\begin{pmatrix} 1 & 0,0,\ldots,t_i,0\ldots & 0 \\ 0 & I_n & 0 \\ 0 & 0 & 1 \end{pmatrix} \cdot \chi^{-1}(m) = \begin{pmatrix} 1 & \bar{x}+t\bar{e}_i & x^0+p_it \\ 0 & I_n & \bar{p} \\ 0 & 0 & 1 \end{pmatrix} \tag{693}$$
$$\to \chi_{*m}(\xi_{A_i}) = \partial_{x^i} - p_i\partial_{x^0} = X_i,$$

$$\begin{pmatrix} 1 & 0 & 0 \\ 0 & I_n & 0,0,\ldots,t_j,0\ldots \\ 0 & 0 & 1 \end{pmatrix} \cdot \chi^{-1}(m) = \begin{pmatrix} 1 & \bar{x} & x^0 \\ 0 & I_n & \bar{p}+t\bar{f}_j \\ 0 & 0 & 1 \end{pmatrix} \tag{694}$$
$$\to \chi_{*m}(\xi_{B_j}) = \partial_{p_j} = P_j.$$

The pull-back of the contact form $\theta = dx^0 + p_l dx^l$ from P to $\mathcal{H}_n$ defines the one-form $\theta_{\mathcal{H}}$ on $\mathcal{H}_n$:

$$\theta_{\mathcal{H}} = \chi^*(\theta) = -dc + b_i da^i.$$

The Reeb vector field of this form is

$$\xi_{\mathcal{H}} = -\xi_c = -\partial_c,$$

and we have

$$\chi_*(\xi_{\mathcal{H}}) = \xi$$

for the Reeb vectors of contact manifolds $(\mathcal{H}_n, \theta_{\mathcal{H}})$ and (P, θ).

The kernel $D_{\mathcal{H}}$ of this one-form (a distribution of codimension one on $\mathcal{H}_n$) is at each point g generated by the values of the vector fields ξ_{A_i}, ξ_{B_j} of the left translations, and is therefore **right invariant**. As a result, the distribution $D_{\mathcal{H}}$ defines the **right invariant contact structure on $\mathcal{H}_n$** (given as the kernel of the form $\theta_{\mathcal{H}}$).

Considering the **right translations** on the group $\mathcal{H}_n$, corresponding to the one-dimensional Lie subalgebras of $\mathfrak{h}_n$ with generators Z, A_i, B_j we find that their generators have the form

$$\eta_C = \partial_c, \qquad \eta_{A_i} = \partial_{a_i}, \qquad \eta_{B_j} = \partial_{b_j} + a_i\partial_c \tag{695}$$

and it is easy to check that the form θ **is invariant under the flow of these (therefore contact) vector fields**. Thus, the form $\theta_{\mathcal{H}}$ is **right invariant**. The diffeomorphism χ sends these vectors into

$$\chi_*(\eta_C) = \xi; \qquad \chi_*(\eta_{A_i}) = \partial_{x^i}; \qquad \chi_*(\eta_{B_j}) = \partial_{p_j} + x^j\partial_{x^0}.$$

Comparing this result with the description of the Killing vector field of the metric G we see that these vector fields form the nilradical of the Lie algebra of the Killing vector fields of the metric G.

Remark 1. *The distribution $D_{\mathcal{H}}$ is the direct sum of two n-dimensional distributions*

$$D_{\mathcal{H}} = D_a \oplus D_b$$

in obvious notations. The distributions D_a and D_b are integrable having as basis at each point values of pairwise commuting vector fields $\xi_{\bar{a}}$ and $\xi_{\bar{b}}$, respectively. Let A and B, respectively, denote the abelian subgroups of $\mathcal{H}_n$ of matrices of the form

(686) with $c = \bar{b} = 0$ and $c = \bar{a} = 0$, respectively. Then integral manifolds of the distributions D_A and D_B are orbits of the left translations by the subgroup A and by the Lie subgroup B, respectively.

The metric G is transferred under the diffeomorphism χ into the metric $G_{\mathcal{H}}$ on the Heisenberg group. This metric is constant in the right invariant (non-holonomic) frame $(\xi_C, \xi_{A_i}, \xi_{B_j})$ and is, therefore, right invariant by itself.

As a result we have proved the following:

THEOREM 4. *The diffeomorphism χ defined by*

$$\chi : g = \begin{pmatrix} 1 & \bar{x} & x^0 \\ 0 & I_n & \bar{p} \\ 0 & 0 & 1 \end{pmatrix} \to m = \begin{pmatrix} -x^0 \\ \bar{p} \\ \bar{x} \end{pmatrix}$$

determines an isomorphism of the "thermodynamical metric contact manifold" (P, θ, G) with the Heisenberg group $\mathcal{H}_n$ endowed with the right invariant contact from $\theta_{\mathcal{H}}$ and the right invariant metric $G_{\mathcal{H}}$ of signature $(n+1, n)$.

Remark 2. *Recall (see [**A19**]) that the automorphism group $Aut(\mathcal{H}_n)$ of the Heisenberg group has the Lie group $Aut_0(\mathcal{H}_n) = Sp(n, \mathbb{R}) \times A^{2n+1}$ as its connected component of unity where A^{2n+1} is the abelian group of dimension $2n+1$. This group acts on the space of all right invariant contact one-forms on the group $\mathcal{H}_n$. Since right invariant one-forms on $\mathcal{H}_n$ are defined by their values at the unit of the group $e \in \mathcal{H}_n$ and since the automorphism group $Aut(\mathcal{H}_n)$ leaves $e \in \mathcal{H}_n$ fixed, it is sufficient to study the action of this group on the set of elements $\nu \in \mathfrak{h}_n^*$.*

The following result for the left invariant contact structures (with the sketch of the proof) was sent to the author in a letter by M. Goze ([**A9**]). We reformulate this result for the right invariant contact structures due to the obvious duality between left and right translations.

Let $\{X_i, Y_j, Z\}$ be a standard basis of $\mathfrak{h}_n$ with the only non-trivial brackets being $[X_i, Y_i] = Z$, $i = 1, \ldots, n$. Let $\{\alpha_i, \beta_j, \omega\}$ be the dual basis in $\mathfrak{h}_n^*$. Then, extending this basis to the coframe of right invariant vector fields we get the relations

$$d\alpha_i = d\beta_j = 0, \ d\omega = \sum_i \alpha_i \wedge \beta_i.$$

Hence $\omega \wedge (d\omega)^n \neq 0$ and, therefore, the one-form ω defines the right invariant contact structure on $\mathcal{H}_n$.

PROPOSITION 4. *Let $\omega_1 = a\omega + \sum_i a_i\alpha_i + \sum_j b_j\beta_j$ be a contact form in $\mathfrak{h}_n^*$. The group $Aut(\mathcal{H}_n)$ acts transitively on the set of right invariant contact structures with the isotropy group of ω being the intersection of the group $Aut(\mathcal{H}_n)$ with the group of ω-conformally contact diffeomorphisms of $\mathcal{H}_n$.*

It follows that the contact structure $\theta_{\mathcal{H}}$ of the thermodynamical phase space is the typical representative of the $Aut(\mathcal{H}_n)$-conjugacy class of right invariant contact structures on the Heisenberg group defined by a choice of the canonical basis $\{X_i, Y_j, Z\}$ of the Lie algebra $\mathfrak{h}_n$. Therefore, $\theta_{\mathcal{H}}$ is unique up to an automorphism of the group $\mathcal{H}_n$.

Remark 3. *After the isomorphism of the TPS (P, θ, G) with $(\mathcal{H}_n, \theta_{\mathcal{H}}, G_{\mathcal{H}})$ is established, many properties of the metric G can be obtained from the corresponding results for invariant metrics on Lie groups.*

A.9. Geodesics of the Metric G

In this section we study the structure of geodesic curves of the metric G. We will be using the frame (680) (ξ, P_l, X_i) (so we will actually work on the Heisenberg group $\mathcal{H}_n$) to shorten the calculations and then reformulate the results in the holonomic Darboux basis $(\partial_{x^0}, \partial_{x_i}, \partial_{p_j})$.

Let Z_i be a frame in $T(P)$. Let $\hat{\Gamma}^i_{jk}$ be the Christoffel coefficients of the metric G in the basis Z_i given by $\nabla_{Z_j} Z_i = \hat{\Gamma}^k_{ji} Z_k$. Then, for any vector fields X, Y with components $X = X^i Z_i$ and $Y = Y^j Z_j$ we have

$$\begin{aligned}(\nabla_X Y = X^s \nabla_{Z_s} Y) &= X^s (Z_s Y^j) Z_j + X^s Y^j \nabla_{Z_s} Z_j \\ &= (Z \cdot Y^j) Z_j + X^s \hat{\Gamma}^i_{sk} Y^k Z_i \\ &= (X \cdot Y^k + \hat{\Gamma}^k_{ij} X^i Y^j) Z_k.\end{aligned}$$

Let $\gamma(t)$ be a curve in P. Decompose its velocity $\gamma'(t)$ in a frame Z_i of $T(P)$ as $\gamma'(t) = \hat{\gamma}^k(t) Z_k$. Then we have

$$\begin{aligned}\nabla_{\gamma'(t)} \gamma'(t) &= [\hat{\gamma}^k(t)(Z_k \cdot \hat{\gamma}^l(t)) + \hat{\Gamma}^l_{ks} \hat{\gamma}^k(t) \hat{\gamma}^s(t)] Z_l \\ &= [\frac{\partial}{\partial t} \hat{\gamma}^l(t) + \hat{\Gamma}^l_{ks} \hat{\gamma}^k(t) \hat{\gamma}^s(t)] Z_l.\end{aligned}$$

As a result, the curve $\gamma(t)$ is a geodesic of the metric G if and only if

(696) $$\dot{\hat{\gamma}}^l(t) + \hat{\Gamma}^l_{ks} \hat{\gamma}^k(t) \hat{\gamma}^s(t) = 0, \qquad l = 1, \ldots, 2n+1.$$

Here $\dot{f}(t)$ is the derivative of the (scalar) function $f(t)$ by the parameter t.

Applying these considerations to the decomposition of the velocity

(697) $$\gamma'(t) = \hat{\gamma}^0 \xi + \hat{\gamma}^i X_i + \hat{\gamma}_j P_j$$

with respect to the basis (ξ, P_l, X_i) (of right invariant vector fields on the group $\mathcal{H}_n$) and using expressions of proposition 2 for the covariant derivatives of elements of this basis we get the equations for components of velocity $\gamma'(t)$ of geodesics in the form

(698) $$\begin{cases} \dot{\hat{\gamma}}^0(t) = 0, \\ \dot{\hat{\gamma}}^i(t) - \hat{\gamma}^0(t) \hat{\gamma}^i(t) = 0, \\ \dot{\hat{\gamma}}_j(t) + \hat{\gamma}^0(t) \hat{\gamma}_j(t) = 0. \end{cases}$$

Solving this system we get for the velocity of a geodesic

(699) $$\gamma'(t) = c\xi + c^i e^{ct} X_i + c_j e^{-ct} P_j$$

with arbitrary constants c, c^i, c_j.

We now introduce the three-dimensional Lie subalgebra $\mathfrak{h}_\gamma$ of the Heisenberg algebra $\mathfrak{h}_n$ as the linear span

(700) $$\mathfrak{h}_\gamma =< \xi, X_\gamma = c^i X_i, P_\gamma = c_j P_j > .$$

Then one has:

PROPOSITION 5. (1) *For all t we have $\gamma'(t) \in R_{\gamma(t)\,*}\mathfrak{h}_\gamma$,*

(2) *A geodesic $\gamma(t)$ lies in the three-dimensional affine subspace*

$$P_\gamma = \gamma(0) + \mathfrak{h}_\gamma \subset P.$$

(3) *Subalgebra $\mathfrak{h}_\gamma$ is abelian if $q_\gamma = < c^i, c_j >= 0$. It is the three-dimensional Heisenberg algebra $\mathfrak{h}_1$ if $q_\gamma = < c^i, c_j > \neq 0$.*

(4) *The restriction of the metric G to the subalgebra $\mathfrak{h}_\gamma$ in the basis ξ, X_γ, P_γ has the form*

$$G_{\mathfrak{h}_\gamma} = \begin{pmatrix} 0 & 0 & 0 \\ 0 & 0 & q_\gamma \\ 0 & q_\gamma & 0 \end{pmatrix}.$$

PROOF. The first two statements were obtained above, the third follows from the bracket relation $[P_i, X_j] = -\delta_{ij}\xi$. The last one is proved by a direct calculation. □

Substituting expressions for the basic vectors X_i, P_j into the formula (699) for the velocity of a geodesic $\gamma(t) = (\gamma^0(t), \gamma_j(t), \gamma^i(t))$ in (x^0, p_j, x^i) coordinates we get

(701) $$\gamma'(t) = \dot\gamma^0\partial_{x^0} + \dot\gamma^i\partial_{x^i} + \dot\gamma_j\partial_{p_j} = ((c - \gamma_i(t)c^i e^{ct})\partial_{x^0} + c_j e^{-ct}\partial_{p_j} + e^{ct}c^i\partial_{x^i}.$$

Integration yields:

PROPOSITION 6. (1) *For $c \neq 0$,*

(702) $$\begin{cases} \gamma^0(t) = \gamma^0(0) + (c + \frac{q_\gamma}{c})t - \frac{<\gamma^0(0), c^i>}{c}(e^{ct} - 1), \\ \gamma_j(t) = \gamma_j(0) - \frac{c_j}{c}(e^{-ct} - 1), \\ \gamma^i(t) = \gamma^i(0) + \frac{c^i}{c}(e^{ct} - 1). \end{cases}$$

(2) *For the limit case $c = 0$,*

(703) $$\begin{cases} \gamma^0(t) = \gamma^0(0) - < \gamma_i(0), c^i > t - q_\gamma \frac{t^2}{2}, \\ \gamma_j(t) = \gamma_j(0) + c_j t, \\ \gamma^i(t) = \gamma^i(0) + c^i t. \end{cases}$$

(3) *For all c*

(704) $$\theta(\gamma'(t)) = c.$$

In particular, a geodesic $\gamma(t)$ is **horizontal** *if and only if $c = 0$.*

(4) *For the norm of velocity vector $\gamma'(t)$ we have*

$$G(\gamma'(t), \gamma'(t)) = c^2 + q_\gamma.$$

PROOF. The first two statements are obtained by a direct calculation. We now prove the third statement. For $c \neq 0$,

$$\begin{aligned} \theta(\gamma'(t)) &= (dx^0 + p_i dx^i)((c - \gamma_i(t)c^i e^{ct})\partial_{x^0} + c_j e^{-ct}\partial_{p_j} + e^{ct}c^i\partial_{x^i}) \\ &= (c - \gamma_i(t)c^i e^{ct}) + \gamma_i(t)e^{ct}c^i \\ &= c. \end{aligned}$$

For $c = 0$, $\gamma'(t) = -\gamma_i(t)c^i\partial_{x^0} + c_j\partial_{p_j} + c^i\partial_{x^i}$. As a result $\gamma_i(t) = \gamma_i(0) + c_i t$ and $\dot\gamma^0(t) = -(\gamma_i(0) + c_i t)c^i = -q_\gamma t - < c^i, \gamma_i(0) >$. Calculating

$$\theta(\gamma'(t)) = -\gamma_i(t)c^i + \gamma_i(t)c^i = 0$$

we get the result. The fourth statement follows from (699) and the commutation relations in the frame (680). □

Remark 4. *Since the metric $G_{\mathcal{H}}$ and its Levi-Civitá connection on the group $\mathcal{H}_n$ are right invariant, the right translation of a geodesic is a geodesic. As a result, one may restrict the study to the geodesics that start at the unit of the group $\mathcal{H}_n$.*

In order to present a real thermodynamical process, a curve $\gamma(t)$ in the space P should satisfy a second law of thermodynamics that states that the entropy is non-decreasing along the process. Therefore, it is interesting to see under which conditions a geodesic $\gamma(t)$ of the metric G satisfies this condition. Assuming that the choice of a canonical chart (x^0, p_j, x^i) is such that x^0 is the entropy of our system we see that along a geodesic with $c \neq 0$,

$$\frac{dx^0}{dt} = \frac{q_\gamma}{c} + [c - \frac{q_\gamma}{c} - < c^i, \gamma_i(0) >]e^{ct}.$$

As a result we proved the following:

PROPOSITION 7. (1) *If $c > 0$, then $\frac{dx^0}{dt} \geqq 0$ for all $t \geqq 0$ if and only if*

$$\begin{cases} \text{If } q_\gamma \leqq 0, \text{ then } c- < c^i, \gamma_i(0) > \geqq 0, \\ \text{If } q_\gamma > 0, \text{ then } c- < c^i, \gamma_i(0) > \geqq q_\gamma. \end{cases}$$

(2) *If $c < 0$, then $\frac{dx^0}{dt} \geqq 0$ for all $t \geqq 0$ if and only if*

$$q_\gamma \leqq 0 \text{ and } c- < c^i, \gamma_i(0) > \geqq 0.$$

(3) *If $c = 0$, then $\frac{dx^0}{dt} \geqq 0$ for all $t \geqq 0$ if and only if*

$$q_\gamma \leqq 0 \text{ and } < c^i, \gamma_i(0) > \leqq 0.$$

A.10. Symplectization of the Manifold (P, θ, G)

Let $\tilde{P}$ be the standard $(2n+2)$-dimensional real vector space $\mathbb{R}^{2n+2}$ with the coordinates (p_i, x^j), $i, j = 0, \ldots, n$, endowed with the one-form

(705) $$\tilde{\theta} = \sum_{i=0}^{n} p_i dx^i$$

and the standard symplectic structure

(706) $$\omega = d\theta = \sum_i dp_i \wedge dx^i.$$

We consider the **embedding** of the space $(P, \theta = dx^0 + p_l dx^l)$ into $\tilde{P}$

(707) $$J : (x^0, x^i, p_j) \to (x^0, x^i; p_0 = 1, p_l, \ l = 1, \ldots, n)$$

as the affine subspace $p_0 = 1$.

It is easy to see that

PROPOSITION 8. (1) *The pull-back by J of the 1-form $\tilde{\theta}$ coincides with the contact form θ:*

$$J^*(\tilde{\theta}) = \theta.$$

(2) *The symplectic manifold $(\tilde{P} = \{(p, x) \in \mathbb{R}^{2n+2} | p_0 > 0\}, \omega)$ is the standard* **symplectization** *of (P, θ) (see* [**A1, A15**]*) and J is the section of the symplectization bundle $\pi : \tilde{P} \to P$.*

(3) *The symmetrical tensor*

$$\tilde{G} = (\tilde{G}_{ij}) = \begin{pmatrix} 0_{n+1\times n+1} & I_{n+1} \\ I_{n+1} & p_i p_j \end{pmatrix} \tag{708}$$

determines the pseudo-Riemannian metric of signature $(n+1, n+1)$ *in* $\tilde{P}$.

(4) *The restriction of the metric* $\tilde{G}$ *to the image of the embedding* J *coincides with the metric* G*:*

$$J^*\tilde{G} = G.$$

(5) *There is a bijection between the Legendre submanifolds of the contact manifold* (P, θ) *and the homogeneous Lagrange submanifolds of the symplectic manifold* $(\tilde{P}, \omega)$ *under the action*

$$(x^i, p_j) \to (x^0, \lambda p_0, x^1, p_1, \ldots, x^n, p_n)$$

of R^+ *on the manifold* $\tilde{P}$. *This correspondence is defined by the intersection of a homogeneous Lagrangian submanifold* $\tilde{K}$ *with the image of the embedding* J *and by the action of the dilatation group on the image of a Legendre submanifold* $K \subset P$ *under the embedding* J.

PROOF. Almost all the statements of this proposition follow simply from the construction or are known ([**A2, A15**]). The determinant of the matrix $\tilde{G}_{ij}$ of the metric $\tilde{G}$ is equal to $(-1)^{n+1}$ which proves its non-degeneracy. $\square$

A.11. Properties of the Metric $\tilde{G}$

The basic properties of the metric $\tilde{G}$ in the space $\tilde{P}$ are given in the following two statements proved in [**A22**]:

PROPOSITION 9. (1) *The Ricci Tensor of the metric* $\tilde{G}$ *is*

$$Ric(\tilde{G}) = \begin{pmatrix} 0_{(n+1)\times(n+1)} & \frac{n+2}{2} I_{n+1} \\ \frac{n+2}{2} I_{n+1} & \frac{n+2}{2} p_i p_j \end{pmatrix} = \frac{n+2}{2}\tilde{G}. \tag{709}$$

(2) *The scalar curvature of the metric* $\tilde{G}$ *is*

$$R(\tilde{G}) = \mathrm{tr}\left(\tilde{G}^{-1}\frac{n+2}{2}\tilde{G}\right) = (n+1)(n+2). \tag{710}$$

(3) *The metric* $\tilde{G}$ *is the indefinite Einstein metric of the scalar curvature* $R(\tilde{G}) = (n+1)(n+2)$.

THEOREM 5. *The Lie algebra* $\mathfrak{iso}_{\tilde{G}} \simeq \mathfrak{sl}(n+2, \mathbb{R})$ *of the Killing vector fields of the metric* $\tilde{G}$ *is (as a vector space) the linear sum*

$$\mathfrak{iso}_{\tilde{G}} = \mathfrak{q} \oplus \mathfrak{d} \oplus \mathfrak{x}$$

of the Lie subalgebras $\mathfrak{q}, \mathfrak{d}$ *and* $\mathfrak{x}$ *given as follows:*

(1)

$$\mathfrak{q} = < Q^i_j = x^i \partial_{x^j} - p_j \partial_{p_i} >,$$

with the commutation relations

$$[Q^i_j, Q^p_k] = \delta^p_j Q^i_k - \delta^i_k Q^p_j.$$

Thus the subalgebra $\mathfrak{q}$ *is isomorphic to* $\mathfrak{gl}(n+1, \mathbb{R})$.

(2) *The abelian subalgebra*

$$\mathfrak{d} =< D^i = \frac{x^i}{2}Q + (1 - \frac{1}{2}(x^l p_l))\partial_{p_i}, >$$

where $Q = \sum_i Q^i_i = \sum_i (x^i \partial_{x^i} - p_i \partial_{p_i})$ *is the generator of the hyperbolic rotation* $H_t : (p, x) \to (e^t p, e^{-t} q)$.

(3) *The abelian subalgebra*

$$\mathfrak{x} =< X_s = \frac{\partial}{\partial x^s} > .$$

The generators Q^i_j, X_j, D^j *satisfy the following commutation relations*

$$[Q^i_j, X_s] = -\delta^i_s X_j, \qquad [Q^i_j, D^s] = \delta^s_j D^i, \qquad [X_s, D^i] = \frac{1}{2}Q^i_s + \frac{1}{2}\delta^i_s Q. \tag{711}$$

The vector fields Q^i_j, X_i, D^j *are Hamiltonian with Hamiltonian functions*

$$H_{Q^i_j} = -x^i p_j, \qquad H_{X_k} = -p_k, \qquad H_{D^s} = x^s(1 - \frac{< \bar{x}, \bar{p} >}{2}).$$

A.12. Constitutive Hypersurface and its Lift to $\tilde{P}$

For dimensional reasons (cf. [**A5**]), the fundamental thermodynamical constitutive equation (law) $x^0 = \phi(x^i)$ of any material is **homogeneous of order one**, i.e. the following condition is fulfilled for all $\lambda \neq 0$:

$$\phi(\lambda x^i) = \lambda \phi(x^i) \tag{712}$$

(cf. [**A5**]). In other words, the action of the one-parameter group Λ of transformations

$$D_\lambda : (x^0, p_l, x^i) \to (\lambda x^0, p_l, \lambda x^i), \ \lambda \in R^* \tag{713}$$

leaves the constitutive Legendre surface Σ_ϕ of a real material invariant.

As a result the surface Σ_ϕ lays in the canonical quadric (hyperbolic paraboloid)

$$\mathcal{C} := \{m = (x^0, p_l, x^i) | x^0 + \sum_{i=1}^{n} p_l x^l = 0\}. \tag{714}$$

The intersection of the contact distribution D in $T(\mathcal{C})$ with the fibers of the tangent bundle $T(\mathcal{C})$ determines the subbundle $D_{\mathcal{C}}$.

Along the hypersurface $\mathcal{C}$ one has

$$0 = dx^0 + \sum_{i=1}^{n} p_i dx^i + \sum_i x^i dp_i = \theta + \sum_i x^i dp_i. \tag{715}$$

As a result, on the distribution $D_{\mathcal{C}}$ we have

$$\sum_i x^i dp_i = 0 \tag{716}$$

which represents the abstract Gibbs-Duhem equation (cf. [**A5**]).

Remark 5. *In the Heisenberg group* $\mathcal{H}_n$ *the one-parameter transformation group* D_λ *acts by automorphisms on* $\mathcal{H}_n$. *One can show that the hypersurface* $\mathcal{C}$ *transferred to the Heisenberg group is related to the symmetrical form of the 2-cocycle* $R^{2n}_{\bar{a},\bar{b}} \to R_z$ *obtained from the cocycle* ω_+ *by adding a proper co-boundary (cf. section A.8).*

There exists a natural lift of the constitutive hyperquadric $\mathcal{C}$ to the space $\tilde{P}$ as the homogeneous hypersurface

$$\tilde{\mathcal{C}} = \{(p_i, x^i) | x^0 p_0 + \sum_{i=1}^{n} p_i x^i = 0\}. \tag{717}$$

The hyperquadric $\mathcal{C}$ is invariant under the action of $\Lambda = \mathbb{R}^*$: $(x^i, p_i) \to (x^i, \lambda p_i)$ as well as under the hyperbolic rotations $(x^i, p_i) \to (\lambda x^i, \lambda^{-1} p_i)$.

Polarizing the coordinates (x^i, p_j), i.e. introducing new coordinates

$$\xi_i = x^i + p_i, \qquad \eta_j = x^j - p_j,$$

we rewrite the equation of $\tilde{\mathcal{C}}$ as

$$\sum_{i=0}^{n} (\xi_i^2 - \eta_i^2) = 0.$$

From this we see that the hyperquadric $\tilde{\mathcal{C}}$ is a cone of signature $(n+1, n+1)$ in the space $\tilde{P}$.

A.13. Hyperbolic Rotations and the Projectivization of $\tilde{P}$

In this section we construct a natural compactification of the TPS P endowed with the extension of the contact structure θ and that of the indefinite metric G.

Consider the action of the one-parameter group R in the space $\tilde{P}$ acting by the one-parameter group HR of **hyperbolic rotations**

$$g^t : (p_l, x^i) \to (e^t p_l, e^{-t} x^i). \tag{718}$$

We obviously have:

LEMMA 1. (1) *The one-form $\tilde{\theta}$ is invariant under the action (718) of the group HR.*
(2) *The metric $\tilde{G}$ is invariant under the action of the group HR.*

PROPOSITION 10. *The space $\hat{P}$ of orbits of the points $\tilde{P} \setminus \{0\}$ under the action of the group HR is canonically isomorphic to the projective space $P_{2n+1}(\mathbb{R})$.*

PROOF. Cover the space $\mathbb{R}^{2n+2}$ with the open subsets of two types:

(1) The sets of the first type are $\hat{U}_j = \{m | p_j \neq 0\}$. Associate with these sets the affine domains $U_j \equiv \mathbb{R}^{2n+1}$ of the projective space $P_{2n+1}(\mathbb{R})$ with the coordinates
$$\left(x^i p_j, \frac{p_l}{p_j}\right).$$
(2) The sets of the second type are $\hat{V}_k = \{m | x^k \neq 0\}$. Associate with these sets the affine domains $V_k \equiv \mathbb{R}^{2n+1}$ of the projective space $P_{2n+1}(\mathbb{R})$ with the coordinates
$$\left(\frac{x^i}{x^k}, p_l x^k\right).$$

Notice that on the intersections $U_{j_1} \cap U_{j_2}$ we have the following relations between the corresponding affine coordinates:

$$x^i p_{j_2} = x^i p_{j_1} \cdot (\frac{p_{j_2}}{p_{j_1}}), \qquad \frac{p_k}{p_{j_2}} = \frac{p_k}{p_{j_1}} \cdot (\frac{p_{j_1}}{p_{j_2}}).$$

On the intersections $U_j \cap V_k$, we have these relations between the corresponding affine coordinates:

$$x^l p_j = \frac{x^l}{x^k} \cdot (x^l p_j), \qquad \frac{p_l}{p_j} = p_l x^k \cdot (\frac{1}{x^k p_j}).$$

Finally, on the intersections $V_{j_1} \cap V_{j_2}$ we have as relations between the corresponding affine coordinates:

$$\frac{x^k}{x^{j_2}} = \frac{x^k}{x^{j_1}} \cdot (\frac{x^{j_1}}{x^{j_2}}), \qquad p_l x^{j_2} = p_l x^{j_1} \cdot (\frac{x^{j_2}}{x^{j_1}}).$$

This shows that the affine coordinates of all affine charts are related by the transition functions **invariant under the action of hyperbolic rotations**. Thus, they are glued together to the standard projective space $P_{2n+1}(\mathbb{R})$. □

Combining lemma 2 and the previous construction we get the following:

PROPOSITION 11. (1) *The projections $\hat{\theta}$ of the one-form $\tilde{\theta}$ and that of the metric $\hat{G}$ of $\tilde{G}$ endow the projective space $\hat{P}$ with the contact structure and the metric of signature $(n+1, n)$.*

(2) *The composition J of the embedding $j : P \to \tilde{P}$ with the projection $\tilde{P} \to \hat{P}$ define the* **compactification** *$(\hat{P}, \hat{\theta}, \hat{G})$ of the TPS (P, θ, G) with the contact structure and Mrugala metric G.*

A.14. Group Action of $\mathcal{H}_n$ and the "Partial Orbit Structure" of $\hat{P}$

In this section we consider the lift of the action of the group $\mathcal{H}_n$ on P to the space $\tilde{P}$ discussed in section A.8 and the action of subgroups of $\mathcal{H}_n$ on the cells of smaller dimension of the standard CW-structure of the projective space $\hat{P}$.

The differential operators X_i, P_j, ξ of the canonical frame (680) also act in the space $\tilde{P}$ with the same commutation relations. This action generates the action of the Lie group $\mathcal{H}_n$ on the space $\tilde{P}$ leaving hyperplanes $p_0 = \text{const.}$ invariant.

We introduce the sequence L_k, $k = 1, \ldots, n$, of subgroups of the Heisenberg group $\mathcal{H}_n$ defined by the condition

(719) $$L_{n-k} = \{g(\bar{a}, \bar{b}, c) | b_1 = \cdots = b_k = 0\}.$$

These subgroups form the series

(720) $$\mathcal{H}_n \supset L_{n-1} \supset L_{n-2} \supset \cdots \supset L_0.$$

It is easy to see that

(721) $$L_{n-k} \simeq \mathbb{R}^k \times \mathcal{H}_{n-k}$$

is the product of the k-dimensional abelian group $\mathbb{R}^k$ and the Heisenberg group $\mathcal{H}_{n-k}$.

The right invariant vector fields on $\mathcal{H}_n$ tangent to (and generated by) the subgroup L_{n-k} are (in terms of the isomorphism of section A.8)

$$\xi, \quad X_i, i = 1, \dots, n, \quad P_j, j = k+1, \dots, n.$$

In the space $\tilde{P}$ consider the affine planes V_k defining the cells of the standard cell structure of the projective space $\hat{P} = P_{2n+1}(\mathbb{R})$ with respect to the (hyperbolically) homogeneous coordinates of $\hat{P}$ given by

$$V_k = \{(x^i, p_j | p_0 = p_1 = \dots = p_{k-1} = 0, \ p_k = 1\}, \tag{722}$$

with $k = 0, 1, \dots, n$. The projective space $\hat{P}$ is obtained in the following way. At first we glue the smaller cells V_1, V_2, V_3 consecutively to the cell $V_0 = j(P)$ as in the standard construction of projective space. Then, after n such steps the last glued cell will be V^{n+1} which, by construction, is canonically diffeomorphic to the product $R^{n+2} \times H^1$. We consider this cell V^{n+1} as the open cell of P^{n+1} and close it to P^{n+1} by adding to it the n-dimensional projective space $P_n(\mathbb{R})$. This copy of $P_n(\mathbb{R})$ is obtained by the action of hyperbolic rotations (usual dilatations here, because all the $p_j = 0$) on the subspace $V_{n+1} = \{(x^i, p_j = 0 \mid j = 0, \dots, n\}$.

The projective space $\hat{P}$ is the union of cells V^k, $k = 2n+1, \dots, 0$:

$$\begin{aligned} P &\simeq V^{2n+1} \cup V^{2n} \cup \dots \cup V^{n+1} \cup V^n \cup \dots V^0 \\ &= \mathcal{H}_n \cup R \times \mathcal{H}_{n-1} \cup \dots \cup R^{n-2} \times \mathcal{H}_1 \cup R^n \cup \dots \cup R^0. \end{aligned}$$

Now it is easy to see that each cell V_k is **canonically diffeomorphic** to the group L_{n-k} whose action on V_k is induced by the action of the Heisenberg group H_n on the space $\tilde{P}$ considered above.

Even though the action of $\mathcal{H}_n$ on P cannot be extended to the compactification $\hat{P}$, a coherent action of the subgroups of the series (720) produces the partial cell structure of $\hat{P}$ starting with the projective subspace $V_{n+1} \simeq P_n(\mathbb{R})$.

The restriction of the one-form $\tilde{\theta}$ to the cell V_k has the form

$$\theta_k = \tilde{\theta}|_{V_k} = dx^k + \sum_{i=k+1}^{n} p_i dx^i.$$

Therefore, this form determines the canonical contact structure on the cell $V_k \simeq L_{n-k} \simeq \mathbb{R}^k \times \mathcal{H}_{n-k}$, called Heisenberg factor; it is zero on the first factor.

The restriction of the Mrugala metric G to the cell V_k, in variables

$$(p_k, \dots, p_n; x^0, \dots, x^{k-1}; x^k, \dots, x^n),$$

has the form

$$G_k = \tilde{G}|_{V_k} = \begin{pmatrix} 0_{(n-k)\times(n-k)} & 0_{(n-k)\times k} & I_{n-k} \\ 0_{k\times(n-k)} & 0_{k\times k} & 0_{k\times(n-k)} \\ I_{n-k} & 0_{(n-k)\times(n-k)} & p_i p_j \end{pmatrix}. \tag{723}$$

Thus this metric is zero on the first abelian factor of the cell $V_k \simeq L_{n-k}$ and coincides with the metric G of R. Mrugala on the Heisenberg factor $\mathcal{H}_{n-k}$ of the cell V_k.

Combining these arguments we get the following:

THEOREM 6. (1) *The restriction of the action of the Heisenberg group $\mathcal{H}_n$ on the space $P \simeq V_0$ (embedded in $\hat{P}$) to the subgroup $L_{n-k} \simeq R^k \times \mathcal{H}_{n-k}$ of the form (719) extends to the action of this subgroup on the cell $\pi(V_k) \subset \hat{P}$ and determines the diffeomorphism of the group L_{n-k} with V_k and with its image $\pi(V_k) \subset \hat{P}$.*

(2) *Restrictions of the one-form $\tilde{\theta}$ and the metric $\tilde{G}$ to the cell V_k endow the Heisenberg factor $\mathcal{H}_{n-k}$ of V_k with the contact structure θ_k and the Mrugala metric G_k and are both zero on the abelian factor $\mathbb{R}^k$.*

Remark 6. *Every cell V_k represents the thermodynamical phase space of an abstract thermodynamical system with $n+1$ extensive and $n-k$ intensive variables. This corresponds to a situation where the thermodynamical potential $\phi(x^i)$ depends on $x^{k+1}, \dots, x^n$ but not on the first k extensive variables x^i. As a result the participation of factors $x^i, i = 0, \dots, k-1$, in the processes is "switched out" and they become parameters only.*

Remark 7. *Let Σ_ϕ be a Legendre submanifold of TPS (P, θ) of the form (676). Consider its embedding $\hat{\Sigma}_\phi = \pi \circ j(\Sigma_\phi) \subset \hat{P}$. If ϕ is* **regular** *at the origin $0 \in P$, then the closure of $\hat{\Sigma}_\phi$ in $\hat{P}$ is the union of Σ_ϕ and its complement lying in $\pi(V_n) \simeq P_n$. This complement is the projective space P_{n-1} embedded into V_n as $\{(x^0 = d\phi(0)(\bar{y}), x^i = y^i)\}$.*

EXAMPLE 2. Consider the case $n = 2$, i.e. take P^5 to be five-dimensional with the contact form $\theta = dU - SdT + pdV$ (a one-component homogeneous system, per 1 mole). The hypersurface $\mathcal{C}$ in section A.12 has, in this case, the well-known form $U - ST + pV = 0$. Its lift to $\tilde{P}$ - $\tilde{\mathcal{C}}$ has the form $p_0 U - ST + pV = 0$.

The intersection of this quadric with the plane $p_0 = 0$ is the (degenerate) quadric $\mathcal{C}_1 : pV = ST$. Fixing the value of S, say, and taking $S = R -$ const (Ridberg constant!) determines the cell $\simeq V_1$ that projects onto the cell V_1 of the compact space $\hat{P}$. The image of the quadric $\mathcal{C}_1$ under this projection determines in the three-dimensional H_1-factor of the cell V_1 the surface $pV = RT$ given by the **constitutive equation of mono-atomic ideal gas**.

The hypersurface $\mathcal{C}$ is the submanifold containing all the constitutive (equilibrium) surfaces of all thermodynamical systems with the TPS P^5. Closures of these surfaces in $\hat{P}$ contain points from cells of smaller dimension V_1 and V_2. Thus the equation of a mono-atomic ideal gas appears here as the equation of the surface formed by the limit points in V_1 of all possible constitutive surfaces in $(P^5, \theta = dU - SdT + pdV)$.

Remark 8. *The construction of a compact manifold $\hat{P}$ in terms of a series of subgroups (720) of a Lie group $\mathcal{H}$ represents a way to represent a manifold in terms of a Lie group with the open dense orbit isomorphic to the group $\mathcal{H}$ itself and a natural Whitney stratification in terms of extensions of subgroup actions on the cells of smaller dimension. Removed generators (here P_i, $i = 1, \dots, k$) determine the projections from cells of higher dimension to the cells in their closure (cf.* [**A26**]*).*

Bibliography

[A1] V. Arnold, *Mathematical Methods of Classical Mechanics*, Springer, 1989.

[A2] V. Arnold, A. Givental, *Symplectic Geometry*, in: "Dynamical Systems IV", Springer, 1988.

[A3] L. Benayoun, *Methodes geometriques pour l'etude des systemes thermodynamiques et la generation d'equations d'etat*, These, INPG, Grenoble, 1999.

[A4] D. Blair, *Riemannian Geometry of Contact and Symplectic Manifolds*, Birkhauser, Boston, 2002.

[A5] H. Callen, *Thermodynamics*, Whiley, 1960.

[A6] C. Caratheodory, *Unterschungen uber die Grundlagen der Thermodynamyk*, Gesammelte Mathematische Werke, B.2., Munchen, 1955, S. 131-177.

[A7] L. Diosi, P. Salamon, *From Statistical Distances to Minimally Dissipative Processes*, in "Thermodynamics of Energy Conversion and Transport" eds. S. Sieniutycz, A. de Vos., Springer, 2000, pp.286-318.

[A8] J.W. Gibbs, *The Scientific Papers*, vol.1, Dover Publ., 1961.

[A9] M. Goze, private communication, July 2005.

[A10] M. Goze, P.Piu, *Classification des Metriques Invariantes a Gauche sur le Groupe de Heisenberg*, Rendiconti del circolo matematico di Palermo, Ser.II, T. XXXIX (1990), pp.299-306.

[A11] H. Heemeyer, *Pseudoriemannsche Faserbundel und ihre Anwendung in der allegemeinrealitivistischen Nichtgleichgewichtsthermodynamik*, Diss. Tech. Univ. Berlin, Wissenschaft und Technik Verlag, Berlin, 1995.

[A12] R. Hermann, *Geometry, Physics and Systems*, Dekker, N.Y., 1973.

[A13] A. Kirillov, *Local Lie algebras*, Uspekhi Mat. Nauk, v.31, n4, pp.57-76, 1976 (in Russian).

[A14] S. Kobayashi, K. Nomizu, *Foundations of Differential Geometry*, Wiley (Interscience), N.Y., Vol.I, 1963; Vol.II, 1969.

[A15] P. Libermann., C.-M. Marle, *Differential Geometry and Analytical Mechanics*, D. Reidel, 1987.

[A16] R. Mrugala, J. Nulton, J. Schon, P. Salamon *Contact structure in thermodynamical theory*, Reports of Mathematical Physics, v.29, No.1, 1991, pp.109-121.

[A17] R. Mrugala, *On a Riemannian Metric on Contact Thermodynamic Spaces*, Reports of Mathematical Physics, v.38, No.3, 1996, pp.339-348.

[A18] R. Mrugala, *Geometrical Methods in Thermodynamics*, in "Thermodynamics of Energy Conversion and Transport" eds. S. Sieniutycz, A. de Vos., Springer, 2000, pp.257-285.

[A19] A. Onishchik, E. Vinberg, (Eds), *Lie Groups and Lie Algebras III*, EMS, v.41, Springer-Verlag, Berlin, 1994.

[A20] M. Postnikov, *Geometry VI. Riemannian Geometry*, EMS, v.91, Springer-Verlag, Berlin, 2001.

[A21] S. Preston, *On the indefinite metrics compatible with the sympectic and contact structures*, in preparation.

[A22] S. Preston, J. Vargo, *The indefinite metric of R.Mrugala and the geometry of the thermodynamical phase space*, ArXive math.DG/0509267, 12 Sept.2005.

[A23] G. Ruppeiner, *Riemannian geometry in thermodynamic fluctuation theory*, Reviews of Modern Physics, v.67, n.3, 1995, pp.605-659.

[A24] G. Ruppeiner, *Thermodynamical Curvature: Origin and Meaning*, in "Nonequilibrium Theory and Extremum Principles, eds. S. Sieniutycz, P. Salamon, Taylor and Francis, N.Y., 1990, pp.129-174.

[A25] W. Schempp, *Harmonic analysis on the Heisenberg Nolpotent Group with applications to Signal Theory*, Pitman Res. Notes in Math., v.147, Longman scientific and Technical, 1986.

[A26] R. Thom, *Ensembles et morphismes stratifies*, Bull. Amer. Math. Soc., (N.S0, 75, 1969, 240-284.

[A27] F. Weinhold, *Metric Geometry of equilibrium thermodynamics*, p. I-V,Journal of Chemical Physics, v.63, n.6, 2479-2483, 2484-2487, 2488-2495, 2496-2501, 1976, v.65, n.2, pp.559-564, 1976.

[A28] J. A. Wolf, A. Gray, *Homogeneous spaces defined by Lie group automorphisms. II.* J. Differential Geometry, v.2, 1968, pp.115–159.

[A29] S. Preston, J. Vargo, *The indefinite metric of R. Mrugala and the geometry of the thermodynamical phase space*, in "Proceedings of THERMOCONN 2005", Atti dell'Accademia Peloritana dei Pericolanti, Messibna, It., Classe di Scienze Fisiche, Matematiche e Naturali Vol. LXXXVI, C1S0801019 (2008) - Suppl. pp.1-12.

Bibliography

[1] Ralph Abraham and Jerrold E Marsden. *Foundation of Mechanics*. Benjamin/Cummings, Reading, 1978.

[2] Ralph Abraham, Jerrold E. Marsden, and Tudor Ratiu. *Manifolds, Tensor Analysis, and Applications*. Number 2 in Global Analysis Pure and Applied B. Addison-Wesley, London, Amsterdam, Don Mills (Ontario), Sydney, Tokyo, 1983.

[3] Erik Alfsen and Frederic Shultz. *State Spaces of Operator Algebras*. Birkhäuser, Basel, Boston, Berlin, 2001.

[4] Syed Twareque Ali, Jean-Pierre Antoine, and Jean-Pierre Gazeau. *Coherent States, Wavelets and their Generalizations*. Springer, New York, Berlin, Heidelberg, 2000.

[5] Michael Artin. *Algebra*. Birkhäuser, Basel, Boston, Berlin, 1993.

[6] H. Atmanspacher, T. Filk, and H Römer. Weak quantum theory: Formal framework and selected application. In Theo M. Nieuwenhuizen Guillome Adenier, Andrei Yu. Khrennikov, editor, *Proceedings of "Quantum Theory: Reconsideration of Foundations - 3." Växjö (Sweden) June 6-11, 2005*, pages 34–46, Melville, New York, 2006. American Institute of Physics.

[7] Heinz Bauer. *Maß- und Integrationstheorie*. de Gruyter, Berlin, 1992.

[8] R. Beattie and H. P. Butzmann. *Convergence Structures and Applications to Functional Analysis*. Kluwer Academic Publishers, Dordrecht, 2002.

[9] M. Berger and B. Gostiaux. *Differential Geometry: Manifolds, Curves, and Surfaces*. GTM **115**. Springer, New York, 1988.

[10] Marcel Berger. *Lectures on Geodesics in Riemannian Geometry*. Tata Institute of Fundamental Research, Bombay, 1965.

[11] Ernst Binz. Two natural metrics and their covariant derivatives on a manifold of embeddings. *Monatshefte für Mathematik*, 89:275–288, 1980.

[12] Ernst Binz, Reinhard Honegger, Sonja Pods, and Alfred Rieckers. Heisenberg Algebra Bundles and Field Theoretic Weyl Quantization. (in preparation).

[13] Ernst Binz, Reinhard Honegger, and Alfred Rieckers. Construction and Uniqueness of the C^*-Weyl Algebra over a General Pre-Symplectic Space. *J. Math. Phys.*, 45(7), 2004.

[14] Ernst Binz, Reinhard Honegger, and Alfred Rieckers. Field Theoretic Weyl Quantization as a Strict and Continuous Deformation Quantization. *Annales Henri Poincaré 5*, pages 327 – 346, 2004.

[15] Ernst Binz, Reinhard Honegger, and Alfred Rieckers. Infinite Dimensional Heisenberg Group Algebra and Field-Theoretic Strict Deformation Quantization. *Intern. J. of Pure and Appl. Math., Vol. 38 No.1*, pages 43 – 78, 2007.

[16] Ernst Binz and Sonja Pods. A Heisenberg Algebra bundle of a Vector Field in Three-Space and its Weyl Quantization in Quantum Theory. In *AIP Proceedings*, volume 810, Melville, New York, 2006.

[17] Ernst Binz, Sonja Pods, and Walter Schempp. Heisenberg groups — a unifying structure of signal theory, holography and quantum information theory. *J. of Appl. Math. and Comp.*, 11(1–2):1–57, 2003.

[18] Ernst Binz, Jedrzej Sniatycki, and Hans Fischer. *The Geometry of Classical Fields*. Dover Publication Inc., Mineola, New York, 2006.

[19] Arno Bohm. *Quantum Mechanics, Foundations and Applications*. Springer, New York, Berlin, 1993.

[20] David Bohm and Basil Hiley. *The Undivided Universe*. Routledge, London, 1995.

[21] Max Born and Emil Wolf. *Principle of Optics*. Cambridge University Press, 1994.

[22] Helmut Boseck. *Grundlagen der Darstellungstheorie*. Deutscher Verlag der Wissenschaften, Berlin, 1973.

[23] Karl-Heinz Brenner. A Discrete Version of the Wigner Distribution Function. pages 307–310, 1983.

[24] Karl-Heinz Brenner. *Phasendarstellung in Optik und Signalverarbeitung.* PhD thesis, Universität Erlangen-Nürnberg, 1983.

[25] Karl-Heinz Brenner and Wodkiewicz K. Time Dependent Physical Spectrum of Light and Wigner Distribution Function. *Opt. Comm.*, 43(2):103–106, 1982.

[26] Karl-Heinz Brenner and Xiyuan Liu. New Description of the Inhomogeneous Wave for Plane Wave Expansion in Absorbing Media. *EOS Topical Meeting on Advanced Optical Imaging Techniques, 29.06.-01.07.2005*, 2005.

[27] Karl-Heinz Brenner and A.W. Lohmann. Wigner Distribution Function Display of Complex 1d Signals. *Opt. Comm.*, 42(5):310–314, 1982.

[28] Karl-Heinz Brenner and J Ojeda-Castaneda. The Ambiguity Function Applied to Partially Coherent Imagery. *Optica Acta*, 31(2):213–223, 1984.

[29] Claude Cohen-Tannoudji, Bernard Diu, and Franck Laloë. *Quantum Mechanics*, volume 1. John Wiley & Sons, New York, 1977.

[30] T. M. Cover and J. A. Thomas. *Elements of Information Theory.* John Wiley & Sons, New York, 1991.

[31] Maurice de Gosson. *Maslow Classes, Metaplectic Representations and Lagrangian Quantization*, volume 95 of *Mathematical Reserch.* Akademie Verlag, Berlin, 1997.

[32] Maurice de Gosson. *Symplectic Geometry – Weyl – Moyal – Calculus, and Quantum Mechanics.* Birkhäuser Verlag, Basel, Boston, Berlin, 2006.

[33] Serge de Gosson. *Multioriented Symplectic Geometry and the Extension of Path Intersection, Indices, Thesis.* PhD thesis, University of Växjö (Sweden), 2005.

[34] Manfredo P. do Carmo. *Differential Forms and Applications.* Springer, Berlin, Heidelberg, New York, 1994.

[35] Samuel Eilenberg and Norman Steenrod. *Foundation of Algebraic Topology.* Princeton University Press, Princeton, New Jersey, 1952.

[36] D. E. Evans and J. T. Lewis. *Dilatations of Irreducible Evolutions in Algebraic Quantum Theory.* Dublin Institute for Advanced Studies, Dublin, 1977.

[37] Gerald B. Folland. *Harmonic Analysis in Phase Space.* Princeton University Press, 1989.

[38] Gerald B. Folland. *Fourier Analysis and its Applications.* Brooks / Cole, Pacific Grove, 1992.

[39] Daniel S. Freed and Karen K. Uhlenbeck. *Instantons and Four-Manifolds*, volume 1 of *Mathematical Sciences Research Institute Publications.* Springer, 1984.

[40] A. Frölicher and A. Kriegl. *Linear Spaces and Differentation Theory.* John Wiley & Sons, Chicester, 1988.

[41] A. Gerrard and J. M. Burch. *Introduction to Matrix Methods in Optics.* John Wiley & Sons, 1975.

[42] Marvin J. Greenberg. *Lectures on Algebraic Topology.* Benjamin, 1971.

[43] Werner Greub. *Linear Algebra.* Springer, 4 edition, 1975.

[44] Werner Greub. *Multilinear Algebra.* Springer, 2 edition, 1978.

[45] Werner Greub, Stephen Halperin, and Ray Vanstone. *Connections, Curvature, and Cohomology II.* Number 47 in Pure and Applied Mathematics. Academic Press, New York, 1973.

[46] Werner Greub, Stephen Halperin, and Ray Vanstone. *Connections, Curvature, and Cohomology. Vol. I & II.* Number 47 in Pure and Applied Mathematics. Academic Press, New York, 1973.

[47] Karl-Heinz Gröchenig. *Foundations of Time-Frequency Analysis.* Birkhäuser, Boston, 2000.

[48] Hans-Christoph Grunau. *Abbildungsgrad und Fixpunktsätze.* www-ian.math.uni-magdeburg.de/home/grunau/grunaupublic.html.

[49] Victor Guillemin and Alan Pollack. *Differential Topology.* Prentice-Hall, New Jersey, 1974.

[50] Victor Guillemin and Shlomo Sternberg. *Symplectic Techniques in Physics.* Cambridge University Press, 1991.

[51] J. Gutknecht. *Die* C^∞_Γ*-Struktur aus der Diffeomorphsimengruppe einer kompakten Mannigfaltigkeit.* PhD thesis, ETH Zürich, 1977.

[52] Eugene Hecht. *Optics.* Addison-Wesley Pub Co Inc, Reading, Mass, New York, London, Amsterdam, Don Mills (Ontario), Sydney, Tokio, third edition, 1998.

[53] Wolfgang Hein. *Einführung in die Struktur- und Darstellungstheorie der klassischen Gruppen.* Springer, Berlin, Heidelberg, 1990.

[54] F. Hirzebruch. *Topological Methods in Algebraic Geometry.* Springer, Berlin, Heidelberg, New York, third edition, 1978. Second, Corrected Printing.
[55] Reinhard Honegger and Alfred Rieckers. *Photons in Fock Space and Beyond.* (to appear).
[56] Sze-Tsen Hue. *Homotopy Theory.* Academic Press, New York, 1959.
[57] H. H. Keller. *Differential Calculus in Locally Convex Spaces.* Springer, Berlin, Heidelberg, 1974.
[58] Aleksandr A. Kirillov. *Elements of the Theory of Representations.* Springer, New York, 1974.
[59] N. P. Landsman. *Mathematical Topics Between Classical and Quantum Mechanics.* Springer, New York, 1998.
[60] Jerrold E. Marsden and Tudor S. Ratiu. *Introduction to Mechanics and Symmetry.* Texts in Applied Mathematics 17. Springer, Berlin, Heidelberg, New York, 1994.
[61] James R. Munkres. *Elements of Algebraic topology.* Addison-Wesley, Menlo Park, London, Amsterdam, 1984.
[62] Gregory L. Naber. *Topology, Geometry, and Gauge Fields: Foundations.* Number 25 in Texts in Applied Mathematics. Springer, New York, 1997.
[63] Gregory L. Naber. *Topology, Geometry, and Gauge Fields: Interactions.* Number 141 in Texts in Applied Mathematics. Springer, New York, 2000.
[64] Johann von Neumann. *Mathematische Grundlagen der Quantenmechanik (Nachdruck).* Springer, Berlin, Heidelberg, New York, 1968.
[65] Michael A. Nielsen and Isaac L. Chuang. *Quantum Computation and Quantum Information.* Cambridge University Press, 2000.
[66] Richard S. Palais. *Seminar on the Atiyah-Singer Index Theorem.* Princeton Univ. Pr, 1966.
[67] Stephen Parrott. *Relativistic Electrodynamics and Differential Geometry.* Springer, New York, Berlin, Heidelberg, 1987.
[68] Sonja Pods. *Die Theorie der Heisenberggruppen zur Beschreibung einer Informationsübertragung insbesondere bei der Bildgebung in der Magnetresonanztomographie.* PhD thesis, Universität Mannheim, Logos–Verlag, Berlin, 2003.
[69] A. Pressley and G. Segal. *Loop Groups.* Oxford Science Publication. Clarendon Press, 1990.
[70] I. Raeburn and D. P. Williams. *Morita Equivalence and Continuous Trace C^*-Algebras,* volume 60. AMS Mathematical Surveys and. Monographs, Boston, 1991.
[71] Walter Johannes Schempp. *Harmonic Analysis on the Heisenberg Nilpotent Lie Group with Applications to Signal Theory.* Pitman Research Notes in Mathematics Series **147**. Longman Scientific & Technical, 1986.
[72] Walter Johannes Schempp. *Magnetic Resonance Imaging, Mathematical Foundations and Applications.* Wiley–Liss, 1998.
[73] Wolfgang P. Schleich. *Quantum Optics in Phase Space.* Wiley-VCH, Berlin, Weinheim, 2001.
[74] Günter Schwarz. *Hodge Decomposition — A Method of Solving Boundary value Problems,* volume 1607 of *Lecture Notes in Mathematics.* Springer, Berlin, Heidelberg, 1995.
[75] Jedrzej Sniatycki. *Geometric Quantization and Quantum Mechanics.* Applied Math. Sciences **30**. Springer, New York, 1980.
[76] Jedrzej Sniatycki. Geometric Quantization. Unpublished preprint, 1998.
[77] H. Spanier Edwin. *Algebraic Topology.* Series in Higher Mathematics. McGraw–Hill, 1966.
[78] Shlomo Sternberg. *Group Theory and Physics.* Cambridge University Press, 1994.
[79] Norbert Straumann. *Thermodynamik,* volume 265 of *Lecture Notes in Physics.* Springer, 1986.
[80] Masamichi Taksaki. *Theory of Operator Algebras I.* Springer, New York, Berlin, Heidelberg, 1979.
[81] Audrey Terras. *Fourier Analysis on Finite Groups and Applications.* London Mathematical Society Student Texts 43. Cambridge University Press, 1999.
[82] François Trèves. *Topological Vector Spaces, Distributions and Kernels.* Pure and Applied Mathematics 25. Academic Press, New York, 1967.
[83] Varadarajan. V.S. *Lie Groups, Lie Algebras and their Representation.* Springer, New York, Berlin, Heidelberg, Tokyo, 1984.
[84] Varadarajan. V.S. *Geometry of Quantum Theory.* Springer, New York, Berlin, Heidelberg, Tokyo, second edition, 1985.
[85] F. W. Warner. *Foundations of Differentiable Manifolds and Lie Groups.* Springer, New York, 1983.
[86] Joachim Weidmann. *Linear Operators in Hilbert Spaces.* Springer, New York, 1980.

[87] Hermann Weyl. *The Theory of Groups and Quantum Mechanics.* Dover Publication, Inc., New York, 1950.

[88] Bernardo Wolf, Kurt. *Geometric Optics on Phase Space.* Springer, Berlin, Heidelberg, 2004.

Index

Titles in This Series

151 **Ernst Binz and Sonja Pods,** The geometry of Heisenberg groups: With applications in signal theory, optics, quantization, and field quantization, 2008

150 **Bangming Deng, Jie Du, Brian Parshall, and Jianpan Wang,** Finite dimensional algebras and quantum groups, 2008

149 **Gerald B. Folland,** Quantum field theory: A tourist guide for mathematicians, 2008

148 **Patrick Dehornoy with Ivan Dynnikov, Dale Rolfsen, and Bert Wiest,** Ordering braids, 2008

147 **David J. Benson and Stephen D. Smith,** Classifying spaces of sporadic groups, 2008

146 **Murray Marshall,** Positive polynomials and sums of squares, 2008

145 **Tuna Altinel, Alexandre V. Borovik, and Gregory Cherlin,** Simple groups of finite Morley rank, 2008

144 **Bennett Chow, Sun-Chin Chu, David Glickenstein, Christine Guenther, James Isenberg, Tom Ivey, Dan Knopf, Peng Lu, Feng Luo, and Lei Ni,** The Ricci flow: Techniques and applications, Part II: Analytic aspects, 2008

143 **Alexander Molev,** Yangians and classical Lie algebras, 2007

142 **Joseph A. Wolf,** Harmonic analysis on commutative spaces, 2007

141 **Vladimir Maz′ya and Gunther Schmidt,** Approximate approximations, 2007

140 **Elisabetta Barletta, Sorin Dragomir, and Krishan L. Duggal,** Foliations in Cauchy-Riemann geometry, 2007

139 **Michael Tsfasman, Serge Vlăduţ, and Dmitry Nogin,** Algebraic geometric codes: Basic notions, 2007

138 **Kehe Zhu,** Operator theory in function spaces, 2007

137 **Mikhail G. Katz,** Systolic geometry and topology, 2007

136 **Jean-Michel Coron,** Control and nonlinearity, 2007

135 **Bennett Chow, Sun-Chin Chu, David Glickenstein, Christine Guenther, James Isenberg, Tom Ivey, Dan Knopf, Peng Lu, Feng Luo, and Lei Ni,** The Ricci flow: Techniques and applications, Part I: Geometric aspects, 2007

134 **Dana P. Williams,** Crossed products of C^*-algebras, 2007

133 **Andrew Knightly and Charles Li,** Traces of Hecke operators, 2006

132 **J. P. May and J. Sigurdsson,** Parametrized homotopy theory, 2006

131 **Jin Feng and Thomas G. Kurtz,** Large deviations for stochastic processes, 2006

130 **Qing Han and Jia-Xing Hong,** Isometric embedding of Riemannian manifolds in Euclidean spaces, 2006

129 **William M. Singer,** Steenrod squares in spectral sequences, 2006

128 **Athanassios S. Fokas, Alexander R. Its, Andrei A. Kapaev, and Victor Yu. Novokshenov,** Painlevé transcendents, 2006

127 **Nikolai Chernov and Roberto Markarian,** Chaotic billiards, 2006

126 **Sen-Zhong Huang,** Gradient inequalities, 2006

125 **Joseph A. Cima, Alec L. Matheson, and William T. Ross,** The Cauchy Transform, 2006

124 **Ido Efrat, Editor,** Valuations, orderings, and Milnor K-Theory, 2006

123 **Barbara Fantechi, Lothar Göttsche, Luc Illusie, Steven L. Kleiman, Nitin Nitsure, and Angelo Vistoli,** Fundamental algebraic geometry: Grothendieck's FGA explained, 2005

122 **Antonio Giambruno and Mikhail Zaicev, Editors,** Polynomial identities and asymptotic methods, 2005

121 **Anton Zettl,** Sturm-Liouville theory, 2005

120 **Barry Simon,** Trace ideals and their applications, 2005